Radioactive Waste Confinement: Clays in Natural and Engineered Barriers

Geological Society books refereeing procedures

The Society makes every effort to ensure that the scientific and production quality of its books matches that of its journals. Since 1997, all book proposals have been refereed by specialist reviewers as well as by the Society's Books Editorial Committee. If the referees identify weaknesses in the proposal, these must be addressed before the proposal is accepted

Once the book is accepted, the Society Book Editors ensure that the volume editors follow strict guidelines on refereeing and quality control. We insist that individual papers can only be accepted after satisfactory review by two independent referees. The questions on the review forms are similar to those for *Journal of the Geological Society*. The referees' forms and comments must be available to the Society's Book Editors on request.

Although many of the books result from meetings, the editors are expected to commission papers that were not presented at the meeting to ensure that the book provides a balanced coverage of the subject. Being accepted for presentation at the meeting does not guarantee inclusion in the book.

More information about submitting a proposal and producing a book for the Society can be found on its website: www.geolsoc.org.uk.

It is recommended that reference to all or part of this book should be made in one of the following ways:

Norris, S., Bruno, J., Van Geet, M. & Verhoef, E. (eds) 2017. *Radioactive Waste Confinement: Clays in Natural and Engineered Barriers*. Geological Society, London, Special Publications, **443**.

Toprak, E., Olivella, S. & Pintado, X. 2017. Coupled THM modelling of engineered barriers for the final disposal of spent nuclear fuel isolation. *In*: Norris, S., Bruno, J., Van Geet, M. & Verhoef, E. (eds) *Radioactive Waste Confinement: Clays in Natural and Engineered Barriers*. Geological Society, London, Special Publications, **443**, 235–251. First published online September 26, 2016, https://doi.org/10.1144/SP443.19

GEOLOGICAL SOCIETY SPECIAL PUBLICATION NO. 443

Radioactive Waste Confinement: Clays in Natural and Engineered Barriers

EDITED BY

S. NORRIS
Radioactive Waste Management, UK

J. BRUNO
Amphos 21, Spain

M. VAN GEET
ONDRAF/NIRAS, Belgium and

E. VERHOEF
COVRA, The Netherlands

2017
Published by
The Geological Society
London

THE GEOLOGICAL SOCIETY

The Geological Society of London (GSL) was founded in 1807. It is the oldest national geological society in the world and the largest in Europe. It was incorporated under Royal Charter in 1825 and is Registered Charity 210161.

The Society is the UK national learned and professional society for geology with a worldwide Fellowship (FGS) of over 10 000. The Society has the power to confer Chartered status on suitably qualified Fellows, and about 2000 of the Fellowship carry the title (CGeol). Chartered Geologists may also obtain the equivalent European title, European Geologist (EurGeol). One fifth of the Society's fellowship resides outside the UK. To find out more about the Society, log on to www.geolsoc.org.uk.

The Geological Society Publishing House (Bath, UK) produces the Society's international journals and books, and acts as European distributor for selected publications of the American Association of Petroleum Geologists (AAPG), the Indonesian Petroleum Association (IPA), the Geological Society of America (GSA), the Society for Sedimentary Geology (SEPM) and the Geologists' Association (GA). Joint marketing agreements ensure that GSL Fellows may purchase these societies' publications at a discount. The Society's online bookshop (accessible from www.geolsoc.org.uk) offers secure book purchasing with your credit or debit card.

To find out about joining the Society and benefiting from substantial discounts on publications of GSL and other societies worldwide, consult www.geolsoc.org.uk, or contact the Fellowship Department at: The Geological Society, Burlington House, Piccadilly, London W1J 0BG: Tel. +44 (0)20 7434 9944; Fax +44 (0)20 7439 8975; E-mail: enquiries@geolsoc.org.uk.

For information about the Society's meetings, consult *Events* on www.geolsoc.org.uk. To find out more about the Society's Corporate Affiliates Scheme, write to enquiries@geolsoc.org.uk.

Published by The Geological Society from:
The Geological Society Publishing House, Unit 7, Brassmill Enterprise Centre, Brassmill Lane, Bath BA1 3JN, UK

The Lyell Collection: www.lyellcollection.org
Online bookshop: www.geolsoc.org.uk/bookshop
Orders: Tel. +44 (0)1225 445046, Fax +44 (0)1225 442836

British Library Cataloguing in Publication Data

A catalogue record for this book is available from the British Library.
ISBN 978-1-78620-273-4
ISSN 0305-8719

Distributors
For details of international agents and distributors see:
www.geolsoc.org.uk/agentsdistributors

Typeset by Nova Techset Private Limited, Bengaluru & Chennai, India
Printed and bound by CPI Group (UK) Ltd, Croydon CR0 4YY

Contents

Mass transfer

Bentonite evolution

Gas transfer

Radioactive waste confinement: clays in natural and engineered barriers – introduction

SIMON NORRIS

Radioactive Waste Management, Building 587, Curie Avenue, Harwell Campus, Didcot, Oxon, OX11 0RH, UK and Corresponding Editor, Scientific Committee, Brussels 2015

simon.norris@nda.gov.uk

There is general agreement internationally (Nuclear Energy Agency, OECD 2008) that geological disposal provides the safest long-term management solution for higher-activity radioactive waste. Many countries (e.g. Canada, Finland, France, Switzerland, Sweden, UK and USA) have chosen to dispose of all or part of their radioactive waste in facilities constructed at an appropriate depth in stable geological formations. The development of a repository (sometimes also referred to as a geological disposal facility) on a specific site requires a systematic and integrated approach, taking into account the characteristics of (i) the waste to be emplaced, (ii) the enclosing engineered barriers and (iii) the host rock and the geological setting of the host rock. Three main rock types are usually considered for geological disposal: crystalline rocks, salt and clays. Each type includes bedrock formations with a relatively broad spectrum of geological properties. The engineered barriers contain different types of materials, such as metals, concrete and natural materials, such as clay.

This Special Publication highlights the importance of clays and clayey material in the development of almost all national geological disposal systems (for further information on the uses of clay proposed by a range of national waste management programmes, see, for example, ANDRA 2016; Bundesamt für Strahlenschutz 2016; COVRA 2016; Nagra 2016; NWMO 2016; ONDRAF/NIRAS 2016; Ontario Power Generation 2016; Posiva 2016*a*; PURAM (RHK Kft.) 2016; SKB 2016; UK Government 2016*a*, *b*; United States Department of Environment 2016). Clays exhibit many interesting properties, which are exploited in the development of most geological disposal systems. Clays are used both as host rock and as material for engineered barriers. Whatever their use, clays present various characteristics that make them high-quality barriers to the migration of radionuclides and chemical contaminants towards the surface environment. As host rocks, clays are, in addition, hydrogeologically, geochemically and mechanically stable over geological time-scales, i.e. millions of years.

The disposal system as a whole

The disposal system consists of the waste, the engineered barriers and the host rock (Figs 1 & 2).

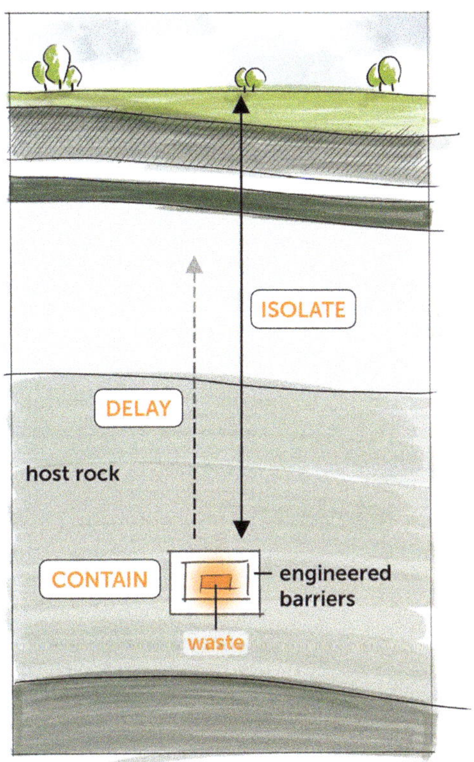

Fig. 1. Illustrative diagram of geological disposal. Geological disposal provides protection of people and the environment without human intervention being necessary once the facility is closed (source: ONDRAF/NIRAS 2016). Note that the depth at which a disposal facility could be constructed is dependent on, for example, the national strategy, the geology and the disposal concept. In the case of the UK, the facility will be located at a depth between 200 and 1000 m below the surface (UK Government 2016*a*).

From: NORRIS, S., BRUNO, J., VAN GEET, M. & VERHOEF, E. (eds) 2017. *Radioactive Waste Confinement: Clays in Natural and Engineered Barriers*. Geological Society, London, Special Publications, **443**, 1–8.
First published online March 1, 2017, https://doi.org/10.1144/SP443.26

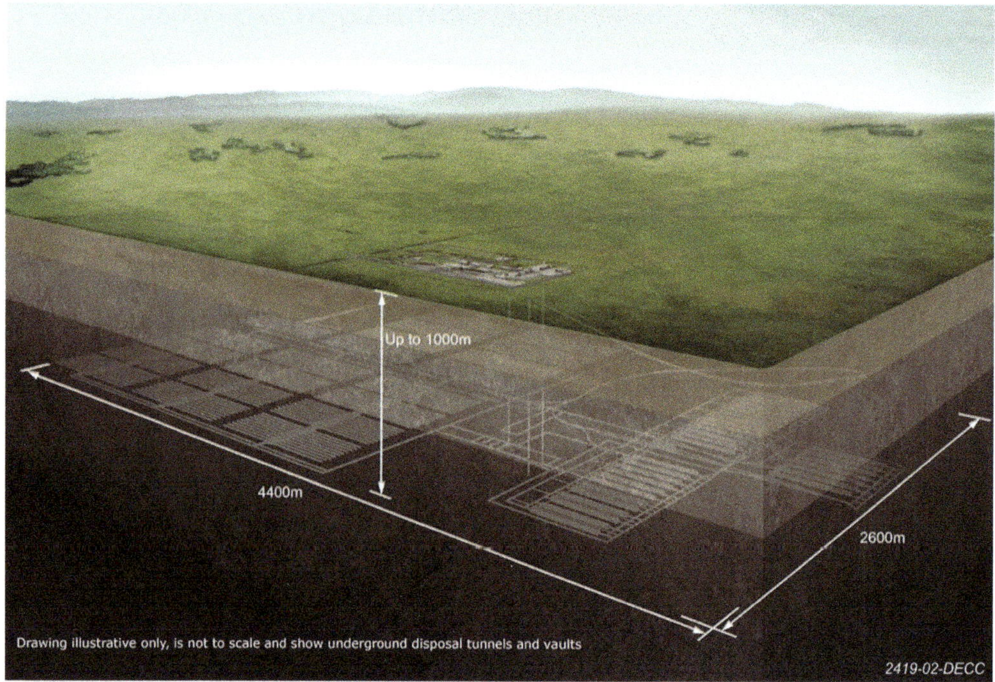

Fig. 2. Artist's impression of one potential layout of a geological disposal facility for UK higher activity wastes (© UK Crown copyright 2014). The precise layout and design will depend on the inventory for disposal and the specific geological characteristics at the site in question.

Each of these elements fulfils, separately or in complementary fashion, multiple safety functions (e.g. isolation, containment). The 'defence-in-depth' principle requires multiple levels of protection, which are designed to enhance safety through their diversity and redundancy. It is the whole system that must be taken into account in a safety case, and not each system component separately.

Clays as host rocks

France (ANDRA 2016) and Switzerland (Nagra 2016) have chosen to dispose of their high-level and long-lived intermediate-level radioactive waste in indurated clays – in the Callovo-Oxfordian and Opalinus Clay formations, respectively. In Switzerland, Opalinus Clay is also the proposed host rock for low- and intermediate-level waste. In Belgium, the technical solution recommended for the long-term management of high-level and long-lived low- and intermediate-level waste is geological disposal in poorly indurated clay (Boom Clay or Ypresian clays) (ONDRAF/NIRAS 2016). In the Netherlands, clays are considered as a potential host rock for the disposal of all types of radioactive waste (COVRA 2016). Other countries also consider the use of clays as host rock (e.g. NWMO 2016; UK Government 2016*b*).

The research and development studies performed internationally over several decades have highlighted the favourable properties of clays in relation to geological disposal (for further information see the work of the respective national waste management organisations, e.g. ANDRA 2016; Bundesamt für Strahlenschutz 2016; COVRA 2016; Nagra 2016; NWMO 2016; ONDRAF/NIRAS 2016; Ontario Power Generation 2016; Posiva 2016*a*; PURAM (RHK Kft.) 2016; SKB 2016; UK Government 2016*a*, *b*; United States Department of Environment 2016). These favourable properties can be listed as follows:

- *very little water movement* – thanks to their low permeability, there is practically no water movement in clays. Radionuclide and chemical contaminant transport via this medium is thus strongly delayed;
- *diffusive transport* – given the limited water movement, transport in clays is essentially diffusive, which means species migrate primarily under the influence of their concentration gradient, and migration under the influence of pore-water movement is minimal;

- *retention capacity* – clays have a strong retention capacity for many radionuclides and chemical contaminants. Their migration through clays is thus considerably delayed;
- *buffer effect* – clays display a significant buffer effect with regard to chemical perturbations. The thickness of the clay that is chemically perturbed by the disposal facility is, therefore, very limited;
- *self-sealing capacity* – clays show a high capacity for self-sealing. Any fractures and fissures that occur, in particular those created by excavation activities, close quite rapidly;
- *stability* – the selected clay host rocks and, therefore, their favourable properties have remained unchanged over millions of years. The migration of natural chemical species through these clay host rocks has remained diffusive during at least the last million years;
- *vertical homogeneity* – radionuclide and chemical contaminant transport properties are very homogeneous almost throughout the entire thickness of the selected clay host rocks;
- *lateral continuity* – clays are present within simple geological structures, with a significant lateral continuity, which facilitates their large-scale characterization.

Clays may also contribute to the safety of disposal systems whose host rock is not clay, by being present in their geological environment. For example, in Germany, the disposal facility for low- and intermediate-level waste is under construction in an old iron mine (limestone) located under a clay formation (Bundesamt für Strahlenschutz 2016); in Canada, the host rock proposed for the disposal of low- and intermediate-level waste is limestone situated beneath a sequence of clay formations (Ontario Power Generation 2016).

Clays as engineered barriers

The favourable properties of clay (low permeability, self-sealing, stability) also make it a material of choice for engineered barriers. Clay is mainly planned for use as buffer, backfill or sealing material.

- Buffer material – the empty space between the disposal package and the host rock is filled with clay. For instance, bentonite, a swelling clay, is used as buffer material filling the voidage between the disposal packages and the host rock in the disposal facility designs selected in, for example, Canada (NWMO 2016), Finland (Posiva 2016*a*), Sweden (SKB 2016) and Switzerland (Nagra 2016) (Fig. 3).
- Backfill material – clay (for instance in the form of blocks or pellets) is used to fill excavated spaces (placement rooms, access ways), sometimes in combination with other materials.
- Sealing material – clay, sometimes in combination with other materials, is used to isolate parts of the disposal facility. Seals are works of limited dimensions with specific purpose placed at key locations of the disposal facility. For instance, in France (ANDRA 2016), the seals aim to limit water flow within the underground facilities.

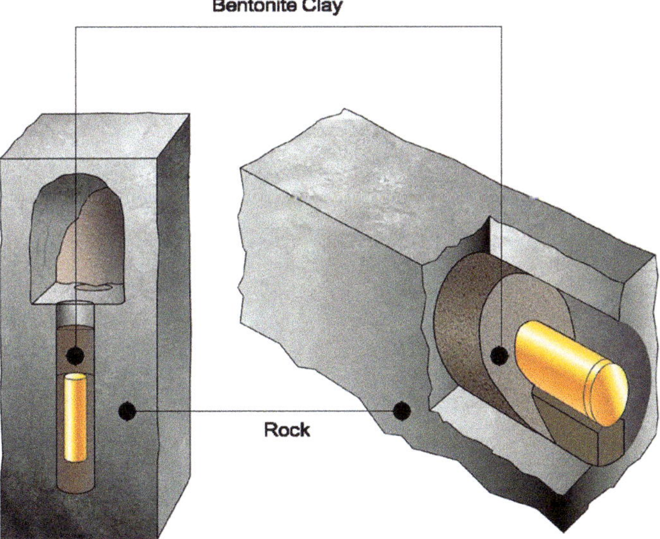

Fig. 3. Examples of placement of disposal packages in a repository. Voidage is filled with bentonite (source: NWMO 2016).

Table 1. Use of clays as buffer, backfill or sealing materials in operational and planned national disposal facilities

	Buffer	Backfill	Sealing
Operational facilities			
United States of America (United States Department of Environment 2016) Waste types: long-lived low- and intermediate-level military waste Host rock: salt			x
Hungary (PURAM (RHK Kft.) 2016) Waste types: short-lived low- and intermediate-level waste Host rock: crystalline rock (granite)			x
Applications submitted for disposal facility construction licence			
Canada (Ontario Power Generation 2016) Waste types: low- and intermediate-level waste Host rock: limestone overlain by clay			x
Sweden (SKB 2016) Waste type: irradiated fuel Host rock: crystalline rock (granite)	x	x	x
Finland (Posiva 2016*a*) Waste type: irradiated fuel Host rock: crystalline rock (granite)	x	x	x
Host rock selected			
France (ANDRA 2016) Waste types: long-lived intermediate- and high-level waste Host rock: clay		x	x
Switzerland (Nagra 2016) Waste types: irradiated fuel and long-lived low- and intermediate-level waste Host rock: clay	x	x	x
Host rock not yet selected			
Canada (NWMO 2016) Waste type: irradiated fuel	x	x	x
United Kingdom (UK Government 2016(a)) Waste type: intermediate- and high-level, irradiated fuel	x	x	x

The geological disposal facilities in operation, under development or being planned in many countries thus contain, or are proposed to contain, clay, whatever the selected host rock. Table 1 lists some of these facilities, sorted according to the progress of the national geological disposal programme. The use of clay as buffer, backfill or in sealing is indicated.

Multidisciplinary research and this volume

Organizations responsible for implementing geological repositories, such as ANDRA, Bundesamt für Strahlenschutz, COVRA, Nagra, NWMO, ONDRAF/NIRAS, Ontario Power Generation, Posiva, PURAM (RHK Kft.), SKB, the UK Government and the United States Department of Environment, are undertaking significant programmes of research – nationally and internationally – to advance understanding of clays and their contribution to ensuring safe long-term management of long-lived radioactive waste. Multidisciplinary approaches, including geology, mineralogy, geochemistry, rheology, the physics and chemistry of clay minerals and assemblages, are required in order to provide a detailed characterization of the geological host formations considered for the disposal of radioactive waste and to assess the behaviour of engineered and natural barriers when submitted to various types of perturbations induced by such facilities. Existing and proposed experimental programmes in underground research laboratories are designed to evaluate the performance of the natural barrier and the impact of repository-induced disturbances upon the confinement properties of clay-rich geological formations. Interpreting the subsequent scientific results, modelling the long-term behaviour of radioactive waste repositories and carrying out safety assessment exercises all further contribute to the knowledge base.

This Geological Society, London, Special Publication contains 25 papers of scientific studies presented at the 6th conference on 'Clays in natural and engineered barriers for radioactive waste confinement', hosted by ONDRAF/NIRAS (the Belgian National Radioactive Waste Management Agency) and held in Brussels, Belgium, in 2015. The papers cover a range of outputs from the Brussels meeting and provide insight for the reader of the range of clay-related work currently being undertaken internationally in relation to natural and engineered barriers for radioactive waste confinement.

Since 2002, national waste management organizations have developed this conference into the most important event for all kinds of scientists from all over the world dealing with the disposal of radioactive waste. These conferences, each attracting a large number of papers and attendees from around the world, have contributed significantly to the outstanding scientific level of research relevant to argillaceous media in the context of the disposal of radioactive waste.

The 2015 meeting covered aspects of clay characterization and behaviour considered at various temporal and spatial scales relevant to the confinement of radionuclides in clay, from basic phenomenological process description to the global understanding of the performance and safety at repository and geological scales. Special emphasis was put on the modelling of processes occurring at the mineralogical level within the clay barriers.

The 25 papers published in this Special Publication have been classified according to different areas within the field of disposal of radioactive waste research. The assignment of a study to one of the topic areas was not always easy because many studies provide information of, and for, different aspects of disposal research. With that caveat, the papers in this volume consider research into argillaceous media under the following topic area headings: large-scale geological characterization; general strategy for clay-based disposal systems; geomechanics; mass transfer; bentonite evolution; and gas transfer.

Large-scale geological characterization

Vandersteen *et al.* (2016) discuss data analysis and modelling in relation to regional aquifer hydrogeochemistry in the confined aquifer system below the Boom Clay (NE Belgium).

In their paper, **Gravesen *et al.* (2016)** report on an assessment of Palaeogene and Neogene clay deposits in Denmark as possible host rocks for disposal of low- and intermediate-level radioactive waste.

General strategy for clay-based disposal systems

Approaches to evaluate and underpin the technical feasibility of the Belgian disposal concept are discussed by **Doudou *et al.* (2016*a*)**.

Outputs from the European Commission DOPAS Project (Posiva 2016*b*), which considered the full-scale demonstration of plugs and seals, including constituent clay materials and as will be used in a repository, are discussed in the next three papers. **Doudou *et al.* (2016*b*)** consider approaches to compliance assessment, **Svoboda *et al.* (2016)** report on the DOPAS Czech Experimental Pressure and Sealing Plug (EPSP) experiment, as constructed at the Josef Underground Laboratory in granitic rock, while **Trpkošová *et al.* (2016)** discuss laboratory models for the EPSP experiment in the DOPAS Project relating to the saturation of bentonite pellets.

Geomechanics

Papers on the geomechanical behaviour of clays in this Special Publication focus on clays investigated as part of French and Swiss studies, utilizing access to a number of underground research facilities.

Shao *et al.* (2016*a*) report on quantitative characterization of the excavation damage zone fracture network in the Meuse/Haute-Marne (Bure) Underground Research Laboratory, constructed in Callovo-Oxfordian clay in France, focusing on an *in-situ* experiment and numerical interpretation of a helium injection test. Further work in relation to this underground research laboratory is discussed by **Guayacán-Carrillo *et al.* (2016)**, focusing on convergence analysis of an unsupported microtunnel.

Su *et al.* (2016) report on work undertaken at the Tournemire experimental station, also in France, on the laboratory and numerical investigation of the mechanical behaviour of the Tournemire argillite.

Moving to the Swiss national programme, **Alcolea *et al.* (2016)** report on a pragmatic approach to abstract the excavation damage zone around tunnels of a geological radioactive waste repository. This is applied to the HG-A experiment in the Mont Terri rock laboratory, a research facility near St-Ursanne in the Canton of Jura, in which investigations are being carried out in the Opalinus Clay – a proposed host rock for the future deep geological disposal of radioactive waste in Switzerland.

Again based on Mont Terri studies, **Shao *et al.* (2016*b*)** consider the influence of different supports on the properties of the excavation damaged zone along the Full-scale Emplacement tunnel in the rock laboratory.

Mass transfer

It is important to understand how groundwater and other chemical species, as might derive from waste disposed in a repository, could potentially move through natural and engineered clay barriers. As well as understanding mass transfer at the very small scale, it is also important to consider how a very detailed conceptual model for the movement of groundwater and other chemical species, for example, can be upscaled for inclusion in a larger-scale model of an evolving repository.

M'Jahad *et al.* (2016) present a characterization of transport and water retention properties of damaged Callovo-Oxfordian claystone, while **Jean-Baptiste *et al.* (2016)** present an analysis of helium and $^{40}Ar/^{36}Ar$ vertical distribution in porewaters of the Eastern Paris Basin (Meuse/Haute-Marne), discussing constraints that can be derived in relation to transport processes through the sedimentary sequence.

Kolomá & Červinka (2016) report on a study of ^{85}Sr transport through a column filled with crushed granite in the presence of bentonite colloids, while **Aertsens *et al.* (2016)** propose an improved model for through-diffusion experiments, with application to strontium and tritiated water (HTO) diffusion in Boom Clay (as relevant to the Belgian national programme) and compacted illite. **Meng & Pfingsten (2016)** discuss multispecies random walk simulations in radial symmetry – the model concept, benchmark and application to HTO, ^{22}Na and ^{36}Cl diffusion in clay; **Bildstein *et al.* (2016)** report on a study to gain insight into corrosion processes from numerical simulations of an integrated iron-claystone experiment.

Toprak *et al.* (2016) present coupled thermo-hydro-mechanical modelling of engineered barriers for the final disposal of spent nuclear fuel isolation and, moving to a larger scale, **Schaedle *et al.* (2016)** combine high-resolution two-phase with simplified single-phase simulations in order to optimize the performance of performance assessment/safety assessment simulations for a deep geological repository for radioactive waste.

Bentonite evolution

Bentonite clay will be emplaced in a repository as part of the engineered barrier system. Over the time-scale of interest typically considered in the evaluation of the performance of a repository (up to 1 million years post-closure), it is possible that the bentonite will evolve. It is important to understand the extent of possible bentonite evolution, and whether or not any related change to properties of the bentonite affect its ability to perform as part of a system of natural and engineered barriers.

Kröhn (2016) considers conceptual models and mathematical descriptions relating to bentonite re-saturation. **Dolder *et al.* (2016)** discuss the alteration of MX-80 bentonite backfill material by high-pH cementitious fluids under lithostatic conditions, presenting an experimental approach using core infiltration techniques. Potential chemical erosion of bentonite is considered by **Reijonen & Marcos (2016)**, who draw on evidence from nature to consider if information can be derived for use in the development of understanding of bentonite evolution in a repository.

Gas transfer

As well as a consideration of how groundwater could move through a clay (see the eight papers in the mass transfer section), it is important to understand how gas – as derived from, for example, the corrosion of waste containers, radiolysis of groundwater – could interact with clay in the engineered barrier system and host rock.

Bénet *et al.* (2016) model water and gas flow through an excavation damaged zone in the Callovo-Oxfordian argillites in the framework of a single porosity model, while **Smutek *et al.* (2016)** report on the gas permeability, breakthrough behaviour and re-sealing ability of Czech Ca–Mg bentonite. Finally, **Jacops *et al.* (2016)** report on a study to measure diffusion coefficients of dissolved helium and argon in three potential clay host formations: Boom Clay (of relevance to the Belgian national programme), Callovo-Oxfordian Clay (of relevance to the French national programme) and Opalinus Clay (of relevance to the Swiss national programme).

The collection of different topics presented in this Special Publication demonstrates the diversity of geological repository research, which will continue as national programmes progress towards their respective licence applications for repository implementation. This continued research will allow uncertainties relating to, for example, the behaviour of an evolving repository system to be reduced; will increase safety margins; and will help with disposal system optimization. As such, geological repository research is a challenging and engaging task for current scientists and hopefully will provide an interesting and long-term career path that is attractive to young scientists.

The 7th conference on 'Clays in natural and engineered barriers for radioactive waste confinement' will be held in Davos, Switzerland, in September 2017, hosted by the Swiss National Radioactive Waste Management Agency, Nagra.

References

Aertsens, M., Van Laer, L., Maes, N. & Govaerts, J. 2016. An improved model for through-diffusion experiments: application to strontium and tritiated water (HTO) diffusion in Boom Clay and compacted illite. *In*: Norris, S., Bruno, J., Van Geet, M. & Verhoef, E. (eds) *Radioactive Waste Confinement: Clays in Natural and Engineered Barriers*. Geological Society, London, Special Publications, **443**. First published online June 24, 2016, https://doi.org/10.1144/SP443.9

Alcolea, A., Kuhlmann, U. *et al*. 2016. A pragmatic approach to abstract the excavation damaged zone around tunnels of a geological radioactive waste repository: application to the HG-A experiment in Mont Terri. *In*: Norris, S., Bruno, J., Van Geet, M. & Verhoef, E. (eds) *Radioactive Waste Confinement: Clays in Natural and Engineered Barriers*. Geological Society, London, Special Publications, **443**. First published online June 24, 2016, https://doi.org/10.1144/SP443.8

ANDRA 2016. http://www.andra.fr/

Bénet, L-V., Blaud, É. & Wendling, J. 2016. Modelling of water and gas flow through an excavation damaged zone in the Callovo-Oxfordian argillites in the framework of a single porosity model. *In*: Norris, S., Bruno, J., Van Geet, M. & Verhoef, E. (eds) *Radioactive Waste Confinement: Clays in Natural and Engineered Barriers*. Geological Society, London, Special Publications, **443**. First published online June 24, 2016, https://doi.org/10.1144/SP443.6

Bildstein, O., Lartigue, J-É., Schlegel, M.L., Bataillon, C., Cochepin, B., Munier, I. & Michau, N. 2016. Gaining insight into corrosion processes from numerical simulations of an integrated iron-claystone experiment. *In*: Norris, S., Bruno, J., Van Geet, M. & Verhoef, E. (eds) *Radioactive Waste Confinement: Clays in Natural and Engineered Barriers*. Geological Society, London, Special Publications, **443**. First published online June 22, 2016, https://doi.org/10.1144/SP443.2

Bundesamt für Strahlenschutz 2016. http://www.endlager-konrad.de/EN/topics/nwm/repositories/konrad/konrad.html

COVRA 2016. http://www.covra.nl/

Dolder, F., Mäder, U., Jenni, A. & Münch, B. 2016. Alteration of MX-80 bentonite backfill material by high-pH cementitious fluids under lithostatic conditions – an experimental approach using core infiltration techniques. *In*: Norris, S., Bruno, J., Van Geet, M. & Verhoef, E. (eds) *Radioactive Waste Confinement: Clays in Natural and Engineered Barriers*. Geological Society, London, Special Publications, **443**. First published online June 24, 2016, https://doi.org/10.1144/SP443.10

Doudou, S., Harvey, E.J., Richardson, P.J., Wickham, S.M., Van Marcke, P., Raymaekers, D. & Wacquier, W. 2016*a*. Approaches to evaluate and underpin the technical feasibility of the Belgian disposal concept. *In*: Norris, S., Bruno, J., Van Geet, M. & Verhoef, E. (eds) *Radioactive Waste Confinement: Clays in Natural and Engineered Barriers*. Geological Society, London, Special Publications, **443**. First published online June 24, 2016, https://doi.org/10.1144/SP443.11

Doudou, S., White, M.J., Johnson, M., Bosgiraud, J.-M. & Grahm, P. 2016*b*. DOPAS full-scale experiments: approaches to compliance assessment. *In*: Norris, S., Bruno, J., Van Geet, M. & Verhoef, E. (eds) *Radioactive Waste Confinement: Clays in Natural and Engineered Barriers*. Geological Society, London, Special Publications, **443**. First published online August 26, 2016, https://doi.org/10.1144/SP443.16

Gravesen, P., Pedersen, S.A.S., Nilsson, B. & Binderup, M. 2016. An assessment of Palaeogene and Neogene clay deposits in Denmark as possible host rocks for final disposal of low- and intermediate-level radioactive waste. *In*: Norris, S., Bruno, J., Van Geet, M. & Verhoef, E. (eds) *Radioactive Waste Confinement: Clays in Natural and Engineered Barriers*. Geological Society, London, Special Publications, **443**. First published online June 22, 2016, https://doi.org/10.1144/SP443.3

Guayacán-Carrillo, L.-M., Sulem, J., Seyedi, D.M., Ghabezloo, S., Noiret, A. & Armand, G. 2016. Convergence analysis of an unsupported micro-tunnel at the Meuse/Haute-Marne Underground Research Laboratory. *In*: Norris, S., Bruno, J., Van Geet, M. & Verhoef, E. (eds) *Radioactive Waste Confinement: Clays in Natural and Engineered Barriers*. Geological Society, London, Special Publications, **443**. First published online December 20, 2016, https://doi.org/10.1144/SP443.24

Jacops, E., Maes, N., Bruggeman, C. & Grade, A. 2016. Measuring diffusion coefficients of dissolved He and Ar in three potential clay host formations: Boom Clay, Callovo-Oxfordian Clay and Opalinus Clay. *In*: Norris, S., Bruno, J., Van Geet, M. & Verhoef, E. (eds) *Radioactive Waste Confinement: Clays in Natural and Engineered Barriers*. Geological Society, London, Special Publications, **443**. First published online June 22, 2016, https://doi.org/10.1144/SP443.1

Jean-Baptiste, P., Lavielle, B., Fourre, E., Smith, T. & Pagel, M. 2016. Vertical distribution of helium and $^{40}Ar/^{36}Ar$ in porewaters of the Eastern Paris Basin (Bure/Haute-Marne): constraints on transport processes through the sedimentary sequence. *In*: Norris, S., Bruno, J., Van Geet, M. & Verhoef, E. (eds) *Radioactive Waste Confinement: Clays in Natural and Engineered Barriers*. Geological Society, London, Special Publications, **443**. First published online December 21, 2016, https://doi.org/10.1144/SP443.25

Kolomá, K. & Červinka, R. 2016. Study of ^{85}Sr transport through a column filled with crushed granite in the presence of bentonite colloids. *In*: Norris, S., Bruno, J., Van Geet, M. & Verhoef, E. (eds) *Radioactive Waste Confinement: Clays in Natural and Engineered Barriers*. Geological Society, London, Special Publications, **443**. First published online July 8, 2016, https://doi.org/10.1144/SP443.14

Kröhn, K-P. 2016. Bentonite re-saturation: different conceptual models – similar mathematical descriptions. *In*: Norris, S., Bruno, J., Van Geet, M. & Verhoef, E. (eds) *Radioactive Waste Confinement: Clays in Natural and Engineered Barriers*. Geological Society, London, Special Publications, **443**. First published online July 6, 2016, https://doi.org/10.1144/SP443.12

Meng, S. & Pfingsten, W. 2016. Multispecies random walk simulations in radial symmetry: model concept, benchmark, and application to HTO, ^{22}Na

and ^{36}Cl diffusion in clay. *In*: Norris, S., Bruno, J., Van Geet, M. & Verhoef, E. (eds) *Radioactive Waste Confinement: Clays in Natural and Engineered Barriers*. Geological Society, London, Special Publications, **443**. First published online August 18, 2016, https://doi.org/10.1144/SP443.15

M'Jahad, S., Davy, C.A., Skoczylas, F. & Talandier, J. 2016. Characterization of transport and water retention properties of damaged Callovo-Oxfordian claystone. *In*: Norris, S., Bruno, J., Van Geet, M. & Verhoef, E. (eds) *Radioactive Waste Confinement: Clays in Natural and Engineered Barriers*. Geological Society, London, Special Publications, **443**. First published online December 22, 2016, https://doi.org/10.1144/SP443.23

Nagra 2016. http://www.nagra.ch/en

Nuclear Energy Agency (NEA), Organisation for Economic Co-operation and Development 2008. 'Moving Forward with Geological Disposal – A Collective Statement by the NEA Radioactive Waste Management Committee', http://bit.ly/1jzKJfw

NWMO 2016. http://www.nwmo.ca/

ONDRAF/NIRAS 2016. http://www.niras.be/geologischeberging

Ontario Power Generation 2016. http://www.opg.com/generating-power/nuclear/nuclear-waste-management/Deep-Geologic-Repository/Pages/Deep-Geologic-Repository.aspx

Posiva 2016*a*. http://www.posiva.fi/en

Posiva 2016*b*. http://www.posiva.fi/en/dopas

PURAM (RHK Kft.) 2016. http://www.rhk.hu/en/our-premises/nrwr/

Reijonen, H.M. & Marcos, N. 2016. Chemical erosion of the bentonite buffer: do we observe it in nature? *In*: Norris, S., Bruno, J., Van Geet, M. & Verhoef, E. (eds) *Radioactive Waste Confinement: Clays in Natural and Engineered Barriers*. Geological Society, London, Special Publications, **443**. First published online June 24, 2016, https://doi.org/10.1144/SP443.13

Schaedle, P., Kaempfer, T., Pépin, G., Wendling, J. & Brommundt, J. 2016. Combining high-resolution two-phase with simplified single-phase simulations in order to optimize the performance of PA/SA simulations for a deep geological repository for radioactive waste. *In*: Norris, S., Bruno, J., Van Geet, M. & Verhoef, E. (eds) *Radioactive Waste Confinement: Clays in Natural and Engineered Barriers*. Geological Society, London, Special Publications, **443**. First published online June 22, 2016, https://doi.org/10.1144/SP443.4

Shao, H., Göthling, S., Liu, W., Hesser, J., Morel, J. & Sönnke, J. 2016*a*. Quantitative characterization of the excavation damaged zone fracture network in the Meuse/Haute-Marne Underground Research Laboratory: *in situ* experiment and numerical interpretation of helium injection test. *In*: Norris, S., Bruno, J., Van Geet, M. & Verhoef, E. (eds) *Radioactive Waste Confinement: Clays in Natural and Engineered Barriers*. Geological Society, London, Special Publications, **443**. First published online October 13, 2016, https://doi.org/10.1144/SP443.21

Shao, H., Paul, B., Wang, X., Hesser, J., Becker, J., Garitte, B. & Müller, H. 2016*b*. The influence of different supports on the properties of the excavation damaged zone along the FE tunnel in the Mont Terri Underground Rock Laboratory. *In*: Norris, S., Bruno, J., Van Geet, M. & Verhoef, E. (eds) *Radioactive Waste Confinement: Clays in Natural and Engineered Barriers*. Geological Society, London, Special Publications, **443**. First published online September 26, 2016, https://doi.org/10.1144/SP443.18

SKB 2016. http://www.skb.com

Smutek, J., Hausmannova, L. & Svoboda, J. 2016. The gas permeability, breakthrough behaviour and re-sealing ability of Czech Ca–Mg bentonite. *In*: Norris, S., Bruno, J., Van Geet, M. & Verhoef, E. (eds) *Radioactive Waste Confinement: Clays in Natural and Engineered Barriers*. Geological Society, London, Special Publications, **443**. First published online June 22, 2016, https://doi.org/10.1144/SP443.5

Su, X., Nguyen, S., Haghighat, E., Pietruszczak, S., Labrie, D., Barnichon, J.-D. & Abdi, H. 2016. Characterizing the mechanical behaviour of the Tournemire argillite. *In*: Norris, S., Bruno, J., Van Geet, M. & Verhoef, E. (eds) *Radioactive Waste Confinement: Clays in Natural and Engineered Barriers*. Geological Society, London, Special Publications, **443**. First published online October 12, 2016, https://doi.org/10.1144/SP443.20

Svoboda, J., Pacovský, J., Dvořáková, M., Hanusová, I., Večerník, P. & Trpkošová, D. 2016. DOPAS EPSP experiment. *In*: Norris, S., Bruno, J., Van Geet, M. & Verhoef, E. (eds) *Radioactive Waste Confinement: Clays in Natural and Engineered Barriers*. Geological Society, London, Special Publications, **443**. First published online August 26, 2016, https://doi.org/10.1144/SP443.17

Toprak, E., Olivella, S. & Pintado, X. 2016. Coupled THM modelling of engineered barriers for the final disposal of spent nuclear fuel isolation. *In*: Norris, S., Bruno, J., Van Geet, M. & Verhoef, E. (eds) *Radioactive Waste Confinement: Clays in Natural and Engineered Barriers*. Geological Society, London, Special Publications, **443**. First published online September 26, 2016, https://doi.org/10.1144/SP443.19

Trpkošová, D., Večerník, P., Gondolli, J., Havlová, V., Svoboda, J. & Hanusová, I. 2016. Laboratory experiments on bentonite pellet saturation. *In*: Norris, S., Bruno, J., Van Geet, M. & Verhoef, E. (eds) *Radioactive Waste Confinement: Clays in Natural and Engineered Barriers*. Geological Society, London, Special Publications, **443**. First published online November 3, 2016, https://doi.org/10.1144/SP443.22

UK Government 2016*a*. http://www.gov.uk/government/collections/geological-disposal-facility-gdf-for-high-activity-radioactive-waste

UK Government 2016*b*. http://www.gov.uk/government/uploads/system/uploads/attachment_data/file/517677/ngs-guidance.pdf

United States Department of Environment 2016. http://www.wipp.energy.gov/index.htm

Vandersteen, K., Leterme, B. & Gedeon, M. 2016. Regional aquifer hydrogeochemistry in the confined aquifer system below the Boom Clay (NE Belgium): data analysis and modelling. *In*: Norris, S., Bruno, J., Van Geet, M. & Verhoef, E. (eds) *Radioactive Waste Confinement: Clays in Natural and Engineered Barriers*. Geological Society, London, Special Publications, **443**. First published online June 22, 2016, https://doi.org/10.1144/SP443.7

Regional aquifer hydrogeochemistry in the confined aquifer system below the Boom Clay (NE Belgium): data analysis and modelling

K. VANDERSTEEN*, B. LETERME & M. GEDEON

Institute for Environment, Health and Safety, Belgian Nuclear Research Centre (SCK•CEN), Boeretang 200, 2400 Mol, Belgium

**Correspondence: katrijn.vandersteen@sckcen.be*

Abstract: Geochemical data from the deep aquifer system below the Boom Clay in NE Belgium have been collected and reinterpreted. The data were obtained between 1980 and 2014 within the framework of radioactive waste disposal studies. Currently, groundwater in the deep aquifer system mainly ranges between $NaHCO_3$ type and NaCl type. Because of the low groundwater velocity in this aquifer system, re-equilibration with the mineralogical composition of the host formations generally occurs. The main geochemical indicators point out that the current formation waters are a mixture between the original saline marine porewaters and freshwater recharge. SE–NW gradients of increasing ion concentrations are observed and can be explained in agreement with the pattern of natural groundwater flow. Calcite dissolution and cation exchange are still ongoing in this freshening aquifer system. A low sulphate content in the deep aquifer system indicates sulphate reduction associated with organic matter degradation. Inverse modelling along a flow path using PHREEQC generally confirms the reactions that have been derived in the data analysis.

SCK•CEN has been involved for more than 30 years in the Belgian research programme on the long-term management of high-level and/or long-lived radioactive (Class B and C) waste co-ordinated by ONDRAF/NIRAS. One aspect of the research deals with the regional hydrogeology and geochemistry of the aquifer systems surrounding the Boom Clay in NE Belgium.

The confined aquifers between the Boom Clay and the Ypresian clays in NE Belgium include, from top to bottom, parts of the Oligocene aquifer system and the Ledo-Paniselian–Brusselian aquifer system, which are separated by the Bartoon aquitard system (Fig. 1). The aquifer system below the Boom Clay crops out in the south.

The results of previous studies on the geochemistry of the deep aquifers (Oligocene and Ledo-Paniselian–Brusselian aquifer systems) of NE Belgium can be found in Blommaert *et al.* (1988, 1994), Beaucaire *et al.* (2000) and Pitsch & Beaucaire (2000). Blommaert *et al.* (1988, 1994) classified groundwater below the Boom Clay as $NaHCO_3$- to NaCl-type waters. The change from $NaHCO_3$- to NaCl-type waters reflects the increasing salinity with depth in these northwards-sloping aquifers. Pitsch & Beaucaire (2000) noted that, in terms of alkalinity, most of the groundwater samples of the Oligocene aquifer system are in equilibrium with the host rock. Spatial variations in the hydrogeochemistry were also noted, with mineralization and carbonate concentrations found to increase from south to north. Chloride follows the same trend, with a less pronounced increase in the aquifer central part, indicating a slower flushing of southern recharge freshwater in the distal parts of the aquifer (Pitsch & Beaucaire 2000). A similar south–north gradient was observed for sulphate concentrations. Pitsch & Beaucaire (2000) recognized the influence of mixing freshwater with seawater on the measured ions. These previous studies were based on limited datasets. The first attempt at geochemical modelling of the Oligocene aquifer in NE Belgium was undertaken by Beaucaire *et al.* (2000). While carbonates and silicates were included in the model, cation exchange and organic matter were not considered. Walraevens *et al.* (2007) implemented a one-dimensional (1D) geochemical model to represent the freshening of both the Bartonian Clay and the underlying Ledo-Paniselian aquifer along a flow pathway in NW Belgium. According to this model, the hydrogeochemical variations in the Ledo-Paniselian aquifer are the result of a spatial separation of the marine cations being sequentially desorbed by the recharging freshwater. Then, water compositions of the porewater leaving the Bartonian Clay were used to freshen the Ledo-Paniselian aquifer, in which the model included cation exchange and sulphate reduction (Walraevens *et al.* 2007).

This study presents a synthesis and re-evaluation of available hydrogeochemistry data, in support of modelling the regional groundwater hydrogeochemistry in the aquifers between the Boom Clay and the Ypresian clays in NE Belgium. First, the available hydrogeochemistry data are presented. The objective is then to identify the main geochemical processes that have led to the current water

From: NORRIS, S., BRUNO, J., VAN GEET, M. & VERHOEF, E. (eds) 2017. *Radioactive Waste Confinement: Clays in Natural and Engineered Barriers*. Geological Society, London, Special Publications, **443**, 9–28.
First published online June 22, 2016, https://doi.org/10.1144/SP443.7

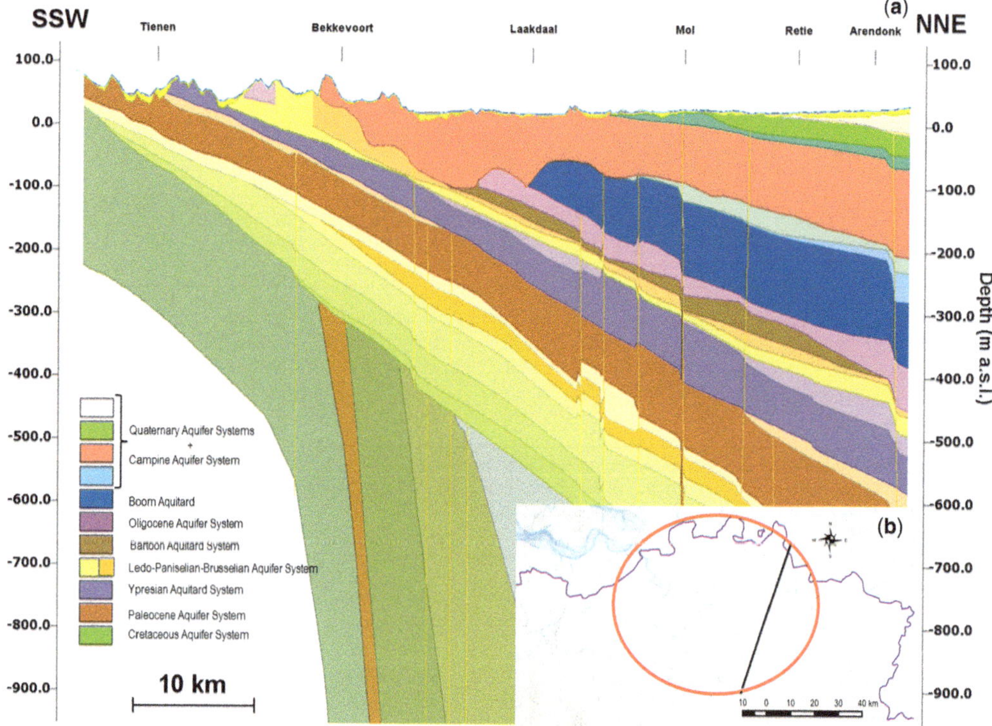

Fig. 1. (**a**) SSW–NNE hydrogeological profile in the Campine according to the 3D geological model of VITO (Matthijs *et al.* 2013). (**b**) The scope of the hydrogeochemical research is limited to the area marked in the circle.

quality in the confined aquifers below the Boom Clay in NE Belgium. This constitutes a first step in the prediction of future groundwater chemistry in the studied aquifers.

Geological and hydrogeological context

In NE Belgium, several marine transgressions and regressions took place during the Neogene (24–2.5 Ma). At the end of the Neogene, the sea drew back from this area after having deposited shallow-marine and estuarine sands and some clay. From then on, the connate marine porewater in the aquifers was gradually diluted as the aquifers were flushed with fresh recharge water. Water–rock interactions, including cation exchange, began to play a role in the deep aquifer system. This led to changes in the hydrogeochemistry over time. In the regional context of the Campine, the main geological units influencing and bounding the studied hydrogeological system are shown in Figure 2a. The Campine Basin, characterized by stacked units of sand with various amounts of clay, is situated between the shallow bedrock of the Brabant Massif in the south and west, and the rapidly subsiding Roer Valley Graben in the east and north. In the NW, the Campine Basin merges into the southern North Sea Basin. This leads to the assumption that, in general, groundwater flows in a SE–NW direction under natural hydraulic gradient conditions (Fig. 2a). However, today, groundwater flow in the confined aquifer system below the Boom Clay is considerably influenced by groundwater extraction situated close to where the aquifers crop out in the south, causing the groundwater flow, even in the very north of Belgium, to be directed towards areas characterized by high groundwater extraction (Vandersteen *et al.* 2014*b*). So far, we have found no evidence for other phenomena, such as the presence of relict groundwater pressures from, for example, basin subsidence or sea-level changes. The fact that these deep aquifers are influenced, even in the northern part of the study area, by the rather recent groundwater extraction in the southern part of the aquifer, implies that the hydraulic conductivity is still significantly large and that propagation of pore-pressure changes in the aquifer will be relatively fast, making the occurrence of relict groundwater pressure seem unlikely.

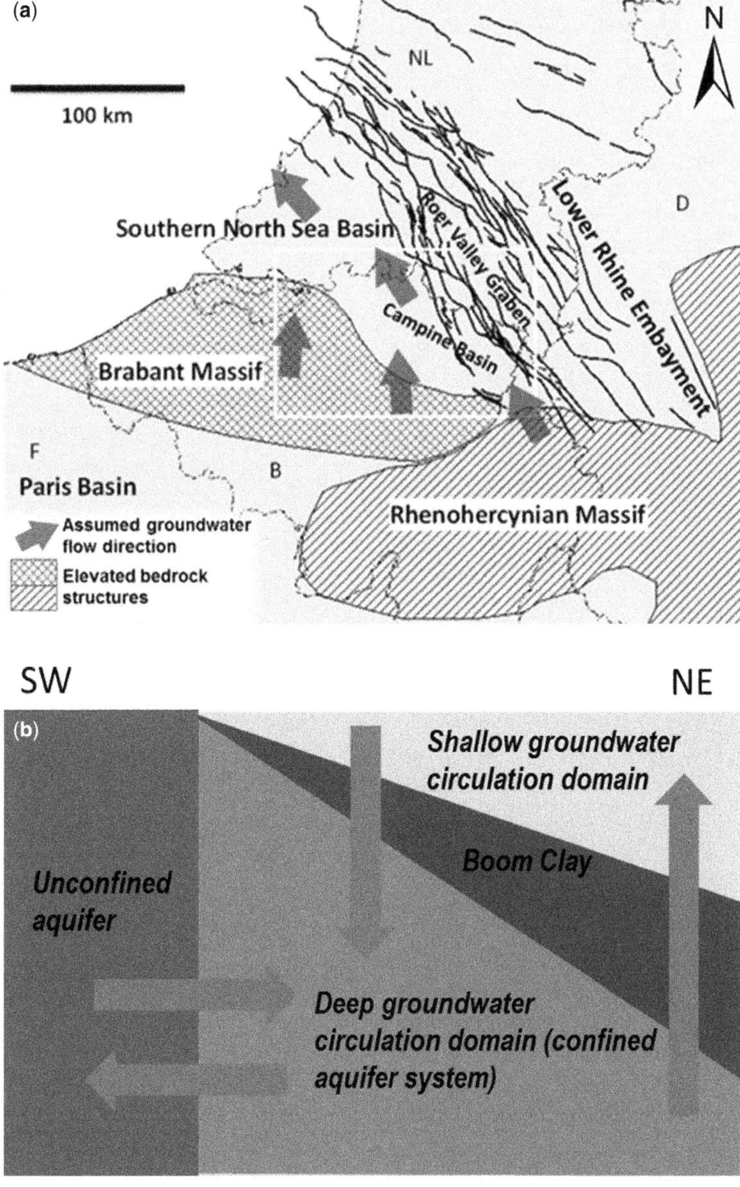

Fig. 2. (**a**) Structural geological framework in Western Europe (after Beerten *et al.* 2013) (the white rectangle denotes the study area) with an indication of the assumed natural groundwater flow direction. (**b**) Schematic overview of groundwater circulation in the deep aquifers of NE Belgium under natural conditions.

Water balance calculations with a regional groundwater flow model for the confined aquifers in NE Belgium (Vandersteen *et al.* 2012, 2014*b*) indicate that the system is hydraulically closed under natural conditions (Vandersteen *et al.* 2013). The lack of a groundwater outlet in the deep aquifers causes the groundwater in these aquifers to rise up into the shallow groundwater circulation domain (including the unconfined parts of the deep aquifer system), where it discharges into the rivers (Fig. 2b). The water balance calculations also indicate that, within the deep aquifer system, water is exchanged between the Oligocene and the Ledo-Paniselian–Brusselian aquifer systems: about 50% of the water that enters the Oligocene aquifer system originates from the Ledo-Paniselian–Brusselian aquifer system, while about the same amount of water circulates back to the Ledo-Paniselian–Brusselian aquifer system from the Oligocene aquifer system (Vandersteen *et al.* 2013). Modelled horizontal Darcy velocities in the Oligocene and Ledo-Paniselian–Brusselian aquifer systems are estimated at around 4×10^{-12} and 8×10^{-11} m s^{-1}, respectively (Vandersteen *et al.* 2012). The groundwater age in the northern part of the deep aquifer system (NE Belgium) was roughly estimated to be up to 850 ka in the north, based on solute breakthrough simulated by the transport model using the flow field of the regional groundwater flow model (Rogiers *et al.* 2012). However, ^{14}C analysis in the PHYMOL project (synthesized in Marivoet *et al.* 2000) resulted in an estimated groundwater age for the confined aquifer system of up to only 47 ka. The real groundwater age probably lies in-between, because of the presence of modelling uncertainties and because the age dating method used in the PHYMOL project does not account for the mixing of groundwater and, implicitly, assumes a piston flow model that, for asymptotic decay methods, leads to a considerable underestimation of the groundwater age (Bethke & Johnson 2008).

Lithology

The Ledo-Paniselian–Brusselian aquifer system mainly consists of fine to coarse glauconitic and calcareous sands, containing limestone benches, and

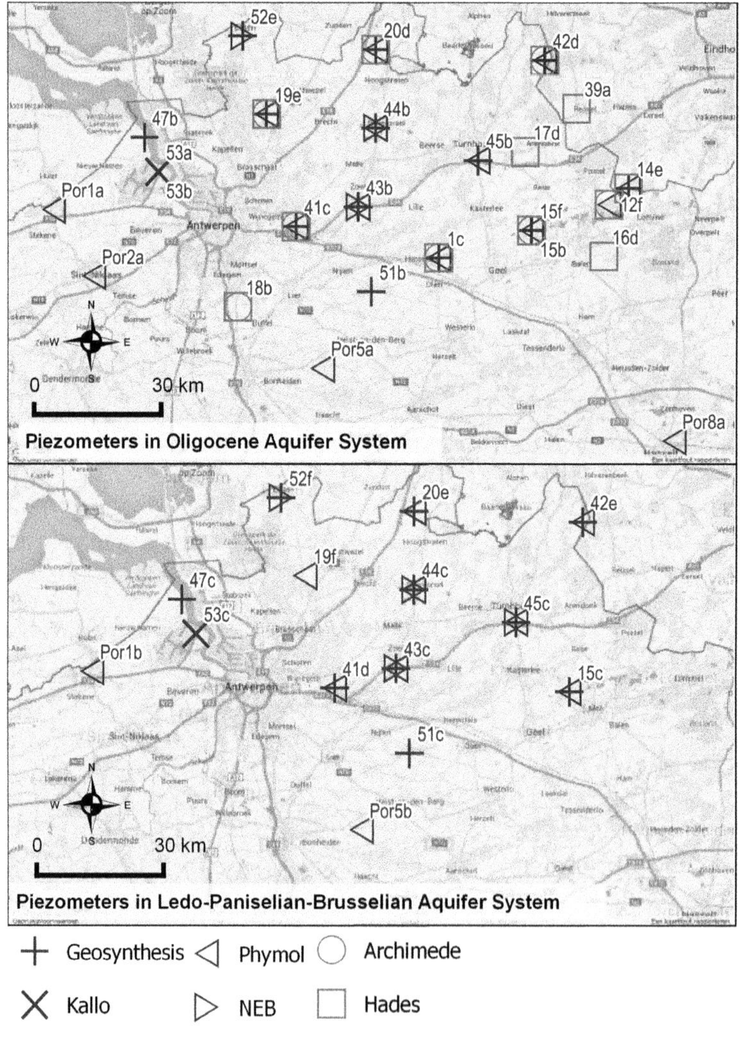

Fig. 3. Location of the piezometers providing hydrogeochemical information on the groundwater composition.

Table 1. *Detailed mineralogy of the Doel-2B sample taken at a depth of 151.7 m below the surface in the Oligocene aquifer system, determined by FT-IR**

Silica minerals		Feldspar		Carbonate		Clay		Others	
Quartz	44.5	K-Feldspar	6.0	Calcite	0.3	Illite	7.0	Muscovite	0.0
Chert	3.9	Na-Feldspar	5.4	Dolomite	0	Smectite	5.7	Biotite	0.0
Opal	1.1	Ca-Feldspar	0.0	Ankerite	0	Kaolinite	1.1	Pyrite	0.5
				Aragonite	0	Chlorite	0.0	Gypsum	0.0
				Siderite	0	Glauconite	24.3	Anhydrite	0.0
				Magnesite	0.3			Barite	0.2
				High-Mg calcite (HMC)	0			Hematite	0.0
								Fluorite	0.0
								Celestite	0.0
Total	**49.4**		**11.4**		**0.6**		**38.1**		**0.7**

*FT-IR (Fourier transform infrared spectroscopy). Results are given in wt%. All values below 1 wt% should be taken as being indicative only. *Source*: Schlumberger-Doll Research (1999).

locally thin marl and clay lenses. Lagrou *et al.* (2004) reported high carbonate concentrations for the Brussels and Lede sands, attributed to the abundance of calcareous microfossils and other skeletal grains. The Oligocene aquifer system consists of sandy clay and clayey sands intercalated with thin discontinuous clay layers. Unfortunately, very limited research has been carried out on the mineralogy of the deep aquifers. One core sample from the ON-Doel-2B borehole (at site 47 in Fig. 3, situated in the Oligocene aquifer system) has been characterized by Schlumberger-Doll Research (1999). The sample is mainly composed of silica minerals (49.4%), clay (38.1%) and feldspar (11.4%) (Table 1).

Regional hydrochemistry

Data collection and representation

Figure 3 provides an overview of all locations where deep aquifer hydrogeochemical data were collected

Table 2. *Overview of data available and studies related to the groundwater chemistry of the aquifers below the Boom Clay at the regional scale*

References	Location	Analyses	Sampling dates
HADES: Blommaert *et al.* (1988, 1994)	Twelve piezometers in the Oligocene aquifer system	pH, major ions, stable isotopes	1981–86
ARCHIMEDE: Griffault *et al.* (1996); Beaucaire *et al.* (2000)	Eight piezometers in the Oligocene aquifer system	pH, alkalinity, major ions, stable isotopes	1992–93
PHYMOL: Philippot *et al.* (2000); Pitsch & Beaucaire (2000)	Fifteen piezometers in Oligocene and 12 in Ledo-Paniselian-Brusselian aquifer system	pH, alkalinity, major ions, TOC* Stable isotopes	1997–98 1997–98
NEB: Wemaere *et al.* (2004a, b, 2005a, b)	Three piezometers in the Oligocene and four in the Ledo-Paniselian-Brusselian aquifer system	pH, major ions, TIC*, TOC*	1996
GEOSYNTHESIS: Labat (2011); Vandersteen *et al.* (2014a, b)	Fourteen piezometers in the Oligocene and 10 in the Ledo-Paniselian-Brusselian aquifer system	Major and minor ions, alkalinity	2008–09, 2013, 2014
KALLO: Labat *et al.* (2014)	Two piezometers in the Oligocene and one in Ledo-Paniselian-Brusselian aquifer system	Major and minor ions, stable isotopes	2010

*TIC, total inorganic carbon; TOC, total organic carbon.

in the framework of radioactive waste disposal research, while Table 2 gives a summary of the data availability and source. In the 1980s, groundwater samples were usually derived from the piezometric network available at that time. These analyses, grouped under the HADES project, were summarized by Blommaert *et al.* (1988, 1994). This dataset includes hydrogeochemical analyses and stable isotopes of groundwater samples from the Oligocene aquifer system (Fig. 3; Table 2). A decade later, more hydrogeochemical data were collected from the Oligocene aquifer system in the framework of the ARCHIMEDE project (Griffault *et al.* 1996; Beaucaire *et al.* 2000). Again, chemical composition and stable isotopes were analysed. The next geochemical sampling campaign was acquired from the European PHYMOL project (Marivoet *et al.* 2000; Philippot *et al.* 2000; Pitsch & Beaucaire 2000). It focused on sampling the aquifers below the Boom Clay. During the two regional site characterization campaigns NEB96 and NEB05, a limited amount of hydrogeochemical analyses was performed on aquifer groundwater samples extracted from the installed piezometers in the Zoersel (Wemaere *et al.* 2004*b*), Rijkevorsel (Wemaere *et al.* 2004*a*), Turnhout (Wemaere *et al.* 2005*a*) and Essen boreholes (Wemaere *et al.* 2005*b*). In 2009, geochemical and stable isotope data were collected at Kallo from deep aquifer groundwater samples (Labat *et al.* 2014). The recent sampling campaign, GEOSYNTHESIS (2008–09, 2013, 2014), yielded analytical results of groundwater composition from the existing, well-functioning, piezometers, with a screen located in aquifers below the Boom Clay

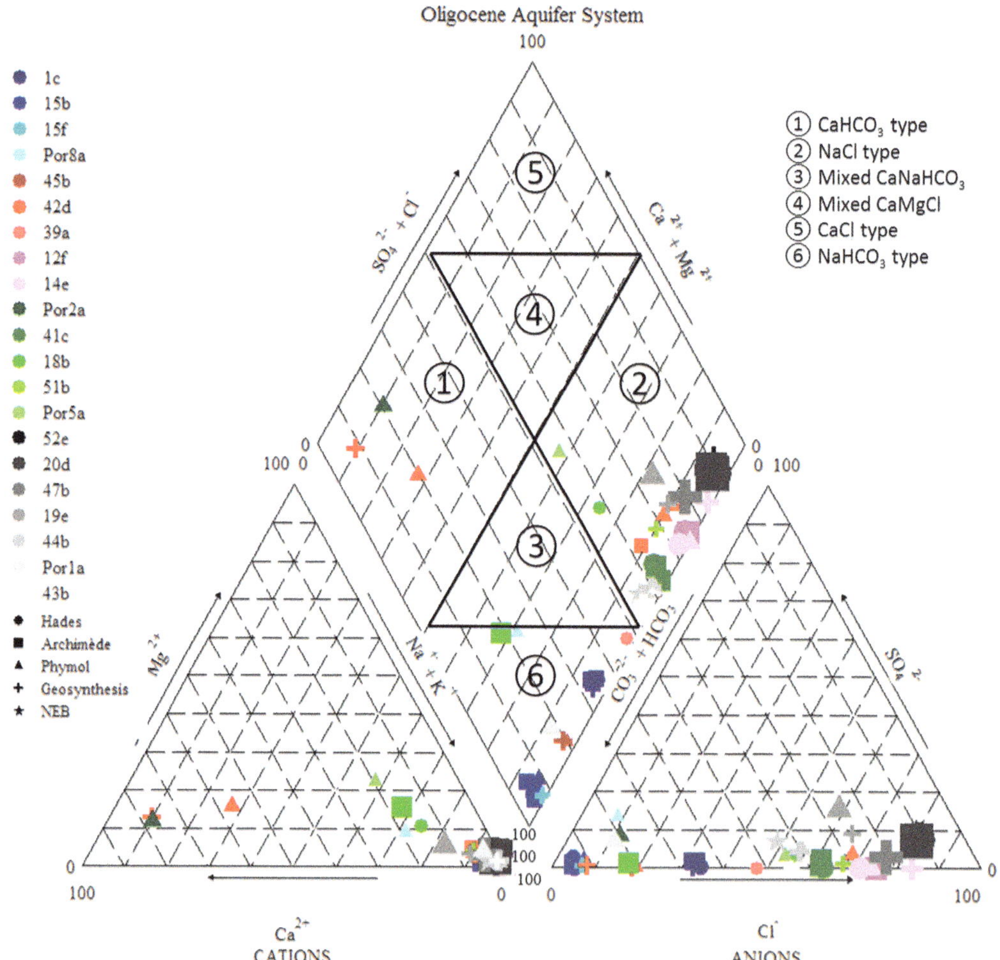

Fig. 4. Piper diagram of all measurements for the Oligocene aquifer system for the different measurement campaigns. The symbol size is proportional to the total dissolved solid content (TDS).

(Labat 2011; Vandersteen & Labat 2014; Vandersteen *et al.* 2014*c*).

In most cases, HCO_3^- and CO_3^{2-} ions were not measured directly but were calculated in this study from pH and alkalinity measurements using the PHREEQC software (Parkhurst & Appelo 1999). Samples that showed an electrical imbalance of more than 5% were discarded. The evolution in space of the deep aquifer general hydrochemistry, as well as the variability over different measurement campaigns, is illustrated in a Piper diagram (Piper 1953) (Fig. 4). The Piper diagram is a common way of classifying water samples by plotting the relative abundance of cations and anions. In this way, the nature of a given sample and its relationship to others is easily visualized. Each measurement campaign (HADES, ARCHIMEDE, NEB, PHYMOL and GEOSYNTHESIS) is represented by a different symbol. Complementary to the Piper diagrams, the Stuyfzand classification system (Stuyfzand 1986, 1989) provides additional information on the aquifer general geochemistry. This classification system refers to the general model of a freshening aquifer described by Appelo (1994). In this model, the displacing ion is Ca^{2+}, which has a higher affinity for the cation exchanger than the displaced saltwater ions Na^+, K^+ and Mg^{2+}. The freshening of the aquifer thus produces fronts, whereby first Na^+, then K^+ and, finally, Mg^{2+} are displaced (Appelo 1994). According to the Stuyfzand classification system,

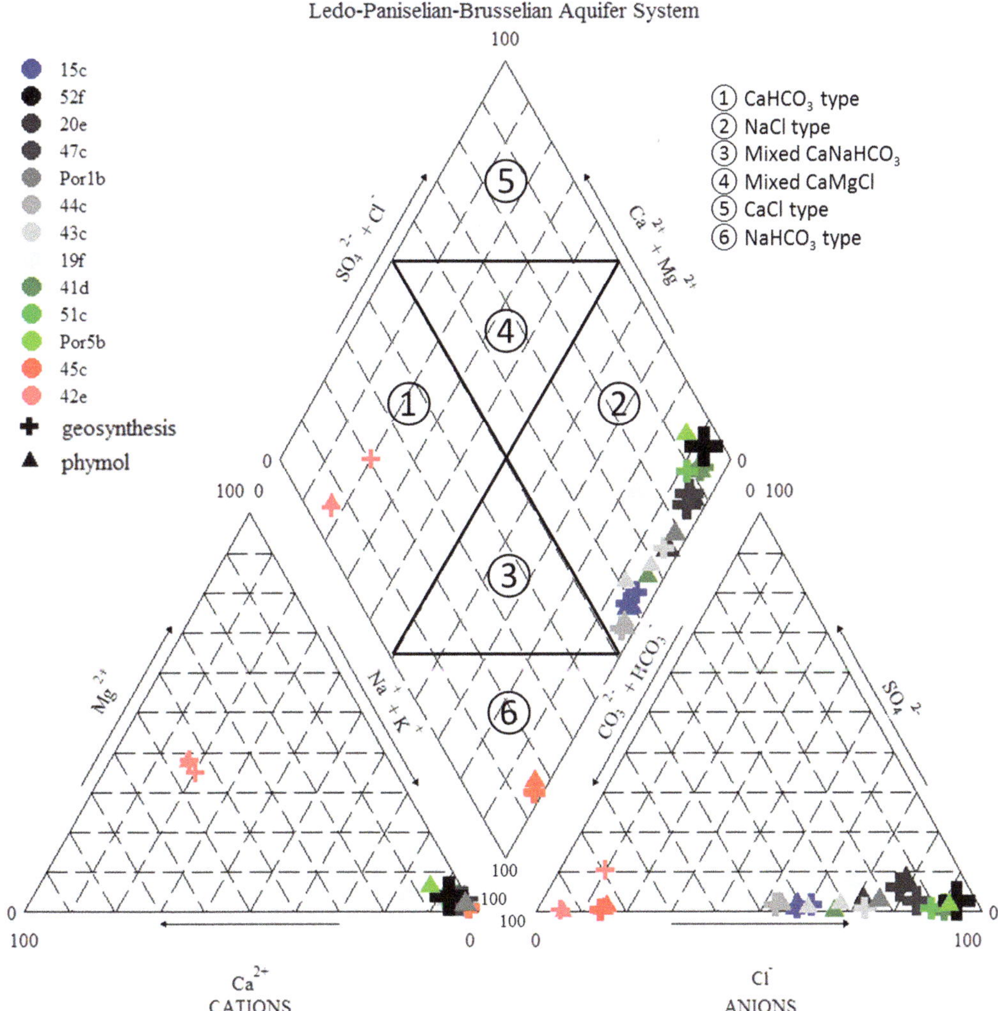

Fig. 5. Piper diagram of all measurements for the Ledo–Paniselian–Brusselian aquifer system for the different measurement campaigns. The symbol size is proportional to the total dissolved solid content (TDS).

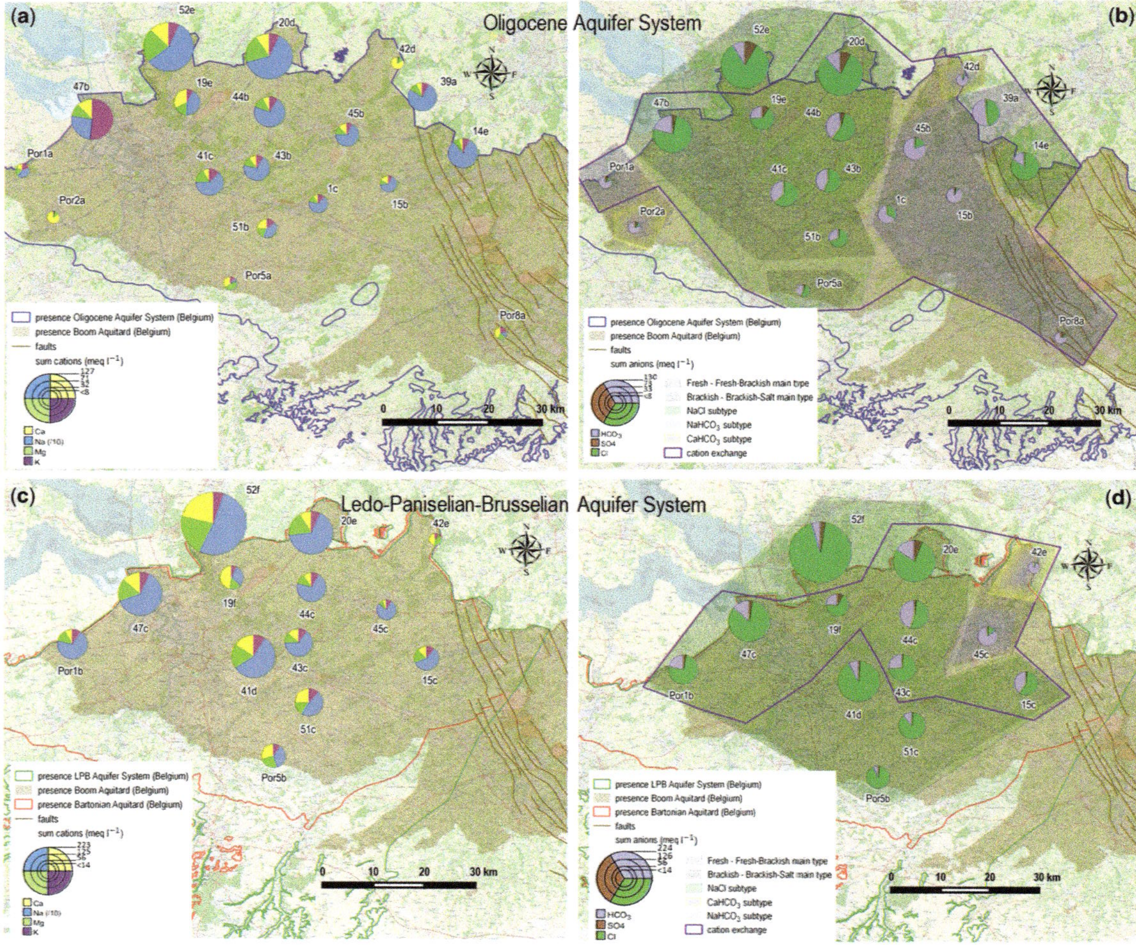

Fig. 6. Sum of major cations/anions (Ca, Na, Mg, K, HCO_3^-, SO_4^{2-}, Cl^- in meq l^{-1}) from the GEOSYNTHESIS project supplemented with data from previous campaigns, in the Oligocene (**a**) & (**b**) and the Ledo–Paniselian–Brusselian (**c**) & (**d**) aquifer systems. The circle radius is proportional to the square root of the total cation/anion content (meq l^{-1}). Note that the proportion of Na^+ in the pie chart is based on its concentration divided by 10 for clarity. The main type, subtype and class based on the Stuyfzand (1986) classification system are given for both aquifer systems.

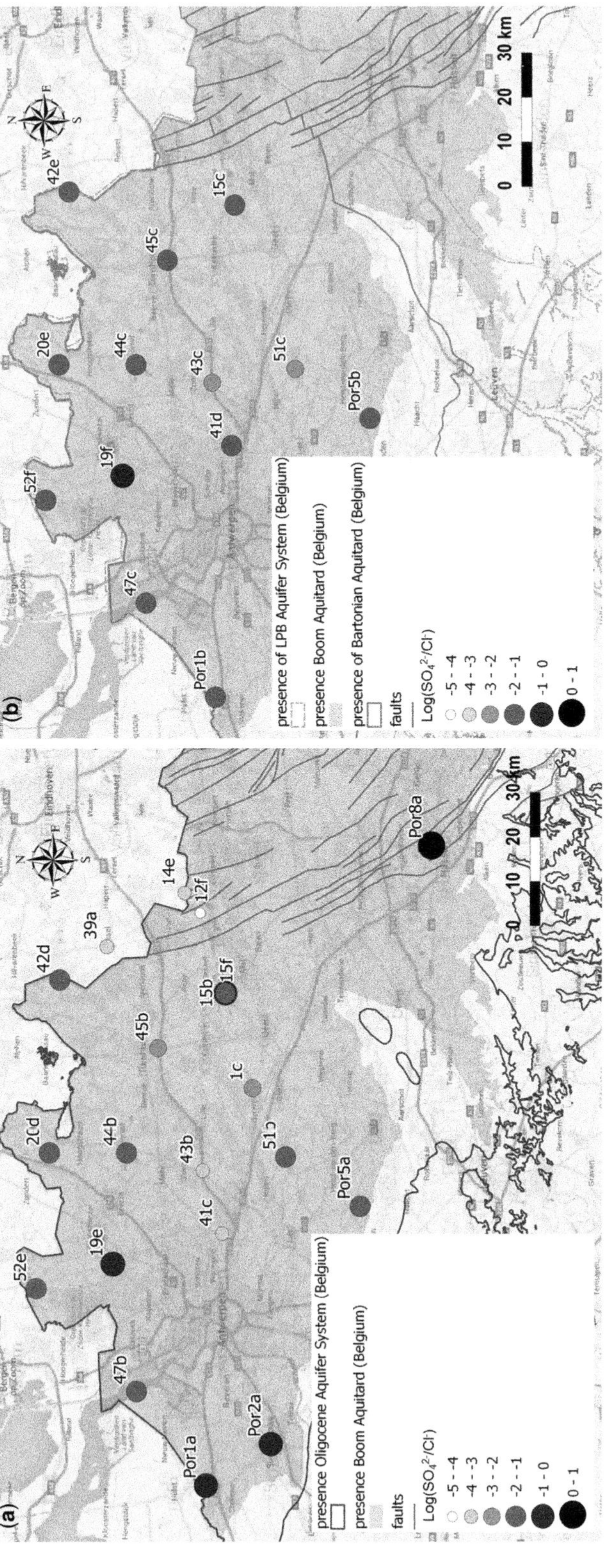

Fig. 7. Ratio of SO_4^{2-}/Cl^- ((meq l^{-1})/(meq l^{-1})) in (**a**) the Oligocene and (**b**) the Ledo–Paniselian–Brusselian aquifer systems.

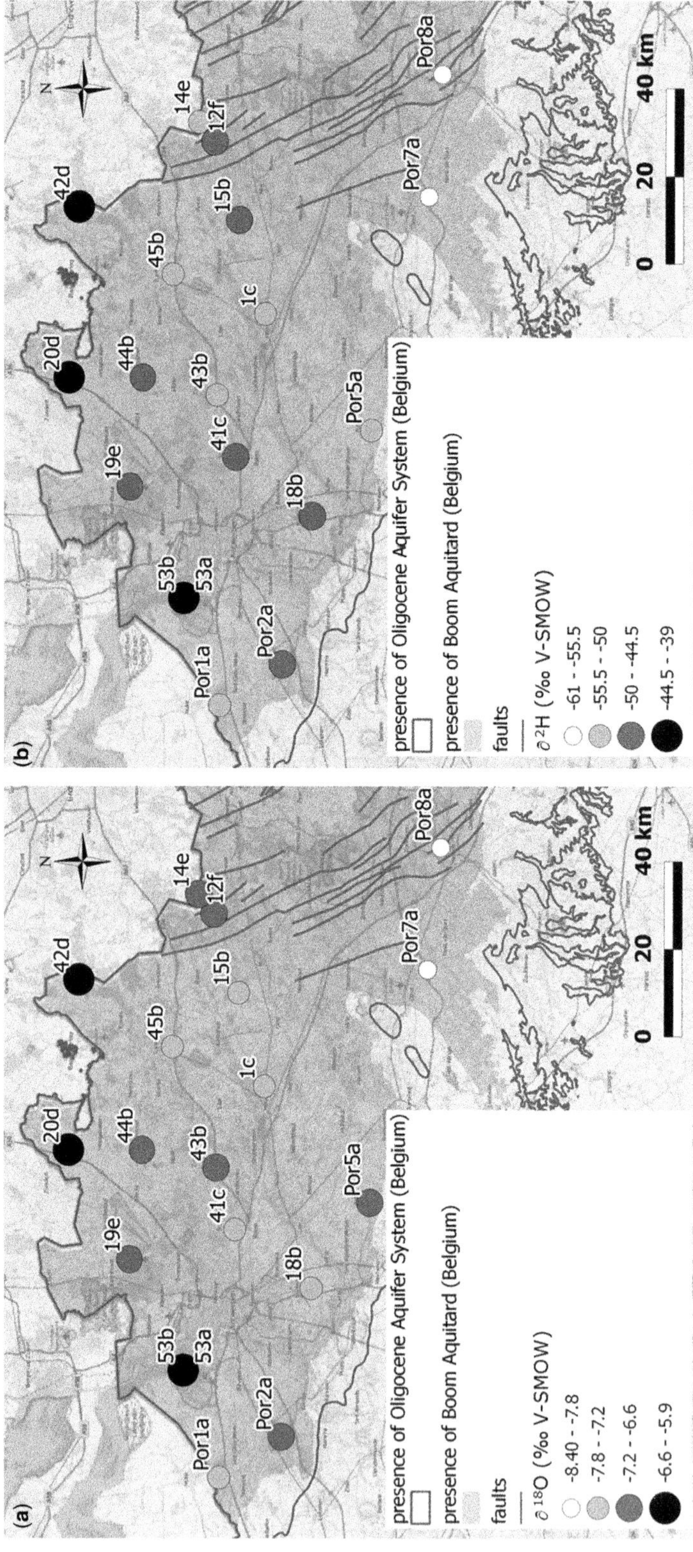

Fig. 8. **(a)** $\delta^{18}O$ and **(b)** $\delta^{2}H$ (‰ SMOW) in the Oligocene aquifer system.

the main groundwater type is determined by the chloride content: it ranges from very oligohaline (<5 mg l^{-1} of Cl^-) to hyperhaline ($>2 \times 10^4$ mg l^{-1} of Cl^-). The main cation and anion define the groundwater subtype. The classification system also takes cation exchange into account.

Data analysis

Major ions. It is clear from Figure 4 that most of the examined samples in the Oligocene aquifer system are of the NaCl or $NaHCO_3$ type. The different measurement campaigns show similar results, except for a couple of piezometers. For piezometer 42d, measurement campaigns ARCHIMEDE and PHYMOL of February 1998 show a NaCl-type water, while samples from PHYMOL from March 1998 and GEOSYNTHESIS campaigns are classified as a $CaHCO_3$-type water. This was noted by Pitsch & Beaucaire (2000), who suggested that this was related to recent Neogene water intrusion into the deeper Oligocene aquifer system through a fault associated with the Roermond earthquake in 1992, or the existence of a hydraulic connection along the piezometer hole (a well-completion artefact). The sampling campaign of February 1998 yielded NaCl-type water, which makes a well artefact a more logical explanation. For piezometer 18b, the sample from the HADES campaign shows a NaCl-type water, while the sample from the ARCHIMEDE campaign corresponds to a $NaHCO_3$ type. No recent samples have been analysed, which precludes a definitive classification for 18b. Figure 4 also suggests a correlation of the Cl^- content with the concentration of total dissolved solids (TDS).

In general, the different measurement campaigns show similar results for the groundwater chemistry within the Ledo-Paniselian–Brusselian aquifer system, as indicated in Figure 5. The largest deviation can be seen for sample 41d, where one of the two PHYMOL samples has a lower Cl^- content than the other PHYMOL sample and that from the GEOSYNTHESIS campaign. For sample 42e, one of the two samples of the GEOSYNTHESIS campaign has a higher sulphate and chloride content, and lower (bi)carbonate content, compared to the other two samples (GEOSYNTHESIS and PHYMOL campaigns). No clear explanation was found for these differences. An increase in TDS with Cl^- content, observed for the Ledo-Paniselian–Brusselian aquifer system, reflects the variations seen in Figure 5.

Figure 6 shows selected results of major regional ion geochemistry for the Oligocene and Ledo-Paniselian–Brusselian aquifer systems. Most data are taken from the latest GEOSYNTHESIS measurement campaign as it has the most complete dataset, supplemented with data from previous campaigns. Figure 6a presents the total cation concentration (meq l^{-1}) of groundwater samples from the Oligocene aquifer system. The proportions of Ca^{2+}, Na^+ (divided by a factor 10 for clarity), K^+ and Mg^{2+} are also displayed. Figure 6b presents the total anion concentration and proportional pie charts of HCO_3^- and CO_3^{2-}, SO_4^{2-}, and Cl^- from the Oligocene aquifer system. The groundwater main type and subtype are also indicated (these types are identical to the groundwater type defined in the Piper diagram), as well as the presence of cation exchange according to the Stuyfzand (1986) classification. A SE–NW gradient of increasing total anion and cation concentrations is clearly visible, which is in accordance with the groundwater flow direction. However, piezometer 19e shows a quite low ion concentration, despite its position in the NW of the domain. In piezometer 42d, an unusually low ion concentration is present, which might be related to the presence of a fault, providing a pathway for the intrusion of relatively dilute Neogene water in the Oligocene aquifer system, or, more likely, to a well-completion artefact. In piezometers 14e and 39a (east), a high total ion concentration is present, which is, at first sight, not in accordance with the groundwater flow direction. A lower groundwater velocity, reducing the recharge by freshwater into the deep aquifer, in this region might explain these higher values. This lower groundwater velocity is possibly caused by the deeper burial of the aquifer in the fault system, compared to the situation west of the fault system. Deeper burial can lead to more compaction and lower hydraulic conductivity, thus a lower groundwater velocity. Piezometers Por1a, Por2a, Por5a and Por8a have a low total ion concentration (Fig 6a, b) as these piezometers lie close to the Boom Clay outcrop and the connate marine porewater is already substantially diluted by fresh recharge water. When looking at the main water type and subtype indicated in Figure 6b, the saline groundwater types (brackish and brackish-salt) are situated towards the centre and west of the domain. Exceptions are piezometers 14e and 39a, which also

Table 3. *Type and number of stable isotope analyses performed in the different projects and aquifers*

Aquifer system	Project	$\delta^{18}O$	δ^2H	$\delta^{13}C$
Oligocene	HADES	6	0	13
	ARCHIMEDE	15	15	8
	PHYMOL	20	21	24
	KALLO	2	2	
Ledo-Paniselian–Brusselian	HADES	3	0	8
	ARCHIMEDE	0	0	0
	PHYMOL	24	24	18
	KALLO	1	1	

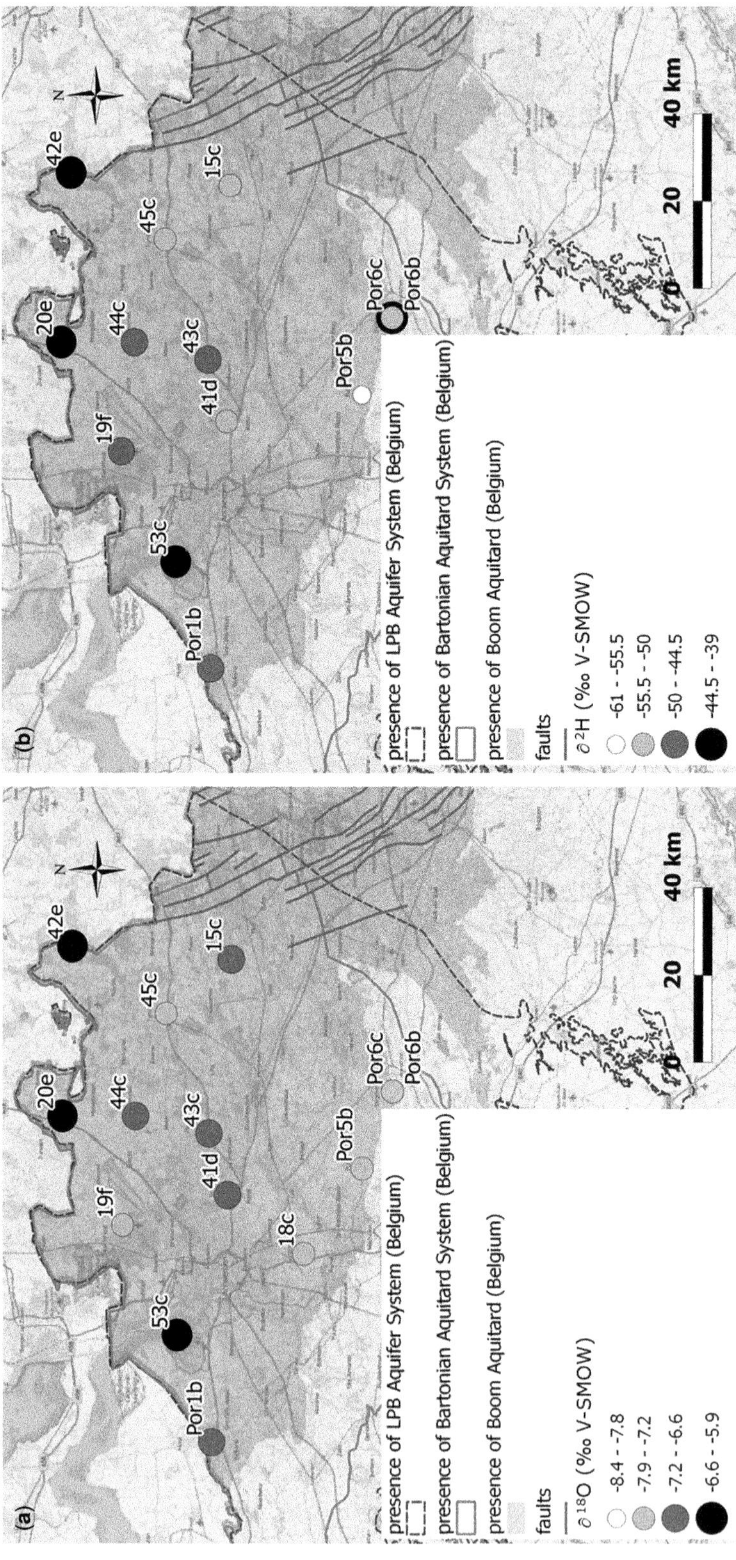

Fig. 9. **(a)** δ^{18}O and **(b)** δ^{2}H (‰ SMOW) in the Ledo–Paniselian–Brusselian aquifer system.

have this type of saline groundwater, but they are situated in the east. The fresh and fresh-brackish main types are found in the east of the domain and near the Boom Clay outcrop. The groundwater subtype (according to the Stuyfzand classification) indicates that $NaHCO_3$-type water is present in the east and near the outcrop in the west, while NaCl-type water is found in the centre and west of the domain. An exception is piezometer 14e, which also indicates a NaCl-type water. Piezometers 42d and Por2a have $CaHCO_3$-type water, indicative of fresh groundwater. The purple line in Figure 6b indicates the presence of cation exchange processes. The related area covers almost the entire domain, except for piezometers Por2a and 52e. The analysis of water types in the Oligocene aquifer system is clearly in accordance with the groundwater flow of a freshening aquifer.

Figure 7a shows the ratio of sulphate to chloride concentrations in the Oligocene aquifer system. Near the Boom Clay outcrop, the sulphate proportion is higher and the ratio is similar to fresh recharge water, this as a result of the oxidation of pyrite by oxidizing water. In the centre and east of the aquifer, the ratio is close or equal to zero, which reflects sulphate reduction (associated with the degradation of organic matter). Towards the northwest, the ratio increases, approaching the seawater ratio (0.13).

Figure 6c, d presents the total cation and anion concentration, and proportional pie charts, of the major cations and anions from the Ledo-Paniselian–Brusselian aquifer system. Similar to the Oligocene aquifer system, the SE–NW gradient of the increasing total ion concentrations is visible, although the trend is less clear in this case. The main exception is the low ion concentration in piezometer 42e. At piezometer 41d, a higher than expected ion concentration is present, the cause of which is unknown. The brackish or brackish-saline main type of water, which coincides with the NaCl-subtype water, is situated over the largest part of the domain (Fig. 6d). The fresh and fresh-brackish main types are found in the east of the domain, at piezometers 42e and 45c. Piezometer 42e has a $CaHCO_3$-subtype water, indicative of fresh groundwater (similar to piezometer 42d situated in the Oligocene aquifer system), while piezometer 45c has $NaHCO_3$-subtype water. The purple line in Figure 6d indicates the presence of cation exchange processes, influencing the centre of the domain. As in the Oligocene aquifer system, the analysis of water-type variation in the groundwater chemistry in the Ledo-Paniselian–Brusselian aquifer system is consistent with the groundwater flow in a progressively freshening aquifer.

Figure 7b shows the ratio of sulphate to chloride concentrations in the Ledo-Paniselian–Brusselian aquifer system, demonstrating a trend similar to the Oligocene aquifer system (Fig. 7a).

Stable isotopes. As already indicated in Table 2, stable isotope data of groundwater ($\delta^{18}O$, δ^2H and $\delta^{13}C$) have been collected in the framework of the HADES, ARCHIMEDE, PHYMOL and KALLO projects. Table 3 further details the available stable isotope analyses.

Figures 8 and 9 show the $\delta^{18}O$ and δ^2H values for the Oligocene and the Ledo-Paniselian–Brusselian aquifer systems, respectively. The isotope values at each point are averaged from several measurements. The relative concentration of heavy isotopes ($\delta^{18}O$ and δ^2H) seems to increase from south to north in the Oligocene aquifer system (Fig. 8). This probably reflects a mixture of fresh recharge water with older, deeper saline water (for seawater $\delta^{18}O = \delta^2H = 0$). The same interpretation has previously been made by Philippot *et al.* (2000) in the PHYMOL project and was also shared by Beaucaire *et al.* (2000), who described the saline front as coming from the NW of the aquifer. This pattern of mixing between a meteoric end member (in the south) against a saline end member (in the north) is also found for the Ledo-Paniselian–Brusselian aquifer system (Fig. 9). The $\delta^{18}O$ and δ^2H data for both aquifer systems are plotted together with the global meteoric water line (GMWL) in Figure 10. The equation for GMWL is given by:

$$\partial^2 H = 8\partial^{18} O + 10. \quad (1)$$

In Figure 10, we use a local variant of the GMWL, derived from local weather station data, having a

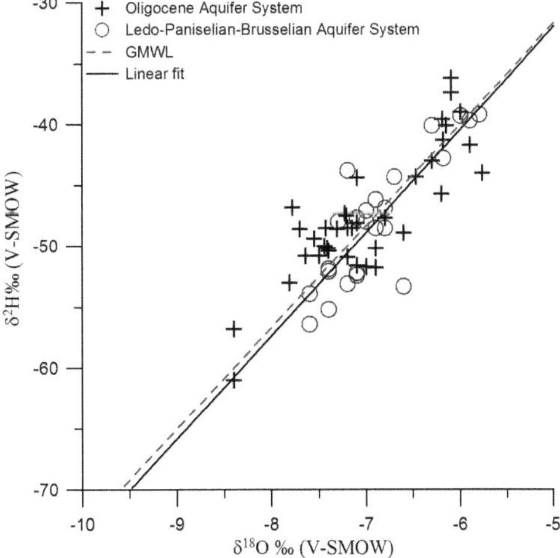

Fig. 10. Plot of δ^2H v. $\delta^{18}O$ for the Oligocene and the Ledo–Paniselian–Brusselian aquifer systems, in relation to the (local variant of the) general meteoric water line (GMWL) and a linear fit.

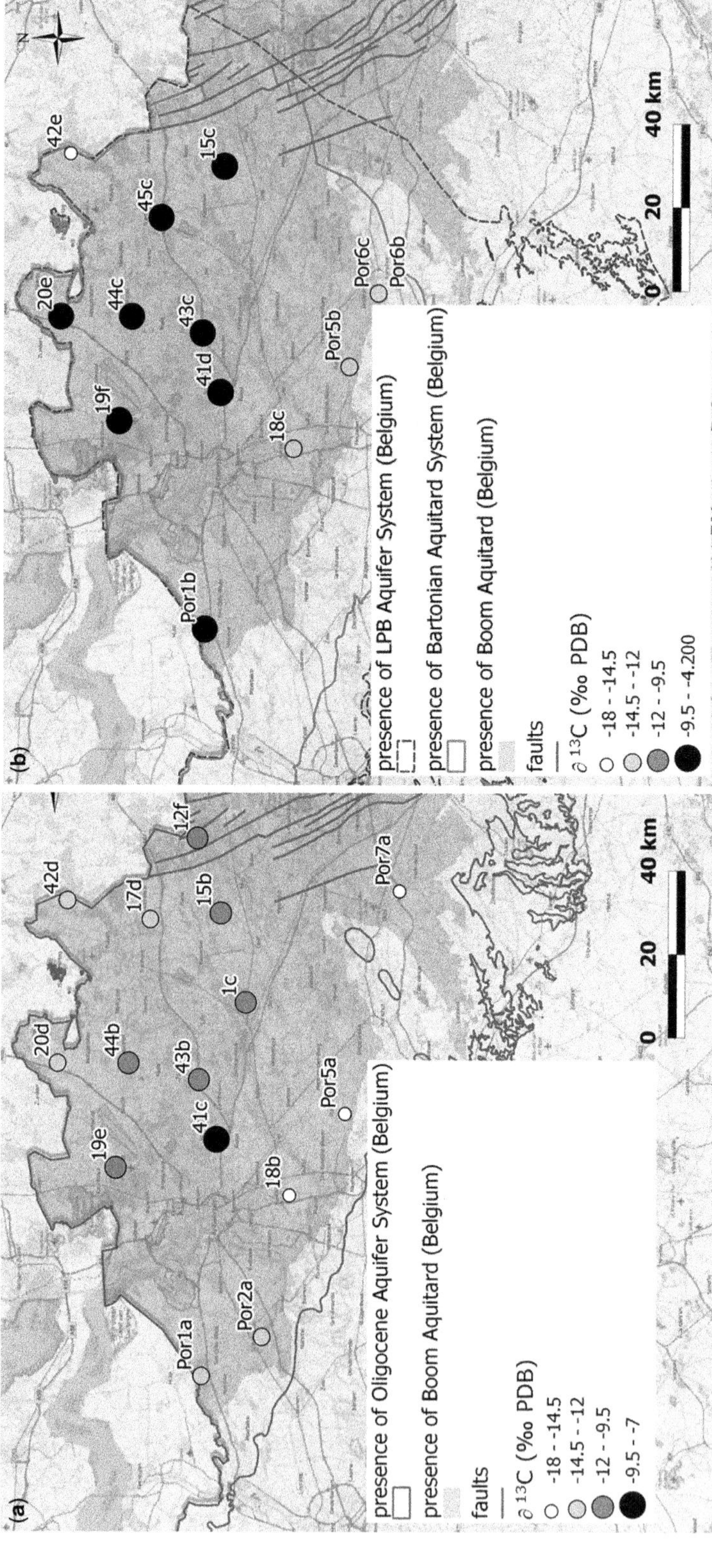

Fig. 11. $\partial^{13}C$ isotope values in **(a)** the Oligocene and **(b)** the Ledo–Paniselian–Brusselian aquifer systems (‰ PDB).

slightly different slope and intercept compared to equation (1) (Philippot *et al.* 2000).

The linear fit for the $\delta^{18}O$ and $\delta^{2}H$ values for the two aquifer systems coincides with the GMWL (Fig. 10). There is no visible difference between values from the Oligocene and the Ledo-Paniselian–Brusselian aquifer systems, which possibly points to the mixing of water from the two aquifer systems, as suggested by Philippot *et al.* (2000) and which is also confirmed by the groundwater modelling, as explained earlier in the text (Vandersteen *et al.* 2013).

Figure 11 shows that in the centre and the north of both aquifer systems, the $\delta^{13}C$ values are close to −12‰, which is the common value for CO_2-closed systems (Appelo & Postma 2006). In a system closed for CO_2 gas, carbonic acid is consumed and not replenished as dissolution of carbonate minerals (calcite) proceeds. In the Oligocene aquifer system, $\delta^{13}C$ concentrations vary in between −16.5 and −7.8‰. In general, concentrations are lower in the Boom Clay outcrop area and in the SE (piezometers 18a, Por5a, Por7a and Por8a). In the centre and the west of the domain, higher values point to renewed carbonate dissolution that is most likely to be due to cation exchange: that is, the loss of Ca^{2+} from solution into the cation exchange sites (in this case, clay minerals) (Appelo & Postma 2006). However, Walraevens (1990) proposed another explanation for the $\delta^{13}C$ variation in the NW Belgium part of the Ledo-Paniselian aquifer system. In that study, it was suggested that high $\delta^{13}C$ values could also be due to higher values of $\delta^{13}C$ in soil-CO_2 (i.e. $\delta^{13}C > -25‰$) when the cold and dry climate of Pleistocene times caused the vegetal activity to be very low. In the Ledo-Paniselian–Brusselian aquifer system, measured $\delta^{13}C$ values are between −17.6 and −4‰. Similar to the Oligocene aquifer system, $\delta^{13}C$ are lower near the outcrop of the clay layers, and higher towards the centre and the north. Almost all $\delta^{13}C$ values point to a system that is closed to CO_2. In piezometer 42e, an anomaly is present, which could possibly be explained by Neogene water intrusion. $\delta^{13}C$ values in the Ledo-Paniselian–Brusselian aquifer system are, in general, higher than those in the Oligocene aquifer system. This points to increased $CaCO_3$ dissolution in the Ledo-Paniselian–Brusselian aquifer system, related to the higher groundwater velocities in this aquifer system (Vandersteen *et al.* 2014*b*).

Modelling

The aim of this modelling exercise is to unravel the reactions that have led to the observed water quality in the deep aquifers below the Boom Clay in NE Belgium. The geochemical modelling was carried out in PHREEQC and the eponymous geochemical database was used. First, we calculated the saturation indices for each solution in the Oligocene and the Ledo-Paniselian–Brusselian aquifer systems, based on the ion measurements, and calculated alkalinity (Table 4). For both aquifer systems, all water samples are close to saturation or supersaturated for carbonates (saturation index (SI) of between −0.28 and 1.74 for calcite; between −0.61 and 3.4 for dolomite) and quartz (SI between −0.8 and 0.61), which is in accordance with the mentioned low flow rates in the deep aquifers as calculated by the groundwater model. The mass balance errors are generally small (between −8 and 5%).

Secondly, we use the inverse modelling tool within PHREEQC, which aims to identify the reactions that account for observed water compositions along a flowline. This type of modelling considers two or more water compositions as starting and ending compositions along a flow path. Inverse modelling is then used to calculate the moles of minerals and gases that must enter or leave the solution to account for the observed differences in composition. In this approach, time and flow path are implicitly taken into account (but not explicitly, as the geochemical model is not coupled to a dynamic groundwater movement and mixing model). Inverse modelling leads to sets of mineral and gas mole transfers that account for differences in composition between waters within specified compositional uncertainty limits (Parkhurst & Appelo 2013). The mineral composition of the aquifer material and water end members along a flow path are required input. We use the following end members: (i) seawater (Parkhurst & Appelo 2013); (ii) fresh recharge water; and (iii) Boom Clay porewater sampled from piezometers from the HADES URF at

Table 4. *Minimum and maximum values for saturation indices (SI) in Oligocene and Ledo-Panisselian–Brusselian aquifer systems*

Aquifer system	Percentage error	SI calcite	SI dolomite	SI quartz
Oligocene	−8.05 to 5.17 ($n = 19$)	0–1.74 ($n = 19$)	−0.26 to 3.4 ($n = 19$)	−0.8 to 0.31 ($n = 12$)
Ledo-Paniselian–Brusselian	−3.22 to 4.39 ($n = 10$)	−0.28 to 1.39 ($n = 10$)	−0.61 to 3.04 ($n = 10$)	−0.02 to 0.61 ($n = 10$)

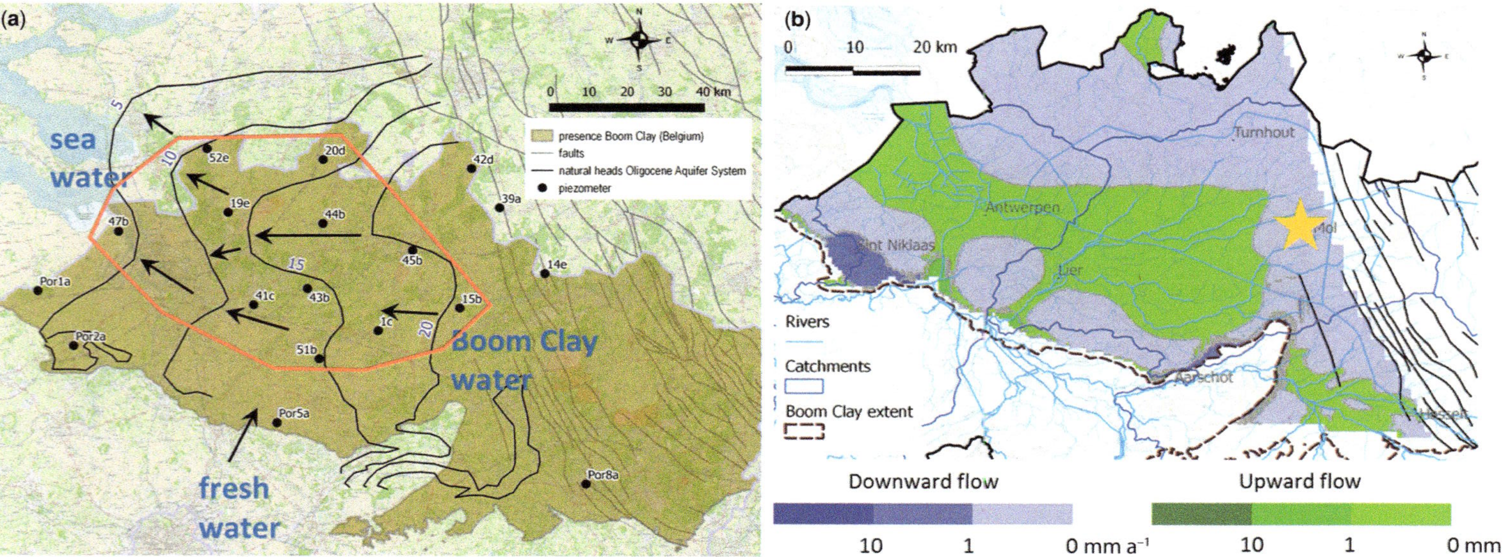

Fig. 12. (**a**) Water end members (blue labels) in relation to the general horizontal groundwater flow under natural conditions (black arrows) and the location of the piezometers in the Oligocene aquifer system. The area is delineated by red points of the piezometers that are selected for the inverse modelling. (**b**) Vertical flow through the Boom Clay under assumed natural gradient conditions, as calculated by the groundwater model (Vandersteen *et al.* 2013). The piezometer at the Mol site is marked with a star.

Table 5. *Reference Boom Clay porewater composition (De Craen* et al. *2004)*

pH	Alkalinity (meq l^{-1})	Ca (mg l^{-1})	Cl (mg l^{-1})	K (mg l^{-1})	Mg (mg l^{-1})	Na (mg l^{-1})	SO_4 (mg l^{-1})	Fe (mg l^{-1})	C (mg l^{-1})	Si (mg l^{-1})
8.5	15.1	2	26	7.2	1.6	359	2.2	0.2	181.3	3.4

Mol (near piezometer 15) (De Craen *et al.* 2004), which is consistent with the groundwater flow direction under natural conditions (Fig. 12a, b) as calculated from the groundwater model (Vandersteen *et al.* 2013). The reference Boom Clay porewater composition was calculated by cation exchange and mineral dissolution reactions that are calibrated against measured piezometer water compositions (Table 5) (De Craen *et al.* 2004). For the fresh recharge water end member, we use the water composition at piezometer Por2a that is located near the outcrop of the Boom Clay at the Wase cuesta, which is characterized by a high downwards flux through the Boom Clay (Fig. 12b). Piezometer Por2a has $CaHCO_3$-type water, indicative of fresh groundwater. We use the same end members for both aquifer systems, since considerable exchange is assumed between the two aquifer systems (Vandersteen *et al.* 2013). Owing to the low groundwater velocities and the large time period it takes for the groundwater to move, the current Boom Clay water composition might not be the same as that of past situations. However, the Boom Clay water that is used in the geochemical modelling reflects the geochemical processes that have resulted in its current water composition and therefore we consider it representative. The geochemical modelling results will, however, only allow qualitative inferences on the geochemical reactions and cannot be used to determine, for example, reaction rates (Glynn & Plummer 2005). Following the results of the $\delta^{13}C$ data analysis, we consider the system to be closed to CO_2. As mineral phases, we include the cation exchange complexes (CaX_2, NaX, KX and MgX_2), calcite, magnesite, quartz, K-feldspar, albite, kaolinite and organic matter in line with the measured mineralogical composition. Clay minerals have been considered in the model through the use of the cation exchange complexes. Other possible reactions with clay minerals have not been included in the model mainly because of uncertainties on the precise composition of the clay minerals. We force the model to dissolve calcite, NaX and KX, while precipitation of CaX_2 and MgX_2 is included.

For all considered piezometers, one or more models could be found under the given constraints, explaining the geochemical composition of the end members. The results are summarized in Figure 13, showing, besides the seawater fraction per piezometer, the phase mole transfers as calculated by the inverse geochemical modelling for the piezometers of the Oligocene and the Ledo-Paniselian–Brusselian aquifer systems. Positive values of mole transfer indicate dissolution, while negative values point to precipitation. We only take into account the models that are reduced to the minimum number of phases that can satisfy all of the constraints within the specified uncertainty limits. These models contain the most essential geochemical reactions. The results of the inverse geochemical modelling correctly represent increasing seawater dilution from NW to SE. Only in the case of piezometer 41d (Fig. 13b), do both data and model show a higher than expected salt concentration. Throughout the study area, calcite dissolution, silicate dissolution, sulphate reduction by organic matter and cation exchange can be found for most piezometers, which is in accordance with the reactions deduced from the data analysis. Cation exchange is visible through the replacement of Na^+, K^+ and Mg^{2+} (positive mole transfer in Fig. 13), which would originally have occupied cation exchange sites when the rocks were saturated with the early connate marine porewater, by Ca^{2+} in the aquifer (negative mole transfer in Fig. 13). For piezometers 1c, 41d, 15c and 51c, cation exchange reactions are absent in the modelling results (Fig. 13a). For piezometers 51c and 41d, this is in accordance with the results of the Stuyfzand classification visualized in Figure 6, which show that, for these piezometers, no cation exchange is present at piezometers 52e, 41d, 52f and 51c. For piezometer 52e and 52f, however, the current chemical water composition cannot be explained without taking into account cation exchange. As the ranges in possible phase mole transfer for the different piezometers are, in most cases, quite broad for the cation exchange parameters, as well as for calcite and silicates, no inferences about the spatial variability of these parameters can be made. For the sulphate reduction, Figure 13a shows greater reduction in the central area (piezometers 41c, 43b, 51b, 1c, 45b and 47b) in the Oligocene aquifer system, which is in accordance with the spatial variability of the SO_4^{2-}/Cl^- ratio of the data (Fig. 7).

Conclusions

Over the last 30 years, several studies have been undertaken in the deep aquifer system of NE

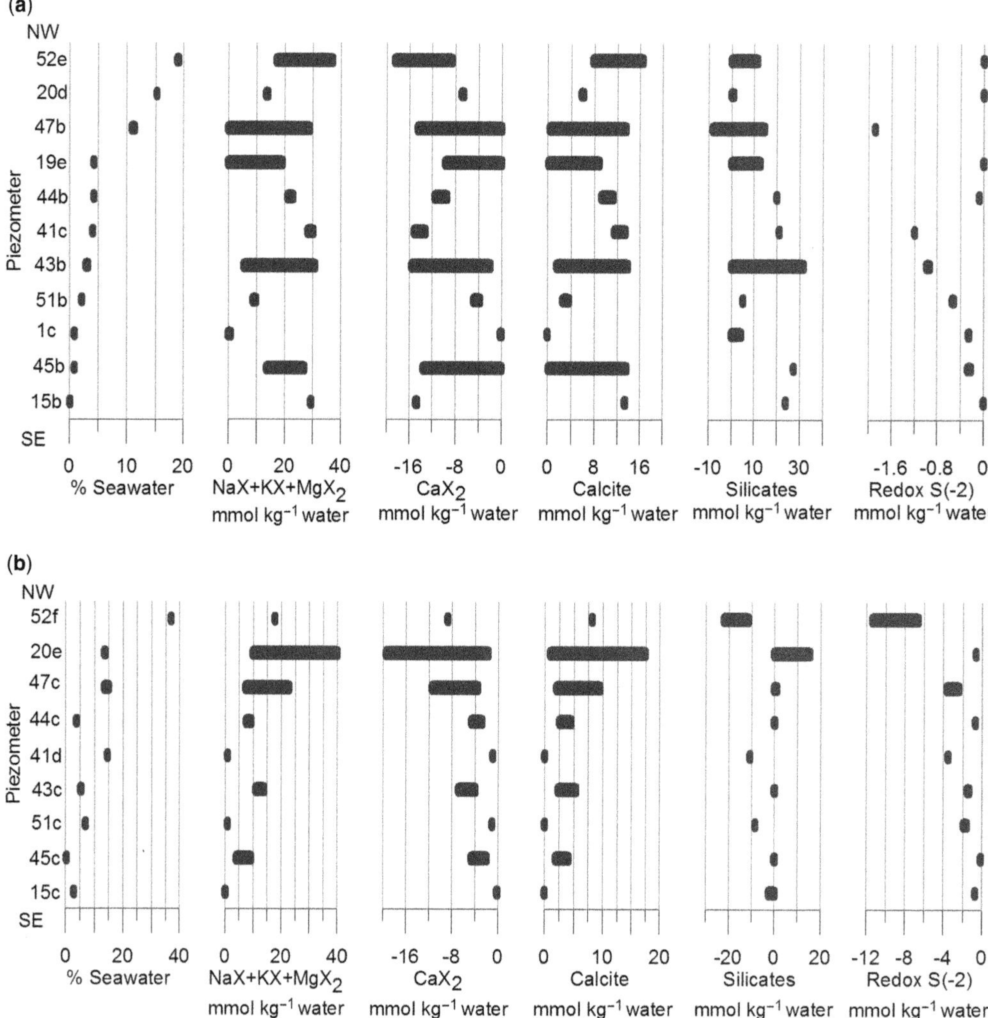

Fig. 13. Seawater fraction (%) and phase mole transfer (mmol kg^{-1} water) for (**a**) the Oligocene aquifer system and (**b**) the Ledo–Paniselian–Brusselian aquifer system. Positive values indicate dissolution and negative values point to precipitation. The piezometers are ranged from NW to SE. Consistency exists in cation exchange, calcite dissolution, silicate dissolution and sulphate reduction. The bar width represents the range in phase mole transfer for the inverse models that were found.

Belgium in the context of the geological disposal of nuclear waste research. These campaigns yielded similar results overall. Geochemical data analysis reflected an aquifer freshening process, with mainly $NaHCO_3$- and NaCl-type waters found. Also, the data are consistent with the direction and velocity of the groundwater flow as determined from the groundwater flow model. Evidence was found for calcite dissolution, sulphate reduction and cation exchange.

Inverse geochemical modelling using PHREEQC confirmed that the current geochemical composition of the deep aquifer system is the result of mixing three end-member waters (seawater, fresh recharge water and Boom Clay porewater) in combination with different water–rock interactions. Dilution of the original marine-derived porewater increases progressively from NW to SE.

More accurate data, however, are needed to include in the geochemical modelling in order to reduce model uncertainty in the geochemical model. More accurate data on aquifer mineralogy and stable isotopes could be used to further constrain the inverse model, while data on the cation-exchange capacity of the aquifer systems would enable forward geochemical modelling.

This work is performed in close co-operation with, and with the financial support of, NIRAS/ONDRAF, the Belgian Agency for Radioactive Waste and Fissile Materials, as part of the programme on geological disposal of high-level/long-lived radioactive waste that is carried out by ONDRAF/NIRAS. Dr Koen Beerten, Dr Mieke De Craen, Dr Miroslav Honty, Dr ir Diederik Jacques, Dr Bart Rogiers and two anonymous reviewers are acknowledged for giving valuable comments to the document.

References

Appelo, C.A.J. 1994. Cation and proton exchange, pH variations, and carbonate reactions in a freshening aquifer. *Water Resources Research*, **30**, 2793–2805.

Appelo, C.A.J. & Postma, D. 2006. *Geochemistry, Groundwater and Pollution*, 2nd edn. A.A. Balkema, Leiden.

Beaucaire, C., Pitsch, H., Toulhoat, P., Motellier, S. & Louvat, D. 2000. Regional fluid characterization and modelling of water–rock equilibria in the Boom clay Formation and in the Rupelian aquifer at Mol, Belgium. *Applied Geochemistry*, **15**, 667–686.

Beerten, K., De Craen, M. & Wouters, L. 2013. Patterns and estimates of post-Rupelian burial and erosion in the Campine area, north-eastern Belgium. *Physics and Chemistry of the Earth, Parts A/B/C*, **64**, 12–20, https://doi.org/10.1016/j.pce.2013.04.003

Bethke, C.M. & Johnson, T.M. 2008. Groundwater age and groundwater age dating. *Annual Review of Earth and Planetary Sciences*, **36**, 121–152, https://doi.org/10.1146/annurev.earth.36.031207.124210

Blommaert, W., de Marmol, P. & Patyn, J. 1988. *Hydrochemical research within the HADES project (Disposal of High Radioactive Waste into the Boom Clay) 1982–86*. SCK•CEN, Mol, Belgium (accessible through Royal Institute of Natural Sciences).

Blommaert, W., de Marmol, P. & Patyn, J. 1994. Hydrogeology and hydrochemistry of the lower and upper sand aquifers. *In*: Beaufays, R., Blommaert, W. et al. 1994. *Characterization of the Boom Clay and its Multilayered Hydrogeological Environment*. Final Report. EC Contract No. FI1W/0055. European Commission, Luxembourg.

De Craen, M., Wang, L., Van Geet, M. & Moors, H. 2004. *Geochemistry of Boom Clay Pore Water at the Mol Site*. SCK-CEN-BLG-990. SCK-CEN, Mol, Belgium.

Glynn, P.D. & Plummer, L.N. 2005. Geochemistry and the understanding of ground-water systems. *Hydrogeology Journal*, **13**, 263–287, https://doi.org/10.1007/s10040-004-0429-y

Griffault, L., Merceron, T. et al. 1996. *Participation au projet 'Archimède – Argile'. Acquisition et Regulation de la chimie des eaux en milieu argileux pour le projet de stockage de déchets radioactifs en formation géologique*. Final Report. EC Contract No. FI2W-CT92-0117. European Commission, Luxembourg.

Labat, S. 2011. *Short Note on the Chemistry of Aquifers*. Note EHS/W&D/11/MDC/R-19. SCK•CEN, Mol, Belgium.

Labat, S., Marivoet, J. & Maes, T. 2014. *Kallo Boreholes: Technical Aspects and Hydrogeological Investigations*. SCK•CEN-ER-224. SCK•CEN, Mol, Belgium.

Lagrou, D., Dreesen, R. & Broothaers, L. 2004. Comparative quantitative petrographical analysis of Cenozoic aquifer sands in Flanders (N Belgium): overall trends and quality assessment. *Materials Characterization*, **53**, 317–326. https://doi.org/10.1016/j.matchar.2004.07.012

Marivoet, J., Van Keer, I. et al. 2000. *A Palaeohydrogeological Study of the Mol Site (PHYMOL Project)*. Final Report EUR 19146. European Commission, Luxembourg.

Matthijs, J., Lanckacker, T., De Koninck, R., Deckers, J., Lagrou, D. & Broothaers, M. 2013. *Geologisch 3D lagenmodel van Vlaanderen en het Brussels Hoofdstedelijk Gewest – versie 2, G3Dv2*. VITO Report 2013/R/ETE/43. VITO, Mol, Belgium.

Parkhurst, D.L. & Appelo, C.A.J. 1999. *User's Guide to PHREEQC (Version 2) – A Computer Program for Speciation, Batch-Reaction, One-Dimensional Transport, and Inverse Geochemical Calculations*. United States Geological Survey, Water-Resources Investigations Report, **99-4259**.

Parkhurst, D.L. & Appelo, C.A.J. 2013. *Description of Input and Examples for PHREEQC Version 3 – A Computer Program for Speciation Batch-Reaction, One-Dimensional Transport, and Inverse Geochemical Calculations*. United States Geological Survey, Techniques and Methods, **6-A43**, http://pubs.usgs.gov/tm/06/a43

Philippot, A.C., Michelot, J.-L. & Marlin, C. 2000. *A Palaeohydrogeological Study of the Mol Site, Belgium (PHYMOL Project)*. Rapport spécifique n°3: Analyse des isotopes et des gaz nobles. EC Contract No. FI4W-CT96-0026. European Commission, Luxembourg.

Piper, A.M. 1953. *A Graphic Procedure in the Geochemical Interpretation of Water Analysis*. United States Geological Survey, Washington DC.

Pitsch, H. & Beaucaire, C. 2000. *A Palaeohydrogeological Study of the Mol Site, Belgium (PHYMOL Project). Topical Report 2: Hydrogeochemistry*. EC Contract No. FI4W-CT96-0026. European Commission, Luxembourg.

Rogiers, B., Beerten, K., Gedeon, M. & Vandersteen, K. 2012. A comparison between hydrogeological modelling results and literature groundwater ages for shallow and deep aquifers in northern Belgium. Paper presented at GDAT 2012, 15–17 October 2012, Rennes, France.

Schlumberger-Doll Research 1999. Mineralogy and Chemistry Analysis Project (Doel-1A and Doel-2B). Report for ONDRAF/NIRAS.

Stuyfzand, P.J. 1986. A new hydrochemical classification of watertypes: principles and application to the coastal dunes aquifer system of The Netherlands. *In*: Boekelman, R.H., van Dam, J.C., Evertman, M. & ten Hoorn, W.H.C. (eds) *Proceedings of the 9th Salt Water Intrusion Meeting*, 12–16 May 1986,

Delft. Delft University of Technology, Delft, 641–655.

STUYFZAND, P.J. 1989. A new hydrochemical classification of water types. *In*: *Proceedings of the Baltimore Symposium, May 1989, Regional Characterization of Groundwater Quality. International Association of Hydrological Sciences* (IAHS), Publications, **182**, 89–98.

VANDERSTEEN, K. & LABAT, S. 2014. *Analysis of Geochemical Measurements for Well 15 (Major Anions & Cations). Sampling Campaign September 2014.* Note 14/KVa/N-93. SCK•CEN, Mol, Belgium.

VANDERSTEEN, K., GEDEON, M. & ROGIERS, B. 2012. *Transient Model of the Confined Aquifers below the Boom Clay: 2011 Update*. External Report of the Belgian Nuclear Research Centre, SCK•CEN-ER-199. SCK•CEN, Mol, Belgium.

VANDERSTEEN, K., GEDEON, M. & LETERME, B. 2013. *Hydrogeology of North-East Belgium.* Status Report 2012. External report of Belgian Nuclear Research Centre, SCK•CEN-ER-236. SCK•CEN, Mol, Belgium.

VANDERSTEEN, K., GEDEON, M. & BEERTEN, K. 2014*a*. A synthesis of hydraulic conductivity measurements of the subsurface in Northeastern Belgium. *Geologica Belgica*, **17**, 196–210.

VANDERSTEEN, K., GEDEON, M., MARIVOET, J. & WOUTERS, L. 2014*b*. Regional groundwater flow modelling of the confined aquifers below the Boom Clay in NE Belgium. *In*: NORRIS, S., BRUNO, J.ET AL. (eds) *Clays in Natural and Engineered Barriers for Radioactive Waste Confinement*. Geological Society, London, Special Publications, **400**, 163–177, https://doi.org/10.1144/SP400.22

VANDERSTEEN, K., LABAT, S. & LETERME, B. 2014*c*. *Analysis of Geochemical Measurements for Well 15 & 42 (Major Anions & Cations). Sampling Campaign October 2013*. Note 14/KVa/N-16. SCK•CEN, Mol, Belgium.

WALRAEVENS, K. 1990. Hydrogeology and hydrochemistry of the Ledo-Paniseliaan semi-confined aquifer in East- and West-Flanders. *Academiae Analecta*, **52**, 13–66.

WALRAEVENS, K., CARDENAL-ESCARCENA, J. & VAN CAMP, M. 2007. Reaction transport modelling of a freshening aquifer (Tertiary Ledo-Paniselian Aquifer, Flanders, Belgium). *Applied Geomchemistry*, **22**, 289–305.

WEMAERE, I., MARIVOET, J., LABAT, S., MAES, T. & BEAUFAYS, R. 2004*a*. *Rijkevorsel Borehole of the Hydro/96neb Campaign: Technical Aspects & Hydrogeological Investigations. Geological Disposal of Conditioned High-Level & Long-Lived Radioactive Waste.* SCK•CEN-R-3930. SCK•CEN, Mol, Belgium.

WEMAERE, I., MARIVOET, J., LABAT, S., MAES, T. & BEAUFAYS, R. 2004*b*. *Zoersel Borehole of the Hydro/96neb Campaign: Technical Aspects & Hydrogeological Investigations*. Geological Disposal of Conditioned High-Level & Long-Lived Radioactive Waste. SCK•CEN-R-3892. SCK•CEN, Mol, Belgium.

WEMAERE, I., MARIVOET, J., LABAT, S., MAES, T. & BEAUFAYS, R. 2005*a*. *Turnhout Borehole of the Hydro/96neb Campaign: Technical Aspects & Hydrogeological Investigations*. Geological Disposal of Conditioned High-Level & Long-Lived Radioactive Waste. SCK•CEN-R-4124. SCK•CEN, Mol, Belgium.

WEMAERE, I., MARIVOET, J., LABAT, S., BEAUFAYS, R. & MAES, T. 2005*b*. *The Weelde Boreholes of the Hydro/96neb Campaign: Technical Aspects & Hydrogeological Investigations*. Geological Disposal of Conditioned High-Level & Long-Lived Radioactive Waste. SCK•CEN-R-4187. SCK•CEN, Mol, Belgium.

An assessment of Palaeogene and Neogene clay deposits in Denmark as possible host rocks for final disposal of low- and intermediate-level radioactive waste

PETER GRAVESEN*, STIG A. SCHACK PEDERSEN, BERTEL NILSSON & MERETE BINDERUP

Geological Survey of Denmark and Greenland (GEUS), Øster Voldgade 10, 1350 Copenhagen, Denmark

**Correspondence: pg@geus.dk*

Abstract: In Denmark, mapping and preliminary investigations of Palaeogene and Early Neogene clay deposits have been performed over the past 5 years. The goal was to locate potential host rocks for the final disposal of low- and intermediate-level radioactive waste from the Danish Research Centre Risø, which has to be decommissioned within the next 5–8 years. Five areas with low-permeability Paleocene and Oligocene clay formations, situated in northern Jylland, NE Fyn and southern Lolland, not far from the Baltic coast at Femern Belt, have been suggested. The clay formations are between 75 and 150 m thick, and have large lateral distribution. They are covered by thin layers of glacial clayey tills and the shallow depth to the clay formations is attractive, partly because they are easy to access and partly because there are no groundwater aquifers situated above or below the clay deposits. The paper gives an overall review of the characteristics of the six different clay formations within four of the areas.

According to the decision of the Danish Parliament relating to the decommissioning of the Danish Nuclear Research Centre Risø, the Geological Survey of Denmark and Greenland (GEUS) was asked to locate and describe possible geological sites for depositing radioactive waste from the research facility. The Nuclear Research Centre Risø is located north of the city of Roskilde and 30 km west of Copenhagen, the capital of Denmark. In connection with the search for potential sites for the establishment of a repository for low- and intermediate-level radioactive waste, GEUS has mapped and assessed the geological deposits (Gravesen *et al.* 2011*a*). The investigation was based on a screening of existing information and the description of Danish geology to a depth of 300 m below ground surface. The task was performed during 2009–2013. The screening resulted in a selection of six areas, which were recommended for further examination (Fig. 1). Five of the six areas were located on Palaeogene and Early Neogene clays. The sixth locality was located on Precambrian basement rocks.

Potential clay host rocks of the same age and similar lithology have been studied in Belgium, where the Oligocene Boom Clay has been investigated for many years (Decleer *et al.* 1983; Wemaere *et al.* 2008; Gedeon *et al.* 2012; Yu *et al.* 2012). The Oligocene Boom Clay is also considered as a potential host rock in The Netherlands (Vis & Verweij 2014).

This paper provides a review of the criteria for the selection of a candidate locality for a repository and the key characteristics of the Danish clay deposits.

Repository concept and criteria for the selection of sites

During the initiation of the mapping, a set of criteria were defined for the selection of geological sites (Gravesen *et al.* 2011*a*) based on guidelines from the International Atomic Energy Agency (IAEA 2014). The set of criteria addresses the demands for the rocks of interest to host a repository and to act as a barrier to prevent low and intermediate-level radioactive waste from accessing the environment.

The following types of repositories may be established for up to 10 000 m^3 of waste after more detailed examination in a subsequent project design phase (Gravesen *et al.* 2011*b*):

- Concept A – a near-surface repository (on the ground surface or down to a maximum of 30 m below ground surface).
- Concept B – a near-surface repository in combination with a borehole to contain some parts of the long-lived waste.
- Concept C – an intermediate-depth repository (30–100 m below ground surface).

From: Norris, S., Bruno, J., Van Geet, M. & Verhoef, E. (eds) 2017. *Radioactive Waste Confinement: Clays in Natural and Engineered Barriers*. Geological Society, London, Special Publications, **443**, 29–38.
First published online June 22, 2016, https://doi.org/10.1144/SP443.3

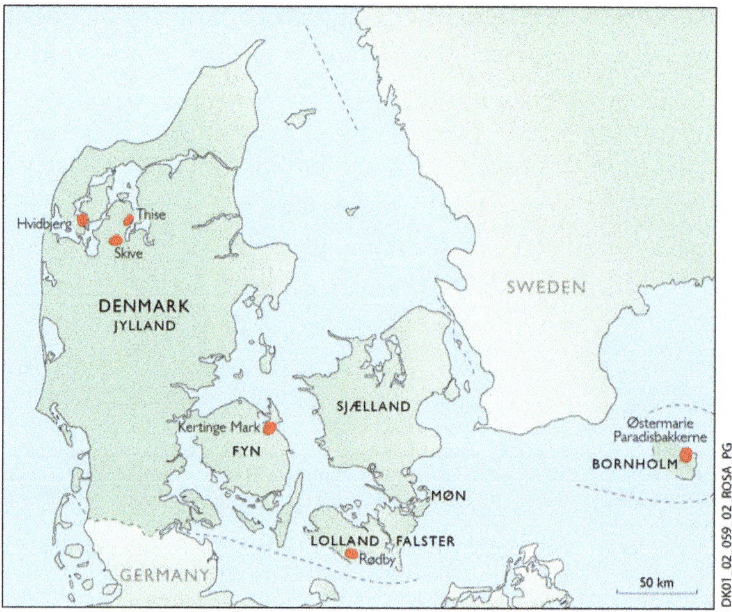

Fig. 1. Map of Denmark with the six selected localities. The localities have Palaeogene–Neogene clays as potential host rocks for a repository, except for the locality on Bornholm where Precambrian basement rocks have been chosen.

Geological requirements

The terrain within the repository site finally chosen should be predominantly flat, non-dipping and geotechnically stable. Deposits from the surface down to the greatest possible depth should be homogeneous and of low permeability. Only a very few localities in Denmark contain Palaeogene and Neogene clay outcrops so, at the surface, clayey tills should dominate, but this cover of clayey till should be as thin as possible over most of the site because the best protection for the repository is in the older clays. It should be possible for the Palaeogene and Neogene clay deposits to enclose or underlie a repository, which is best achieved if there are dense, low-permeability layers of great thickness and significant lateral extent within the sites.

A near-surface repository concept (Concept A) should be a combination of clayey till at the ground surface and with more low-permeability strata immediately below. The aim is to gain access to the particularly low-permeability layers as quickly as possible. If the repository is placed at the surface level, the clayey till should be as thin as possible. The intermediate-depth repository concept (Concept C) should always be in the low-permeability strata.

If part of the waste inventory is placed in boreholes, these will always be located in the low-permeability layers (Concept B). The depth of the borehole will depend on the depth and lateral extent of the geological host-rock layers.

The screening resulted in the selection of 22 potential areas, which on demand were reduced to six areas by an evaluation based on a priority-rating table (Gravesen *et al.* 2011*a*). All of the areas were described in detail (Gravesen *et al.* 2013). It is suggested that a future waste storage facility will cover between 20 000 and 30 000 m^2. All of the six areas are much larger, varying in size from 5 to 20 km^2, so it will be possible to place the site optimally within the area.

The main geological criteria were to find thick sequences of low-permeability sediments and rocks that have a sufficient horizontal extent within the focused areas. The final six areas with the highest priority rating were, furthermore, selected on details of geological conditions, such as a low thickness of clayey till above the Palaeogene and Neogene clays, and there being no possible presence of glaciotectonic or deep tectonic features. The tectonic features were evaluated based on fieldwork, boreholes and geophysical surveys. Finally, the vulnerability of groundwater aquifers was assessed.

Geological setting of the Danish land areas

The geological map of the Danish bedrock is shown in Figure 2. In northern Denmark, the

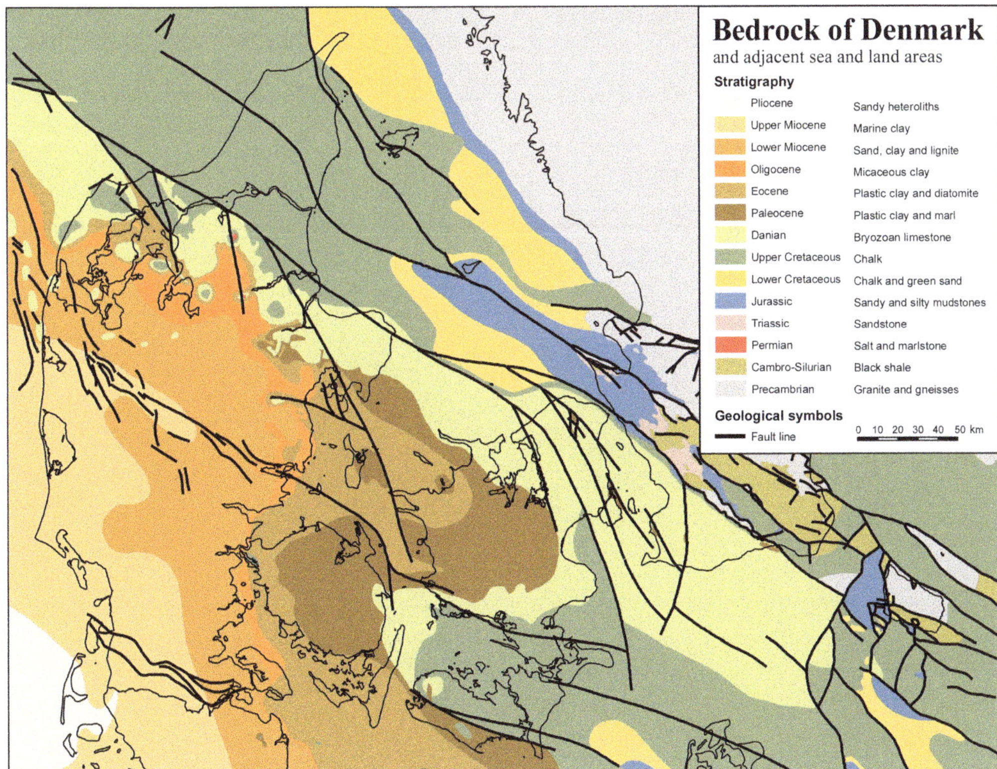

Fig. 2. Geological map of the bedrock in Denmark (modified from Håkansson & Pedersen 1992). The formations become increasingly younger towards the west, where they link up with the depression in the North Sea. The framework of fault zones is mainly related to the Sorgenfrei–Tornquist Zone, which is a SE–NW-striking wrench fault system separating the soft sedimentary bedrock in Denmark from the basement rock in Sweden.

geological setting is related to the main tectonics of the Norwegian–Danish Basin (Danish Subbasin). A very strong structural element in this setting is the Sorgenfrei–Tornquist Zone, which is a NW-striking fault zone separating the Scandinavian Basement to the north and east from the up to 10 km deep basin to the south and west. The main fault activity in the Sorgenfrei–Tornquist Zone took place about 60 myr ago, but earthquakes in southern Scandinavia are still concentrated along this zone. The five clay localities are located outside this zone, in areas of low seismic activity and earthquakes (Gravesen *et al.* 2013). Towards the south, the Ringkøbing-Fyn High forms a basement feature that separates the Danish Basin from the North German Basin (Sorgenfrei & Buch 1964).

In the following, remarks on the Cretaceous, Danian, Palaeogene, Neogene and Quaternary deposits are presented, because these deposits occur immediately below or above the investigated clays.

Cretaceous deposits

The uppermost Cretaceous marine chalk forms the northern part of the large North European carbonate platform (Fig. 2). The Upper Cretaceous Maastrichtian chalk crops out on the pre-Quaternary surface in the Kattegat Platform and in the Baltic area. The chalk is exposed in pits in northern Denmark, in the centres of some dome structures formed by saltdiapirs and in the glaciotectonic complex at Møns Klint in SE Denmark.

Danian limestones

The Cretaceous–Tertiary boundary forms the lower boundary of the Danian Limestone, which is dominantly present in the subsurface of Denmark (Fig. 2). The Danian Limestone constitutes a bryozoan reef limestone interbedded with chert. In few localities, coral limestone is present, and the limestone succession grades up into a hard, massive limestone. The best-known locality for the Danian

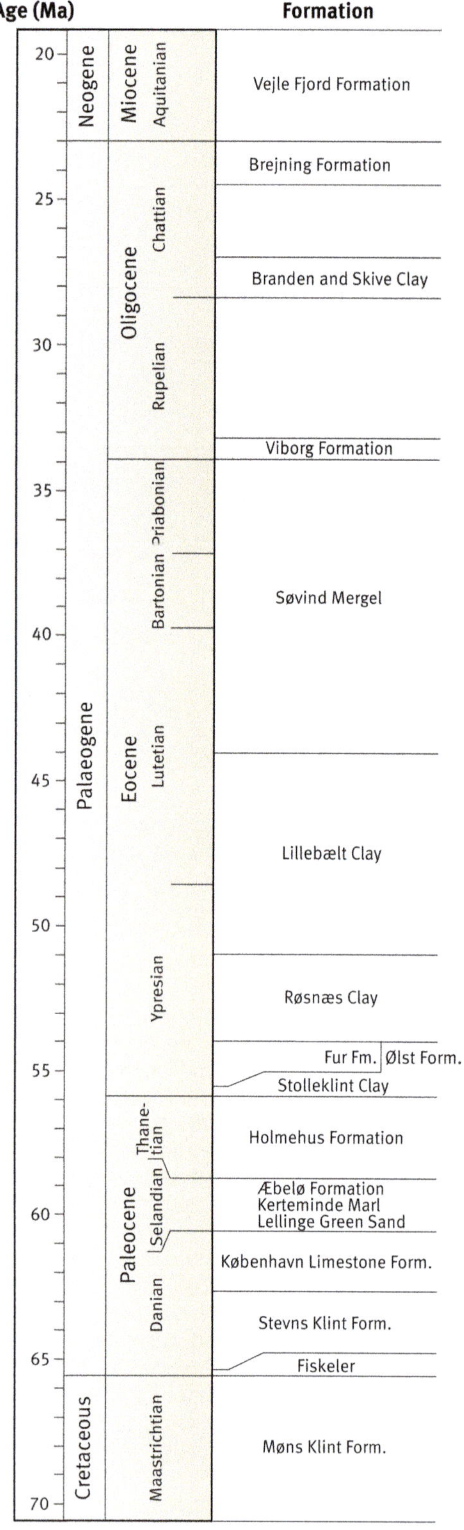

Fig. 3. Stratigraphic scheme for the Danish Palaeogene and Early Neogene. Note that the scheme only has an equivalent timescale; information about the thickness is not given. Thus, the Søvind Marl is a relatively thin formation (*c.* 10 m) deposited over a long period. In contrast, the Fur Formation was deposited over a period of less than 2 myr, but has a thickness of 60 m.

Limestone is Stevns Klint (central eastern Denmark), where the Fish Clay with a high iridium content, contributing to the interpretation of the meteorite impact hypothesis for the mass extinction at the boundary, is exposed (Surlyk *et al.* 2006).

Palaeogene and Neogene deposits

The Danian Limestone represents the lowermost Palaeogene unit overlain by marine greensand deposits. The Paleocene, Eocene and Oligocene consist of fully marine clay-rich formations (Figs 2 & 3), which represent a temperate marine environment. In only the northern part of Denmark, a succession of clayey diatomite and volcanic ash beds occur in the thick clay sequence (Pedersen & Surlyk 1983). The ash layers represent the volcanic activity related to the early break-up of the Laurentien continent and the formation of the North Atlantic Ocean on the Paleocene–Eocene boundary (Larsen *et al.* 2003).

The marine Late Oligocene clay and silt beds are known from the eastern and central part of Denmark, and the clays consist of fine-grained and silty, often micaceous, sediments at the Neogene boundary. The Neogene is dominated by a large delta system building out from the NE European–Baltic platform towards the North Sea during Miocene times (Rasmussen *et al.* 2010). These deposits dominate the western part of central Denmark, and they continue into the Norwegian–Danish Basin in the North Sea. Apart from the delta system, parts of the Miocene in central Jylland consist of thick sandy successions containing lignite deposits. Towards the west, marine Miocene deposits containing clays, silts and fine-grained sands reach thicknesses of up to 350–700 m (Rasmussen *et al.* 2010).

Quaternary deposits

The Quaternary Period is characterized by repeated glacial events, where glacial advances resulted in glaciodynamic successions (Pedersen 2012). Denmark is a lowland, with hills generally less than 150 m high, where the surface deposits are dominated by the Weichselian successions. The deposits are characterized by proglacial meltwater successions, typically deformed by glaciotectonic thrusting and folding, and capped by basal clayey till, except for the outwash plains, which are located in the areas beyond the ice front. During the 100 kyr of the Weichselian glaciation, the glaciodynamic events were caused by ice advances from the Baltic, Norway, Sweden and, again, from the Baltic (Pedersen 2014). During Holocene time, the eustatic sea-level rise resulted in deposition of marine fine sand and clay in the northern part of the country. After the Atlantic transgression of 8–7 ka, glacio-isostatic rebound brought these deposits to the surface.

Stratigraphy of Tertiary clay units

In Denmark, the Palaeogene and lowermost Neogene successions are dominated by clay deposits, except for the oldest Danian limestone deposits and the lowermost Selandian greensand limestone, sand and sandy marl. The lithostratigraphy for the whole Palaeogene and lowermost Neogene is illustrated in Figure 3. Several of the clay-dominated formations have been included in the focused investigation: the Paleocene Kerteminde Marl, the Æbelø and Holmehus formations, the Oligocene Brejning Formation, the Branden and Skive clays, and the Miocene Vejle Fjord Formation. Several other clay-rich Eocene and Early Oligocene formations are not included because they do not fulfil the criteria for selection (Fig. 3). The formations are not well exposed but are known from many boreholes recorded in the GEUS database (Gravesen & Fredericia 1984). All formations are lithologically described in detail, with age determination based on foraminifers, dinoflagellates and rare macrofossils (e.g. see Gry 1935; Heilmann-Clausen 1985; Heilmann–Clausen *et al.* 1985; Friis 1994; Nielsen *et al.* 1994; Rasmussen *et al.* 2010). An important feature for the stratigraphic frame is that all formations consist of marine sediments.

The Palaeogene and Neogene formations are almost all covered by Quaternary deposits, of which most are of Weichselian or Holocene age. The prominent sediments are clayey tills, meltwater sand and gravel, marine mud, sand and gravel, and aeolian sand.

Below the Quaternary deposits, the older deposits demonstrate an undulating subcrop as a result of the subsidence of the basins between the Cambrian and the Cenozoic (Fig. 2). The total thickness of the Palaeogene (excluding Danian) and Early Neogene deposits is approximately 200–300 m in the Danish Sub-basin and 250–330 m in the North German Basin (Dinesen *et al.* 1977; Nielsen *et al.* 1994).

Characteristics of the formations

Kerteminde Marl Formation (Paleocene, Selandian)

The Kerteminde Marl Formation consists of olive grey or grey fine-grained clay with silt lenses and often a high content of calcium carbonate (between 40 and 70%). Re-sedimented limestone fragments from the underlying chalk and limestone deposits occur. Cemented and chertified horizons can also be found. The clastic grain-size distribution is approximately 1% sand, 30–40% silt and 60–70% clay. The small sand fraction contains quartz, glauconite and pyrite. The clay fraction is dominated by smectites (78%) with subordinate illite (22%), while kaolinite is almost absent (Tank 1963; Nielsen 1994).

The formation is exposed at only a few localities in Jylland, but abundant boreholes show that the formation is distributed from the central part of Sjælland towards Fyn and eastern part of Jylland (e.g. Heilmann-Clausen 1985, 1995; Thomsen & Heilmann-Clausen 1985; Nielsen *et al.* 1994). The formation is also known in the North Sea area as the North Sea Marl (Clausen & Huuse 2002). Towards the northern part of Jylland, and towards the south on Lolland and Langeland, the formation is missing, probably because of erosion or non-deposition (Sheldon *et al.* 2012). The palaeogeographical development of the Danian Limestone, the Lellinge Greensand Formation and the Kerteminde Marl Formation/North Sea Marl deposits demonstrated an extensive marine area from the North Sea towards Sjælland and further to the east (Clausen & Huuse 2002; Clemmensen & Thomsen 2005).

The borehole data include sediment samples and geophysical logging data (Klitten 2003). These data demonstrate that the thickness can reach more 50 m in the central part of Sjælland. The Kerteminde Marl Formation has a low permeability and contains no exploitable groundwater, although the underlying Lellinge Greensand Formation is an important aquifer in the central and eastern part of Sjælland (Gry 1935; Klitten 2003) (Fig. 4).

Æbelø Formation (Paleocene, Selandian)

This formation consists of silty and fine-grained grey, dark grey or black clay, which is often layered. The clay is non-calcareous or very slightly calcareous with silicified horizons. The sand content is low (0.25–1.25%), while the clay content is between 50 and 75% (Nielsen 1994). The sand fraction mainly consists of quartz, glauconite and other authigenic minerals in an amount up to 10%. The clay fraction is dominated by smectites (*c.* 80%), while kaolinite

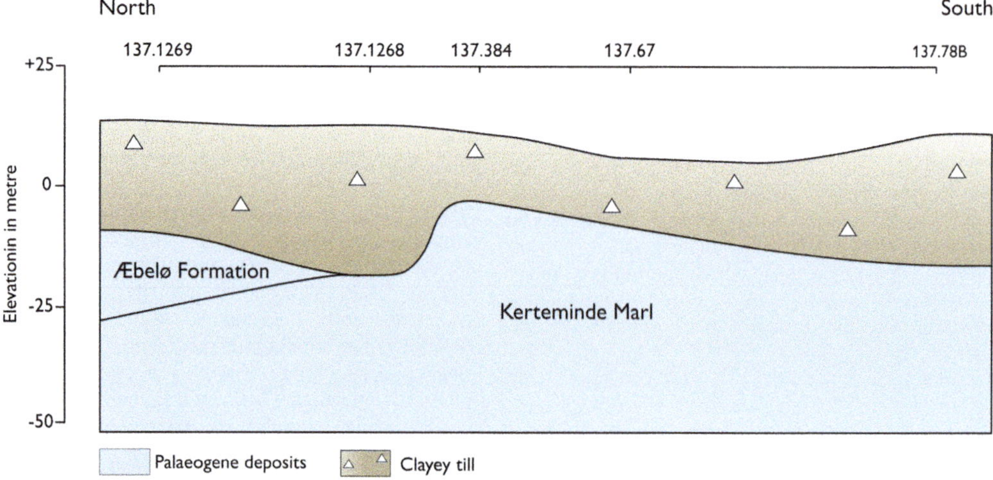

Fig. 4. Geological cross-section from the Kertinge Mark area on eastern Fyn. Note that the numbers at the top of the figure are references to the borehole identification numbers in the borehole database Jupiter at the Danish Geological Survey (GEUS). Boreholes 137.1268 and 137.1269 are investigation boreholes carried out to test the site and document that it qualifies to be one of the selected sites. The irregular geological boundary is interpreted as being due to glacial erosion. Vertical exaggeration is approximately ×20.

is only subordinate (Nielsen 1994). The formation is widespread in the Danish area, being documented from many boreholes, but is often described as the grey, slightly calcareous clay younger than the Kerteminde Marl (Bøggild 1918; Dinesen *et al.* 1977) (Fig. 4). Only two outcrops are known from Jylland and the island of Æbelø (Heilmann-Clausen 1980, 1985). The palaeogeographical evolution described by Clemmensen & Thomsen (2005) shows that the formation was deposited in a sea of larger extent than that of the Kerteminde Marl in the North Sea area, reaching, for example, the Lolland area. However, the sea seems to have been restricted towards the east (Sheldon *et al.* 2012).

Because of its small grain size and the layering, the formation has very low permeability. The formation is between 16 and 57 m thick in western Denmark, but is some 35–40 m thick on Lolland and up to 60 m thick on Falster, perhaps increasing in thickness up to a total of 120 m (the Holmehus Formation and Æbelø Formation together).

Holmehus Formation (Paleocene, Selandien–Thanetian)

The Holmehus Formation consists of green-grey, green, brown or black very-fine-grained plastic clay. The clay is mainly laminated, but is sometimes faintly horizontally laminated and often disturbed by bioturbation. Sideritic and phosphatic concretions occur. The clay fraction varies between 50 and 70%, with 30–40% silt. The rest is quartz and feldspar sand grains. The clay fraction can be up to 80% in northern Jylland and in Store Bælt (Nielsen *et al.* 1986; Nielsen 1994). Most of the clay is non-calcareous, but thin calcareous layers do occur. At Lolland, the clay mineralogy comprises 70–80% smectite (but 45–50% on average), 20–25% illite, 2–4% kaolinite and very little chlorite (Pedersen 1992). The clay is classified as a combination of Na and Ca bentonite, and has a high plasticity. In northern Jylland and Store Bælt, smectite (often up to 80%) is the dominant clay mineral, but it includes some illite (approximately 15–20%) and sporadic kaolinite (Nielsen *et al.* 1986, 1994; Nielsen 1994).

The formation has a very low permeability because of its very fine grain size. Percolation tests have measured hydraulic conductivities of the clay to be as low as 2.5×10^{-12} m s^{-1} (Pusch *et al.* 2015).

The formation is widely distributed in the Danish Basin and is primarily known from borehole information, because outcrops are scarce and only found in Jylland and Æbelø (Bøggild 1918; Heilmann-Clausen *et al.* 1985). The formation was deposited in a fully marine environment. The thickness is between a few metres and 40 m in Jylland, and up to 45 m on Lolland and Falster (Pedersen 1992) (Fig. 5).

Branden Clay–Skive Clay (Oligocene, Chattian)

The Branden Clay and Skive Clay are two low-permeability informal Oligocene units consisting of fine-grained clay of the same age (Heilmann-Clausen

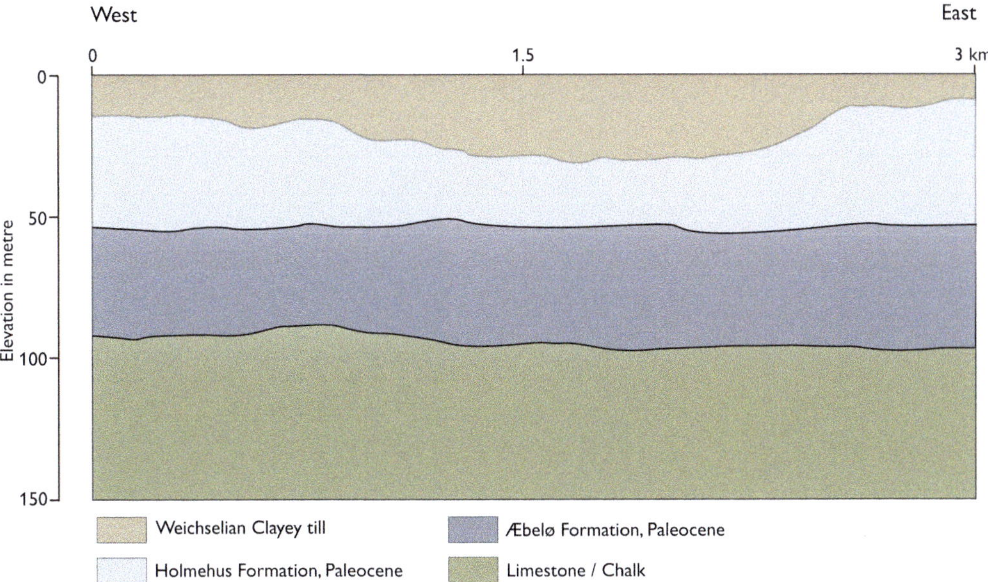

Fig. 5. Geological cross-section from the Rødbyhavn area on southern Lolland (after Pedersen 1992). The interpretation of the geology is based on a seismic section and documentation of the stratigraphy from one deep drill hole and four shallow wells. Vertical exaggeration is approximately ×10.

& Surlyk 2012). The clays are green-grey and grey-green and often mica, glauconite, pyrite and silt lamination occur. Large calcareous septarian concretions occur in the clays. The units start with a clay layer dominated by glauconite. The clays have a varied calcium carbonate content, but are often non-calcareous. The Branden Clay consists of 50% clay and 50% silt (Friis 1994). The clay mineralogy is 20–50% smectite, 30–60% illite and the rest is kaolinite (Friis 1994). The clay mineralogy of the Skive Clay is 53% smectite, 26% illite and 21% kaolinite.

Because of the grain size and structures, the two units have a low permeability. The Branden Clay unit appears to have a limited distribution in northern Jylland and the thickness is from 45 m to approximately 90 m, while the Skive Clay unit is more widely distributed in northern Jylland and with a greater thickness. They are known from a few outcrops and several boreholes (Fig. 6). They were deposited in a fully marine environment.

Brejning Formation (Oligocene, Chattian)

The Brejning Formation consists of low-permeability greenish-brown, brown to black, glauconitic micaceous clay. The clay can contain pebbles and sideritic concretions. In northern Jylland, the formation consists of 50% clay, 40% silt, and coarser material of glauconite and pyrite. In the upper part of the formation, increased organic matter, silt and sand occur (Rasmussen *et al.* 2010). The clay fraction is dominated by 36–40% illite, 30–40% smectite, and 30% kaolinite and gibbsite (Friis 1994). In eastern Jylland, however, the composition is different (Rasmussen 1995).

This Oligocene marine formation is widely known in the Jylland area from several exposures in clay pits and coastal cliffs (Larsen & Dinesen 1959; Rasmussen *et al.* 2010). Furthermore, many boreholes reach the formation (Gravesen *et al.* 2013) (Fig. 6). The thickness varies from a few metres to approximately 40–50 m. The marine area seems to have covered large parts of the Danish land area (Rasmussen *et al.* 2010).

Vejle Fjord Formation (Miocene, Aquitanian)

The Vejle Fjord Formation consists of dark brown, brown and grey-brown micaceous clayey silt, very silty clay, and fine-grained sand, often as thin layers with pyrite (Larsen & Dinesen 1959). The clay mineral fraction consists, on average, of 15–30% smectite, 40% illite and 30% kaolinite (Rasmussen 1995).

The formation was deposited in a brackish to fully marine environment in western Denmark (Jylland), restricted by islands on the Ringkøbing-Fyn High. It is known from many coastal exposures along the east coast of Jylland and western Limfjorden, and is also known from abundant boreholes (Fig. 6). The formation is from 18 m to approximately 100 m thick.

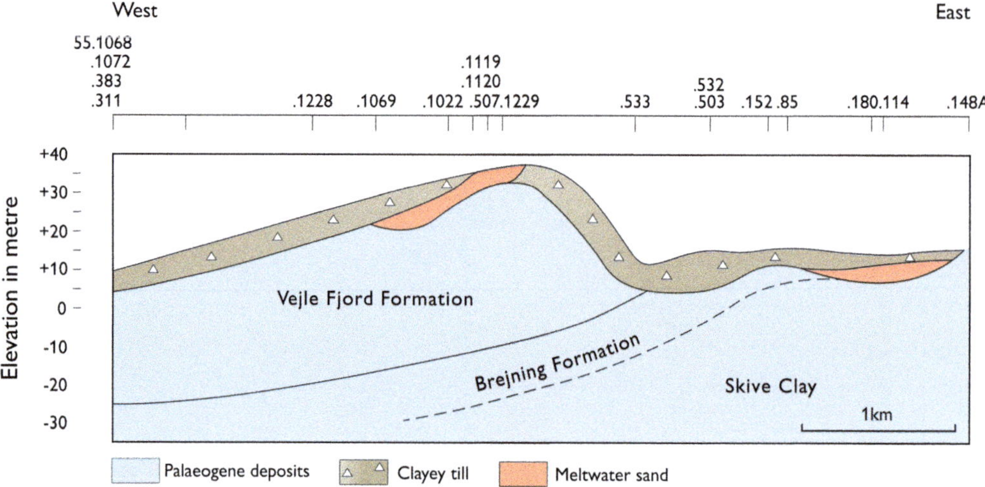

Fig. 6. Geological cross-section from the Skive area in northern Jylland. The numbers at the top of the figure are references to the borehole identification numbers in the borehole database Jupiter at the Danish Geological Survey (GEUS). Boreholes 55.1228 and 55.1229 are investigation boreholes to test the occurrence of the black Oligocene and Miocene clays. Note that the bedding is gently dipping, but appears to be steeper due to the vertical exaggeration (approximate exaggeration is ×25).

Site characteristics

The sites comprising Danish Palaeogene and Neogene formations included in this investigation all consist of marine deposits and are characterized by containing more than one formation at each locality. The potential host-rock successions comprise at least two or more formations in a stratigraphically normal position.

The formations are located relatively close to the ground surface and are mainly covered by glacial clayey till that may be intersected by tectonic fractures and fractures formed by freeze–thaw activities (Klint & Gravesen 1999). Although there is a possibly that the uppermost layers might have been disturbed by glaciotectonic displacements, sedimentological relationships and the stratigraphic succession seem not to show that this has occurred. Limestone–chalk units below the clay unit are situated approximately 100 m below ground surface and normally contain salty groundwater.

The Kerteminde Marl Formation and the Æbelø Formation of Late Paleocene age are found on NE Fyn (Kertinge Mark area, size *c.* 7 km^2) in a more than 75 m-thick succession overlying Danian Limestone (Fig. 4). Although the areas surrounding the locality are dominated by glaciotectonically disturbed sediment, the layers at the locality are considered as being *in situ* below the glacial basal clayey till.

On Lolland (Rødbyhavn area, size *c.* 20 km^2), there is a gradual transition between the Æbelø Formation and the Holmehus Formation of Late Paleocene age. Therefore, it is relevant to consider these two formations as one host sequence of nearly 80 m in thickness, resting on Cretaceous chalk (Fig. 5).

At northern Jylland (Thise area, size *c.* 14 km^2), the Branden Clay is approximately 100 m thick and overlain by the approximately 40 m-thick Brejning Formation (both of Late Oligocene age).

At Skive (Skive area, size *c.* 18,5 km^2), the Skive Clay is approximately 90 m thick and is overlain by 20 m of the Brejning Formation: above this, the Vejle Fjord Formation is 30 m thick (Fig. 6).

Discussion

Based on desk studies and a few supplementary new geological boreholes at the selected sites, most probably the clay sequences of the described areas meet the requirement of the main demands for host rocks, according to the applicable site suitability criteria:

- The clay formations have a lateral continuity within the rather large selected areas, and a vertical thickness of up to about 100 m;
- They are likely to have this distribution and homogeneity, because they were all deposited in a relatively deep marine environment and because the vertical sequences of different formations are a stratigraphic sequence without breaks.
- There is very little groundwater movement because of the low or very low permeability of

the clays and diffusive water transport in the clays is regarded as very low (therefore, any transport of radionuclides and hazardous materials will be delayed).

- The clays have self-sealing characteristics and fractures created by excavations will close.
- The clays are between approximately 20 and 60 million years old and have been stable over this period.
- Only the uppermost parts of the clay deposits may have been disturbed by glacial activity during the Pleistocene (last 2.6 million years). The preliminary investigations, however, show no glacial disturbances.

In Denmark, the investigations of clay formations are at an initial stage, but the properties seem to be promising for consideration as a host rock for the waste, cf. research in e.g. France, Switzerland and Belgium (ONDRAF/NIRAS & ANDRA 2015) where comparable lithologies are present. The clay formations studied in Switzerland and France are mainly of Mesozoic age (Jurassic and Cretaceous) (Enssle *et al.* 2011; Vinsot *et al.* 2011), and Cretaceous clay and marl formations are already used as impermeable layers below the low- and intermediate-level short-lived waste in the ANDRA disposal facility in the Aube District in France. The early Oligocene, Rupelian, Viborg Formation in Denmark is an equivalent to the low permeable Boom Clay (Wemaere *et al.* 2008; Gedeon *et al.* 2012) known from Belgium and the Netherlands, while the Ypresian plastic Røsnæs and Lillebælt Clay Formations are comparable to the Belgium Ypres Clay. The clay types are all deposited in a marine environment and have large horizontal distribution.

Although the studied formations represent different time intervals (Jurassic, Cretaceous, Palaeogene and Neogene) they have the important marine clay characteristics as mentioned above, and the clays from the five selected Danish areas share characteristics from the other clay types in Europe.

Concluding remarks

Palaeogene and Neogene clay deposits in Denmark are considered as possible host sediments for low- and intermediate-level radioactive waste. Although the clay deposits have only been preliminarily investigated, the host-rock requirements for a repository in Denmark are likely to be met by these clay rocks. Future detailed investigations in selected areas will demonstrate whether or not the clays have the optimal characteristics in accordance with investigated clays in France, Switzerland and Belgium, although it is to be noted that the first two of these clays are from different geological settings.

The Danish Parliament funded the investigations. The paper is published with permission of the Geological Survey of Denmark and Greenland.

References

Bøggild, O.B. 1918. *Den vulkanske Aske i Moleret samt en Oversigt over Danmarks ældre Tertiærbjergarter.* Danmarks Geologiske Undersøgelse series, **2**, (33).

Clausen, O.R. & Huuse, M. 2002. Mid-Paleocene palaeogeography of the Danish area. *Bulletin of the Geological Society of Denmark*, **49**, 171–186.

Clemmensen, A. & Thomsen, E. 2005. Palaeoenvironmental changes across the Danian–Selandian boundary in the North Sea Basin. *Palaeogeography, Palaeoclimatology, Palaeoecology*, **219**, 351–394.

Decleer, J., Viaene, W. & Vandenberghe, N. 1983. Relationships between chemical, physical and mineralogical characteristics of the Rupelian Boom Clay, Belgium. *Clay Minerals*, **18**, 1–10.

Dinesen, A., Lieberkind, K. & Michelsen, O. 1977. *A Survey of the Paleocene and Eocene Deposits of Jylland and Fyn.* Geological Survey of Denmark, Series B, **1**.

Enssle, C.P., Cruchaudet, M., Croise, J. & Brommundt, J. 2011. Determination of permeability of the Callovo-Oxfordian clay at the meter to decametre scale. *Physics and Chemistry of the Earth*, **36**, 1669–1678.

Friis, H. 1994. Lithostratigraphy and sedimentary petrography of the Oligocene sediments from the Harre borehole, Denmark. *Aarhus Geoscience*, **1**, 35–43.

Gedeon, M., Wemaere, I. & Labat, S. 2012. Characterization of groundwater flow in the environment of the Boom Clay Formation. *Physics and Chemistry of the Earth*, **36**, 1486–1495.

Gravesen, P. & Fredericia, J. (eds) 1984. *ZEUS Geodatabasesystem. Borearkivet. Databeskrivelse, Kodesystem og Sideregistre.* Danmarks Geologiske Undersøgelse, Series C, **3**.

Gravesen, P., Binderup, M., Nilsson, B. & Pedersen, S.A.S. 2011*a*. Geological characterization of potential disposal areas for radioactive waste from Risø, Denmark. *Bulletin of the Geological Survey of Denmark and Greenland*, **23**, 21–24.

Gravesen, P., Binderup, M. et al. 2011*b*. *Slutdepot for Risøs Radioaktive Affald. Geoviden – Geologi og Geografi*, No. 2, 2011.

Gravesen, P., Nilsson, B., Binderup, M., Larsen, T.B. & Pedersen, S.A.S. 2013. Geology, seismic activity and groundwater conditions at six potential disposal sites for radioactive waste from Risø, Denmark. *Bulletin Geological Survey of Denmark and Greenland*, **28**, 13–16.

Gry, H. 1935. *Petrology of the Paleocene Sedimentary Rocks of Denmark.* Geological Survey of Denmark, Series 2, **61**.

Håkansson, E. & Pedersen, S.A.S. 1992. *Map of Bedrock Geology of Denmark.* Varv and Geological Survey of Denmark and Greenland, Copenhagen.

HEILMANN-CLAUSEN, C. 1980. Paleocene plastic clay from the Vejle Fjord area. *Bulletin of the Geological Society of Denmark*, **29**, 47–52.

HEILMANN-CLAUSEN, C. 1985. *Dinoflagellate Stratigraphy of the Uppermost Danian to Ypresian in the Viborg 1 Borehole, Central Jylland, Denmark.* Geological Survey of Denmark, Series A, **7**.

HEILMANN-CLAUSEN, C. 1995. Palæogene aflejringer over Danskekalken. *In*: NIELSEN, O.B. (ed.) *Danmarks geologi fra Kridt til i dag. Aarhus Geokompendier 1*. Geologisk Institut, Aarhus Universitet, Aarhus, 69–114.

HEILMANN-CLAUSEN, C. & SURLYK, F. 2012. 10. Koralrev og lerhav. *Naturen i Danmark. Geologien. Gyldendal*, 181–226.

HEILMANN-CLAUSEN, C., NIELSEN, O.B. & GERSNER, F. 1985. Lithostratigraphy and depositional environments in the Upper Paleocene and Eocene of Denmark. *Bulletin of the Geological Society of Denmark*, **33**, 287–323.

IAEA 2014. *Near Surface Disposal Facilities for Radioactive Waste. Specific Safety Guide No. SSG-29*. International Atomic Energy Agency, Vienna.

KLINT, K.E.S. & GRAVESEN, P. 1999. Fractures and biopores in Weichselian clayey till aquitards at Flakkebjerg, Denmark. *Nordic Hydrology*, **30**, 267–284.

KLITTEN, K. 2003. Log-stratigrafi for Selandien Lellinge Grønsand formationen og Kerteminde Mergel formationen. *Geologisk Tidsskrift*, **2**, 20–22.

LARSEN, G. & DINESEN, A. 1959. *Vejle Fjord Formation ved Brejning: Sedimenterne of foraminiferfaunaen (Oligocæn–Miocæn)*. Danmarks Geologiske Undersøgelse, Series 2, **82**.

LARSEN, L.M., FITTON, J.G. & PEDERSEN, A.K. 2003. Paleogene volcanic ash layers in the Danish basin: composition and source areas in the North Atlantic Igneous Province. *Lithos*, **71**, 47–80.

NIELSEN, O.B. 1994. Lithostratigraphy and sedimentary petrography of the Paleocene and Eocene sediments from the Harre borehole, Denmark. *Aarhus Geoscience*, **1**, 15–34.

NIELSEN, O.B., BAUMANN, J., DEYU, Z., HEILMANN-CLAUSEN, C. & LARSEN, G. 1986. Tertiary deposits in Store Bælt. *Geoskrifter*, **24**, 235–253.

NIELSEN, O.B., FRIIS, H. & KORSBECH, U. 1994. Lithology and lithostratigraphy of the Harre borehole, Denmark. *Aarhus Geoscience*, **1**, 4–14.

ONDRAF/NIRAS & ANDRA 2015. *Clays in Geological Disposal Systems*. Brochure published on the occasion of the 6th International Conference 'Clays in Natural and Engineered Barriers for Radioactive Waste Confinements', Brussels, Belgium.

PEDERSEN, A.D. 1992. *Bentonitprojekt Lolland. Fase 3: Supplerende feltundersøgelser samt analyser og vurdering af leret ved Rødbyhavn*. Storstrøms amt. Miljøkontoret.

PEDERSEN, G.K. & SURLYK, F. 1983. The Fur Formation, a late Paleocene ash-bearing diatomite from northern Denmark. *Bulletin of the Geological Society of Denmark*, **32**, 43–65.

PEDERSEN, S.A.S. 2012. Glaciodynamic sequence stratigraphy. *In*: HUUSE, M., REDFERN, J., LE HERON, D.P., DIXON, R.J., MOSCARIELLO, A. & CRAIG, J. (eds) *Glaciogenic Reservoirs and Hydrocarbon Systems*. Geological Society, London, Special Publications, **368**, 29–51, https://doi.org/10.1144/SP368.2

PEDERSEN, S.A.S. 2014. Architecture of glaciotectonic complexes. *Geosciences*, **4**, 269–296.

PUSCH, R., KASBOHM, J., HOANG-MINH, T., KNUTSSSON, S. & NGUYEN-THANH, L. 2015. Holmehus clay – A Tertiary smectitic clay of potential use for isolation of hazardous waste. *Engineering Geology*, **188**, 39–47.

RASMUSSEN, E.S. 1995. Vejle Fjord Formation: mineralogy and geochemistry. *Bulletin of the Geological Society of Denmark*, **42**, 57–67.

RASMUSSEN, E.S., DYBKJÆR, K. & PIASECKI, S. 2010. Lithostratigraphy of the upper Oligocene–Miocene succession of Denmark. *Bulletin of the Geological Survey of Denmark and Greenland*, **22**, 1–92.

SHELDON, E., GRAVESEN, P. & NØHR-HANSEN, H. 2012. Geology of the Femern Bælt area between Denmark and Germany. *Bulletin of the Geological Survey of Denmark and Greenland*, **26**, 13–16.

SORGENFREI, Th. & BUCH, A. 1964. *Deep Tests in Denmark 1935–1959*. Geological Survey of Denmark, Series 3, **36**.

SURLYK, F., DAMHOLT, T. & BJERAGER, M. 2006. Stevns Klint, Denmark: uppermost Maastrichtian chalk, Cretaceous–Tertiary boundary, and lower Danian bryozoan mound complex. *Bulletin of the Geological Society of Denmark*, **54**, 1–48.

TANK, R.W. 1963. Clay mineralogy of some lower Tertiary (Paleogene) sediments from Denmark. Geological Survey of Denmark, Series 4, **9**.

THOMSEN, E. & HEILMANN-CLAUSEN, C. 1985. The Danian–Selandian boundary at Svejstrup with remarks on the biostratigraphy of the boundary in western Denmark. *Bulletin of the Geological Society of Denmark*, **33**, 341–362.

VINSOT, A., DELAY, J., VAISSIERE, R. DE LA & CHRUCHAUDET, M. 2011. Pumping tests in a low permeability rock: results and interpretation of a four-year long monitoring of water production flow rates in the Callovo-Oxfordian agillaceous rock. *Physics and Chemistry of the Earth*, **36**, 1679–1687.

VIS, G.-J. & VERWEIJ, J.M. 2014. *Geological and Geohydrological Characterization of the Boom Clay and its Overburden*. OPERA-PU-TNO411 Report. Central Organization for Radioactive Waste (COVRA), Nieuwdorp.

WEMAERE, I., MARIVOET, J. & LABAT, S. 2008. Hydraulic conductivity variability of the Boom Clay in north-east Belgium based on four core drilled boreholes. *Physics and Chemistry of the Earth*, **33**, 24–36.

YU, H.-D., CHEN, W.-Z., JIA, S.-P., CAO, J.-J. & LI, X.-L. 2012. Experimental study on the hydro-mechanical behavior of Boom clay. *International Journal of Rock Mechanics & Mining Sciences*, **53**, 159–165.

Approaches to evaluate and underpin the technical feasibility of the Belgian disposal concept

S. DOUDOU[1]*, E. J. HARVEY[1], P. J. RICHARDSON[1], S. M. WICKHAM[1], P. VAN MARCKE[2], D. RAYMAEKERS[2] & W. WACQUIER[2]

[1]*Galson Sciences Ltd, 5 Grosvenor House, Melton Road, Oakham LE15 6AX, UK*

[2]*ONDRAF/NIRAS, Avenue des Arts 14, 1210 Brussels, Belgium*

**Correspondence: sd@galson-sciences.co.uk*

Abstract: In 2009, ONDRAF/NIRAS, the Belgian Agency for Radioactive Waste and Enriched Fissile Materials, established a six-year research, development and demonstration (RD&D) programme to explore the feasibility of its concept for the geological disposal of Category B and C radioactive waste in Belgium. This programme, generally referred to as the 'B&C Techno project', aims to acquire evidence and build arguments in order to demonstrate that the proposed disposal system can be constructed, operated and progressively closed, taking into account long-term safety requirements and operational safety requirements.

Galson Sciences Ltd (GSL) has been supporting the co-ordination of the activities involved in this project. This support has included synthesis of research outputs, and the development of various tools and methods to demonstrate that the Belgian geological disposal concept is feasible to implement. Two examples of studies conducted by GSL for ONDRAF/NIRAS as part of the B&C Techno project to substantiate feasibility are described in this paper:

- A review of other national geological disposal concepts to identify transferable experience relating to the feasibility of constructing, and assembling comparable disposal system components.
- Development of a storyboard illustrating the different steps involved in disposing of Category B and C waste.

In 2009, ONDRAF/NIRAS, the Belgian Agency for Radioactive Waste and Enriched Fissile Materials, established a 6 year research, development, and demonstration (RD&D) programme to explore the feasibility of its concept for the geological disposal of Category B and C radioactive waste in Belgium. This programme, generally referred to as the 'B&C Techno project', aims to acquire evidence and build arguments in order to demonstrate that the proposed disposal system can be constructed, operated and progressively closed, taking into account long-term safety requirements and operational safety requirements. In particular, it aims to demonstrate that there are no fundamental flaws or 'showstoppers' relating to plans for the construction and operation of the disposal facilities. The outcomes from this project will support ONDRAF/NIRAS' first Safety and Feasibility Case (SFC1).

The project brings together international experts from a range of waste management organizations, and supporting research and consultancy organizations, to explore, at the conceptual design level, the anticipated feasibility of activities in three distinct areas:

- fabrication and assembly of disposal waste packages, along with construction and operation of the associated fabrication and 'post-conditioning' facilities on the surface;
- construction of the repository (i.e. excavation and construction of underground installations);
- repository operations, including waste transfer and emplacement, and closure (backfilling and sealing) of the repository.

The programme of work has been carried out in accordance with ONDRAF/NIRAS' Feasibility Strategy and Feasibility Assessment Methodology for the geological disposal of radioactive waste (Wacquier *et al.* 2011). This strategy and methodology centres on the development of arguments to support a hierarchy of Feasibility Statements (the latest versions of which are provided in (section 5 of ONDRAF/NIRAS 2013). These cover all the steps of the disposal process, from retrieval from interim storage facilities, through to transport to the disposal site, packaging in disposal containers, transfer and emplacement underground, to the backfilling and sealing of the repository.

Galson Sciences Limited (GSL) has been supporting the co-ordination of activities involved in the B&C Techno project. This support has included synthesis of research outputs, and the development of various tools and methods to demonstrate that

From: Norris, S., Bruno, J., Van Geet, M. & Verhoef, E. (eds) 2017. *Radioactive Waste Confinement: Clays in Natural and Engineered Barriers*. Geological Society, London, Special Publications, **443**, 39–48.
First published online June 24, 2016, https://doi.org/10.1144/SP443.11

the Belgian geological disposal concept is feasible to implement. RD&D activities undertaken to substantiate feasibility focus on issues of particular relevance to the Belgian concept, and aim to avoid repeating work where transferable evidence is available from elsewhere.

This paper summarizes selected outcomes from the B&C Techno project that advance confidence in the feasibility of the Belgian geological disposal concept across the three areas listed above. It then discusses two examples of studies conducted by GSL for ONDRAF/NIRAS under the B&C Techno project to substantiate feasibility:

- A review of other national geological disposal concepts to identify transferable experience relating to the feasibility of constructing, and assembling comparable disposal system components. This enables identification of key gaps in knowledge and, hence, the prioritization of ONDRAF/NIRAS' ongoing RD&D activities.
- Development of a storyboard illustrating the steps involved in disposing of Category B and C waste, from retrieval from interim storage facilities, through to the transport to the disposal site, to the backfilling and sealing of the repository.

The Belgian disposal concept

A geological disposal facility (GDF) is proposed to be sited in poorly indurated clay (Boom Clay or Ypresian clays). RD&D in the period 2009–15 focused on a reference repository depth of 230 m. The GDF will accommodate both Category B (long-lived low- and intermediate-level waste (LL–LILW)) and Category C (high-level waste (HLW) and spent nuclear fuel (SNF)) wastes. The waste disposal containers will be emplaced in a series of horizontal disposal galleries lined with pre-cast concrete wedge blocks. The repository sections for the Category B and C wastes will be at one level, but spatially separated. Access to the underground facility will be via three shafts: two service shafts for operator access and an air inlet at either end of the repository; and a central larger waste and air outlet shaft. The reference layout of the Belgian GDF is shown in Figure 1.

Vitrified HLW canisters and SNF assemblies will be disposed of in the GDF within a 'supercontainer'. The supercontainer consists of a carbon-steel overpack containing the primary waste container(s), filler and buffer, with or without an envelope. In the case of SNF, a canister-filling material is used inside the primary waste package canister (made of stainless steel), and a cast-iron insert fills the voids between the SNF canisters and the carbon-steel overpack. The supercontainer provides radiological shielding and complete containment of the Category C waste during the thermal phase.

The waste disposal package for the Category B waste is the 'monolith B'. This monolith is a cylindrical concrete shell with a flat base, allowing its emplacement on the disposal gallery support structure. The monolith consists of a concrete caisson and lid that serve as a disposal package, filling mortar immobilizing the primary waste packages and filling the void space, and the primary waste packages placed within the caisson.

The existing primary waste packages are currently stored in interim storage facilities at Belgoprocess and will need to be transported to a post-conditioning facility. The handling and transport of nuclear waste packages are already routinely carried out in the nuclear industry in Belgium and worldwide. The feasibility of the retrieval, handling and transport of primary waste packages from interim storage facilities to the post-conditioning facility is therefore considered to be sufficiently demonstrated and proven.

After the primary waste packages are placed in disposal packages in the post-conditioning facility, these disposal packages will be transported from the post-conditioning facility to their final position in the underground repository by means of a transport cart that will be moved by a battery-operated locomotive. At the crossing between the access gallery and the disposal gallery, the disposal package will be turned by a turntable.

After waste emplacement, the galleries and shafts in the underground repository need to be backfilled to avoid them acting as water-conducting features, and to limit the voids in the repository. Seals will be installed in the disposal galleries, at both ends of the access galleries and in the shafts. The seals will contribute to isolation by creating a physical barrier that prevents access to the waste. Seals will also divide the repository into compartments, thereby limiting the impact of disruptive events or processes affecting one part of the repository. Both the backfill material and the seal design have to take into account the potential requirement for waste retrievability. The backfill and seal therefore have to be designed and installed so that it is possible to safely dismantle them to provide access to, and allow the retrieval of, the disposal waste packages from the disposal galleries. No decisions on the design of the seal (bentonite seal, concrete seal or a combination of both) have yet been made.

Selected outcomes from the B&C Techno project

Selected main outcomes from the programme of work include:

- Determination of the performance of the Category B monolith under various accident

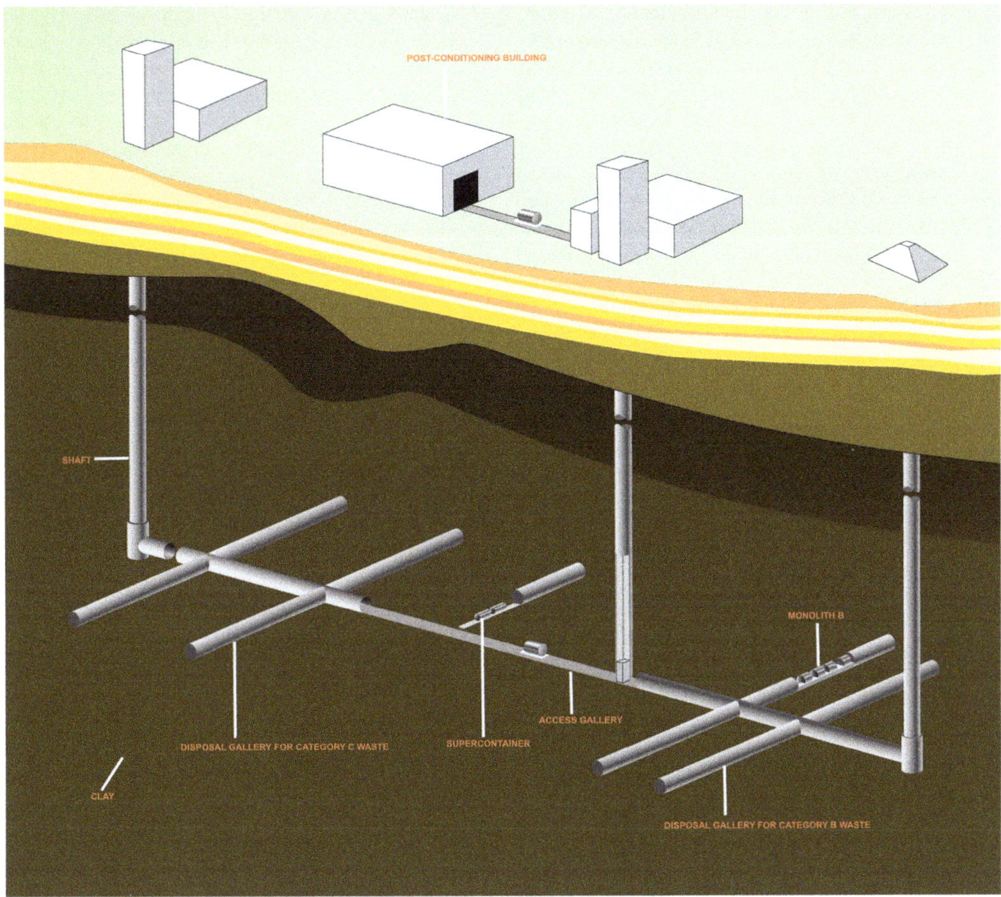

Fig. 1. A simplified view of the ONDRAF/NIRAS disposal concept for Category B and C wastes illustrating the separation between the regions for the two waste types (ONDRAF/NIRAS 2013, fig. 14).

scenarios, through simulation of impact damage resulting from various drop configurations (Tractebel Engineering 2014). Three drop configurations from a height of 50 cm were considered:

- the first with the monolith falling on its flat base;
- the second with impact on the edge of the base;
- the third with impact on a corner of the base of the monolith.

The overall conclusions were that no significant tensile failures (cracking) were observed for the three drop configurations. However, compression failure (crushing) was observed for the bottom drop (first) and corner drop (third) cases. This damage could be reduced by increasing the compressive strength and strain (ductility) of the monolith concrete through the use of more fibre reinforcement. It should be noted that the drop test configurations considered were quite severe and their probability of occurrence is low.

- Design of a device to tilt the supercontainer (proposed for packaging Category C waste) from a vertical to a horizontal orientation following assembly and prior to transfer (Tractebel Engineering 2013). The tilting device aims to perform this operation in a reversible manner (i.e. from vertical to horizontal and from horizontal back to a vertical position, if required). The tilting operation is performed using an overhead crane to place a vertical supercontainer on a pallet into the tilting device. The tilting device then rotates the supercontainer and the pallet into a horizontal position, before a trolley is driven underneath the supercontainer to unload it from the pallet and the tilting device.
- Development of a concrete recipe for the supercontainer concrete buffer, and testing its

construction feasibility and performance in half-scale tests (EURIDICE 2011). Self-compacting concrete formulations are preferred for use in the supercontainer buffer. Two half-scale tests were carried out in 2009 and 2013 to test the performance of the concrete and the construction feasibility of the buffer. The first half-scale test focused on fabricating the concrete buffer and investigating the impacts of inserting a hot overpack, simulated by means of a heater, in the buffer. An array of superficial small cracks appeared on the surface of the buffer after insertion of the heat source. The formation of these cracks is further examined in a second half-scale test. The analysis of this test is ongoing.

- Design of a repository ventilation system for repository construction and operation (Herold *et al.* 2013*b*; Antea-BG 2014*a*). The proposed ventilation system during construction would be installed in three phases: Phase 1 for ventilation of the shafts; Phase 2 for ventilation of the access gallery; and Phase 3 for ventilation of disposal galleries. Phases 1 and 2 use blowing ventilation, whilst Phase 3 uses a double air flux system (air blowing and air suction). During repository operation, the general ventilation concept includes a continuous airflow inside the main gallery. The service shafts will function as the air intake shafts, while the waste shaft will function as the outlet. The airflow in the access gallery will be generated by a fan at the top of the waste shaft, which works by suction. The disposal galleries will be ventilated by an auxiliary ventilation system. The main components of the auxiliary system are ducts in the disposal galleries, the main gallery and the waste shaft, and an exhaust fan at the top of the waste shaft. In the same manner as the main fan in the waste shaft is used to generate airflow in the access gallery, the fan in the auxiliary system also works by suction to generate airflow in the disposal galleries.
- Development of a shaft hoisting system and an underground transport system for transporting the supercontainers and monoliths from the surface to their disposal positions in the repository (Herold *et al.* 2013*a*). The multi-cable hoisting system assumes a maximum payload weight of 80 tonnes based on the mass of the heaviest assembled supercontainer. Safety features are included throughout the design of the hoisting system. Underground transport of the disposal packages is achieved using a trolley and a battery-powered locomotive, and employs a rail–wheel configuration. The rail system consists of parallel grooved rails in which the wheels of the trolley and locomotive travel. Instead of traditional rail wheels, a cart with castor type wheels that rest inside the rails installed in the access gallery would be used. In the disposal galleries, no grooved rails are envisaged: instead, hydraulically damped telescopic wheels will be fitted to the sides of the locomotive to ensure that it remains centred in the waste transport channel.
- Development of functional requirements for the repository backfill, and research to develop backfill recipes that fulfil these requirements (Engelhardt 2014). The main function of the backfill is to fill void space and provide stability to the galleries. However, its strength must be low enough to enable retrieval of waste packages. Furthermore, the backfill needs to provide a chemical environment consistent with the design concept (i.e. it should not disturb the alkaline corrosion-protective environment and should not cause the formation of mobile radionuclide complexes). A high-porosity material is preferred as it can provide a storage volume for gas generated in the repository and, consequently, limit the gas pressure build-up. Research is ongoing to develop and optimize the appropriate backfill formulation that fulfils these requirements.

Review of disposal concepts

The objective of this study (Doudou & Richardson 2013) was to review the designs and concepts for geological disposal under consideration by waste management organizations in countries with established repository programmes, and focus on the engineering and technical feasibility of fabricating or constructing components similar to those envisaged in the Belgian geological disposal concept for Category B and C wastes, with a view to identifying the extent to which RD&D work in those countries helps to substantiate the feasibility of implementing the Belgian concept.

Disposal concepts in the following nine countries were reviewed: Canada, Finland, France, Germany, Japan, Sweden, Switzerland, the UK and the USA. The specific components of disposal concepts considered were: the SNF canister, the SNF overpack insert, the overpack, the concrete container for LL–LILW, the repository layout, the waste transport methods, the ventilation approach, the backfilling concept and the repository sealing method.

The review focused on disposal system components that are common to both the Belgian concept and these international concepts. Little or no consideration was given to those aspects of the Belgian disposal concept that are unique, such as the use of a cementitious buffer within a supercontainer or disposal in a poorly indurated clay host rock (the

Boom Clay), as it is assumed that relatively little transferable experience relating to these components is available from other national programmes. Aspects of feasibility relating to these components are being substantiated through ONDRAF/NIRAS' own ongoing programme of RD&D (ONDRAF/NIRAS 2013).

Transferability of experience from other programmes

The SNF canister. The SNF primary waste canister in the Belgian concept is made from stainless steel. The void between the SNF assemblies and the steel canister will be filled with a material intended to avoid criticality (likely to be limestone sand).

Most other national programmes do not use specific canisters to enclose SNF assemblies in the disposal containers, although guiding frames or structures may be used to hold the assemblies together. A stainless-steel case (or tube) is used in France for some Pressurized Water Reactor (PWR) pre-conditioned assemblies, and prototypes have been built for the long-term storage of SNF assemblies. A carbon-steel basket is used to contain SNF bundles in Canada and a demonstration of the fuel assembly transfer system to the carbon-steel basket is planned for the period 2012–18. In the German concept, the fuel rods are put in a stainless steel can before insertion into the disposal containers.

The SNF overpack insert. A spent fuel overpack insert, made of cast iron, is envisaged in the Belgian concept. This insert will provide mechanical strength to the carbon-steel overpack and improve thermal conductivity inside the package. The SNF insert has either four square channels for uranium oxide (UOX) SNF canisters or one channel for a mixed oxide (MOX) canister. In the case of the UOX assemblies, the insert also provides a way of minimizing criticality by separating the UOX canisters.

Many other geological disposal programmes use similar materials as insert to the overpack for SNF waste: for example, Finland, France, Sweden and, possibly, Switzerland. Demonstration experiments and prototypes of these components have been built in Finland, France and Sweden. Therefore, the feasibility of fabricating the cast-iron insert has been shown and valuable transferable experience exists.

The SNF overpack. A carbon-steel overpack is envisaged in the Belgian concept in order to assure waste containment during the thermal phase. Carbon steel was chosen for its low corrosion rate.

Many countries use copper overpacks in their disposal concepts (Canada, Finland, Sweden and, possibly, Switzerland). However, there are some countries that intend to use steel overpacks for disposal containers, such as France (P235 unalloyed steel) and Switzerland (carbon steel). Canada has an alternative concept, where an overpack made of carbon steel is considered. Prototype steel overpacks have been built in France.

The LL-LILW concrete overpack. The monolith B concept has been developed in Belgium for Category B waste. In the Belgian programme for Category A waste (low-level waste (LLW)) disposal, a similar concrete container (monolith A) is used for near-surface disposal. Some experiments and tests have been carried out and prototypes of monolith A have been constructed. Research experience gained for monolith A can be transferred to monolith B. Many other countries also use concrete overpacks for containers of LL–LILW, although the designs are varied. Countries using or considering use of such overpacks are: Canada, Finland (for short-lived waste), France, Germany, Japan (previously), Sweden, Switzerland and the UK.

In France, LL–LILW concrete containers (similar to monolith B) have been built to demonstrate feasibility, test design variations (e.g. fibrous concrete, reinforced concrete) and also perform drop test experiments to study their mechanical strength. Fire tests have also been performed for cylindrical concrete containers of LL-LILW in Germany. In Japan, experiments were carried out on concrete container prototypes for LL-LILW to test the effects of an accidental fall or an earthquake, and overpressurization due to the generation of gas.

Repository layout. The layout of a repository is inherently inventory dependent and rock-specific, so no two geological repositories will be exactly the same. However, most repository designs incorporate similar components, such as the use of shafts, ramps/drifts and tunnels. No HLW/SNF repository has yet been constructed, although the engineering technologies to be used for the construction and implementation of the layouts have been demonstrated in other repositories (e.g. in salt at the Waste Isolation Pilot Plant (WIPP) in the USA) and in many underground research laboratories (URLs) worldwide in different rock types. Many of these technologies have been developed from civil engineering applications currently in use.

In a similar way to that proposed in Belgium, France, Japan, the UK and, possibly, Switzerland plan to dispose of both HLW and ILW in a co-located facility. There have been many studies of co-location in the UK that the Belgian concept can draw upon (e.g. Solano *et al.* 2013).

It is anticipated that all countries covered here will use shafts for access to the underground repositories, some with a combination of access ramps/tunnels. Many countries have dead end tunnels in their designs, although these may be of different dimensions to the Belgian design for the blind disposal galleries.

As in Belgium, France and Switzerland envisage horizontal-only emplacement of disposal packages. Both vertical and horizontal emplacement options are considered in Canada, Germany and Japan. It should be noted that one of the unique aspects of the Belgian concept is the disposal of LL–LILW in narrow disposal galleries, whereas other countries generally assume large disposal vaults or large-diameter tunnels. This arises from the nature of the poorly indurated clay host rock in Belgium, as it becomes more difficult to support large openings. Significant consideration has been given to this issue under the B&C Techno project as part of demonstrating the feasibility of scaling-up demonstrated excavations at the HADES underground research facility to the required size in the repository itself (e.g. see Antea-BG 2014*b*).

Waste transport. The focus here is on the transfer of waste packages to the underground repository and inside the repository. In the Belgian concept, a waste shaft will be used to transport the disposal containers underground on a transport cart with rubber wheels. A battery-powered locomotive will then be used to transfer the cart with the disposal package to its final position in the disposal gallery.

Many other countries propose to use wheeled vehicles to transport disposal packages underground, some also involving the use of rail systems. The countries that envisage using wheeled vehicles include France, Japan, Sweden, Switzerland and the UK. France and Sweden have built prototypes, and France has demonstrated emplacement methods as part of the ESDRED project.

In the German concept, although a rail system is envisaged, the components of the transport system are very similar to the Belgian concept. Both systems have been developed by DBE Technology in Germany. In the Belgian concept, grooves in the floor are used to guide the transport cart underground instead of steel rails. This was proposed in order to minimize the amount of steel in the underground repository. The German transport system has been tested in surface facilities.

The ventilation system. Many countries, like Belgium, will employ repository ventilation systems that involve the use of shafts, for air intake and air exhaust. Even though the repositories have not yet been constructed, similar ventilation systems are widely used in the mining industry and in URLs worldwide. It is expected that the experience from operating ventilation systems for URLs and mines can be transferred for use in repositories, even though URL systems may be less extensive.

In most countries, the ventilation system of the nuclear waste areas is separated from the construction areas so that potentially contaminated air from the disposal regions does not come in contact with air from construction areas. There is also a desire to keep air from the construction area away from the disposal area to prevent the build-up of dust, which could affect operational activities and/or post-closure performance of the disposal system. It is also noted that a separate (temporary) ventilation system is often required for the construction area, as a higher capacity is necessary. In the Belgian concept, the repository construction and operation phase will not be carried out simultaneously.

Backfilling materials. In all the countries examined, none propose the use of cementitious grout as backfilling material for HLW or SNF, as is proposed in Belgium. However, many envisage the use of such grout in relation to ILW disposal, with the exception of Canada, where no backfill is envisaged. As an example of such grouts, the UK has developed a patented cement-based grout, the so-called Nirex Reference Vault Backfill (NRVB), which may be used as backfill in the ILW disposal vaults if the Phased Geological Repository Concept (PGRC) is taken forward. Most UK ILW waste forms will be immobilized in cement-based grouts.

Sealing materials. Most countries plan to use a mixture of concrete, clay and/or crushed rock for their sealing and plugging systems. In Belgium, no decision has yet been made on whether bentonite seals, concrete seals or a mixture of both will be used to seal the galleries and shafts.

Many experiments demonstrating the designs, functions and feasibility of constructing plugs and seals have been, and are being, carried out internationally.

Storyboard and accompanying diagrams

As part of its role under the B&C Techno project, GSL has also developed a ‘storyboard’, which describes all of the activities relating to the construction, operation and closure of the geological repository. The storyboard, which is a text-based tool developed in Microsoft Excel, aims to provide a global overview of the scope of the B&C Techno project, and acts as a completeness check of activities to underpin feasibility. It captures all the steps in the Category B and C disposal programme from the retrieval of primary waste containers from storage facilities, through to the closure of the repository.

Over the course of the B&C Techno project, GSL has also developed a series of diagrams to illustrate selected steps in the storyboard. The B&C Techno storyboard and accompanying diagrams have been developed to fulfil a number of objectives, namely:

- To provide a completeness check on activities under the B&C Techno project in order to underpin the feasibility of the proposed disposal concept for Category B and C wastes (Wacquier *et al.* 2011).
- To provide a framework within which B&C Techno activities are undertaken, and a clear link to associated feasibility statements and open questions pertaining to feasibility.
- To help identify responsibilities across different B&C Techno contractors, and to identify cross-cutting RD&D topics.
- To act as a communication tool, enabling assumptions about the B&C Techno disposal concept to be communicated to a variety of audiences.
- To illustrate sections of the feasibility part of the first safety and feasibility case (SFC1) submission.
- To help identify potential hazards and risks in the disposal process.

There is a strong link between the B&C Techno storyboard and the hierarchy of feasibility statements. This can be summarized as follows:

- The feasibility statements set out what activities must be shown to be feasible.
- The storyboard sets out what steps will be followed.

Nevertheless, the storyboard has been developed separately from the feasibility statements, as an independent quality assurance activity that provides a completeness check across the B&C Techno project activities. Within the storyboard, activities associated with implementing the Belgian geological disposal concept have been defined and grouped into three topic areas, consistent with the wider organization of the B&C Techno activities:

- Steps associated with the fabrication and assembly of Category B and C disposal waste packages, and the construction and operation of associated facilities on the surface ('disposal packaging activities').
- Sequential steps associated with excavation and construction of the geological repository ('construction activities').
- Sequential steps associated with the operation, backfilling and sealing of the repository ('operational activities').

The storyboard and initial versions of the accompanying diagrams describe and illustrate activities at a relatively high level. Details of some steps will become clearer as the B&C Techno work programme progresses, and as decisions about the implementation of the disposal concept are made.

Storyboard diagrams developed to date focus on two specific topics:

- Retrieval of Category C primary waste containers from storage facilities and transport to the post-conditioning facility.
- Steps involved in the handling of Category B disposal waste packages, including:
  - retrieval of primary waste packages from interim storage facilities;
  - emplacement of primary waste packages into the transport container;
  - emplacement of the transport container into the transport vehicle;
  - transport of primary waste containers from interim storage facilities to the disposal site;
  - removal of primary waste packages from the transport container at the disposal site;
  - fabrication of the monolith B caisson and lid;
  - transfer of the monolith B caisson and lid to the post-conditioning facility;
  - insertion of primary waste packages into the monolith B caisson;
  - pouring mortar into the monolith B caisson;
  - emplacement of the monolith B lid;
  - transfer of the completed monolith B to an interim storage facility prior to transfer underground;
  - retrieval of the completed monolith B from the interim storage facility and transport it to the waste shaft;
  - transport of monolith B down the waste shaft on the transport cart;
  - transport of monolith B underground to the disposal gallery;
  - emplacement of monolith B in the final disposal location and backfilling of the disposal gallery.

Issues and questions raised during preparation of the storyboard diagrams illustrating the Category B disposal concept have been discussed with ONDRAF/NIRAS and members of the B&C Techno project in order to clarify assumptions. Outstanding uncertainties and questions concerning practical implementation that may exist and affect preparation of the diagrams have also been identified and discussed. Further work that resolves ongoing uncertainties and addresses the remaining questions, as ONDRAF/NIRAS' feasibility programme progresses, will provide the basis for future updates to the storyboard diagrams developed to date.

Example diagrams are presented in this paper, but the full suite of diagrams can be found in Richardson *et al.* (2014). Figure 2 outlines the

Fig. 2. Filling the concrete caisson and emplacement of the lid. First, the caisson is partially filled with mortar, before the prefabricated lid is emplaced and bolted onto it. Then, the remainder of the mortar is pumped into the internal voidage. This process is achieved by means of two holes in the lid – the mortar would be pumped into the first hole until it emerges from the second.

steps involved in filling the caisson with mortar, emplacing the lid and bolting it on to produce the complete monolith B, and Figure 3 illustrates the range of safety systems envisaged at the shaft station.

Conclusions

Galson Sciences Ltd (GSL) has been supporting ONDRAF/NIRAS in its 6 year RD&D programme since 2009 to explore the feasibility of its concept

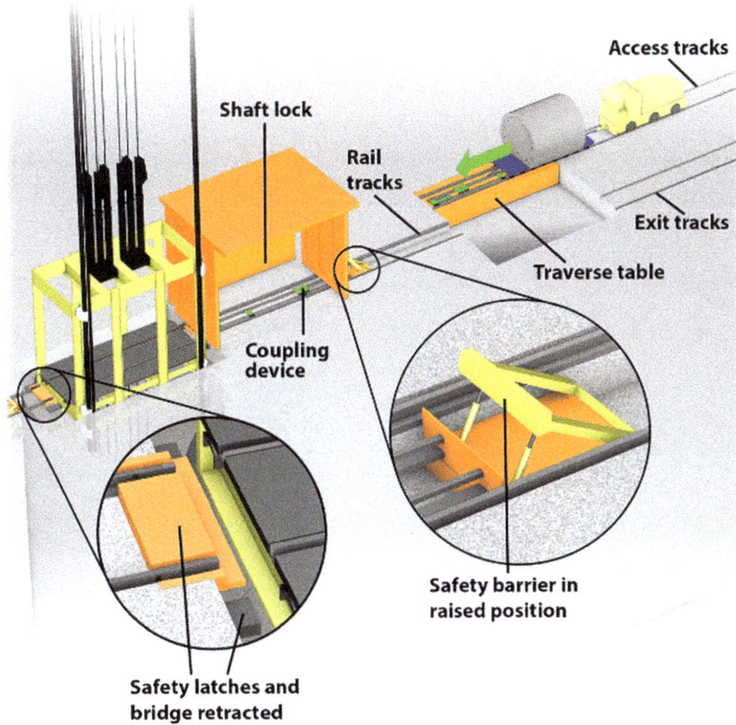

Fig. 3. Arrival of the monolith B and cart at the traverse table. The locomotive pushes the transportation trolley along the grooved rails until it reaches the traverse table, which is the first step in the process of transferring the monolith to the shaft system. This diagram shows the entire system, illustrating the range of safety features envisaged. At this stage, a safety barrier is raised in front of the shaft lock in order to prevent any possible contact with the transport trolley in the event of a failure in the coupling device. The safety latches and the shaft bridge are also retracted at this stage to prevent accidental entry into the shaft cage.

for the geological disposal of Category B and C radioactive waste in Belgium. This support has included synthesis of research outputs, and the development of various tools and methods to demonstrate that the Belgian geological disposal concept is feasible to implement. Two examples of studies conducted by GSL for ONDRAF/NIRAS under the B&C Techno project to substantiate feasibility are:

- A review of other national geological disposal concepts to identify transferable experience relating to the feasibility, and assembling comparable disposal system components. The designs and disposal concepts in nine countries (Canada, Finland, France, Germany, Japan, Sweden, Switzerland, the UK and the USA) have been analysed and compared to the Belgian concept with a view to identifying the extent to which RD&D work in these countries helps to substantiate the feasibility of implementing certain components of the Belgian concept. The focus was on the following components: the SNF canister, the SNF overpack insert, the overpack, the concrete container for LL-LILW waste, the repository layout, the waste transport methods (underground), the ventilation approach, the backfilling concept and the sealing method. The study concluded that transferable experience from the countries reviewed can be used for all the disposal concept components considered, except for the backfill grout foreseen in the galleries containing Category B and C wastes. No other country reviewed uses cement-based materials for backfilling HLW and SNF disposal galleries. However, many countries use such materials for backfilling LL-LILW disposal galleries.
- Development of a storyboard and accompanying diagrams illustrating the steps involved in disposing of Category B and C waste, from retrieval from interim storage facilities through to the transport to the disposal site, to backfilling and the sealing of the repository.

The work described here has been funded by ONDRAF/NIRAS.

References

ANTEA-BG 2014*a*. *B&C Techno Project – Lot 2 – Task 2-210: Construction of Galleries*. A63608 V0, November 2014 Antea Group Belgium, Antwerp.

ANTEA-BG 2014*b*. *B&C Techno Project – Lot 2 – Task 2-140: Feasibility to Construct Large Shafts*. Final Draft, May 2014, Version 1 Antea Group Belgium, Antwerp.

DOUDOU, S. & RICHARDSON, P.J. 2013. *Review of International Disposal Concepts to Substantiate the Belgian Disposal Concept for Category B&C Waste*. 0910-4 Version 1 Draft 3. Galson Sciences Ltd, UK.

ENGELHARDT, H.-J. 2014. *Technical Support for the R&D Feasibility Programme for the Geological Disposal of Category B&C Radioactive Waste*. R&D Study 3-350: Backfilling. Final Report, October 2014. DBE Technology, Germany.

EURIDICE 2011. *Construction and Testing of First Half-Scale Test*. EURIDICE Report EURIDICE, Mol, Belgium.

HEROLD, P., FILBERT, W. & HAVERKAMP, B. 2013*a*. *Technical Support for the R&D Feasibility Programme for the Geological Disposal of Category B&C Radioactive Waste*. R&D Study 3-311: Conceptual Design for the Hoisting System of the Waste Shaft. Draft Report, August 2013. DBE Technology, Germany.

HEROLD, P., NIEDER-WESTERMANN, G. & HAVERKAMP, B. 2013*b*. *Technical Support for the R&D Feasibility Programme for the Geological Disposal of Category B&C Radioactive Waste*. R&D Study 3-360 and 3-370: Ventilation Requirements and Models for the Repository. Final Report, July 2013. DBE Technology, Germany.

ONDRAF/NIRAS 2013. *ONDRAF/NIRAS Research, Development and Demonstration (RD&D) Plan – For the Geological Disposal of High-Level and/or Long-Lived Radioactive Waste Including Irradiated Fuel if Considered as Waste: State-of-the-Art Report as of December 2012*, NIROND-TR 2013-12 E, December 2013 ONDRAF/NIRAS (National Agency for Radioactive Waste and Enriched Fissile Material), Brussels.

RICHARDSON, P.J., HARVEY, E.J., DOUDOU, S. & KENT, C. 2014. *Development of a Storyboard for the Belgian Disposal Concept for Category B&C Waste: Summary of Status*. 0910-5 Version 1. Galson Sciences Ltd, UK.

SOLANO, J.M., HICKS, T.W., WHITE, M.W., WINPENNY, D.B., WATSON, S.P. & THATCHER, K.E. 2013. *Methodology for Determining the Separation Distance between the Disposal Modules of a Co-located Geological Disposal Facility*. 1151b1-3 Version 1. Galson Sciences Ltd, UK.

TRACTEBEL ENGINEERING 2013. *Feasibility Assessment of a Tilting Device for Supercontainer*. Report TIERSDI/4NT/0007754/000/00 Tractebel Engineering, Brussels.

TRACTEBEL ENGINEERING 2014. *Structural Integrity of the Category B monolith*. Report TIERSDI/4NT/0007833/000/00 Tractebel Engineering, Brussels.

WACQUIER, W., VAN HUMBEECK, H., WICKHAM, S. & HARVEY, E.J. 2011. *Feasibility Strategy and Feasibility Assessment Methodology for the Geological Disposal of Radioactive Waste*. ONDRAF/NIRAS SFC1 Level 4 Report: First Full Draft, NIROND-TR 2010-19 E, July 2011. ONDRAF/NIRAS, Belgium.

DOPAS full-scale experiments: approaches to compliance assessment

S. DOUDOU[1]*, M. J. WHITE[1], M. JOHNSON[2], J.-M. BOSGIRAUD[4] & PÄR GRAHM[5]

[1]*Galson Sciences Ltd, 5 Grosvenor House, Melton Road, Oakham LE15 6AX, UK*

[2]*Radioactive Waste Management Limited, Building 587, Curie Avenue, Harwell, Didcot OX11 0RH, UK*

[3]*Present address: Croft Associates Ltd, F4 Culham Science Centre, Abingdon OX14 3DB, UK*

[4]*National Radioactive Waste Management Agency (ANDRA), 1–7 rue Jean-Monnet, 92298 Châtenay-Malabry, France*

[5]*Swedish Nuclear Fuel and Waste Management Co., Oskarshamn, Sweden*

**Correspondence: sd@galson-sciences.co.uk*

Abstract: The Full-Scale Demonstration Of Plugs And Seals (DOPAS) Project is a European Commission programme of work jointly funded by the Euratom Seventh Framework Programme and European nuclear waste management organizations (WMOs). The DOPAS Project aims to improve the industrial feasibility of plugs and seals, the measurement of their characteristics, the control of their behaviour over time in repository conditions, and their hydraulic performance with respect to safety objectives.

Within the project, an approach has been developed and applied to assess the compliance of two of the full-scale experiments to their design bases. The approach involves a review of each requirement in the design basis and the strategy used to demonstrate compliance of the experiment with that requirement. Feedback in the form of proposed updates to the design statements is captured by this approach. Learning points on plugs and seals from the compliance assessment are also noted for consideration in the DOPAS Project outcomes. This developed compliance and assessment approach can be employed as part of, or in conjunction with, other more generic approaches used by WMOs, including monitoring, full-scale testing and the documentation of construction procedures.

The Full-Scale Demonstration Of Plugs And Seals (DOPAS) Project is a European Commission (EC) programme of work jointly funded by the Euratom Seventh Framework Programme and European nuclear waste management organizations (WMOs). The DOPAS Project is running over the period September 2012–August 2016. Fourteen European WMOs, and research and consultancy institutions, from eight European countries are participating in DOPAS. A set of full-scale experiments, laboratory tests and performance assessment studies of plugs and seals in geological repositories are being carried out over the course of the project.

Geological disposal of radioactive waste involves isolation and containment of the waste from the biosphere (IAEA 2011*a*). Containment and isolation can be provided through a series of complementary barriers (e.g. the waste form itself, waste containers, buffer and backfill materials, and the host geology), each of which will be effective over different timescales. The depth of disposal and the characteristics of the host geological environment provide isolation from the biosphere and retardation of migrating radionuclides, and reduce the likelihood of inadvertent or unauthorized human intrusion. Moreover, emplacement at depth in a stable geological formation may significantly reduce the influence of climatic and other surface processes (IAEA 2011*b*).

As part of the backfilling of a repository, specific parts of the underground works will have to be plugged and sealed. The purpose of plugs and seals will depend on the disposal concept, the nature of the geological environment and the inventory to be disposed:

- Plugs and seals may be required during operations to isolate emplaced waste and other engineered barrier system (EBS) components from the rest of the underground excavations.
- Plugs and seals may be required following closure to limit groundwater flow and radionuclide migration.
- Plugs and seals may be required to inhibit inadvertent or unauthorized human access.

DOPAS aims to improve the industrial feasibility of plugs and seals, the measurement of their characteristics, the control of their behaviour over time

From: Norris, S., Bruno, J., Van Geet, M. & Verhoef, E. (eds) 2017. *Radioactive Waste Confinement: Clays in Natural and Engineered Barriers*. Geological Society, London, Special Publications, **443**, 49–58.
First published online August 26, 2016, https://doi.org/10.1144/SP443.16

in repository conditions, and their hydraulic performance with respect to safety objectives. The DOPAS Project is being carried out in seven Work Packages (WPs). WP1 includes project management and coordination. WP2, WP3, WP4 and WP5 address, respectively, the design basis, installation, compliance testing and performance assessment modelling of the five full-scale experiments and laboratory tests. WP6 and WP7 address cross-cutting activities common to the whole project through review and integration of results, and their dissemination to other interested organizations in Europe and beyond.

The DOPAS Project focuses on tunnel, drift and shaft plugs and seals for crystalline, clay and salt rock. This paper focuses on two full-scale experiments of plugs and seals in clay rock and crystalline rock environments:

- Clay rock: the Full-scale Seal (FSS) experiment being undertaken by the French WMO (ANDRA) in a surface facility at St Dizier. FSS is a full-scale demonstration of the reference drift and intermediate-level waste (ILW) disposal vault seal for the French Centre Industriel de Stockage Géologique (CIGEO) repository concept.
- Crystalline rock: the Dome Plug (DOMPLU) experiment related to plugs in horizontal tunnels. DOMPLU is a full-scale demonstration of the reference deposition tunnel plug in the Swedish repository design being undertaken by the Swedish WMO (SKB) at the Äspö Hard Rock Laboratory.

This paper describes the approach followed to assess the compliance of DOMPLU and FSS to their design bases. Other generic strategies used to demonstrate compliance of designs to the design basis are also summarized.

The full-scale experiments

The SKB deposition tunnel plug and the DOMPLU experiment

The KBS-3V method is proposed by SKB for the disposal of spent fuel packaged in copper canisters with cast iron inserts in a crystalline host rock. The long-term safety principles are based on the isolation and containment of radioactive waste through the choice of a stable geological environment at depth, and the use of a multi-barrier system consisting of engineered barriers (canister, buffer, backfill and closure) and the host rock. The canisters are emplaced in vertical deposition holes, containing pre-compacted blocks of bentonite buffer, below horizontal deposition tunnels. The deposition tunnels are backfilled with bentonite blocks and pellets, and closed with a deposition tunnel plug.

Deposition tunnel plugs in the Swedish repository have temporary functions with the objective of supporting the performance of other safety barriers. Their functions during the operational period of the repository are to:

- confine the backfill in the deposition tunnel;
- support saturation of the backfill;
- provide a temporary barrier against water flow that may cause harmful erosion of the bentonite in the buffer and backfill.

The current SKB reference conceptual design for a deposition tunnel plug is described in SKB's report on the design, production and initial state of the backfill and plug in deposition tunnels (SKB 2010), and includes the following components (see Fig. 1):

- Concrete plug: the concrete plug is a dome-shaped structure made of low-pH reinforced concrete. It contains pipes for auxiliary equipment, such as air ventilation pipes, cooling pipes and grouting tubes. The cooling pipes are used to avoid internal cracking due to cement hydration and to pre-stress the concrete dome before contact grouting. The function of the concrete plug is to resist deformation, and to keep the watertight seal, filter and backfill in place.
- Watertight seal: the watertight seal is made of bentonite blocks and pellets in a similar

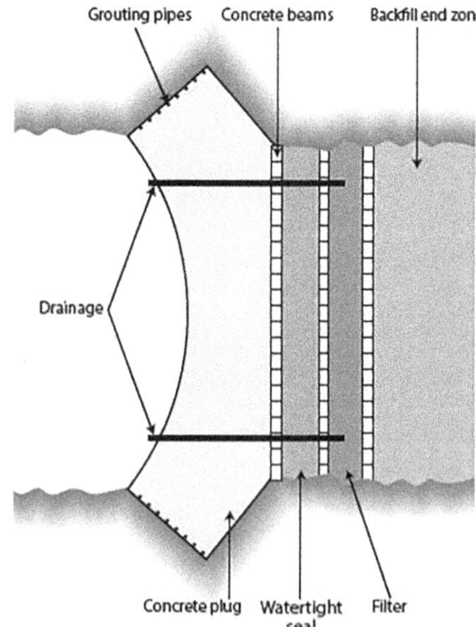

Fig. 1. Schematic diagram of the deposition tunnel plug SKB conceptual design (SKB 2010).

configuration to the backfill. It is 710 mm thick. The functions of the watertight seal are:
  - to seal water leakage paths through small cracks in the concrete plug or between the concrete and the rock surface;
  - to reduce the water pressure acting on the concrete dome so that no unfavourable water pressure is applied in the interface between the rock and the concrete, and so that the water pressure within the backfilled deposition tunnel is equalized.
- Backfill end zone: the part of the backfill closest to the plug in which the density is reduced to mitigate the swelling pressure loads on the plug.
- Filter: the filter is made of sand or gravel. Its function is to collect groundwater leaking from the backfilled deposition tunnel and, if required, drain it to the drainage pipes, so that no water pressure is applied on the concrete plug before it has cured and gained full strength. The filter will also facilitate saturation of the bentonite seal.
- Concrete beams (delimiters): the beams are made of low-pH reinforced concrete. Their function is to facilitate the construction works. The outer beams (towards the concrete plug) are covered with a thin layer of shotcrete to prevent the concrete slurry from mixing with the bentonite during casting of the concrete plug. The function of the outer beams is to keep the watertight seal in place during installation (i.e. acting as an inner formwork for the concrete dome). The inner beams (towards the deposition tunnel) shall keep the backfill in place during installation. The middle beams shall keep the filter in place during installation, and are designed to withstand the development of pressure during the swelling of the watertight seal and/or backfill.
- Drainage pipes: the drainage pipes need to operate throughout the period of operation (up to 100 years), and are made of steel or titanium. They are required to drain the water collected in the filter and to transport it out of the deposition tunnel, which will prevent water pressure being applied on the concrete plug before it has cured and gained full strength.
- Grouting pipes: the grouting pipes are made of steel and may be isolated by geotextile to prevent blocking during pouring. They shall be grouted when the concrete has reached a certain level of strength and shrinkage. The grout shall tighten the contact area between the concrete plug and rock, and contribute to keeping the concrete plug under compression.

DOMPLU represents a detailed iteration of the reference design rather than a fundamental change. A schematic illustration of DOMPLU is provided in Figure 2. The current SKB reference design and DOMPLU design are broadly similar, with the exception of a few modifications intended to test the performance of new materials planned to be introduced as the reference design in the future, or to facilitate experiment implementation. Such modifications include:

- The use of unreinforced concrete instead of reinforced concrete for the concrete dome.
- In DOMPLU, the backfill end zone is redefined as a backfill transition zone, where the swelling pressure from backfill is reduced to a level that is similar to the resulting swelling pressure of the bentonite seal (about 2 MPa). The purpose of introducing a transition zone is to reduce the displacement of the plug system components.
- In DOMPLU, the innermost (towards the backfill) delimiter is considered to be part of the filter. Instead of concrete beams, porous lightweight expanded concrete aggregate (LECA) beams and gravel with high permeability are used. The filter thickness is 600 mm, made up of 300 mm of gravel (with a particle size of 2–4 mm) and 300 mm of LECA beams, compared to a thickness of 700 mm, which is specified in the reference design for the filter.
- The middle delimiter, between the filter and the watertight seal, is composed of a geotextile instead of concrete beams.
- The outer delimiter is composed of low-pH concrete beams, as for the reference design. There is also a double geotextile layer between this delimiter and the concrete dome to prevent adhesion of the delimiter to the concrete dome, and therefore avoid potential cracking of the concrete dome during shrinkage.
- Cooling pipes are made of copper.
- Grouting tubes are made of cross-cut 50 mm plastic drainage tubes.
- The thickness of the watertight seal is 500 mm in DOMPLU, as this is considered sufficient for the timescales of DOMPLU (710 mm is used in the reference design).
- The filter installed dry density is 1400 kg m^{-3} in DOMPLU. A value of 1900 k m^{-3} is considered in the reference design.

ANDRA drift seal and the FSS experiment

In France, high-level waste (HLW) and ILW will be disposed of in a repository referred to as CIGEO. The repository is located in a clay host rock in the Meuse and Haute Marne departments of eastern France. The repository's primary function is to isolate the waste from activities at the surface, and its second function is to confine radioactive substances and to control the transfer pathways that may, in the long term, bring radionuclides into contact with

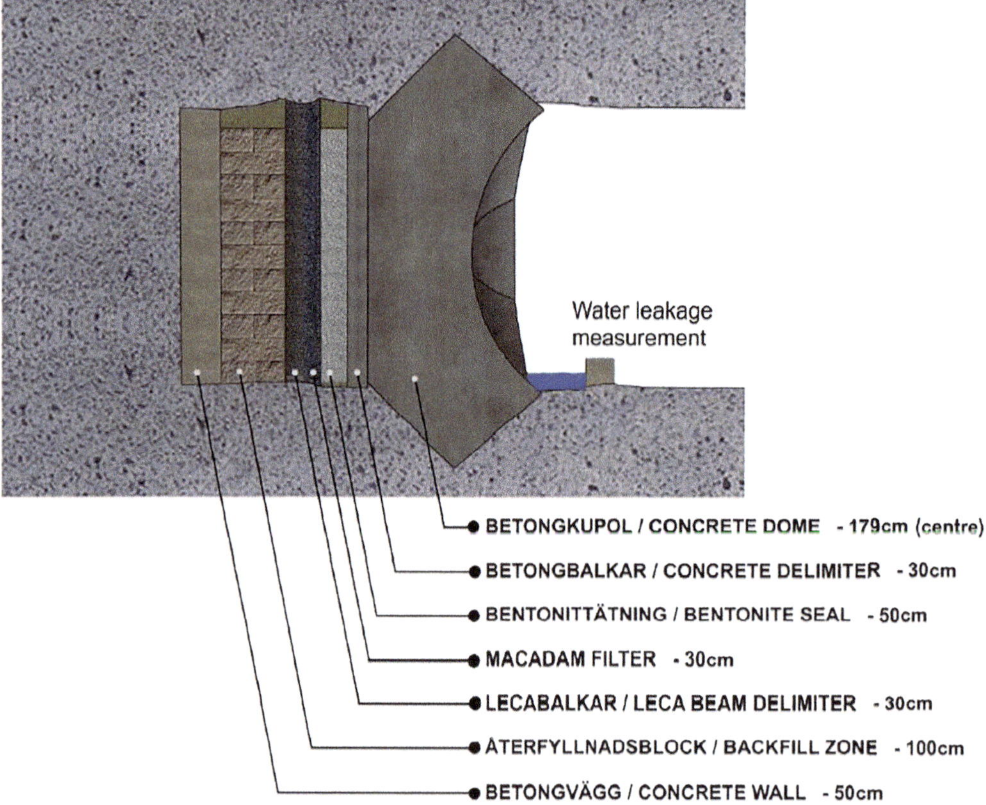

Fig. 2. Schematic diagram of the DOMPLU design (Palmer 2011).

humans and the environment. The principal contribution of the seals in ANDRA's concept is in the second function.

In ANDRA's concept, seals are defined as hydraulic components for closure of large-diameter (several metres) underground installations and infrastructure components. There are three types of seals envisaged in the French reference disposal concept: shaft seals; ramp seals; and drift and ILW disposal vault seals. Each seal consists of a swelling clay core and concrete containment walls (see Fig. 3). The swelling clay core provides the required long-term performance of the seal: the containment walls, however, are included to mechanically contain the bentonite swelling-induced force. The conceptual design of the drift and ILW disposal vault seals is the same.

The safety functions of the drift and ILW vault seals are:

- To limit water flow between the underground works and the overlying formations through the access shafts/ramps.
- To limit the water flow speed within the repository.

The primary difference between the different types of seal (shaft, ramp and drift/ILW disposal vault) is the extent to which the concrete lining is removed before installation of the swelling clay core. Shaft and ramp seals will be located in the upper part of the Callovo Oxfordian Clay, which is more competent than the lower part as it contains more carbonates and, therefore, will generate less damage to the rock during construction (i.e. a reduced excavation damaged zone (EDZ)). As a consequence, complete removal of the lining prior to installation of the swelling clay core can be considered as a reference for shaft and ramp seals: this ensures a good contact between the clay core and the rock, and so a better hydraulic performance. For the drift and ILW disposal vault seals, only partial removal of the lining is envisaged.

The main objective of the FSS experiment is to develop confidence in, and to demonstrate, the technical feasibility of constructing a full-scale drift or

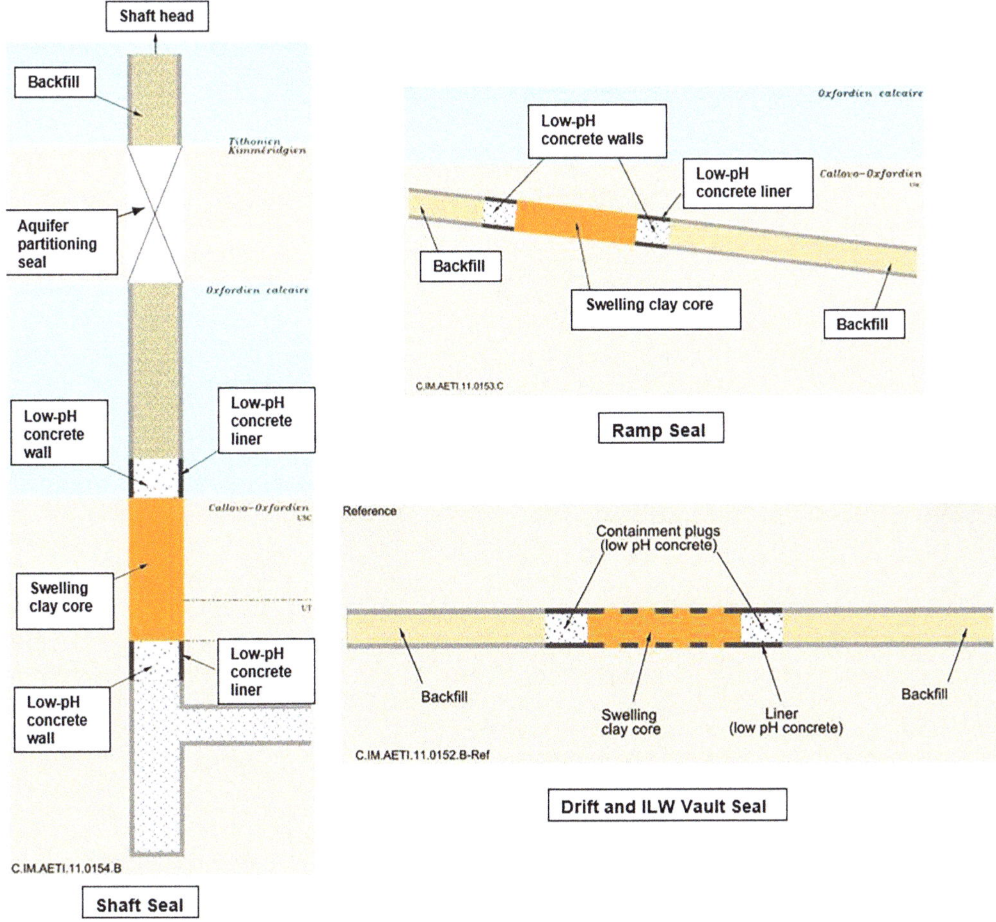

Fig. 3. Conceptual designs for shaft, ramp, and drift and ILW disposal vault seals for the CIGEO reference disposal concept.

ILW disposal vault seal. As the focus of the experiment is on the technical feasibility of seal installation, and not on long-term performance, it can be undertaken in a surface facility. The experiment is housed in a concrete test box, and the environment is controlled such that it mimics the underground environment. Technical feasibility includes demonstrating the ability of the approach used to build significant low pH monoliths (containment walls) and emplace the clay in order for it to be suitable for filling recesses in the clay host rock (i.e. any potential breakouts generated during the removal of the concrete support lining). Therefore, the concrete test box includes recesses that mimic breakouts. As the experiment is focused on the construction and installation of the seal, the materials will not be saturated or otherwise pressurized; complementary experiments are designed to investigate the resaturation process (REM (Resaturation at Metric Scale), part of DOPAS WP5, consists of an 'as close as possible to *in-situ* conditions' resaturation test undertaken in a surface laboratory with the same pellets/powder mixture as used in FSS) and the global performance after resaturation (NSC half-scale *in-situ* seal test in the Meuse/Haute-Marne underground research laboratory (URL)). The conceptual design of the FSS experiment is illustrated in Figure 4.

The main difference between the reference and FSS designs for the ANDRA drift and ILW vault seal is the length of the seal. The real seal underground will be longer than the seal considered in FSS. The FSS experiment will investigate two types of low-pH containment wall (Fig. 4): one self-compacting and one emplaced as shotcrete, to allow

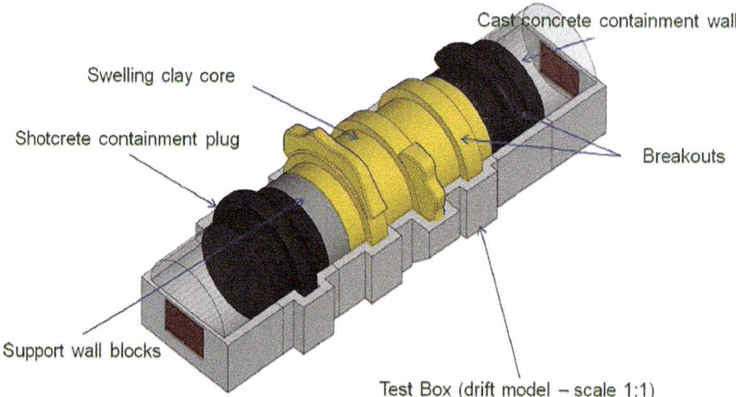

Fig. 4. Conceptual design for the ANDRA FSS experiment.

the preferred method to be selected and incorporated into the reference concept.

Design basis for plugs and seals

The design basis is the set of requirements and conditions taken into account in design (White *et al.* 2014). Requirements are statements on what the design has to do and what it must be like. For a plug or seal this could be a certain strength or a certain hydraulic conductivity of the plug or seal components. Conditions are the boundary conditions imposed on the design. For a plug or seal, this could be the underground environment (e.g. dimensions, air temperature and humidity) or controls on the manner in which the design is implemented (e.g. the time available for construction).

Design bases are generally hierarchical and consist of high-level requirements and low-level requirements. The high-level design basis can be specified and stabilized once the repository concept has been specified and the national regulations developed. This remains fixed. The principal safety functions of a plug or seal are included in the high-level design basis and they are typically described in a qualitative fashion. The lower level, more specific and quantitative design basis is developed through an iterative process involving outputs from full-scale experiments and performance assessment studies.

In DOPAS, a distinction is made between the design basis for the reference repository conceptual design and the design basis for the experiment. The reference design is the one that is used across the national programme (e.g. for licensing). The design basis for the experiment may differ from the design basis for the reference design because the experiments may be testing only specific aspects of a plug or seal, because the testing may involve acceleration of processes or because the experiment may be testing a different design to that of the reference.

In addition, the design basis for specific experiments is developed to a more detailed degree than the design basis for the plugs and seals in the reference repository conceptual designs. This is because the design basis for repository plugs and seals contain high-level requirements and conditions that are fixed, and because outstanding uncertainty does not allow more detailed specification of the reference designs at the present time. In contrast, the design basis for the experiments must be specified in greater detail to allow the experiment to be designed and implemented.

Examples of the most important detailed requirements (i.e. at the lower level of the design basis hierarchy) for the DOMPLU experiment in order to achieve the required safety functions and performance of the deposition tunnel plug are:

- The location of the plug shall be at the tightest section of rock identified in a pilot borehole through water injection tests and core characterization.
- The plug shall resist the sum of the hydrostatic pressure at repository depth and the swelling pressure of the backfill until the main tunnel is filled; 5 MPa water pressure and the swelling pressure from the backfill in the section adjacent to the plug. Currently, the backfill swelling pressure is assumed to be 2 MPa, with the help of backfill transition zone that decreases the load from a high load to 2 MPa.
- The target value for the acceptable leakage through the plug is presently adopted to be 'as low as possible'. Recent calculations predicted an allowed maximum leakage of $0.1\ \mathrm{l\ min^{-1}}$ to prevent loss of bentonite from the buffer and backfill.

Examples of the most important detailed requirements for the FSS experiment in order to achieve the required safety functions and performance of the ILW and vault seal are:

- The seals shall include a swelling core comprised of a clay-based material with a maximum overall hydraulic conductivity of 10^{-11} m s^{-1}, swelling pressure close to but not exceeding the effective mechanical stress of 7 MPa, and a length of two diameters (or 20 m) at least.
- The pH of the concrete shall not exceed a value of 11, and shall ideally lie between 10.5 and 11 at 28 days.
- The temperature at the heart of the concrete and shotcrete of containment walls shall not exceed 50°C.

Compliance demonstration and assessment approaches

To assess the compliance of the DOMPLU and FSS implementations with the requirements in their corresponding design bases, an approach that involves a review of each requirement (or condition) and the strategy used to demonstrate compliance against that requirement (or condition) was described.

The development of the compliance strategy statements in this way allowed for a reconsideration of the design statements and, in certain cases, suggestions for rewording or other types of feedback related to the design basis statements. This is part of the iterative development of the design basis. The requirement-by-requirement review of the design basis allows generic methods for demonstrating compliance to be identified. This approach allows feedback to the specification of the design basis to be captured, and the methodologies and technologies used during each experiment to be evaluated and assessed.

The compliance assessment approach was conducted, in collaboration with the organizations implementing the experiments, by means of tables listing all the design requirements and specifications. For each requirement, the following information was discussed and agreed:

- Requirement justification: the aim of setting a particular requirement.
- Compliance approach: the method used to assess the experiment performance against the requirement.
- Compliance assessment: statement on whether the requirement has been met. This is usually linked to evidence (e.g. a report).
- Feedback to design basis: statement on whether the requirement should be revised based on the assessment and/or based on the outcome of the experiment. A revised requirement could be provided if a revision is proposed, with a new justification.
- Learning points on plugs and seals: provides a method for capturing the learning from the experiment to feed into DOPAS outcomes.

This compliance approach and assessment has been applied to the detailed design requirements for DOMPLU and FSS, including the example requirements above. Table 1 and Table 2 outline the application of the approach to the DOMPLU and FSS example requirements, respectively.

Some of the main outcomes from this compliance assessment exercise, and which are relevant to the design basis work under WP2 of DOPAS, include:

- Compliance of designs with the design basis requires the development of a hierarchy, with compliance focusing on the more detailed quantitative design specifications. Generally, in a design basis hierarchy, top-level requirements dictate the more detailed requirements. Therefore, compliance of the detailed requirements also implies compliance of the top-level requirements.
- Undertaking a compliance assessment is a good process for developing construction procedures and quality assurance requirements. Construction procedures and quality control requirements should be linked to design specifications listed in the design basis.
- Evaluating the compliance of designs with a design specification is an intensive activity. Design bases can include many tens of requirements and discussion of each requirement is undertaken. Therefore, it is recommended that compliance assessments are undertaken by small groups of well-informed staff; ideally from three to five members of staff should be included in the intensive review process. This is considered to be an efficient and cost-effective approach in the long run.

This compliance assessment approach can be applied as part of, or in conjunction with, other more generic approaches used by WMOs. The generic approaches and strategies used by WMOs to demonstrate and assess the compliance of plugs and seals to the design basis, as identified in the DOPAS Project, include:

- Full-scale Testing: full-scale testing is the main strategy adopted by WMOs to compliance demonstration of plugs and seals. For example, DOMPLU represents the main compliance demonstration experiment for deposition tunnel plugs in the SKB programme. Full-scale experiments

Table 1. *Compliance assessment of example requirements for the DOMPLU experiment*

Requirement	Requirement justification	Compliance approach	Compliance assessment	Feedback to design basis	Learning points on plugs and seals
The location of the plug shall be at the tightest section of rock identified in a pilot borehole through water injection tests and core characterization	The aim of the requirement is to reach the best possible conditions for the DOMPLU experiment	Activity plan (AP TD KBP1004-11-041) Characterization – Core drilling and investigation programme for the DOMPLU test tunnel. This specification of the work describes the method to use, and the results expected and data that will be reported	The tightest section of rock was identified and reported in the following documents: BOREMAP data from core characterization (GE038) and core box photography (GE055). Memo from water injection tests (SKB doc id 1348278). Two sections of 7.5 m with hydraulic conductivity below the measurable value. Evaluation and decision of the plug location (SKB doc id 1337821)	The requirement will be different for the Spent Fuel Repository where relevant acceptance criteria for deposition tunnels need to be formulated. This will be based on an understanding of the spatial distribution of hydraulic conductivity within the Forsmark bedrock and the lithological variation underground. Say 'lowest hydraulic conductivity'	A detailed site investigation programme with rock modelling is vital for a final repository. It is important to control the tunnelling work with respect to a feasible plug location
The plug shall resist the sum of the hydrostatic pressure at repository depth and the swelling pressure of the backfill until the main tunnel is filled; 5 MPa water pressure and the swelling pressure from the backfill in the section adjacent to the plug. Currently, the backfill swelling pressure is assumed to be 2 MPa with the help of the backfill transition zone that decreases the load from a high load to 2 MPa	A requirement of 5 MPa gives a safety margin to a worst-case groundwater pressure during operations in the Spent Fuel Repository in Forsmark, located −470 m below sea level, based on the hydrostatic pressure that could exist after saturation of the repository. 2 MPa is a value derived from modelling of the reference deposition tunnel plug system, and includes the swelling pressures developed in the backfill transition zone and the watertight seal, and the related displacement of the filter during operation of the plug	This is a principal objective of the DOMPLU experiment: that is, construct the concrete dome and demonstrate that it can be pressurized. Therefore, the compliance approach was to conduct the demonstration experiment	The DOMPLU experiment did not meet this requirement, as the system has only been pressurized to *c.* 4 MPa within the DOPAS Project, at which time there was significant leakage occurring through the near-field rock and other locations	DOMPLU has shown that the plug system is tighter than the rock. Therefore, greater emphasis should be given in setting stricter requirements on the plug locations, so that no long fractures bypassing the plug system are present	The potential for bypassing the deposition tunnel plug needs to take account of a relatively large volume of near-field rock. When undertaking an experiment, there is a need to fully consider, explain and communicate the representativeness of the experimental site to the final location
The target value for the acceptable leakage through the plug is presently adopted to be 'as low as possible'. Recent calculations predicted a leakage of 0.085 l min^{-1}. However, the DOMPLU experiment aims to achieve values lower than this	Automatic measuring device supplying daily digital leakage data to the plant control room. Measurement based on the mass of water in the measurement device	Not yet checked, but data available. The compliance of the experiment with this requirement should also include analysis of the water breakthrough in the adjacent fracture	The DOMPLU experiment did not meet this requirement, as the system has only been pressurized to *c.* 4 MPa within the DOPAS Project, at which time there was significant leakage occurring through the near-field rock and other locations	The requirement might be revised to account for other routes in addition to 'through the plug'. The performance against this requirement has indicated a possibility of revising the pressure requirements on the deposition tunnel plug	The potential for bypassing the deposition tunnel plug needs to take account of a relatively large volume of near-field rock. When undertaking an experiment there is a need to fully consider, explain and communicate the representativeness of the experimental site to the final location

Table 2. *Compliance assessment of example requirements for the FSS experiment*

Requirement	Requirement justification	Compliance approach	Compliance assessment	Feedback to design basis	Learning points on plugs and seals
The seals shall include a swelling core comprised of a clay-based material with a maximum overall hydraulic conductivity of 10^{-11} m s^{-1}, swelling pressure close to, but not exceeding, the effective mechanical stress of 7 MPa, and a length of two diameters (or 20 m) at least	Performance assessment studies show that seal performance can be achieved with a hydraulic conductivity of 10^{-9} m s^{-1}. 10^{-11} m s^{-1} has been specified because it is an achievable value. The origin of the effective mechanical stress is defined to counterbalance the host rock natural mechanical stress (less than natural stress can result in reactivation of the EDZ, higher than natural stress results in more fractures). The effective mechanical stress is a result of an *in situ* stress of 12 MPa balanced by 5 MPa pore pressure, resulting in 7 MPa. Two diameters is a standard length assumed for seals. Performance assessment calculations showed that 20 m is the minimum necessary length to achieve a lower radionuclide flux through the seal than the host rock	To define additional, more detailed requirements. For hydraulic conductivity: laboratory tests to determine the required density. A design specification is then provided for the design to determine an appropriate dry density. For swelling pressure: metric tests to determine the required density. Once the density is determined, this is set as a (more detailed) requirement on the seal. For length of seal: direct measurement	Pass: design basis includes a requirement for the designers to determine the dry density of the clay core material in order that the hydraulic conductivity and swelling pressure requirements are met	Future iterations of the design basis for the drift and ILW vault seal should include the density for the swelling clay core determined through the work in FSS. The feedback from parallel experiments has demonstrated that resaturation of the EDZ is an important phenomenon for self-healing of the clay (and reduction in permeability and transmissivity) – compression of 3–4 MPa was observed to be enough to achieve self-healing. Rewording of requirement: 'Swelling pressure should be lower than the effective mechanical stress (i.e. several MPa's (3–7 MPa))'	Further understanding of material properties: the required density for swelling clay core to achieve hydraulic conductivity and swelling pressures. It was found to be difficult to achieve the dry density (especially at a large scale) that was initially specified; flexibility in the definition of requirements is required, and there needs to be an iteration between testing and requirement development. Another issue is upscaling from the laboratory scale to full scale
The pH of the concrete shall not exceed a value of 11, and shall ideally lie between 10.5 and 11 at 28 days	At pH < 11, the impact of cement leachate (alkaline plume) on bentonite and argillite performance is acceptable	Three recipes were tested in the laboratory (B50 CEM III/A; B50 CEM I; and B40 CEM III/A)	B50 CEM III/A; B50 CEM I; and B40 CEM III/A gave a pH of 11.3, 11.4 and 12.3, respectively, after 28 days. B50 CEM III/A and B50 CEM I gave a pH of <11 after 90 days	Design basis needs to be changed to reflect the pH of the concrete leachate after 90 days, as the leachate chemistry is still evolving at 28 days. Note that the '28 days' is not critical – it comes from standard practices in the concrete industry	Types of concrete recipes suitable for use alongside bentonite in plugs and seals. Cement and aggregates may change
The temperature at the heart of the concrete and shotcrete of containment walls shall not exceed 50°C	Ettringite formation can occur from 70°C, and lead to expansion and cracking	Temperature sensors placed inside the self-compacting concrete (SCC) and shotcrete containment walls	Pass: The maximum SCC curing temperature reached was 47°C, see p. 20 of D3.11. For shotcrete: include new observations of 65°C	Requirement could be revised to place a temperature constraint on the entire containment wall, and be justified by modelling and interpolation of experimental results. Limit could be increased to 60–65°C. Rewording: 'the curing temperature at the centre of the concrete and...'	Low pH SCC is more likely to be employed than low pH shotcrete for the construction of massive low pH concrete monoliths underground

include demonstration of technical feasibility and appropriate performance.
- Monitoring: WMOs have adopted different approaches to the use of monitoring as part of compliance demonstration strategies for plugs and seals. Monitoring can be used to provide additional information on the performance of plugs and seals during repository operation, and thereby provide confidence in the evolution of the system post-emplacement. WMOs have adopted different approaches to the use of monitoring as part of compliance demonstration strategies for plugs and seals. SKB envisages only minor monitoring of some deposition tunnel plugs and no firm decision to monitor these structures has been made. Detailed plans for monitoring shaft seals have been developed in Germany. Monitoring of shaft seals in the German programme will extend into the post-closure period and may continue for 100 years after closure (as required by regulations). Monitoring in the French programme has so far mainly focused on a demonstration of the retrievability of waste from disposal cells; no firm plans for monitoring drift/vault seals have yet been developed.
- Construction Procedures: WMOs have different views on the use of construction procedures for compliance demonstration. The Finnish WMO (Posiva) regard documentation of construction procedures as one of two elements of the compliance strategy (the other being monitoring). SKB consider construction procedures to be part of quality control. ANDRA's approach is that construction procedures are defined to be consistent with the design, and, therefore, are *a priori* compliant with the design basis.

Other approaches to building confidence in the performance of plugs and seals are generally considered to be part of the design development process rather than compliance demonstration. These include research into long-term material performance, small-scale testing, numerical modelling and the possible use of arguments based on natural analogue information.

Conclusions

The DOPAS Project aims to improve the industrial feasibility of plugs and seals, the measurement of their characteristics, the control of their behaviour over time in repository conditions and their hydraulic performance with respect to safety objectives. A set of full-scale experiments, laboratory tests and performance assessment studies of plugs and seals in geological repositories are being carried out in the course of the project.

Within the project, an approach has been developed and applied to assess the compliance of two of the full-scale experiments (DOMPLU in Sweden and FSS in France) to their design bases. The approach involves a review of each requirement in the design basis and the strategy used to demonstrate compliance of the experiment with that requirement. Feedback in the form of proposed updates to the design statement is captured by this approach. Learning points on plugs and seals from the compliance assessment are also noted for consideration in the DOPAS Project outcomes.

This developed compliance and assessment approach can be employed as part of, or in conjunction with, other more generic approaches used by WMOs. These generic approaches and strategies to demonstrate and assess the compliance of plugs and seals to the design basis include monitoring, full-scale testing and the documentation of construction procedures.

The work and research under DOPAS has received funding from the European Atomic Energy Community's (EURATOM) Seventh Framework Programme FP7/2007-2013 under grant agreement No. 323273, the DOPAS Project. The partners in the project are Posiva, ANDRA, DBE TEC, GRS, NAGRA, RWM, SÚRAO, SKB, CTU, NRG, GSL, BTECH, VTT and UJV. In particular, the following have contributed to the work described in this paper: Behnaz Aghili (SKB), Regis Foin (ANDRA), Jacques Wendling (ANDRA), Aurélien Noiret (ANDRA) and Dean Gentles (RWM).

References

IAEA 2011*a*. *Disposal of Radioactive Waste*. IAEA Safety Standards Series No. SSR-5. International Atomic Energy Agency (IAEA), Vienna.

IAEA 2011*b*. *Geological Disposal Facilities for Radioactive Waste Specific Safety Guide*. IAEA Safety Standards Series No. SSG-14. International Atomic Energy Agency (IAEA), Vienna.

Palmer, S. 2011. *Design Criteria for the Plug Structure in KBS-3V Deposition Tunnels*. SKB Document ID: 1273861, Version 1. Svensk Kärnbränslehantering (SKB), Stockholm.

SKB 2010. *Design, Production and Initial State of the Backfill and Plug in Deposition Tunnels*. SKB Technical Report TR-10-16. Svensk Kärnbränslehantering (SKB), Stockholm.

White, M., Doudou, S. & Neall, F. 2014. *DOPAS Work Package 2, Deliverable D2.1: Design Bases and Criteria, Version 1.1*, February 2014. Galson Sciences Limited Report for the European Commossion.

DOPAS EPSP experiment

JIŘÍ SVOBODA[1]*, JAROSLAV PACOVSKÝ[1], MARKÉTA DVOŘÁKOVÁ[2], IRENA HANUSOVÁ[2], PETR VEČERNÍK[3] & DAGMAR TRPKOŠOVÁ[3]

[1]*Czech Technical University in Prague, Centre of Experimental Geotechnics, Thákurova 7, Prague, Czech Republic*

[2]*SÚRAO, Dlážděná 6, Prague, Czech Republic*

[3]*ÚJV Řež, a. s., Hlavní 130, 250 68 Husinec-Řež, Czech Republic*

**Correspondence: svobodaj@fsv.cvut.cz*

Abstract: The international DOPAS Project (Demonstration Of Plugs And Seals), which involves the participation of 14 European organizations (Posiva, ANDRA, DBE-TEC, GRS, NAGRA, RWA, SÚRAO, SKB, CTU, NRG, GSL, BTECH, VTT and ÚJV Řež), is concerned with the structural solution for sealing plugs to be used in deep geological radioactive waste repositories. The project is funded by the 7th Framework Programme – EURATOM.

The Czech contribution to the DOPAS project (the EPSP experiment – Experimental Pressure and Sealing Plug) consists of an experiment that is being conducted at the Josef Regional Underground Research Centre (Josef URC, Čelina-Mokrsko). The aims of the EPSP experiment are to develop, monitor and verify the functionality of such plugs, and to determine a detailed characterization of the materials from which the plug is constructed. The EPSP experiment is being conducted by a Czech consortium made up of SÚRAO, CTU and ÚJV Řež, a. s.

The programme for the development of a deep geological radioactive waste repository (DGR) in the Czech Republic is based first and foremost on the safe disposal of long-lived highly active radioactive waste. The safety of such repositories will be enhanced by the efficient performance of the plugs and sealing systems that will make up an important part of the overall disposal system. Several types of sealing plugs will be required, the function of which will be to provide for the sealing and closure of individual waste packages not only throughout the period of repository operation, but also following the permanent closure of the facility. Such plugs will have to provide a high level of resistance to the considerable pressure that will be exerted by hydrostatic forces and volumetric changes within the engineered barriers.

The objective of the DOPAS (Demonstration Of Plugs And Seals) international project is to design a sealing plug system for DGR use, provide detailed plans for the design of such plugs, test both the characteristics of the materials to be used and the construction technology, and to install four experimental *in situ* plugs.

This 4 year (2012–16) project is being funded from European Commission financial resources (7th Framework Programme, EURATOM) and the project coordinator is Finland-based Posiva. A total of 14 partners from eight European countries are involved in the project (Posiva, ANDRA, DBE-TEC, GRS, NAGRA, NDA, SÚRAO, SKB, ČVUT, NRG, GSL, BTECH, VTT and ÚJV Řež, a. s.).

In 2012, the construction of a sealing plug commenced at the Äspö underground laboratory (Sweden) in a granitic rock environment containing water with a relatively high level of salinity (*c.* 75 g l^{-1}) (Puigdomeneck 2001) in contrast to Czech granitoids (Pačes *et al.* 2010).

In Finland, a similar plug is undergoing testing at the Onkalo underground laboratory located on the island of Olkiluoto in a migmatitized gneiss rock environment (Dixon *et al.* 2013).

In France, an experiment is underway at the Saint Disier Laboratory focusing on research into sealing systems in claystones, in which it is intended that the French DGR will be constructed (Montes *et al.* 2004).

The Czech EPSP experiment is being conducted in a rock environment consisting of granitoids at the Josef Underground Research Laboratory (URL). The concept of the experiment is based primarily on the use of Czech materials and technology available in the Czech Republic, and the principal aim is to demonstrate the technical viability and functioning of a pressure-resistant plug located in a future DGR.

EPSP project description

The EPSP is aimed at the study of developments concerning the design basis, reference designs and

From: Norris, S., Bruno, J., Van Geet, M. & Verhoef, E. (eds) 2017. *Radioactive Waste Confinement: Clays in Natural and Engineered Barriers*. Geological Society, London, Special Publications, **443**, 59–72.
First published online August 26, 2016, https://doi.org/10.1144/SP443.17

strategies, including compliance issues. The EPSP plug has been designed as a prototype plug for a future Czech deep geological repository. It is expected, therefore, that a similar plug will function over the whole of the operational phase of the repository (i.e. 150 years), with an expected overpressure of up to 7 MPa.

Furthermore, the plug has been designed as a multilayer system consisting of two main structural elements, which ensure the overall stability of the system (i.e. concrete blocks and a sealing element – a bentonite section positioned between the concrete blocks). Fibre shotcrete is being used in the construction of the various elements of the EPSP; the bentonite sealing section will be constructed by means of either compaction or spray technology.

The plug will be tested by means of injecting air/water/a suspension into a pressurizing chamber, followed by the monitoring of the performance of the plug. As a result of the geological conditions within the EPSP experimental drift at the Josef underground laboratory, it was necessary to use grouting in order to lower the permeability of the rock mass prior to the commencement of the EPSP plug experiment.

The technical design of the plug was the responsibility of the Centre of Experimental Geotechnics of the Czech Technical University (CTU), Prague and was based on a structural proposal contained in Reference Design 2011 (Pospíšková *et al.* 2012).

The EPSP experiment run is divided into four stages. The first stage primarily concerned the laboratory verification of the suitability of the materials to be used for plug construction, namely bentonite and a concrete mixture, the quality and detailed characteristics of which received particular attention during work on the design of the plug. Part of this research involves laboratory-scale physical models, which were constructed at ÚJV Řež's laboratories. Two types of models were constructed. The objective of the first model was to obtain data to be used for the calibration of numerical models of the saturation of bentonite. The second type is focused on a laboratory simulation of the interaction processes in the materials to be used, mainly on the concrete–bentonite interface that occurs in the demonstration plug.

The second stage of EPSP consisted of the construction of the *in situ* experiment, which commenced in 2013 with grouting and drilling work in the experimental gallery niche. This was followed by the construction of the plug itself, including the full instrumentation of the experiment and its immediate surroundings.

The third stage consists of pressurization by saturation media (air, water and a bentonite suspension), during which the plug will be exposed to pressure of up to 2.5 MPa (limited by virgin stress of the surrounding rock mass). The experiment and the technology employed will be monitored constantly throughout this period of the experiment.

The final stage of the experiment will concentrate on the evaluation of the data and knowledge obtained from both the *in situ* and laboratory experiments using numerical analysis techniques and modelling. The final output will consist of the verification of the operational safety of the various structural elements of plugs to be used in DGRs, and detailed recommendations for the future design and construction of such plugs.

Josef URL and the plug location

The Josef gallery, opened by the Faculty of Civil Engineering in June 2007, is being used within the DOPAS Project (Fig. 1). The facility is located near the Slapy dam, close to the villages of Čelina and Mokrsko in the Příbram district of Central Bohemia, Czech Republic.

The Josef gallery runs in a NNE direction across the Mokrsko hill rock massif. The total length of the main drift is 1835 m, with a cross-section of 14–16 m^2. The overlying rock thickness is 90–180 m. Two parallel tunnels lead from the entrance portals, each having a length of 80 m and a cross-section of 40 m^2. The main gallery is connected to various exploration workings by a number of insets that follow ore formations and provide access to two further levels. The total length of the galleries is approximately 8 km: 90% of the breakings are not fitted with linings. The end of the main gallery is connected to the ground surface by means of an unsupported 144 m vent.

Geological diversity is one of the great advantages of the Josef URL. There are two basic geological formations, each with a very different history and contact zones (Morávek *et al.* 1992). Moreover, each of the formations exhibits different physical and material properties that change in character towards the contact zone, and which feature a variety of local fracture zones and intrusions. This provides a high level of flexibility with regard to choosing the right place for conducting experiments depending on the conditions required, for example, fracture systems, rock stability, rock strength and mineralogy.

Mokrsko-East and Čelina deposits are hosted by Neoproterozoic rocks of the Jílové Belt (contact metamorphic tuffites), while the Mokrsko-West deposit is largely hosted by amphibolite tonalite of the Variscan Plutonic Complex. The tonalite locally grades into granodiorite or quartz diorite.

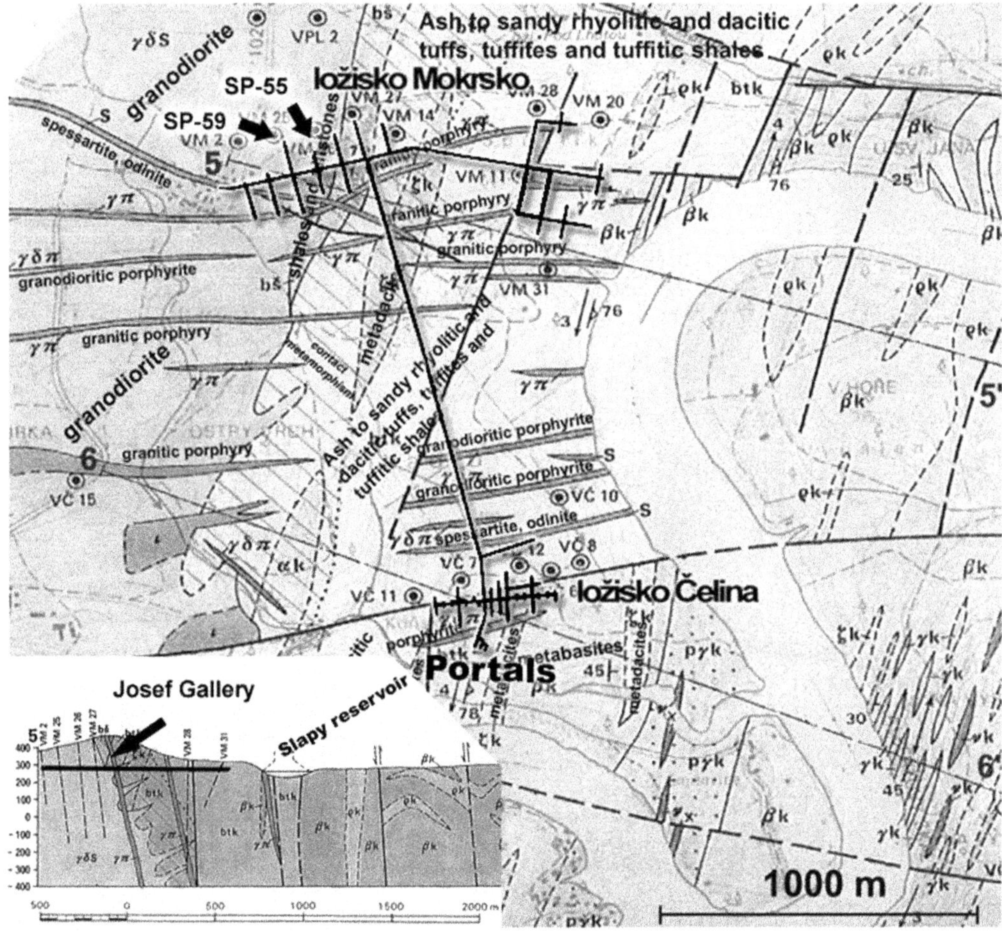

Fig. 1. Geology of the Josef gallery (based on a map published by the Czech Geological Survey (http://www.geology.cz/) in 1991).

Jílové Belt rocks and the tonalite are cross-cut by several types of Variscan dykes: randomly orientated dykes of biotite granodiorite, pegmatite and aplite, east–west-orientated dykes of granite and granodiorite porphyries, and aplites and NW–SE-trending lamprophyre dykes (Zachariáš *et al.* 2014).

The EPSP experiment itself is located inside the M-SCH-Z/SP-59 experimental gallery niche. The niche was reshaped in advance and the surrounding rock has been improved by grouting in order to reduce the water permeability (in order to allow for higher pressures loaded onto the plug itself).

The necessary technology for the experiment is located in a parallel niche, M-SCH-Z/SP-55.

The niches are interconnected by cased boreholes equipped with tubing for pressurization media circulation (four leading into filters and four leading into the pressurization chamber) and for monitoring (five boreholes equipped with sealed lead-through cables).

The experimental gallery niche is traversed by quartz and quartz–carbonate veins, with a maximum thickness of 14 cm. Information on dominant joint systems is recorded in historical mining documentation (Fig. 2), which was subsequently updated according to map source documents owned by Geofond (the map archive maintained by the Czech Geological Survey).

The detailed mineralogical study of the filling of fissures was carried out in niche SP-59 in 2013; the sampling locations are shown in the map (Fig. 2). Six samples were analysed by means of X-ray powder diffraction at the Institute of Chemical Technology, Prague, VŠCHT (X'Pert PRO with Bragg–Brentan geometry, CuK_{α}, 40 kV, 30 mA, High Score Plus) and SEM at the Faculty of Science, Charles University in Prague.

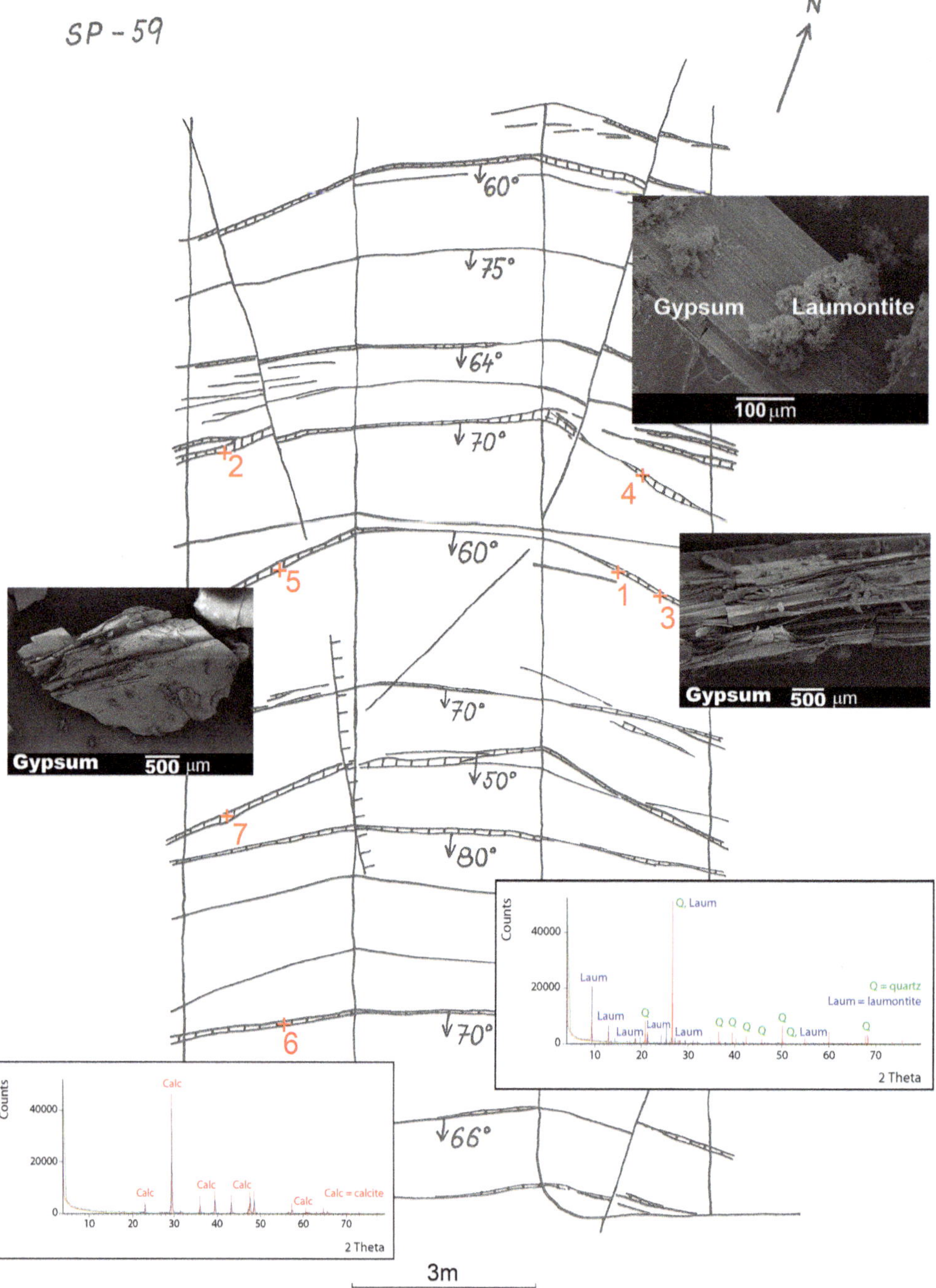

Fig. 2. SP-59 tectonics and mineralogy.

EPSP design overview

The EPSP plug has the following main components: a pressurization chamber, the inner glass fibre-reinforced shotcrete plug, bentonite sealing material, a filter, an external concrete plug and the host-rock environment (Fig. 3). The experiment also comprises a number of auxiliary structures,

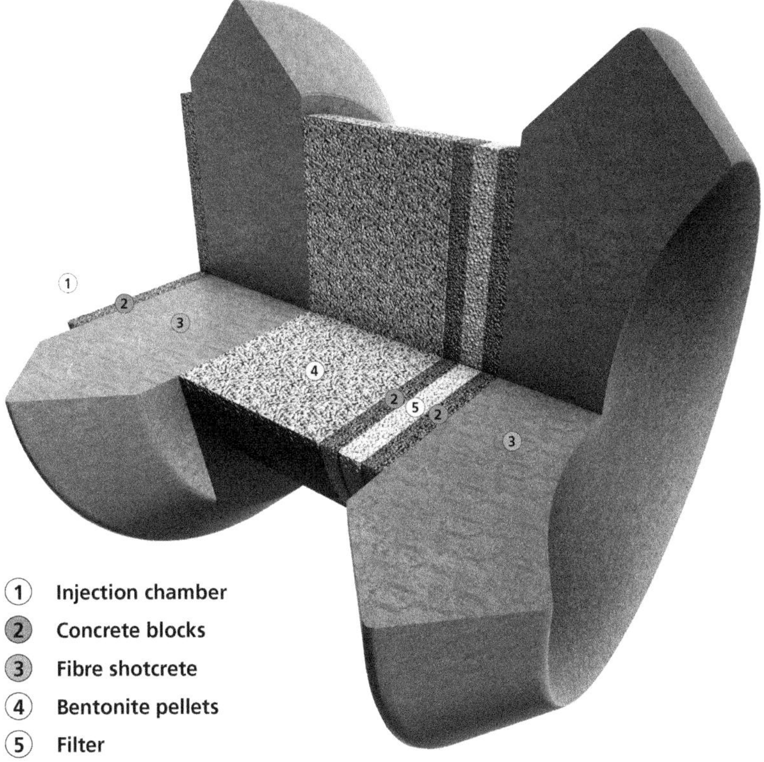

Fig. 3. The EPSP scheme.

such as permeable concrete walls, an extensive monitoring system and a service centre.

The pressurizing media will consist of air, water and a bentonite suspension. The pressurizing medium will be transported to the pressurization chamber via 23 m-long connecting boreholes from the neighbouring service centre niche. The chamber is enclosed by a permeable concrete wall that will serve as support during the construction to the one of the main elements of the EPSP – the inner glass-fibre-reinforced concrete plug.

The inner plug makes up one of the key components of the experiment, and verifying that it works efficiently makes up one of the main objectives of the project. It has two basic functions: static and hydraulic, which means primarily that it must ensure the mechanical stability of the entire system even under high-pressure conditions and, at the same time, must restrict flow through the plug so that the bentonite seal is not damaged through the creation of an erosion channel in the period in which the bentonite sealing zone is undergoing, but has not yet reached, full saturation.

A further major component of the EPSP is the bentonite sealant, the purpose of which is to hydraulically separate/seal the experimental plug. Owing to its exceptional swelling properties (and consequent self-healing capacity) and very low permeability, bentonite is particularly suitable for fulfilling this function. However, its sealing properties depend, in addition to its mineralogical composition, principally on its density. Therefore, in order to achieve the required density level, bentonite will be used in the form of highly compacted pellets. The EPSP experiment also includes the testing of the bentonite installation technology: for example, the use of the shot clay method (Pacovský & Šťástka 2009; Šťástka *et al.* 2014).

The filter has been designed principally for the monitoring of seepage through the plug. Nevertheless, it is connected through the aforementioned boreholes to the service centre niche and can be used, if required, as an alternative pressurizing chamber. The design of the experiment even provides for alternative testing scenarios in which the filter can be used as the pressurizing chamber, enabling the forced saturation of the bentonite sealant, the testing of reverse flow (first by bentonite then by the inner concrete plug) and the testing of the outer concrete plug.

The EPSP is enclosed by an outer concrete plug, the function of which is to mechanically stabilize the entire block.

EPSP construction

Rock mass reshaping and improvement

The works on the EPSP experiment started with niche reshaping. In order to limit further damage to the rock mass, no explosives were allowed on site. Hydraulic wedge splitting was used in combination with non-detonating (GBT) splitting. Using these technologies, the profile of the niche has been adjusted and recesses for concrete plugs excavated.

Once the reshaping was finished, the rock improvement started. The rock properties have been improved by means of grouting, and the resin has been used to lower the rock permeability in order to allow higher pressures to be applied on the plug and to limit unnecessary leakages into the rock mass.

Pressurization chamber

The chamber will serve as a primary point for pressurization media injection. For this purpose, the chamber was equipped with four pressurization tubes perforated at the ends. These tubes are connected to the technology in parallel niche via piping in boreholes.

The rock surface in the area of the pressurization chamber is very uneven, which leads to an excessively high volume in the chamber. Therefore, the volume of the chamber was reduced be filling unnecessary parts by shotcreting (Fig. 4).

This process also served as an important test of the technological set-up for the inner plug (the same machine set-up used).

Once the shape had been adjusted, the surface of the chamber was treated with a waterproofing finish in order to mitigate possible leaks behind the chamber.

As a last step, a permeable separation wall between the chamber and the inner plug was erected. This wall served as a formwork for the shotcreting of the inner plug.

Inner and outer plug

The inner plug was erected in 23 h-long non-stop run. Shotcrete technology was used: the concrete was produced at mixing plant in Prague and subsequently transported in large trucks to the Josef URL (1 – 1.5 h transportation time).

The concrete mix was then gradually reloaded into small trucks, which transported the wet mix from portals to the spraying machine in the

Fig. 4. The chamber before shotcreting.

Fig. 5. The shotcrete wet-mix being reloaded at portals.

experimental niche (a distance of 2 km) (Fig. 5). Two small trucks were alternating loads on the route, with a 40 min turnaround time (e.g. 20–30 min between batches).

The plug was erected in layers – starting in the back on the separation wall and gradually filling the space of the plug with shotcrete (Fig. 6). The short breaks between the batches were used for installation of the instrumentation (Fig. 7).

The outer plug will be built using the same procedure as the inner one.

Bentonite emplacement

The bentonite pellets are going to be emplaced between the inner plug and filter. The lower parts of the bentonite sealing will be compacted using vibration.

The upper part will be emplaced using shot clay technology (Pacovský & Šťástka 2009; Šťástka *et al.* 2014).

The target average dry density after emplacement is 1400 kg m^{-3} as a minimum for both methods.

Filter

The filter will serve as collection point for water that could leak through the EPSP. It will be constructed from gravel held in place by two permeable separation walls.

The complete structure of the filter, with separation walls, has additional functions during the construction of the experiment. First, it has to hold the bentonite in place at the time of emplacement. Therefore, the filter will be raised in stages following the emplacement progress. Once the structure is completed, it will serve as the lost formwork for the outer plug.

Main materials used

Concrete

The shotcrete is an important material in the EPSP plugs. The principal role of the concrete plug components is to fix the middle bentonite section in its initial position. The shotcrete, containing glass fibres, is used instead of iron-based reinforcement. The glass fibres should enhance the physical properties of the concrete in terms of shrinkage and tensile strength, in addition to not bringing more iron into system.

The influence of alkaline leaching on the bentonite barrier will be minimized through the use of concrete with a lowered pH. Highly alkaline leachates

Fig. 6. Shotcreting in progress.

contained in normal pH concretes could negatively influence the properties of the bentonite.

The composition of the concrete:

- cement CEM II/B-M;
- micro-silica;
- aggregates 0–4 and 4–8;
- glass fibres;
- plasticizer;
- retardant;
- accelerator.

Required properties of concrete:

- compressibility strength (ČSN EN 12390-3): >30 MPa;
- flexural strengths (ČSN EN 14488-3): >3 MPa;
- hydraulic conductivity (ČSN CEN ISO/TS 17892-11): $<10^{-8}$ m s^{-1};
- fibre content (pr EN 14488–7): >3 kg m^{-3};
- pH (SKB R-12-02): <11.7.

Bentonite

Following careful consideration of plug construction requirements, factory-produced bentonite (milled, non-activated Ca–Mg bentonite) B75 was selected as the principal material for the bentonite part of the plug (composition shown in Table 1). Based on experience from previous projects, B75 was selected due to it fully complying with the required parameters (hydraulic conductivity and swelling pressure).

The key geotechnical parameter values of the emplaced material are expected to be as follows (Vašíček *et al.* 2014):

- hydraulic conductivity: $<10^{-12}$ m s at a dry density of 1400 kg m^{-3};
- swelling pressure: >2 MPa at a dry density of 1400 kg m^{-3}.

The bentonite will be used in the form of pellets with a dry density of 1900 kg m^{-3} (Fig. 8). The

Table 1. *Silicate analysis of bentonite (Večerník* et al. *2013)*

	SiO_2	Al_2O_3	TiO_2	Fe_2O_3	FeO	MnO	MgO	CaO	Na_2O	K_2O	P_2O_5	CO_2
Content (wt%)	49.83	15.35	2.82	10.9	3.74	0.09	2.88	2.01	0.67	1.05	0.63	3.66

Fig. 7. Installation of the instrumentation.

target average dry density after emplacement is 1400 kg m^{-3} as a minimum.

Monitoring

One of the main objectives of the EPSP is to verify the effectiveness of the sealing plug: therefore, it is essential that the processes taking place within it be accurately determined. Hence, the experiment has been fitted with a comprehensive instrumentation system which has been designed in such a way that it allows for the monitoring of the development of the principal variables at key stages of the experiment.

Fig. 8. Pellets.

The key processes and locations inside the EPSP have been identified, and sensors have been specially selected in order to capture them. Monitoring focuses on water movement inside the experiment and the experiment's response to pressurization (especially the deformation of the plugs):

- Water movement inside the experiment is monitored in terms of water inflow/outflow, water content distribution within the bentonite seal and water (pore) pressure distribution.
- The mechanical response of the plug is monitored by means of strain gauges installed at key locations in the concrete plugs and instrumented rock bolts positioned within the rock. Moreover, contact stress measurement is deployed between the rock and the plug.
- Temperature distribution is monitored as it is important not only during the construction stage (hydration heat) but also during the loading of the experiment as a reference base for sensor compensation.
- The swelling pressure of the bentonite seal is monitored using pressure cells.

In order to obtain good and reliable monitoring results from the various sensors, their position within the EPSP and the quality of their emplacement is crucial. Key locations have been identified and the placement of sensors is focused on those arcas.

Several measurements have been taken in order to ensure the provision of reliable data such as cross-validation (sensors working on different principles are used to measure similar phenomena) and redundancy. Only pretested/calibrated/verified sensors are positioned within the experiment.

A further important aim of the experiment consists of the verification of the suitability of the construction technology selected for particular structures. The monitoring process, therefore, commenced prior to the start of the construction stage and will provide monitoring over the whole of the construction period, which can be divided into the following phases according to the main purpose:

- the preparatory phase (calibration of the system and the collection of data in order to determine the static state of the rock massif);
- construction work up to the completion of the inner plug (monitoring of the inner plug – the development of hydration heat/temperature, the monitoring of deformation and contact stress);
- the testing of the inner plug (inner plug and rock environment response);
- the completion of the construction of the experiment (inspection of the installation of the bentonite sealant and the outer concrete plug);
- trial operation (checking of the functioning of the pressurization equipment and the overall functioning);
- the main experimental programme (comprehensive monitoring of the behaviour of the whole experiment).

The monitoring system employs modern technology that allows the immediate availability of data collected for end users; the data are stored on a continuous basis in the specially designed measurement system database. In addition to the measured data, which are stored by the system in primary units, the database also contains comprehensive information from individual sensors installed within the experiment and the experiment's log. The web interface (Fig. 9), which makes up an indispensable part of the measurement system, allows those involved to obtain a general overview of events occurring within the experiment. The basic information services provided by the interface include: a list of sensors, which enables graphs to be plotted of selected period, the 3D visualization of the current situation, an overview of the overall functioning of the system and the experiment's log.

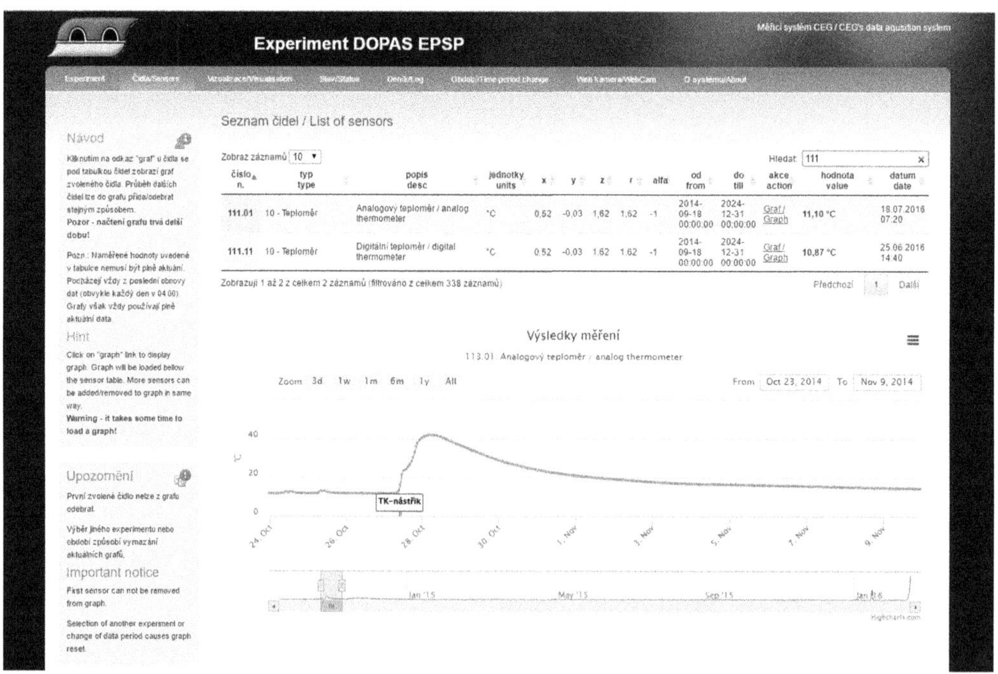

Fig. 9. Monitoring system web interface (temperature progression due to the shotcreting of the pressurization chamber).

Laboratory

The laboratory work will provide data for the subsequent numerical analysis of the behaviour of the EPSP. Three groups of data are being produced: input material parameters, material parameters from verification testing (occasional checking for possible changes as the project progresses) and data from small-scale physical models to be used for the validation of the numerical models.

The improvement of the rock environment in which the experiment is located, and construction of the plug itself, is subcontracted to outside organizations. The laboratory and *in situ* tests have been planned so as to include the checking of the quality of the work performed (e.g. rock permeability, deformation and strength parameters, and concrete behaviour), and to ensure that the relevant parameters concerning the various rock and concrete plug components have been met in full.

Concrete samples were obtained in the testing phase before *in situ* construction and also during the construction of inner concrete part of the EPSP. Both types of samples were tested on leachate pH by the application of methodology described in SKB R-12-02 (Alonso *et al.* 2012). Cylindrical-shaped concrete samples are also used in physical models focused on studies of interactions between concrete and bentonite.

Two types of independent laboratory-scale physical models have been constructed in the laboratories: PHMs, physical hydraulic models; and PIMs, physical interaction models (Fig. 10). The main goal of the first type is to obtain data that will be used for the calibration of numerical models of the bentonite saturation. The second one is focused on a laboratory simulation of the interaction processes that could occur in the demonstration EPSP plug. Those considered are mainly processes on the concrete–bentonite interface. The main reason for performing such an experiment in the laboratory is the lack of *in situ* information, as the EPSP experiment will not be dismantled during the course of the DOPAS Project. Laboratory-prepared synthetic granitic water (SGW: Havlová *et al.* 2010), based on a statistical evaluation of Czech granitic massif groundwaters (depths 20–200 m), is used in the laboratory tests in case conditions as realistic as possible to the real *in situ* experiment are needed. Moreover, large volumes of solution with a defined composition are needed to be prepared repeatedly and reproducibly (Table 2).

The physical hydraulic models (PHMs) consist of a defined number of cylindrically shaped stainless steel chambers 50 mm in length and 80 mm in diameter, each are equipped with sensors in order to record the distribution of water content and pore-water pressure within the bentonite material. Bentonite powder and bentonite pellets were pressed into the nine test chambers in order to reach a final dry bulk density of 1400 kg m^{-3} (an identical bulk density to that assumed for the EPSP experiment) and gradually saturated with synthetic granitic water under pressure. The maximum SGW pressure up to 20 bar is applied in the laboratory tests. At the end of the experiment, samples of the bentonite

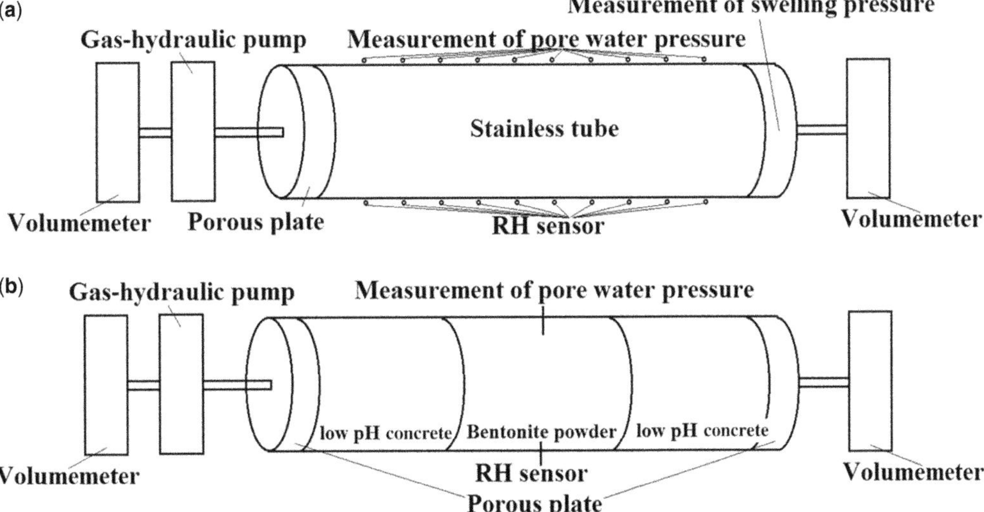

Fig. 10. Scheme of the laboratory physical models: (**a**) the physical hydraulic model (PHM); and (**b**) the physical interaction model (PIM).

Table 2. *Chemical composition of synthetic granitic water (SGW), pH 8.03 (Havlová* et al. *2010)*

	Na^+	K^+	Ca^{2+}	Mg^{2+}	Cl^-	SO_4^{2-}	NO_3^-	HCO_3^-	F^-
Concentration (mg l^{-1})	10.6	1.8	27	6.4	42.4	27.7	6.3	30.4	0.2

material will be dismantled and divided into layers. The water content in each layer then will be determined. It is assumed that the duration of the experiment will be approximately 1 year. Parallel experiments with bentonite pellets are ongoing, using different length configurations in order to describe saturation processes, material homogenization and water distribution.

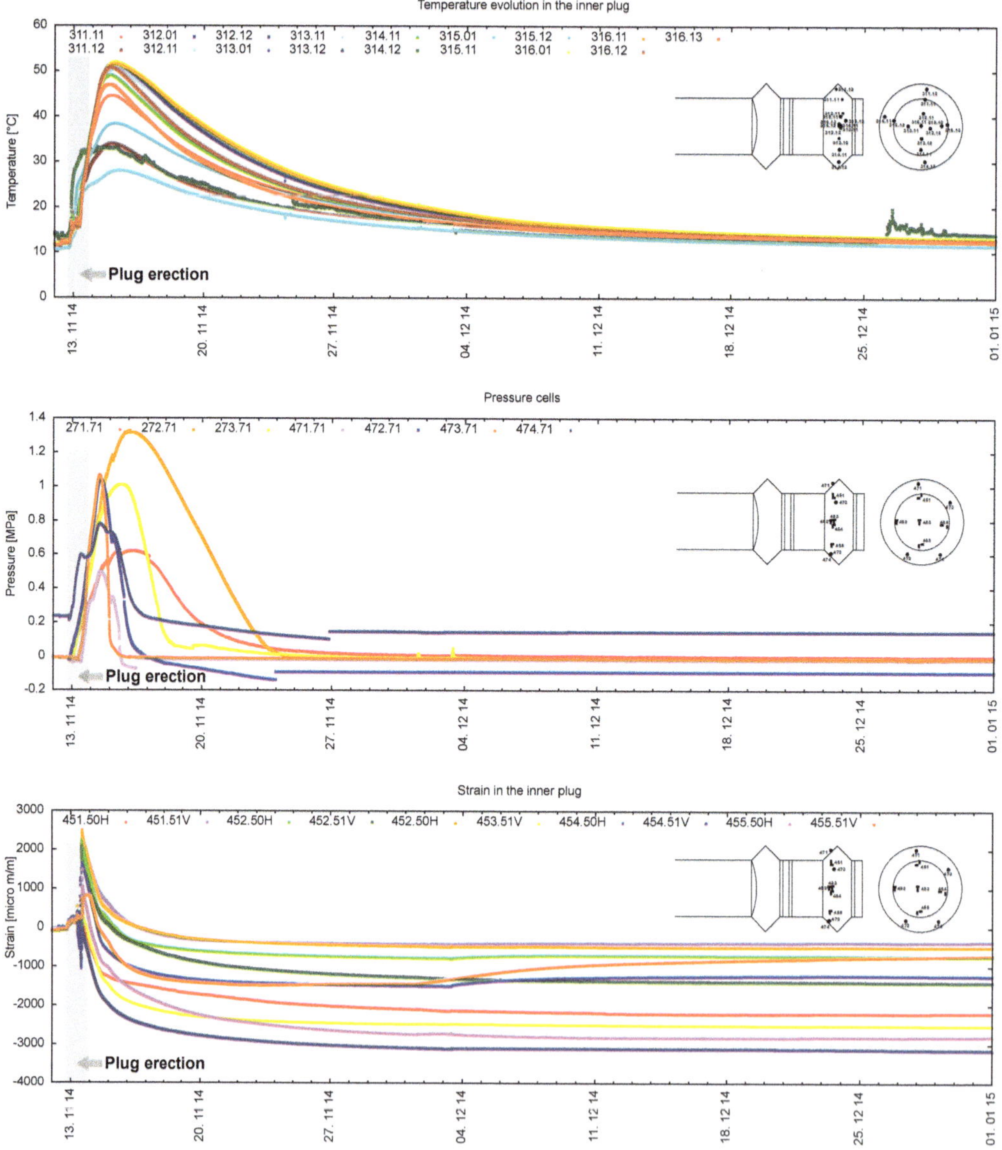

Fig. 11. Measurement in the inner plug.

A laboratory physical interaction model (PIM) consists of a 50 mm-thick bentonite part, surrounded by 50 mm-thick blocks of low-pH concrete from both sides. In order to approach the real plug set-up, the concrete blocks have been sealed in steel chambers using grouting materials, the same as used for grouting the host rock *in situ*. The SGW, under a pressure of up to 20 bar, then infiltrates one side of the experimental cell, sealed with concrete, in order to simulate the same procedure of pressure loading as that in the EPSP.

Results

At the present time, the EPSP experiment is under construction. The rock niche has been reshaped, the rock improved, the pressurization chamber constructed, the inner plug erected, the first stage of the filter constructed, the technology installed and monitoring is running. The next step is the bentonite emplacement.

The production of the slots for the plugs showed that the usage of GBT cartridges is a successful and faster method than hydraulic splitting.

The experience from the construction of the inner plug shows that the shotcrete process is a viable alternative for plug erection from a realistic point of view. The main challenge is logistics and the operation in the confined space of the Josef URL, where size constraints limit the machinery that could be used. In particular, the transportation in the gallery was bottlenecked, limiting the speed of the erection of the plug.

The usage of a wet-mix and tuned concrete mixture greatly reduced the generation of dust. The generation of dust was even lower compared to the traditional wet-mix recipes.

The maximum temperature in the plug due to the hydration process was below 55°C. There was no problem with shotcrete overheating during the emplacement process. The maximum temperature was reached approximately 1 day after shotcreting ended (Fig. 11). No shrinkage cracking has been detected in the plug.

Results of measurements of the concrete leachate pH successfully fall within the set limits. Values of pH obtained for preconstruction testing and for inner EPSP concrete block samples were 11.4. These values are under the predefined limit of 11.7 for the concrete mixture that will be applied in the construction of the EPSP.

The laboratory PIM is still running. The results will be obtained after dismantling the PIM experiment. Analyses will be focused on physical, chemical and mineralogical changes of the bentonite and concrete, especially on surface interfaces. Physical hydraulic models (PHMs) are also running. Measured values are repeatable at the same boundary conditions in PHM experiments. Data obtained during and after dismantling the experiments will be used for verification and calibration of the numerical models of laboratory physical models and the experimental plug.

Conclusions

The EPSP experiment currently underway at the Josef Regional Underground Research Centre (Josef URC) is one of the important international projects that will lead to advances in the knowledge base for the Czech DGR. Emphasis has been placed since the commencement of the project on the use of Czech materials (i.e. B75 bentonite), and technologies that have been under development in the Czech Republic for many years (e.g. sprayed bentonite technology).

The behaviour of the plug will be comprehensively monitored throughout the duration of the experiment. The final assessment of the experiment will involve the use of numerical analysis and modelling techniques. Finally, it is envisaged that the successful completion of the EPSP experiment will contribute towards demonstrating how sealing plug systems behave under real conditions and, thus, will go towards answering one of the many outstanding questions concerning the long-term safety of a future deep geological radioactive waste repository in the Czech Republic.

The research is being funded from the European Union European Atomic Energy Community (EURATOM) Seventh Framework Programme FP7 (2007–13) according to grant agreement No. 323273, the DOPAS Project. The project is also supported by the Czech Republic via funding provided by the Ministry of Education, Youth and Sports – institutional support grant No. 7G13002.

References

Alonso, M., García Calvo, J.L. et al. 2012. *Development of an Accurate pH Measurement Methodology for the Pore Fluids of Low pH Cementitious Materials.* SKB Report, SKB R-12-02. Svensk Kärnbränslehantering (SKB), Stockholm, Sweden.

Dixon, D., Hansen, J., Karvonen, T.H., Korkiala-Tanttu, L., Marcos, N. & Sievänen, U. 2013. *Underground Disposal Facility Closure Design 2012.* POSIVA Working Report 2012-09. POSIVA, Eurajoki, Finland.

Geofond. Map sheet: *Dobříš 1-9/34-24, M-SCH-Z/SP-59.*

Havlová, V., Holeček, J., Vejsada, J., Večerník, P. & Červinka, R. 2010. *Příprava syntetické podzemní vody pro laboratorní experimentální práce. [Preparation of Synthetic Ground water for Experimental Laboratory Work.]* Technical Report of the Ministry of

Industry and Trade of the Czech Republic Project No. FR-TI1/367 [in Czech].

MONTES, H.G., DUPLAY, J., MARTINEZ, L., ESCOFFIER, S. & ROUSSET, D. 2004. Structural modifications of Callovo-Oxfordian argillite under hydration/dehydration conditions. *Applied Clay Science*, **25**, 187–194.

MORÁVEK, P., AICHLER, J. ET AL. 1992. *Zlato v Českém masivu* [Gold in Czech Massif], 1st edn. Czech Geological Survey, Prague, Czech Republic, http://www.geology.cz/1919/historie/publikace/1992-zlato-web.pdf

PAČES, T. ET AL. 2010. *Výzkum procesů pole vzdálených interakcí HÚ vyhořelého jaderného paliva a vysoce aktivních odpadů. Summary Report*. Czech Geological Survey, Praha.

PACOVSKÝ, J. & ŠŤÁSTKA, J. 2009. Development of sprayed backfill technology. *In*: *7th International Conference on Ecosystems and Sustainable Development, ECOSUD 2009. WIT Transactions on Ecology and the Environment, Volume 122*. WIT Press, Southampton, UK, 523–533.

PUIGDOMENECK, I. (ed.). 2001. *Hydrochemical Stability of Groundwaters Surrounding a Spent Nuclear Fuel Repository in a 100 000 Year Perspective*. SKB Technical Report TR-01-28. Svensk Kärnbränslehantering (SKB), Stockholm, Sweden.

POSPÍŠKOVÁ, I., VOKÁL, A., FIEDLER, F., PRACHAŘ, I. & KOTNOUR, P. 2012. *Aktualizace referenčního projektu hlubinného úložiště radioaktivních odpadů v hypotetické lokalitě. Závěrečná zpráva* [Update of reference design of deep repository of radioactive waste at hypothetic locality, Final report]. ÚJV Řež, a. s., Prague, Czech Republic.

ŠŤÁSTKA, J., SVOBODA, J. & SMUTEK, J. 2014. Verification of sprayed clay technology with respect to the geological disposal of radioactive waste. *Acta Geodynamica et Geomaterialia*, **11**, 145–152.

VAŠÍČEK, R., LEVOROVÁ, M., HAUSMANNOVÁ, L., ŠŤÁSTKA, J., VEČERNÍK, P., TRPKOŠOVÁ, D. & GONDOLLI, J. 2014. *Deliverable D3.17 – Interim Results of EPSP Laboratory Testing, DOPAS Project FP7 EURATOM, No. 323273*. Interim DOPAS Report. Czech Technical University in Prague.

VEČERNÍK, P., TRPKOŠOVÁ, D., GONDOLLI, J. & HAVLOVÁ, V. 2013. *Implementation Support and Evaluation of EPSP Experiment in the Project DOPAS. An Annual Report 2013*. ÚJV Řež, a. s., Prague [in Czech].

ZACHARIÁŠ, J., MORÁVEK, P., GADAS, P. & PERTOLDOVÁ, J. 2014. The Mokrsko-West gold deposit, Bohemian Massif, Czech Republic: Mineralogy, deposit setting and classification. *Ore Geology Reviews*, **58**, 238–263.

Laboratory experiments on bentonite pellet saturation

DAGMAR TRPKOŠOVÁ[1]*, PETR VEČERNÍK[1], JENNY GONDOLLI[1], VÁCLAVA HAVLOVÁ[1], JIŘÍ SVOBODA[2] & IRENA HANUSOVÁ[3]

[1]*ÚJV Řež, a. s., Hlavní 130, 250 68 Husinec-Řež, Czech Republic*

[2]*Centre of Experimental Geotechnics, Czech Technical University in Prague, Thákurova 7, 166 29 Prague 6, Czech Republic*

[3]*SÚRAO, Dlážděná 6, 110 00 Prague 1, Czech Republic*

**Correspondence: Dagmar.Trpkosova@ujv.cz*

Abstract: The aim of the DOPAS project is to address the design basis, reference designs and strategies for the plugs and seals to be used in geological disposal facilities. The Czech 'Experimental Pressure and Sealing Plug' (EPSP) experiment has the following objectives: to develop, monitor and verify the functionality of an *in situ* experimental plug, and to determine and describe in detail the materials to be used with respect to experimental plug construction. The functionality of the experimental plug will be verified by means of conducting pressure tests after the construction phase of EPSP. The internal bentonite section of the EPSP will be constructed from bentonite pellets with a dry density of 1850 kg m^{-3}; the pellets will disintegrate following saturation and swelling to form a homogeneous material with a dry density of around 1400 kg m^{-3}. This process leads to the question of the pore space of the bentonite material: that is, whether it will be possible to employ a numerical model for homogeneous compacted bentonite powder porosity instead of that of the double pore/void space of the bentonite pellets used in the experiment. For this reason, a laboratory physical model (PHM – physical hydraulic model) was incorporated into the DOPAS project for the purpose of describing the saturation and disintegration of the bentonite pellets. The aim of this paper is to describe the results of the space distribution of relative humidity in a sample consisting of bentonite pellets with a view to subsequent numerical modelling and to extend the current knowledge concerning the behaviour of the bentonite pellets used in the EPSP experiment. The results show that following water saturation and the swelling of the bentonite the pellets form a homogeneous mass in the same way as does the compacted bentonite powder. After a period of 160 days, the pellets attained 100% relative humidity at a distance of 12.5 cm and swelling pressure recorded at the end of the model was seen to slowly increase over this time period.

The aim of the DOPAS project is to address the design basis, reference designs and strategies for the plugs and seals to be used in geological disposal facilities. According to the Czech deep geological disposal concept, the sealing plugs will be required both to fulfill a mechanical–hydraulic role and to function throughout the entire operational phase of the repository (i.e. 150 years), principally for the purpose of sealing the disposal galleries.

The Czech 'Experimental Pressure and Sealing Plug' (EPSP) experiment is being conducted at the Josef Regional Underground Research Centre (Josef URC), which is operated by the Centre of Experimental Geotechnics of the Czech Technical University in Prague (CEG CTU). The EPSP experiment has the following objectives: to develop, monitor and verify the functionality of an *in situ* experimental plug, and to determine and describe in detail the materials to be employed with respect to experimental plug construction. The demonstration plug has three main construction components: the inner concrete part, which is made from glass-fibre shotcrete with a low-pH leachate; the central bentonite sealing part made from bentonite pellets; and the outer low-pH concrete part (Fig. 1). Local Czech materials were used in the construction of the EPSP.

The functionality of the experimental plug will be verified by means of conducting pressure tests after the construction phase. During the functionality test, saturation of the bentonite part occurs. The tests will involve the successive filling of the experimental injection chamber with water, air and a bentonite suspension under different pressure levels. The response of the plug will be recorded by means of a large number of sensors emplaced within the experiment (thermometers, pressure cells, stress meters and RH sensors). The pressure tests will subsequently be simulated by means of numerical modelling.

The central part of the EPSP will be made from bentonite pellets with a high dry density level

From: NORRIS, S., BRUNO, J., VAN GEET, M. & VERHOEF, E. (eds) 2017. *Radioactive Waste Confinement: Clays in Natural and Engineered Barriers*. Geological Society, London, Special Publications, **443**, 73–83.
First published online November 3, 2016, https://doi.org/10.1144/SP443.22

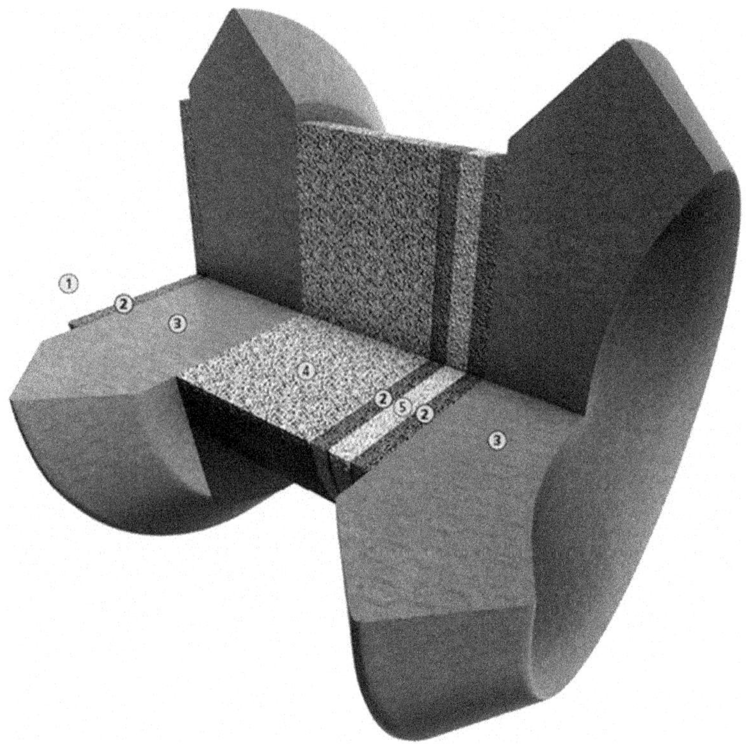

Fig. 1. Cross-section of the 'Experimental Pressure and Sealing Plug' (EPSP) experiment: 1, pressure chamber; 2, concrete brick wall; 3, low-pH glass-fibre shotcrete; 4, bentonite pellets; 5, filter.

(1850 kg m^{-3}). As a result of the saturation of the bentonite, the pellets will disintegrate: that is, they will lose their initial structure and form a visually homogeneous material with a dry density of approximately 1400 kg m^{-3}. This process leads to a question concerning the pore space of the bentonite material: that is, whether it will be possible to employ a numerical model of homogeneous compacted bentonite powder porosity instead of that of the double pore/void space of the bentonite pellets used in the experiment. For this reason, a laboratory physical model (PHM – physical hydraulic model) was incorporated into the DOPAS project for the purpose of describing the saturation and disintegration of the bentonite pellets and the homogenization of the bentonite material.

A large number of authors have already addressed the issue of pellet saturation: for example, Imbert & Villar (2006) studied the saturation of a mixture made up of French Fo–Ca bentonite pellets and bentonite powder (50/50 wt%) for the purpose of describing the hydromechanical response of bentonite pellets following infiltration. Karnland *et al.* (2008) compared various pellet mixtures (pure pellets and a mixture of bentonite and sand) with pure powder in order to assess whether the sealing properties of bentonite change due to the pellet production process. The homogenization of bentonite pellets after saturation was described by Van Geet *et al.* (2005). Current projects aim to determine hydraulic properties only from the incoming water or to determine the water content of a material from subprofiles following the dismantling of the experiment. The laboratory experiments outlined herein, however, aimed to provide a continuous description of the evolution of relative humidity along a number of profiles within a 45 cm-long sample during the bentonite saturation process. Relative humidity was subsequently recalculated to water content by means of a retention curve or determined by means of the weight loss method.

The aim of this paper is to provide a detailed description of the space distribution of water content in a sample consisting of bentonite pellets. The data will subsequently be used for the numerical inverse modelling of the EPSP experiment, which will observe the correspondence between the measured and modelled results depending on the model parameters. The state of the bentonite material and the boundary conditions were chosen with regard to the state of the material in the EPSP experiment in order to obtain model parameters that might be used for simulating the processes underway in the EPSP experiment. The criteria for the calibration of the model describing the bentonite pellets consist of the following variables:

- volume of infiltrated water;
- relative humidity at each observation point;
- retention curve.

This paper addresses the volume of infiltrated water and the relative humidity at each observation point. The retention curve will be determined following the dismantling of the PHM_2 experiment, whereupon the water content of each 'slice' will be measured.

Table 2. *Composition of the synthetic granitic groundwater (SGW)*

Species	Concentration (mg l^{-1})
Na^+	10.6
K^+	1.8
Ca^{2+}	27.0
Mg^{2+}	6.4
Cl^-	42.4
SO_4^{2-}	27.7
NO_3^-	6.3
HCO_3^-	30.4
F^-	0.2
pH	8.0

Materials and methods

The material studied consists of Czech Ca–Mg industrially milled and sifted non-activated bentonite produced by KERAMOST a.s. and code-named B75_2013. This material was selected by the Czech consortium for the sealing of the EPSP as a reference material for potential future use in the planned Czech deep geological repository. The laboratory tests were performed on bentonite pellets that were prepared by the CEG CTU. The chemical and mineralogical composition of the material is presented in Table 1. The pellets consisted of small cylindrical elements (1 cm in diameter and 2–3 cm long). The dry density of the pellets was determined at approximately 1850 kg m^{-3} and the initial water content determined by means of the weight loss method was 12 wt%.

It was decided that synthetic granitic water (SGW) (Havlová *et al.* 2010), specially prepared according to a statistical evaluation of Czech granitic massif groundwater (depths of 20–200 m), would be used as the liquid medium in the laboratory tests, so as to adhere as closely as possible to the conditions in which the *in situ* experiment was being conducted. The main advantage of using SGW consists of the potential for the repeated application of the liquid medium both with a constant composition and in a sufficient amount when required. The chemical composition of SGW is shown in Table 2.

Experiments

A total of five experiments were performed. The first three experiments, code-named Bentonite_powder_60, Bentonite_pellets_60 and Bentonite_pellets_3, were conducted so as to enable the visual confirmation of pellet homogenization following their disintegration. The geometry of the subsequent two experiments, PHM_1 (physical hydraulic model) and PHM_2, was then chosen based on the results of the first three experiments. The main

Table 1. *Chemical and mineralogical composition* of B75_2013*

	Content (wt%)	Minerals	Content (wt%)
SiO_2	51.91	Montmorillonite	75.5
Al_2O_3	15.52	Illite	3.9
TiO_2	2.28	Kaolinite	3.1
Fe_2O_3	8.89	Quartz	8.1
FeO	2.95	Anatase	2.6
MnO	0.11	Calcite	3.1
MgO	2.22	Siderite	1.8
CaO	4.60	Ankerite	0.5
Na_2O	1.21	Cristobalite	1.4
K_2O	1.27		
P_2O_5	0.40		
CO_2	5.15		
Loss of ignition	10.65		
Natural amount of water	6.94 ± 0.43		

*Mineral composition based on chemical composition and X-ray diffraction (XRD) analysis.

aim of the PHM_1 and PHM_2 experiments was to determine a description of the detailed space distribution of relative humidity in a sample consisting of bentonite pellets.

All the experiments were performed in cylindrical stainless steel cells (5 cm in length and 8 cm in diameter) equipped with sensors, according to the requirements of the relevant experiment. All of the experiments were equipped with sensors to record the volume of infiltrating/outflowing water and the pressure at which the water infiltrated. The volume of infiltrated water was measured by means of volume meters equipped with displacement sensors made by the Czech company Datacon, these had a range of −50 to 50 mm; the input water pressure was monitored using pressure sensors also made by Datacon, which have a range from −0.1 to 2.5 MPa. The experimental cells were stacked in order to achieve the required length of the sample and filled with bentonite (powder or pellets) in order to attain a final material dry density level of 1400 kg m^{-3} following homogenization.

Bentonite_powder_60, Bentonite_pellets_60 and Bentonite_pellets_3

The Bentonite_powder_60 experiment consisted of one stainless steel cell filled with compacted bentonite powder. The sample was 5 cm long and the material was saturated with SGW under a pressure of 2 MPa over a period of 60 days. Following saturation and the equilibration of the inflow and outflow of the water, the bentonite sample was sliced into five layers each 1 cm thick. Each layer was then cut into nine parts (Fig. 2): that is, the bentonite sample was divided into a total of 45 small samples, which were subjected to a determination of total water content and the space distribution thereof using the weight loss method.

The Bentonite_pellets_60 experiment had the same geometry and duration of saturation as the Bentonite_powder_60 experiment. However, in this case, the cell was filled with bentonite pellets and the input water pressure was gradually increased to a maximum level of 0.1 MPa. Following saturation and the equilibration of the inflow and outflow of the water, the bentonite sample was divided in the same manner as the Bentonite_powder_60 sample and the 45 sub-samples analysed (Fig. 2) using the weight loss method.

The Bentonite_pellets_3 experiment consisted of a total of three stainless steel cells filled with compacted bentonite pellets. The sample was 15 cm long and the duration of saturation was 3 days only. The input water pressure was gradually increased to a maximum of 0.1 MPa, as in the Bentonite_pellets_60 experiment. Following 3 days of saturation and the equilibration of the inflow and outflow of the water, the sample was sliced into 1 cm-thick layers and the water content of the resultant 15 layers analysed using the weight loss method.

All three experiments included the measurement of the volume of water that infiltrated into the sample and the volume of water issuing from the sample. The rapid water inflow into the void spaces between the pellets led to the correspondingly rapid saturation of the pellets throughout the total profile of all the test cells, as shown in the results of the Bentonite_pellets_60 and Bentonite_ pellets_ 3 experiments (Figs 3 & 4). It is evident that the water content of the samples was the same regardless of the period (3 or 60 days) of saturation.

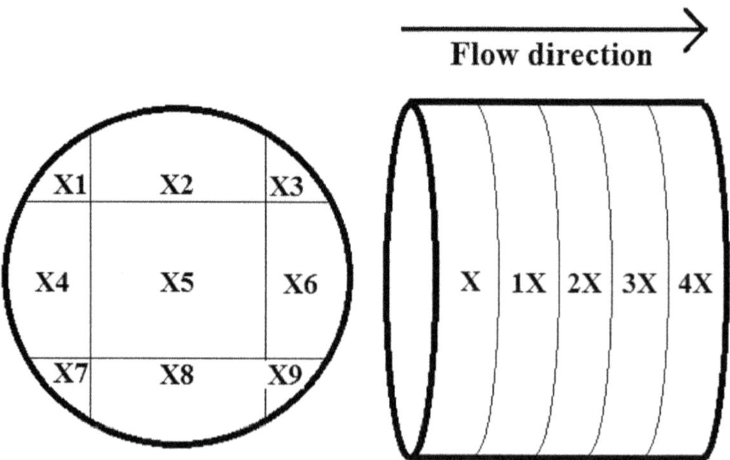

Fig. 2. Labelling the samples during the dismantling experiments.

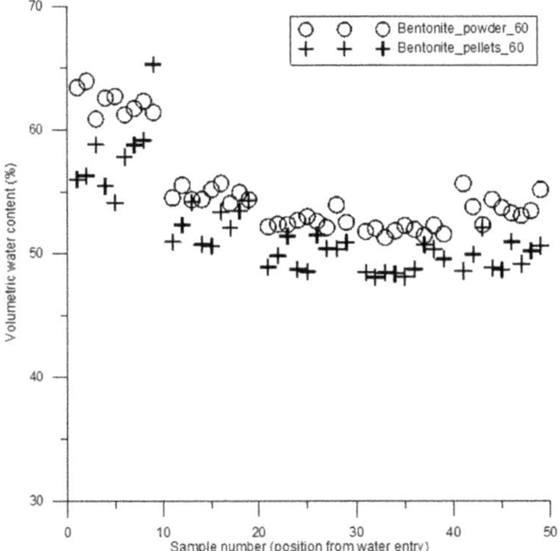

Fig. 3. The x-axis shows the position of the materials from the point of saturation, on this side water flows into the sample in the Bentonite_powder_60 and Bentonite_pellets_60 experiments. The samples denoted 1–9 represent the first 1 cm layer (from the side of water infiltration) and the samples denoted 41–49 represent the fifth layer. The water content of the material in the first layer is influenced by the presence of a water film in the contact zone of the porous frit–bentonite.

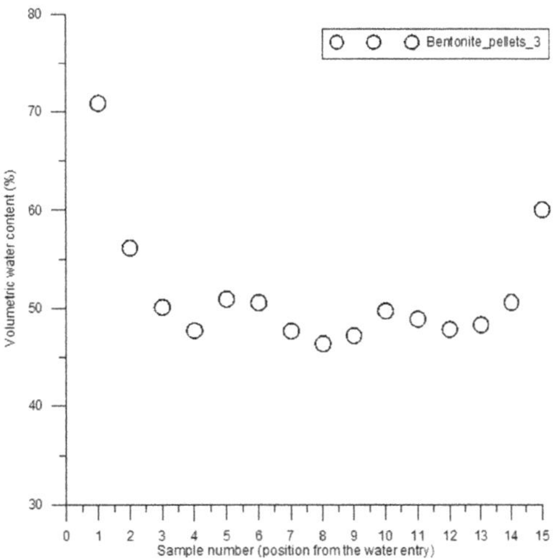

Fig. 4. The x-axis shows the position of materials from the point of saturation, on this side the water flows into the sample in the Bentonite_pellets_3 experiment. The sample denoted as 1 represents the first 1 cm layer (from the side of water infiltration) and the sample denoted as 15 represents the fifteenth layer.

PHM_1 and PHM_2

Both experiments were designed in order to describe the continuous saturation of the bentonite material. The two experiments had the same geometry and saturation conditions (input water pressure): however, they differed in terms of both the duration of the saturation period and the placement of the measuring sensors. The test cells were equipped, in addition to sensors for the measurement of the volume of infiltrating/outflowing water and pressure sensors, with sensors to record the distribution of relative humidity and swelling pressure within the bentonite material. Relative humidity was measured using RH sensors (produced by the Swiss company Rotronic) over a range of 0–100%, the swelling pressure was measured using force sensors produced by the Czech company Lukas-Tenzo, which have a range of up to 50 kN, the volume of infiltrated water was measured using volume meters fitted with displacement sensors produced by the Czech company Datacon, with a range of −50 to 50 mm, and the input water pressure was measured using pressure sensors that are also produced by Datacon and which have a range from −0.1 to 2.5 MPa.

A total of nine cells with a total length of 45 cm were specially designed for the PHM experiments. Initially, it was planned that pellets only would be used in all nine cells, as, for example, in experiments according to Imbert & Villar (2006) and Hoffmann *et al.* (2007). However, since the experiment described herein differed in terms of length (i.e. more than 12 times longer than reported in the cited studies) and saturation took place under pressure, it proved unrealistic to use only bentonite pellets due to the excessively rapid inflow into the void spaces between the pellets and the consequently accelerated saturation of the whole profile of the cell.

It was decided, therefore, that the bentonite pellets would be compacted into eight test cells (the second to ninth test cells) and that the first test cell would be filled with compacted bentonite powder with a dry density of 1400 kg m^{-3}. The target value of the dry density of the bentonite material was identical to that assumed for the EPSP experiment conducted at the Josef URC. The bentonite material in the physical hydraulic model was gradually saturated with synthetic granitic water under pressure in four pressure stages in order to avoid the destruction of the compacted bentonite due to an abrupt increase in pressure. The first pressure level exerted was approximately 0.2 MPa over 2 h, followed by 0.5 MPa over 17 h, 1 MPa over 2 h and a final pressure level of 2 MPa at which the pressure was maintained over the remaining duration of the experiment. The pressure was increased gradually due to the requirement to maintain the sealing contact between the bentonite and the cell wall.

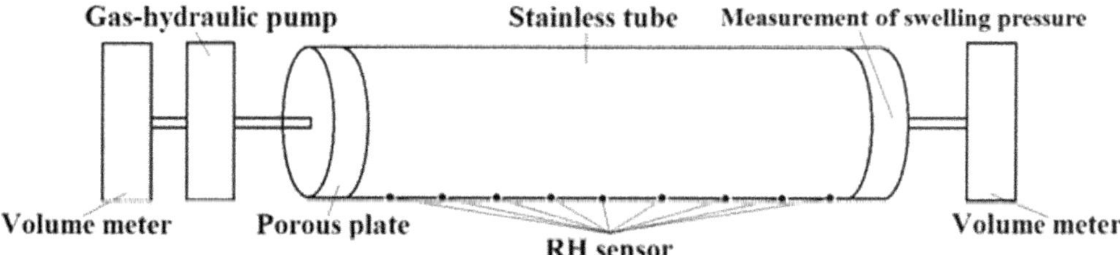

Fig. 5. Scheme of the physical hydraulic models (PHM_1 and PHM_2).

The following parameters were measured during the saturation phase: the input water pressure and the volume of infiltrating water recorded at the inlet of the sample; the evolution of relative humidity recorded at observation points within the cells; and the evolution of the swelling pressure of the bentonite material at the outlet of the experimental apparatus. Observation points were located at distances of 2.5, 7.5, 12.5, 17.5, 22.5, 32.5 and 42.5 cm from the start point of the bentonite sample (i.e. the saturation water inlet). A scheme of the PHM apparatus is shown in Figure 5 and detail of the experimental set-up, including the location of the observation points, is displayed in Figure 6.

The PHM_1 experiment lasted just 8 days owing to reasons outlined below. The PHM_2 experiment was designed based on the PHM_1 experiment: however, in this case, it was planned as a long-term experiment to be conducted over 160 days.

The RH sensors operate within a range of 0–100% relative humidity and, following exposure to the upper limit over long time periods, they may become damaged. Therefore, the sensors were removed from the experimental cell after 100% relative humidity was attained and replaced with stainless steel rods to fill the resulting spaces. With respect to the PHM_1 experiment, the bentonite powder attained 100% RH following a 7-day saturation period. The removal of the sensors, however, was complicated as the bentonite pellets in the other parts of the experimental cell exhibited residual or only slightly increased water content, and, therefore, the swelling pressure of the bentonite inside these cells was not sufficient to seal the contact zone between the bentonite and the cell wall. This resulted in the free pore space surrounding the pellets suddenly becoming filled with water during the removal of the RH sensors from the first cell. Consequently, the experiment was terminated and repeated under the same boundary and initial conditions, but with one significant difference – in the PHM_2 experiment, no sensor was emplaced in the bentonite powder: that is, sensors were emplaced only in the bentonite pellets. In this case, the extraction of the RH sensors was performed with no complications: that is, the intact bentonite powder layer

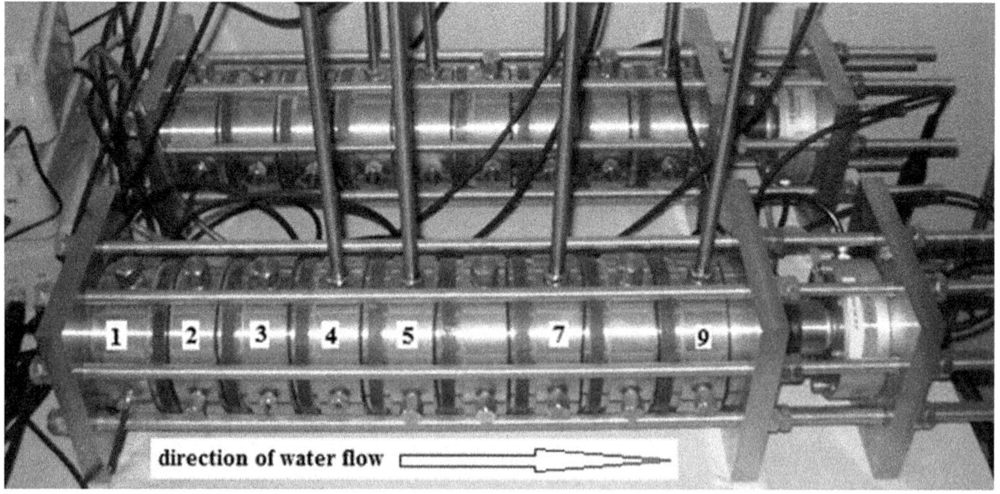

Fig. 6. Physical hydraulic models (PHM_1 and PHM_2) in the laboratory. The model consists of nine cells and the cell numbers represent different observation points.

Fig. 7. Bentonite pellets (**a**) before the commencement of saturation and (**b**) following saturation. It is possible to see the original arrangement of the bentonite pellets on a porous frit.

ensured sufficient sealing ability and so avoided any unexpected abrupt increases in water inflow and the filling of the pore spaces between the bentonite pellets with water.

Results and discussion

The graphs shown in Figures 3 and 4 provide a comparison of the water content of the material following saturation in the Bentonite_powder_60, Bentonite_pellets_60 and Bentonite_pellets_3 experiments. The water content of the material in the first layer is influenced by the presence of a water film in the contact zone of the porous frit–bentonite. The results reveal that the water content of both samples is very similar, even though they differed in terms both of length and the saturation time period. It is clear that the distribution of water content in the pellet samples was extremely

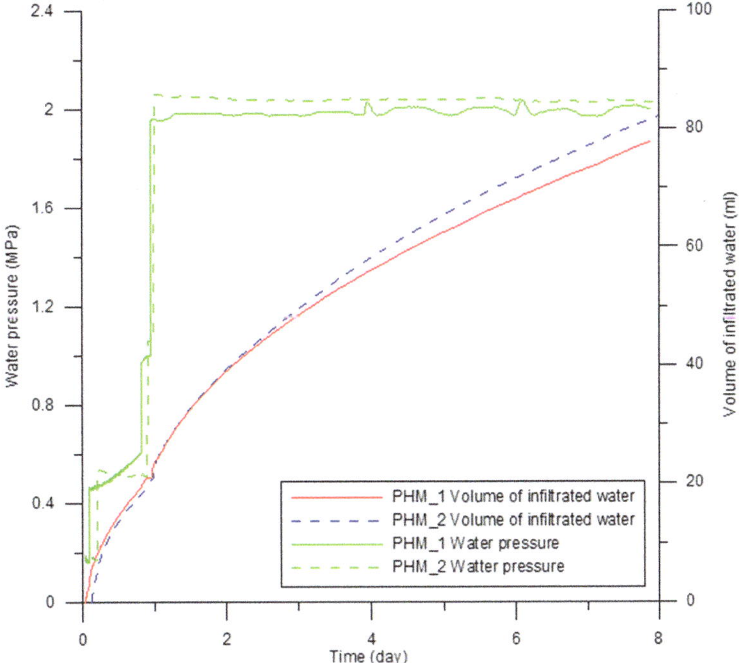

Fig. 8. Comparison of the commencement of the PHM_1 and PHM_2 experiments, the pressure of the water infiltrating into the sample and the volume of water infiltrating into the sample.

rapid; the bentonite pellets after 3 days of saturation exhibited a similar water content as did the bentonite pellets following 60 days of saturation. The state of the bentonite pellets after 3 days of saturation is shown in Figure 7. The pellets have lost their shape, the spaces between the pellets have disappeared and the bentonite has formed a homogeneous mass.

The reason for this phenomenon is as follows: the dry density of the bentonite pellets was approximately 1850 kg m^{-3} and a mixture of pellets (pellets of different length and crushed pellets) was compacted into the stainless steel cells to reach a target bentonite material dry density of 1400 kg m^{-3} following homogenization. It is clear from a comparison of the two density levels that the pellets in the cell were surrounded by connected void spaces (before the commencement of saturation) into which the flow of water was possible; therefore, the flow paths enhanced the ability of water to quickly reach the bentonite material. In other words, saturation was not dependent on the length of the sample but, rather, on the dimensions of the pellets and the void spaces. The bentonite pellet sample was, therefore, fully or almost fully saturated after 3 days (Figs 3 & 4). This, however, will not occur in the case of the EPSP experiment as the water inflow into the plug will be reduced to a minimum by the low level of permeability of the grouted host rock and the concrete part of the plug. The amount of water passing through will be so low that the swelling of the bentonite material and the sealing of the interface layer will be ensured in the void spaces between the pellets. Hence, the PHM_1 and PHM_2 experiments contained a 5 cm-thick layer of powdered bentonite compacted to a dry density of 1400 kg m^{-3} in the first test cells, the aim of which was to protect against the filling of the void spaces between the pellets with pressurized water, thus simulating a layer of low permeability such as concrete.

A comparison of the commencement of the PHM_1 and PHM_2 experiments is shown in Figures 8 and 9: it is clear that the experiments compare well, and it is possible to approximate the progress of relative humidity in the first test cell of PHM_2 from the PHM_1 experiment. Relative humidity in the third test cells of both experiments remained at a residual level over a period of 8 days. In the PHM_2 experiment, conducted over a period of

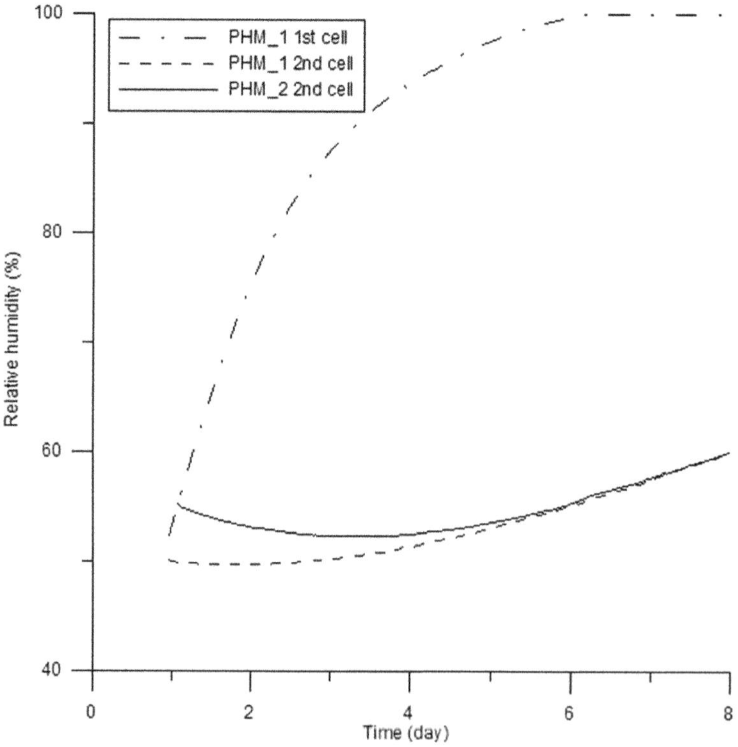

Fig. 9. Comparison of the commencement of the PHM_1 and PHM_2 experiments, the development of relative humidity at the first and second observation points.

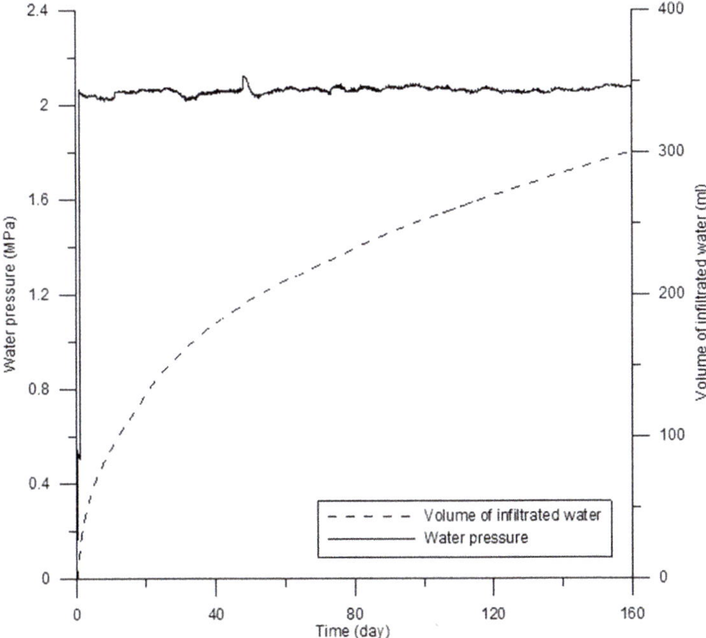

Fig. 10. The pressure and volume of water infiltrating into the PHM_2 sample.

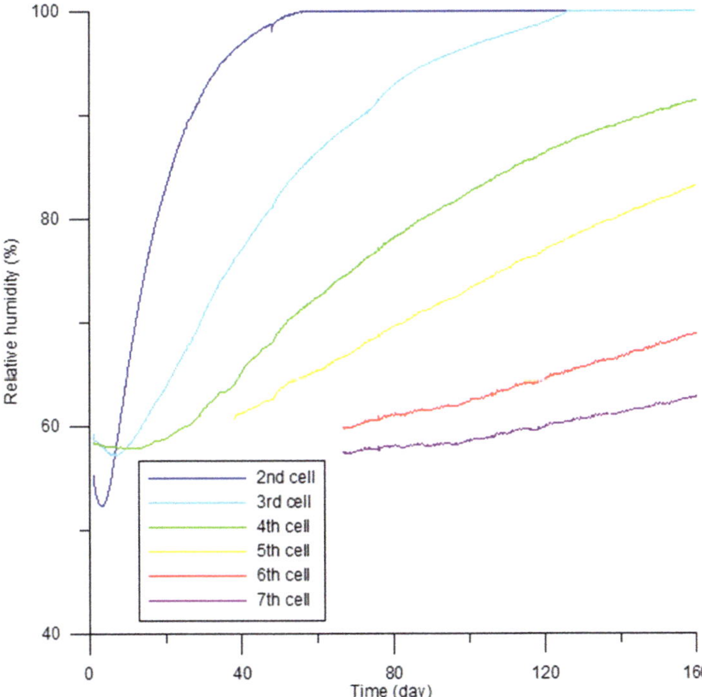

Fig. 11. Development of relative humidity at the PHM_2 observation points. The sensors at the 5th, 7th and 9th observation points were inserted into the sample during the course of the experiment: thus, data from these observation points are not available from the commencement of saturation.

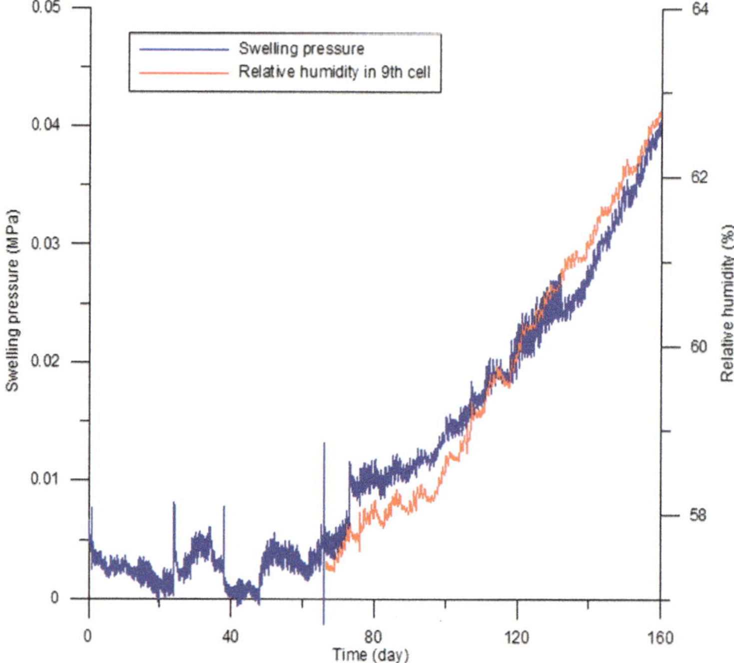

Fig. 12. Development of swelling pressure at the end of the PHM_2 sample.

160 days, relative humidity began to increase from the ninth day.

The results of the PHM_2 experiment are shown in Figures 10–12; water injection took place for 160 days. A decrease in the infiltration of water into the bentonite material is apparent, as is a decreasing rate of saturation with increasing observation point distance from the site of infiltration. While the second observation point showed 100% RH after 56 days, the third observation point showed 100% RH after 126 days. The average saturation rate measured at the second observation point was 1.34 mm/day, whereas that at the third observation point was 0.99 mm/day. It should be emphasized that the development of the progress rate of the saturation front at any given observation point was not linear: indeed, it was found that the rate of progress was retarded, in particular in connection with higher relative humidity values.

Conclusion

The aim of this paper is to describe the results of the experimental determination of the space distribution of relative humidity in bentonite pellet samples during and after hydration. This will later be used as input for numerical modelling of the behaviour of bentonite pellets in the large-scale EPSP experiment. The obtained results revealed that:

- The disintegration rate and saturation of the bentonite pellets is rapid, and pellets form a homogeneous mixture after only 3 days of saturation when water is freely available. After swelling, compacted pellets were found to be as homogeneous as compacted bentonite powder according to a visual comparison and a comparison of water content.
- After 160 days of the experiment, the pellets reached 100% relative humidity at a third cell observation point and the rate of saturation of the bentonite was found to decrease with increasing distance from the infiltration point of the experiment. While the second observation point (at a distance of 7.5 cm from the beginning of the sample) exhibited 100% RH after 56 days, the third observation point (at a distance of 12.5 cm from the beginning of the sample) showed 100% RH after 126 days.
- Swelling pressure monitored at the outlet of the experiment (45 cm from the beginning of the sample) slowly increased during the course of the experiment; the increase began after 60 days. The increase in swelling pressure was found to exhibit the same trend as the evolution of relative humidity at the outlet of the sample.

As full saturation in the overall experiment has not yet been obtained, it cannot yet be concluded that this is general trend for this material and the conditions tested.

The research leading to these results received funding from the European Union's European Atomic Energy Community's (Euratom) Seventh Framework Programme FP7/2007-2013 under grant agreement No 323273, the DOPAS project.

References

HAVLOVÁ, V., HOLEČEK, J., VEJSADA, J., VEČERNÍK, P. & ČERVINKA, R. 2010. *Příprava syntetické podzemní vody pro laboratorní experimentální práce.* [*Preparation of Synthetic Ground Water for Experimental Laboratory Work.*] Technical Report of the Ministry of Industry and Trade of the Czech Republic project FR-TI1/367 [in Czech].

HOFFMANN, C., ALONSO, E.E. & ROMERO, E. 2007. Hydro-mechanical behaviour of bentonite pellet mixtures. *Physics and Chemistry of the Earth, Parts A/B/C*, **32**, 832–849.

IMBERT, CH. & VILLAR, M.V. 2006. Hydro-mechanical response of bentonite pellets/powder mixture upon infiltration. *Applied Clay Science*, **32**, 197–209.

KARNLAND, O., NILSSON, U., WEBER, H. & WERSIN, P. 2008. Sealing ability of Wyoming bentonite pellets foreseen as buffer material – Laboratory results. *Physics and Chemistry of the Earth*, **33**, 472–475.

VAN GEET, M., VOLCKAERT, G. & ROELS, S. 2005. The use of microfocus X-ray computer tomography in characterising the hydration of a clay pellet/powder mixture. *Applied Clay Science*, **29/2**, 73–87.

Quantitative characterization of the excavation damaged zone fracture network in the Meuse/Haute-Marne Underground Research Laboratory: *in situ* experiment and numerical interpretation of helium injection test

HUA SHAO[1], SEBASTIAN GÖTHLING[1], WENTING LIU[1], JÜRGEN HESSER[1], JACQUES MOREL[2] & JÜRGEN SÖNNKE[1]*

[1]*Federal Institute for Geosciences and Natural Resources (BGR), Hanover, Germany*

[2]*French National Agency for Radioactive Waste Management (ANDRA), 55290 Bure, France*

**Correspondence: juergen.soennke@bgr.de*

Abstract: In the underground laboratory of Meuse/Haute-Marne (Bure) in France, different fracture types have been intensively investigated. Within a co-operative project between BGR and ANDRA, geophysical measurements and borehole gas tests were conducted in three well-designed boreholes in the gallery GRM to characterize the fracture structure and to determine the gas tracer velocity within fracture networks.

On the basis of seismic measurements and nitrogen gas interference tests, helium was injected into an interval of borehole OHZ3003, where an unloading joint has been identified. The injection took place in the form of a 10 min 'pulse' with an injection pressure of 2 bar. The other two boreholes, OHZ3002, where an upper part of the shear-mode fractures ('chevron' pattern) dominate, and OHZ3001, where shear-mode fractures (subvertical 'oblique' fractures and 'chevron' pattern) exist, served as observation holes. To maintain the pressure gradient between the injection hole and the observation holes, nitrogen gas was subsequently flushed into the injection hole. Two breakthrough curves of helium concentration and pressure developments in the two observation holes were continuously monitored using two helium leakage detectors and pressure gauges.

To interpret the measured pressure and concentration data, numerical models were constructed. A 3D model was used to simulate nitrogen gas flow and 2D models were applied to simulate a helium transport process. The real volume of the injection interval was considered in the model and the experimental process was simulated. Using the calibrated transport parameter data for a helium tracer from previous studies in the Mont Terri Rock Laboratory, the calculated breakthrough curves agreed well with those obtained from measuring the variation in permeability. The permeability derived from the helium tracer test agrees well with the estimation obtained from the nitrogen gas tests.

After the excavation of underground facility, fractures in the near field of the tunnel are generated to form a so-called 'excavation damaged zone' (EDZ). These fractures are mainly induced by extensile and shear deformation (Nussbaum *et al.* 2007). Depending on the orientation of the excavated tunnel and stress state, the EDZ presents different shapes observed in the underground laboratory Bure (France) (Armand *et al.* 2014). The underground laboratory at the Bure site was excavated by ANDRA (ANDRA 2014) in a 500 m-deep Callovo-Oxfordian clay formation in the Paris Basin in order to characterize the confining properties of the claystone, and to demonstrate the feasibility of the construction and operation of a geological disposal in France.

The EdZ (excavation-disturbed zone) and EDZ (excavation-damaged zone) around an underground opening are becoming increasingly more important because of their altered hydromechanical and geochemical properties. Numerous investigations have been conducted to understand and characterize the EDZ properties of different potential formations for the deep geological disposal of high-level radioactive waste (Tsang *et al.* 2004). A series of experiments aimed at characterizing the EDZ were performed at the Bure site and have been recently documented (Norris *et al.* 2014). Different methods, including geological mapping, structural analysis, and geophysical, geotechnical and geochemical measurements were applied to indicate the coupled hydromechanical and geochemical processes

From: NORRIS, S., BRUNO, J., VAN GEET, M. & VERHOEF, E. (eds) 2017. *Radioactive Waste Confinement: Clays in Natural and Engineered Barriers*. Geological Society, London, Special Publications, **443**, 85–96.
First published online October 13, 2016, https://doi.org/10.1144/SP443.21

involved in the generation and development of fracture systems within the EDZ (Nussbaum *et al.* 2007; Shao *et al.* 2008; Noiret *et al.* 2011; Armand *et al.* 2014; Yildizdag *et al.* 2014). Extensive data from small-scale (borehole) and large-scale (gallery) experiments in the deep underground laboratory have been analysed.

If the tunnel orientation is parallel to the major principal stress, the EDZ and EdZ are characterized by extension (tension, mode 1) and herringbone (shear-mode, mode 2) fractures (Fig. 1). Undefined fractures have also been observed that are oblique to the core axis of drilled boreholes around the niche. *In situ* measurements conducted by ANDRA in the drift showed an increase in permeability of about three or four orders of magnitude (up to 1 m) along the horizontal direction inside of a zone with both fracture modes (the zone shown a with dashed line in Fig. 1). This zone may indicate an EDZ wherein a fracture network is roughly connected and tension mode fractures are dominant (Yildizdag *et al.* 2014). Outside this zone, pure shear fractures were observed and permeability increased, with a maximum of one order of magnitude being measured (the zone shown with a dotted line in Fig. 1). Fractures in this zone may not hydraulically connect with each other. An increase in permeability is, however, observed until 5 m away from the drift. Results from permeability measurements conducted in the drift along the vertical direction are clearly small (de La Vaissière *et al.* 2015).

Within the frame of co-operation between BGR and ANDRA, an experimental programme using geoelectrical methods and ultrasonic borehole measurements, as well as gas permeability and gas tracer measurements, was planned. The objectives of the test programme were: (1) to characterize different type of fractures using different methods; (2) to evaluate quantitatively the damage zone in the vicinity of the opening (EdZ/EDZ) in the Callovo-Oxfordian clay rock; and (3) to check the possible pathways between the different type of fractures and to quantify the transport properties. Based on the results from the geological characterization of the drilling core, geophysical measurements and hydraulic packer tests, well-defined fracture zones/fracture systems were selected for a tracer experiment using helium gas.

In addition to the work of ANDRA on the characterization of the fracture network surrounding a drift using gas and water tests, the focus of this paper is to present the gas tracer experiment in a dipole configuration between three boreholes within the GRM gallery and its associated numerical interpretation technique. The different type of fractures determined in the near field can be characterized quantitatively.

In situ experiments

For more than 10 years, ANDRA (Agency nationale pour la gestion des déchets radioactifs) has carried out numerous experiments in the Meuse/Haute-Marne Underground Research Laboratory (URL) to investigate the feasibility of argillaceous rock as a possible host rock for the disposal of radioactive waste. Within ANDRA's investigation plan, BGR has also been involved in several experiments focusing on the near-field characterization, with the help of geophysical and hydraulic measurements (Fig. 2). The measurements were performed in the boreholes using different orientations and at different depths.

In the programme carried out in 2014, three boreholes in the gallery GRM (Fig. 2), which is parallel to the maximum principal stress, were drilled for combined geophysical and hydraulic measurements. The conceptual model of the fracture pattern

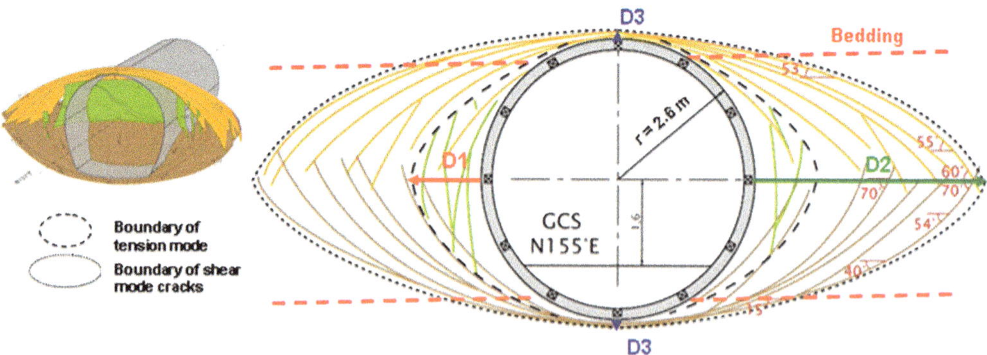

Fig. 1. Generalized EDZ extent around the drift GCS (with the support system) excavated parallel to σ_H at the Meuse/Haute-Marne URL (Noiret *et al.* 2011).

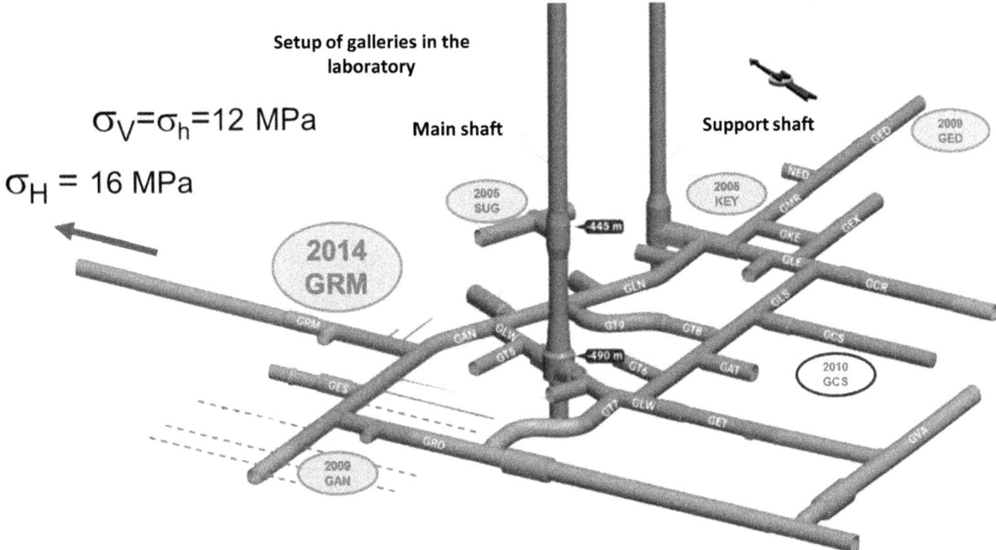

Fig. 2. Layout of the Bure URL (Armand *et al.* 2014) with the BGR hydraulic investigations (year and programme).

in the EDZ as shown in Figure 1 should be valid because the gallery GRM is parallel to the gallery GCS (Fig. 2). All boreholes with a diameter of 86 mm have a length of about 6 m and are denoted by OHZ3001–OHZ3003 (Fig. 2). The distance between each mouth borehole is 0.5 m (Fig. 3, left). Owing to the inclined orientation of OHZ3002 and OHZ3003 (Fig. 3, right), the distance between the deepest part of each borehole is therefore greater than 0.5 m.

Geological and geophysical results

All cores during drilling were mapped and documented (ANDRA 2014), and both tensile and shear fractures were identified. The extension of each fracture zone fits well with the well-established classification at the Bure Site (Fig. 1). In total, 29 fractures in the three boreholes were identified based on the core information (Fig. 4). The deepest extensile fracture was documented at 1.62 m away from the tunnel surface in OHZ3001, at 0.55 m in OHZ3002 and at 1.41 m in OHZ3003. All extensile fractures are within the boundary of the tension mode (Fig. 1). The deepest shear fracture was mapped 3.43 m away from the tunnel surface in OHZ3001, at 4.24 in OHZ3002 and at 4.68 m in OHZ3003.

Shortly after drilling the boreholes, a special BGR ultrasonic borehole probe was used. This method is particularly suited to the detection of fractures in boreholes. With this borehole probe, high-resolution seismic interval velocity measurements using emitted frequencies at 50 kHz were performed along all three boreholes. In total, 44 locations were indicated as anomalies in the seismic signal. Both geological and geophysical information were considered for the last hydraulic characterization, including nitrogen and helium tests.

Nitrogen tests. Before the helium test, all three boreholes were tested using nitrogen gas to determine the permeability and its distribution along the borehole. The permeability measurements were carried out about 2 weeks after the drilling. Before the test, borehole camera logging was performed in all three boreholes to check the borehole quality. In comparison with the previous scan results obtained shortly after the drilling, significant changes concerning breakouts could be observed. Free access for the designed packer, which had a diameter of 84 mm, was not possible in the deeper parts of all three boreholes.

The test intervals were defined taking into account the quality of the borehole wall and the seismic anomalies, which could indicate the existence of an opened fracture. This kind of correlation between seismic results and permeability measurement was often verified in the earlier investigations. Only 80% of the planned 54 locations were possible for the tests. In addition, the results from 9% of the performed tests were not plausible due to possible packer leakage.

For the gas test, a so-called double piston-driven mechanical packer was used. This type of packer can guarantee the unit packer sealing. Generally, an injection pulse of nitrogen with a pressure of

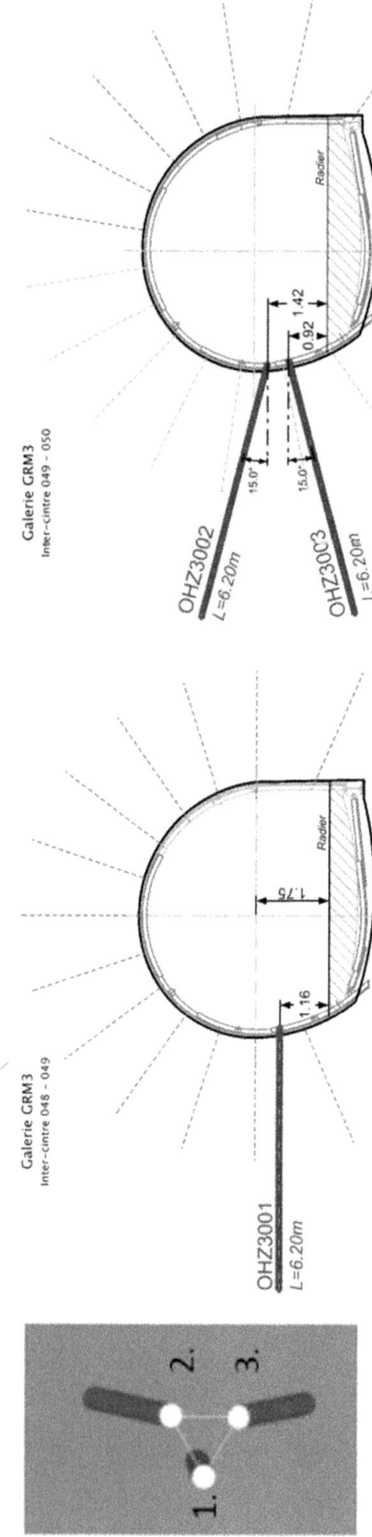

Fig. 3. Borehole configuration of OHZ3001–OHZ3003 with their planning lengths.

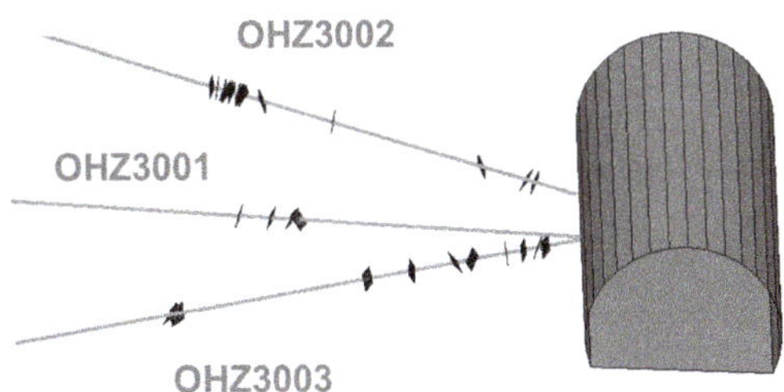

Fig. 4. 3D image of mapped fractures (the tunnel diameter is *c.* 6 m – not scaled).

2 bar was applied in each test interval. In case of a rapid decrease in pressure, a constant rate test of 10 min for the injection of nitrogen gas was attempted. This may indicate a high-permeability zone around the test interval. Almost all test performances went well. During the injection tests in one borehole, the other two boreholes were packed off in order to monitor the pressure interference. As a result, connections were detected from OHZ3003 (1.33–1.43 m from the steel lining or 1.05–1.15 m from the tunnel clay surface, respectively) to OHZ3001, and from OHZ3003 to OHZ3002. These connections were then selected as locations for the helium injection test.

Helium tests. The concentration of helium was measured using helium leakage detectors (type PHOENIX L300i: Fig. 5b). First, as a reference measurement, a concentration of 25 ppm was recorded in the GRM gallery, which is equal to the atmospheric helium concentration. In the boreholes, the concentration of helium was somehow higher than in the gallery. The concentration in OHZ3001 was about 148% higher (62 ppm), in OHZ3002 it was about 28% higher (32 ppm) and in borehole OHZ3003 it was about 52% higher (38 ppm) than the atmospheric content. All of these values can be considered as a zero concentration, however, in comparison with the injected concentration of pure helium, at about 7×10^6 ppm.

As designed, helium was injected into an interval that ran from 1.33 to 1.43 m of borehole OHZ3003. The other two boreholes, OHZ3001 and OHZ3002, were used for the observation of the helium concentration (Fig. 5a). The observation depth stretched from 0.9 m (from the steel lining or 0.64 m from the tunnel clay surface, respectively) to the end of each borehole. Helium gas was injected for a total of 10 min under a pressure of 2 bar. After that, nitrogen gas was flushed into the injection interval of the borehole OHZ3003 for 114 min, also with a pressure of 2 bar. This was made to keep the pressure gradient up between the injection hole and the observation holes. The observation boreholes OHZ3001 and OHZ3002 were sealed with a single packer. The concentrations were measured using the integrated helium leakage detectors.

During the entire test sequence, the temperature and atmospheric pressure in the GRM gallery were monitored (Fig. 6). The experiment confirmed the connectivity between three boreholes, which was detected by nitrogen tests. The connection between OHZ3003 and OHZ3001 is much larger than that between OHZ3003 and OHZ3002. The breakthrough of helium from injection borehole OHZ3003 to the observation hole OHZ3001 took place about 20 s after the die injection of helium started. The rise in the helium concentration in observation hole OHZ3002 took much longer. It began about 270 s after the injection started and increased very smoothly. There was also a large difference between the maximum helium concentrations reached in the two observation boreholes.

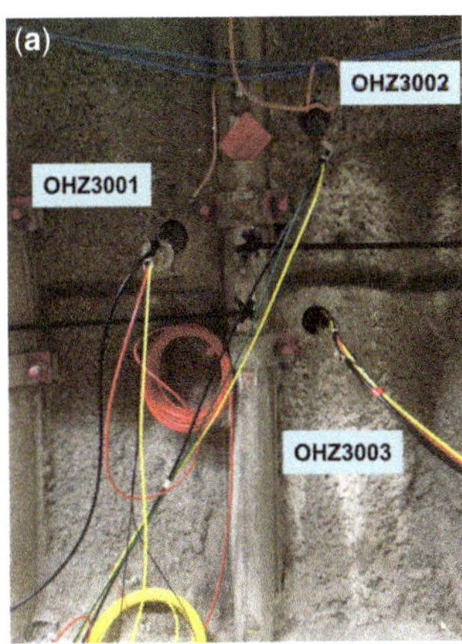

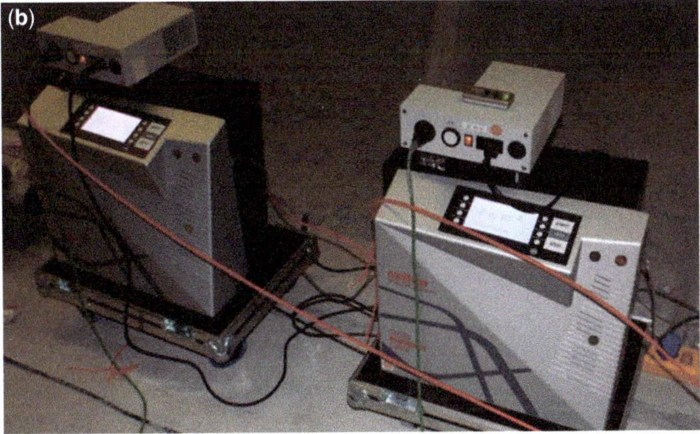

Fig. 5. *In situ* configuration of (**a**) test boreholes and (**b**) helium leakage detectors – PHOENIX L300i.

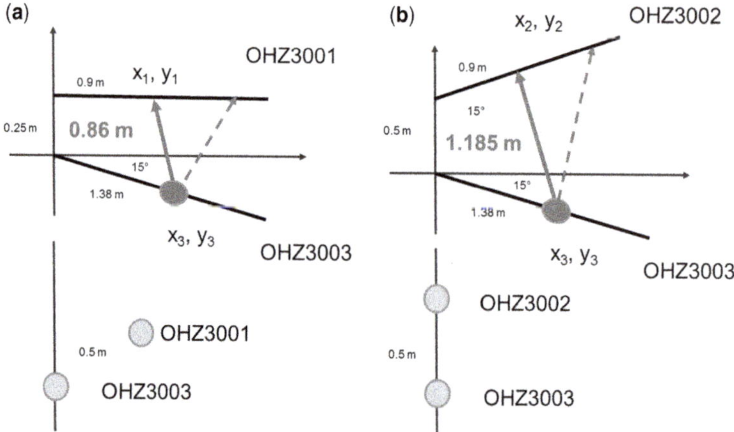

Fig. 6. Possible transport path between boreholes: (**a**) OHZ3003 and OHZ3001; and (**b**) OHZ3003 and OHZ3002).

While borehole OHZ3001 reached a maximum helium concentration of 7×10^6 ppm, the borehole OHZ3002 only reached a maximum concentration of 6.8×10^4 ppm, which may be explained by dilution processes with the nitrogen gas. Furthermore, the maximum concentration in OHZ3001 was reached much earlier than the maximum helium concentration in OHZ3002 (Fig. 7).

Numerical interpretation

To interpret the experimental data from both the nitrogen test and the helium test, numerical models were applied using the FEM program OpenGeoSys (OGS) (Kolditz *et al.* 2014). For the one-phase nitrogen injection test, the module 'air_flow' of the code OGS, which takes account of the properties

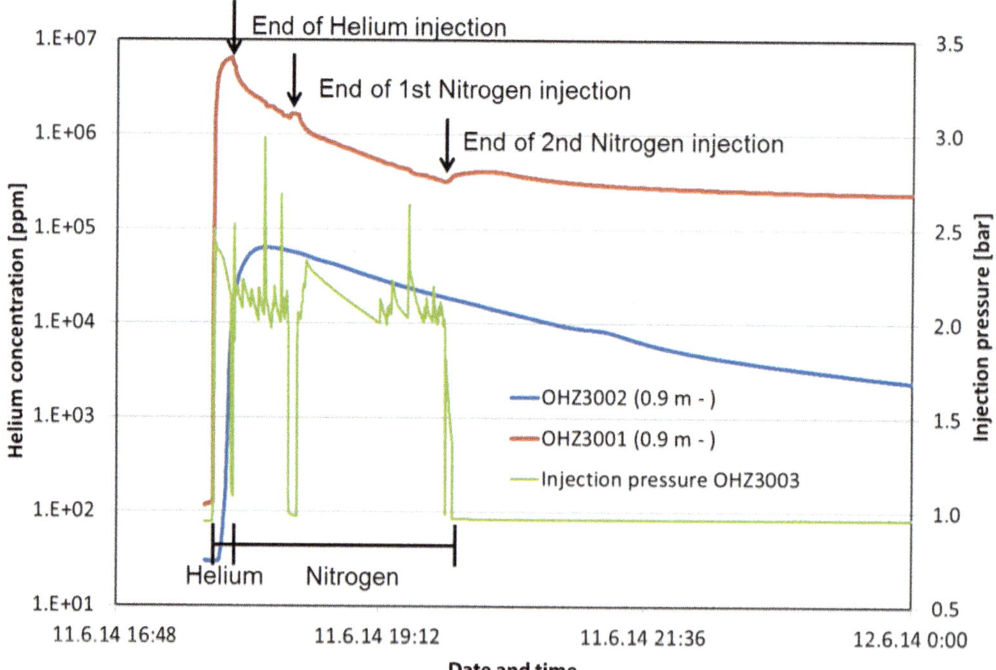

Fig. 7. Measured breakthrough curves in the observation boreholes.

of nitrogen, was used. For the helium test, two modules, 'air_flow' and 'mass_transport', were used thanks to the arbitrary coupling algorithm of the program. The gas mass transport can therefore be described based on the general multiphase flow (equations 1 and 2), as well as on the mass transport (equation 3):

$$\frac{\partial(nS_\alpha\rho_\alpha)}{\partial t} + \text{div}(\rho_\alpha v_\alpha) - \rho_\alpha q_\alpha = 0, \quad (1)$$

where v_α is the gas flow velocity in the 'air_flow' model. In the case of the helium injection test, v_α is the gas flow velocity of two components (nitrogen and helium), which can be calculated by:

$$v_\alpha = -\frac{k_{r\alpha}K}{\mu_\alpha}(\text{grad}(p_\alpha) - \rho_\alpha g), \quad (2)$$

where n is the volumetric porosity, S is the degree of phase saturation, ρ is the fluid density, q is the phase flux, α is the phase (gas or water), $k_{r\alpha}$ is the relative permeability for phase α, μ is the dynamic viscosity, p is the pressure and g the gravity acceleration:

$$\frac{\partial}{\partial t}(nc) + v_\alpha \nabla c - \nabla(nD_s \nabla c) + n\lambda c + q(c - c^*) = 0, \quad (3)$$

where c is the mass fraction of solute per gas mass, v_α is the velocity of phase α (gas), D_s is the diffusion/dispersion tensor, λ is the radioactive decay constant and c^* is the concentration of solute in the source fluid.

This formulation includes dispersion effects according to Fick's first law. The 3D diffusion/dispersion tensor in a (ξ, η, ζ)-coordinate system orientated according to the flow path is written as:

$$D_s = \begin{bmatrix} \alpha_L|v_\alpha| + d_0 & 0 & 0 \\ 0 & \alpha_T|v_\alpha| + d_0 & 0 \\ 0 & 0 & \alpha_T|v_\alpha| + d_0 \end{bmatrix}, \quad (4)$$

where α_L and α_T are the longitudinal and transverse coefficients of mechanical dispersion, and d_0 is the diffusion coefficient.

Numerical model for nitrogen test

For the packer test using nitrogen gas as the injection medium, we used a standard 3D model interpretation based on equations (1) and (2). Owing to the flexible geometry adaptation, the entire system covering the borehole interval hollow, the possible borehole damaged zone (BDZ), the packer itself and all technical lines can be modelled using a finite-element mesh (Fig. 8). Temporary events of a test (e.g. pulse injection and recovery phase) can be simulated. Pressure evolution in the interval can be calculated exactly based on the mass-balance equation of one-phase (water or gas) or two-phase flow (water and gas) according to the fluid existing in the formation and used as the injection medium. In addition, the system leakage (mass loss) can be considered in the model. Therefore, a high resolution of the evaluated value can be obtained.

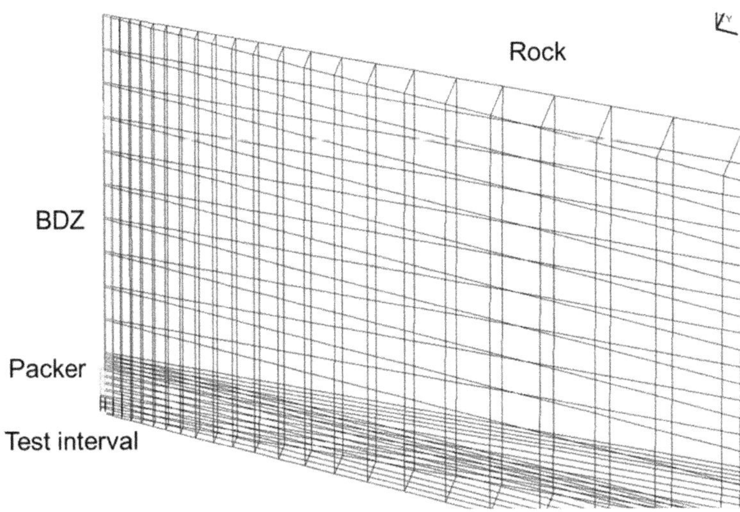

Fig. 8. 3D mesh for the evaluation of packer tests.

Numerical model for the helium test

To simulate the helium tracer test in a dipole configuration, 2D models, including injection and observation boreholes, were developed. The model domain (8 × 5 × 0.1 m) included rock mass and fractures based on the continuum approach. The thickness of the model (0.1 m) reflected the length of the test interval. Possible direct distances for the tracer transport between boreholes were calculated to be about 0.86 m between OHZ3003 and OHZ3001, and 1.18 m between OHZ3003 and OHZ3002 (Fig. 6).

In the injection interval of borehole OHZ3003 (between 1.05 and 1.15 m from the tunnel clay surface), extensile fracture was mapped with a strike of 60° and dip of 80°. In the observation interval of OHZ3001, from 0.9 m to the end of the borehole, both extensile and shear fractures were identified. Because a highly permeable fracture system may determine the flow pattern, a homogeneous model for the extensile fracture zone between OHZ3003 and OHZ3001 was, therefore, used (Fig. 9a).

In another observation borehole, OHZ3002 (from 0.64 m to the end of the borehole), only shear-failure-dominant chevron fractures were

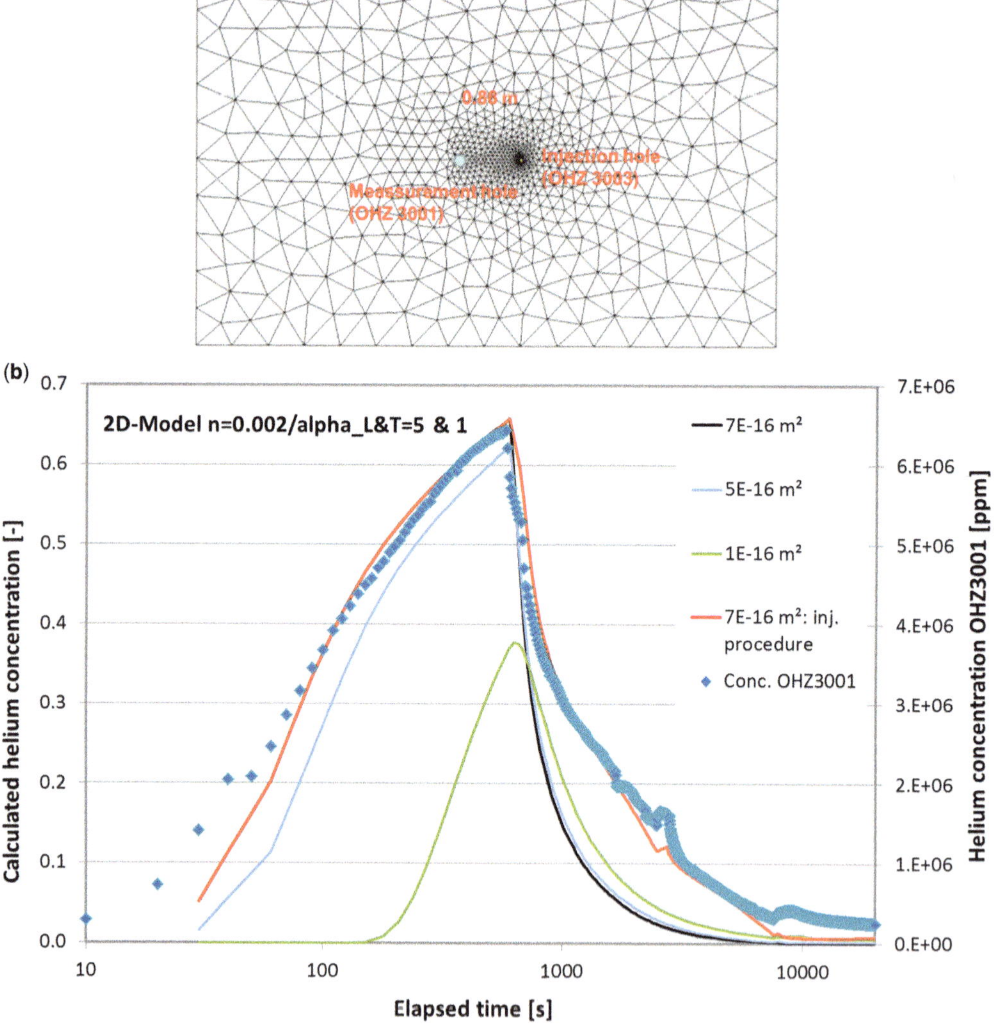

Fig. 9. Homogeneous model for the tracer test between (**a**) OHZ3003 and OHZ3001, and (**b**) its breakthrough interpretation.

observed. The transport pathway was, therefore, conceptualized by an inhomogeneous model with two material groups (Fig. 10a), namely the same properties around the injection borehole as used in the first model (Fig. 9a) and a new material group in the area where borehole OHZ3002 is located. To analyse the injected helium mass, the borehole volume of the injection interval was considered and represented by an additional material group. The atmospheric conditions in the gallery for both pressure and helium concentration were considered as the initial and boundary conditions in the model.

Transport parameters (e.g. longitudinal and transverse dispersion length, and effective porosity for helium in clayey material) were taken from the previous investigation at the URL Mont Terri (Switzerland). Using this calibrated dataset (longitudinal/transverse dispersion 5/1 and effective porosity 0.2%), the breakthrough of helium in the observation boreholes were calculated using the variation in permeability (Figs 9b, 10b).

Results

Both permeability and permeability distribution along the boreholes were evaluated (Fig. 11). High-permeability values induced by packer leakage

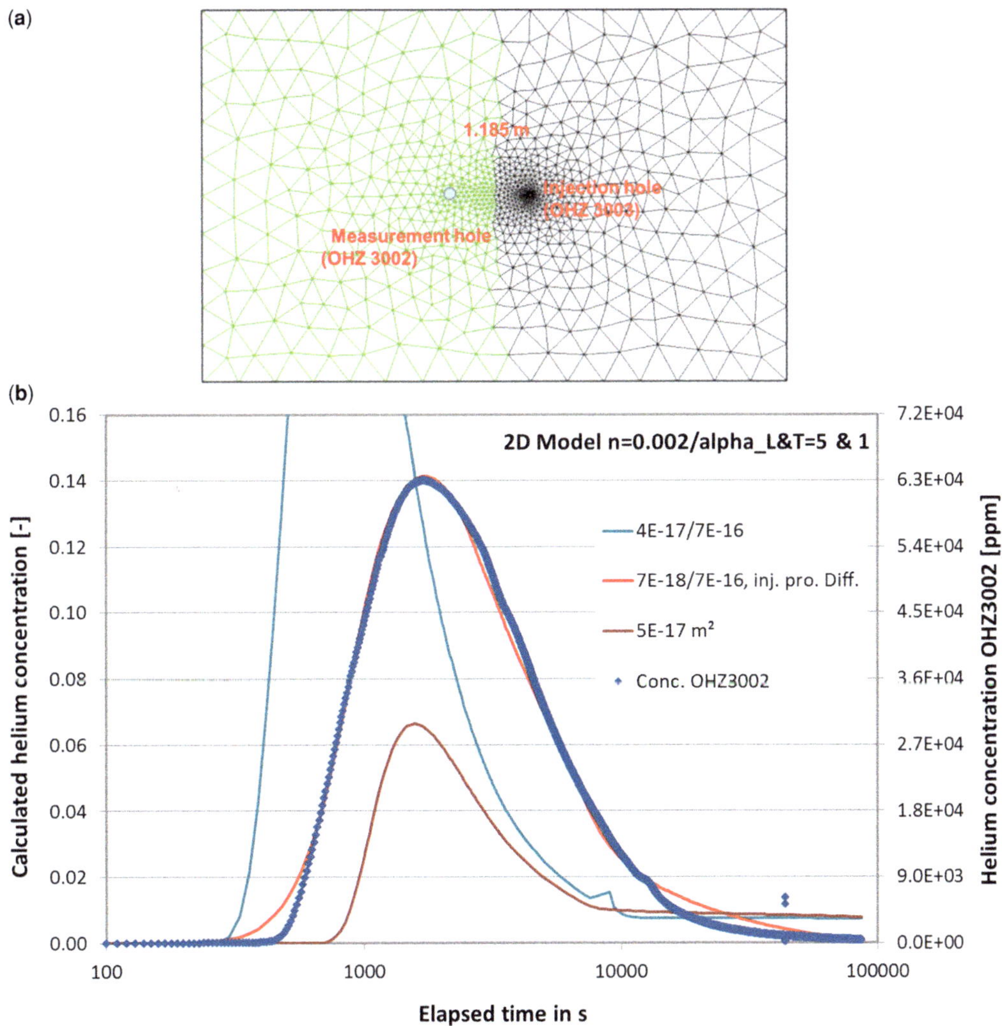

Fig. 10. Inhomogeneous model for the tracer test between (**a**) OHZ3003 and OHZ3002, and (**b**) its breakthrough interpretation.

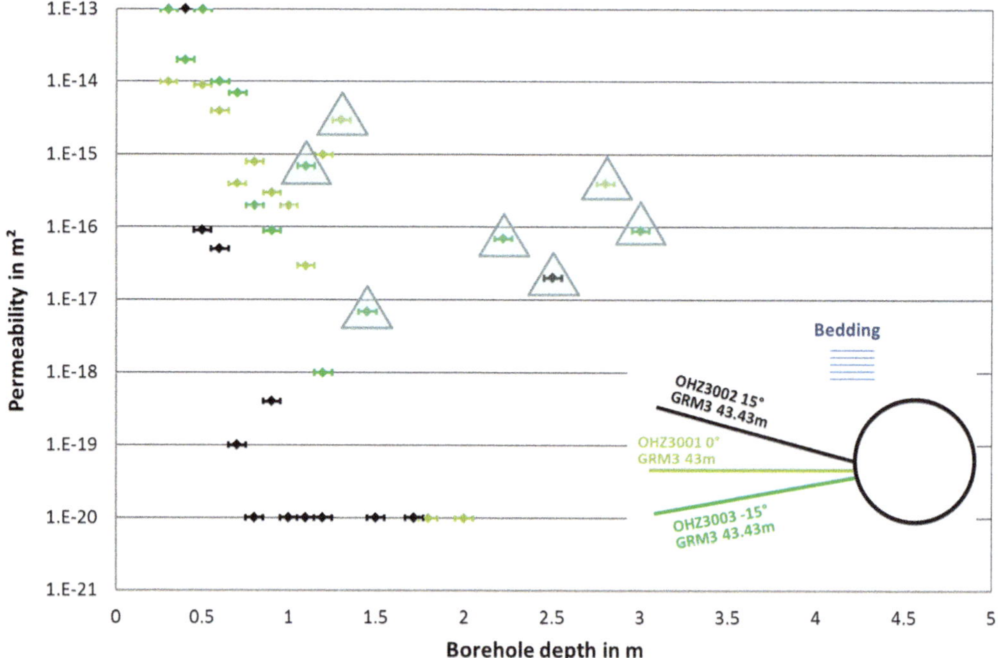

Fig. 11. Permeability distribution along the borehole (values within the triangles indicate the seismic anomaly).

because of breakouts on the borehole wall are not plotted in this figure. As expected, high permeability in all three boreholes was measured to a depth of up to 1.5 m (the zero point is located in the steel line) or 1.2 m from the excavated clay surface. The extent of the high-permeable zone is relative small in borehole OHZ3002, which has an incline angle of 15° upwards. There are exclusively tensile fractures in the area. This agrees well with the conceptual EDZ/EdZ model developed by ANDRA (Fig. 1).

In the first 0.5 m zone of the excavated clay surface, the measured transmissibility using the double-packer system is 1×10^{-14} m³. Assuming that the aperture of a fracture is about 1 mm, which is visible on the tunnel surface, and neglecting the saturated clay matrix permeability, the permeability of such a fracture can be calculated as being greater than 1×10^{-13} m².

Increased permeability at the locations in the deep zone, where only shear fractures were identified by core mapping and anomalies were indicated by seismic measurements, has been measured to be up to four orders of magnitude (about 1×10^{-16} m²). This fact suggests that the shear-failure-induced fractures may also have a high permeability. These locations were marked with triangles in Figure 11.

It is obvious that the permeability of the undisturbed Callovo-Oxfordian formation at the fully saturated state is generally less than 1×10^{-20} m². In the fully saturated state, gas injection tests created a two-phase flow regime in the observed area. Therefore, a two-phase flow model is needed for the evaluation of such tests that takes into account the two-phase flow properties of clay rock, including the relationship between capillary pressure and saturation, and the relationship between relative permeability and saturation (Shao *et al.* 2008).

In the investigated area, which is approximately on the metre-scale, fractures along the tunnel axis may connect with each other and form a transport pathway. Connections between boreholes, which were indicated by the interference tests using nitrogen gas, were quantitatively evaluated using the helium tracer test.

In the injection interval of borehole OHZ3003 (between 1.05 and 1.15 m from the tunnel clay surface), where the tensile fracture was identified, helium gas was injected. The other two boreholes, OHZ3002, where an upper part of the shear fracture dominates, and OHZ3001, where tensile and shear fractures exist, served as observation holes. The tracer transported much faster within the same fracture (e.g. between OHZ3003 and OHZ3001) than within different fracture types from OHZ3003 to OHZ3002. The evaluated parameters are summarized in Table 1. The minimum transport distance in Table 1 is the distance between the midpoints

Table 1. *Parameter data evaluated from the helium tracer tests*

Parameter	Unit	OHZ3003–OHZ3001	OHZ3003–OHZ3002
First arrival time (t_0)	s	20	270
Peak arrival time	s	600	1690
Peak concentration	ppm	6.4E6	6.3E4
Mini. Transport distance (s)	m	0.86	1.185
Transport velocity ($v_0 = s/t_0$)	m s^{-1}	0.043	0.00439
Permeability (evaluated)	m^2	7×10^{-16}	$7 \times 10^{-16}/7 \times 10^{-18}$
Longitudinal/transversal dispersion length	m	5/1	5/1
Effective porosity	%	0.2	0.2
Diffusion coefficient	m^2 s^{-1}	N/A	N/A
Fracture type		Single fracture zone	Fracture network

of both boreholes (Fig. 6). The tortuosity did not vary, and was assumed to be 1 in the model. Owing to the fast breakthrough time, tracer transport within the fracture zone in both cases was dominated by advection rather than diffusion. The diffusion coefficient of the undisturbed clay formation, however, cannot be determined.

Conclusions

As part of the co-operation between ANDRA and BGR, three boreholes were designed and tested by the BGR with the aim of quantitatively characterizing the fracture systems in the near field around an underground facility in the clay formation. Based on the EDZ-concept developed by ANDRA (Fig. 1), different types of fractures (unloading joints, subvertical 'oblique' fractures and chevron-type fractures) were individually quantified using geophysical methods and packer tests with the injection of nitrogen and helium gas.

The nitrogen gas injection tests were performed 13 days after drilling and geophysical borehole measurements. The camera inspection of boreholes showed that the borehole surface was continuously getting worse over this period. In the deep section of up to 4 m, it was not possible to carry out a packer test. In general, the borehole at an inclined angle to the bedding plane has a better quality than those parallel to the bedding plane. To reduce the deteriorating effect over time on the borehole surface, it is worthwhile considering that, in future, the packer test should be carried out immediately after drilling.

The high-permeability zone is limited to a distance of up to 1.2 m from the gallery wall (clay surface). In this zone, both extensile and shear fractures are possible. The unloading fractures may have a permeability that is higher than the 1×10^{-13} m^2 of the immediate near-field zone around the opening. In addition, a single shear feature with a permeability of up to four orders of magnitude higher may exist, even in the deep section down to 3 m, which corresponds to the seismic anomaly detected by seismic measurement. The permeability of the undisturbed Callovo-Oxfordian clay formation can be estimated to be less than 1×10^{-20} m^2 in the saturated state.

On the basis of possible pathways detected using interference tests using nitrogen gas, a helium tracer test was performed in a 'dipole' configuration. Two typical breakthrough curves were registered by the helium leakage detectors – PHOENIX L300i – in two observation boreholes. The fast arrival times of the helium tracer indicated relatively high-permeability pathways.

To interpret the measured pressure and concentration data, two modules, 'air_flow' and 'mass_transport', of the numerical open source code OpenGeoSys were used in a coupled manner. Two-dimensional continuum models were constructed to simulate the tracer transport pathway between boreholes. The experiment process was numerically simulated. This included a helium injection phase and a subsequent nitrogen injection phase during the experiment. The latter was used in order to maintain a pressure gradient between injection and observation boreholes. The volume of the injection borehole was also modelled to retain the mass balance. Using the calibrated transport parameter data, the measured breakthrough curves can be accurately modelled using the variation in permeability. The permeability values derived from the tracer test agree well with those determined from the nitrogen gas tests.

The permeability determined from the helium test in the zone, including extensile fractures between OHZ3003 and OHZ3001, amounted to 7×10^{-16} m^2, which was the same value as that evaluated from the nitrogen test in the injection interval. Maintaining the permeability in the injection area, the permeability from the helium test around the borehole OHZ3002 was evaluated to 7×10^{-18} m^2, which agreed well with the results from the nitrogen tests obtained from 0.65 m to the end of borehole.

As a result, we can conclude that the permeability of extensile fractures can be evaluated at 7×10^{-16} m^2, and the permeability of the connection between different fracture systems (extensile to chevron fractures) may be two orders of magnitude lower. The effective porosity of the helium gas transport in the Callovo-Oxfordian claystone can be determined to be 0.002 from calculations, which is lower than the measured total porosity of 0.2. In such a fracture system, the advection of helium transport is dominant owing to the fast transport velocity of helium. Diffusion is therefore not significant and the diffusion coefficient cannot be determined.

The proposed method contributed mostly to the quantification of the near-field fracture system. In addition, possible connections between different types of fractures (tension fracture and shear fracture) were quantitatively evaluated.

This work was supported by BMWi (Bundesministerium für Wirtschaft und Energie, Berlin). The authors would like to thank all BGR technicians involved in the fieldwork and the ANDRA field supporting team.

References

ANDRA 2014. *Suivi geologique des forages des experimentations dans la galerie GRM forages OHZ3001–OHZ3003.* ANDRA Internal Report D.RP.0GTR. 14.0013 ANDRA (National Agency for Radioactive Waste Management), Châtenay-Malabry, France.

Armand, G., Leveau, F. *et al.* 2014. Geometry and properties of the excavation induced fractures at the Meuse/Haute-Marne URL Drifts. *International Journal of Rock Mechanics and Mining Sciences*, **27**, 21–41.

de La Vaissière, R., Armand, G. & Talandier, J. 2015. Gas and water flow in an excavation-induced fracture network around an underground drift: a case study for a radioactive waste repository in clay rock. *Journal of Hydrology*, **521**, 141–156.

Kolditz, O., Shao, H., Wang, W.-Q. & Bauer, B. 2014. *Thermo-Hydro-Mechanical-Chemical Processes in Fractured Porous Media – Modelling and Benchmarking*. Springer, Heidelberg.

Noiret, A., Armand, G., Cruchaudet, M. & Conil, N. 2011. Mine by experiments in order to study the hydromechanical behavior of the Callovo-Oxfordian claystone at the Meuse Haute-Marne Underground research laboratory (France). Paper presented at the 8th International Symposium on Field Measurements in GeoMechanics, 12–16 September 2011, Berlin, Germany.

Norris, S., Bruno, J. *et al.* (eds) 2014. *Clays in Natural and Engineered Barriers for Radioactive Waste Confinement.* Geological Society, London, Special Publications, **400**, http://sp.lyellcollection.org/content/400/1

Nussbaum, C., Wileveau, Y., Bossart, P., Möri, A. & Armand, G. 2007. Why are the geometries of the EDZ fracture networks different in the Mont Teri and Meuse/Haute-Marne Rock laboratories? Structural approach. Paper presented at the International Conference on Clays in Natural and Engineered Barriers for Radioactive Waste Confinement, 17–18 September 2007, Lille, France.

Shao, H., Schuster, K., Sönnke, J. & Bräuer, V. 2008. EDZ development in indurated clay formations – In situ borehole measurements and coupled HM modelling. *Physics and Chemistry of the Earth, Parts A/B/C*, **33**, (Suppl. 1), S388–S395, https://doi.org/10.1016/j.pce.2008.10.031

Tsang, C.T., Bernier, F. & Davies, C. 2004. Geohydromechanical processes in the Excavation Damaged Zone in crystalline rock, rock salt, and indurated and plastic clays – in the context of radioactive waste disposal. *International Journal of Rock Mechanics & Mining Sciences*, **42**, 109–125, https://doi.org/10.1016/j.ijrmms.2004.08.003

Yildizdag, K., Shao, H., Hesser, J., Noiret, A. & Soennke, J. 2014. Coupled hydromechanical modelling of the mine-by experiment at Meuse/Haute-Marne underground rock laboratory France. *In*: Norris, S. & Bruno, J. *et al.* (eds) *Clays in Natural and Engineered Barriers for Radioactive Waste Confinement.* Geological Society, London, Special Publications, **400**, 265–278, https://doi.org/10.1144/SP400.42

Characterizing the mechanical behaviour of the Tournemire argillite

XUEQING SU[1]*, SON NGUYEN[1], EHSAN HAGHIGHAT[2], STANISLAW PIETRUSZCZAK[2], DENIS LABRIE[3], JEAN-DOMINIQUE BARNICHON[4] & HADJ ABDI[5]

[1]*Canadian Nuclear Safety Commission, Ottawa, Ontario K1P 5S9, Canada*

[2]*McMaster University, Hamilton, Ontario L8S 4L8, Canada*

[3]*CanmetMINING, Natural Resources Canada, Ottawa, Ontario K1A 1M1, Canada*

[4]*Institut de Radioprotection et de Sûreté Nucléaire, 92260 Fontenay-aux-Roses, France*

[5]*University of Ottawa, Ottawa, Ontario K1N 6N5, Canada*

**Correspondence: grant.su@canada.ca*

Abstract: A laboratory programme of uniaxial, triaxial, cyclic and Brazilian tests was conducted to investigate the anisotropic mechanical behaviour of the Tournemire argillite, with different axial loading orientations with respect to the bedding planes (i.e. loading orientation angle, $\theta = 0°$, 30°, 45°, 60° and 90°). The experimental results show that both strength and deformation of the argillite are direction-dependent. Failure occurs in a brittle manner with a sudden collapse of the material strength. The failure mode exhibits localization along distinct failure planes and also depends on the loading orientation. This paper summarizes the experimental results and describes constitutive relationships that were developed in order to simulate the stress–strain behaviour of the Tournemire argillite. A microstructure tensor approach is adopted in order to take into account the anisotropic behaviour of the argillite. The identification procedure for material function and parameters is outlined, and the model is applied to simulate the set of triaxial tests performed at different levels of confining pressure and orientation of the bedding planes. It is demonstrated that the model adequately reproduces the anisotropy, the pre-peak stress–strain response and the onset of material collapse in those tests.

Argillaceous rock formations that are being considered as potential host or cap rocks for the geological disposal of nuclear wastes are usually characterized by the presence of bedding planes, which result in anisotropy of their strength and deformation properties. The anisotropic mechanical behaviour of the argillaceous rocks could affect the extent and characteristics of the damage induced by excavation of a deep geological repository (DGR) in these rock formations, which may have implications on their containment capability for nuclear wastes (Blümling *et al.* 2007; Millard *et al.* 2009; Labiouse & Vietor 2014). Therefore, an understanding of the anisotropic mechanical behaviour of these argillaceous rocks is one of the important aspects in the development of the safety case for a deep geological repository. In Canada, Ontario Power Generation is currently proposing a DGR for the management of its low- and intermediate-level nuclear wastes in an argillaceous limestone, known as the Cobourg limestone, which is overlain by more than 200 m of shale cap rocks. Nuclear Waste Management Organization is currently under a process to select a site for the management of used nuclear fuels in Canada in which the sedimentary rock is considered as one of the potential host rocks. The Canadian Nuclear Safety Commission (CNSC), Canada's nuclear regulator, is responsible for the licensing of future deep geological repositories in Canada. In order to better understand the ability of these argillaceous rock formations to contain and isolate the wastes, and to develop a tool to independently assess the extent and characteristics of damage induced by excavation of a deep geological repository, the CNSC has collaborated with different national and international research organizations and has access to underground research laboratories (URLs) worldwide. One of these URLs is situated in Tournemire, France, in upper Toarcian argillaceous rock (Boisson *et al.* 2001). The Tournemire argillite is characterized by the presence of closely spaced bedding planes and exhibits a strong anisotropic behaviour, so that its strength, as well as its deformation properties, are directionally dependent.

From: Norris, S., Bruno, J., Van Geet, M. & Verhoef, E. (eds) 2017. *Radioactive Waste Confinement: Clays in Natural and Engineered Barriers*. Geological Society, London, Special Publications, **443**, 97–113.
First published online October 12, 2016, https://doi.org/10.1144/SP443.20

Over the last few decades, an extensive research effort has been devoted to studying the mechanical behaviour of anisotropic rocks. Comprehensive references on this topic can be found in a number of review papers (e.g. Amadei 1983; Kwasniewski 1993; Ramamurthy 1993). The transverse isotropy in geomaterials has been examined mainly through triaxial tests, and has been found to be significantly important in the analysis and design of a variety of geotechnical structures, such as foundations, retaining walls and slopes (Casagrande & Carillo 1944; Arthur & Menzies 1972; Oda *et al.* 1978). Other experimentally observed features, like the onset of shear banding and the influence of the intermediate principal stress, which cannot be properly described by isotropic criteria, have also been extensively examined (Gao *et al.* 2010). The strength of sedimentary rocks was found to be strongly affected by the loading direction. A large number of triaxial compression tests were conducted on orientated samples (Donath 1961; McLamore & Gray 1967; Hoek 1968, 1983; Atwell & Sandford 1974; Lerau *et al.* 1981) and the results generally indicate that the maximum strength is associated with specimens in which the direction of major principal stress is either parallel to or perpendicular to the bedding planes, while the minimum strength has been observed for orientations of beddings between 30° and 60°.

Besides experimental studies, an extensive research effort has also been devoted to modelling the mechanical behaviour of anisotropic rocks. Comprehensive reviews on this topic, examining different approaches, are provided by Kwasniewski (1993) and Duveau *et al.* (1998). One of the first attempts to describe the conditions at failure in anisotropic rocks was the work reported by Pariseau (1968), which was an extension of Hill's criterion (Hill 1950). This was followed by more complex tensorial representations (Boehler & Sawczuk 1977; Nova 1980; Amadei 1983). The application of the latter criteria to practical problems is generally difficult due to a large number of independent material functions and/or parameters that appear in the formulation. A simple and pragmatic approach, which incorporates a scalar anisotropy parameter that is a function of a mixed invariant of the stress and the structure orientation tensor, has been developed by Pietruszczak & Mroz (2000, 2001). This approach was later applied to the modelling of sedimentary rocks (Pietruszczak *et al.* 2002; Lydzba *et al.* 2003; Lade 2007).

In this study, a laboratory experimental programme of uniaxial, triaxial, cyclic and Brazilian tests with monitoring of acoustic emissions (AE) was performed on the Tournemire argillite at the CANMET laboratories in Ottawa, Canada, to investigate the mechanical behaviour of the argillite (Abdi & Evgin 2013; Abdi *et al.* 2015). The results of the triaxial tests were modelled using a mathematical framework that employs the microstructure tensor approach. This paper summarizes the experimental results and presents the numerical description of the anisotropic characteristics of the Tournemire argillite.

Description of samples and testing equipment

The Tournemire argillite contains closely spaced bedding planes dipping 5° towards the north and randomly distributed fractures. The rock samples were retrieved in two sets from the Tournemire URL (Fig. 1) in seven boreholes at different angles (0°, 30°, 45°, 60° and 90°) to the sub-horizontal bedding planes (Fig. 2, five boreholes shown only). The first set of samples was drilled between 24 May and 1 June 2011, and the second set was drilled between 5 and 22 December 2011. Boreholes were drilled using dry air (compressed air) as the drilling fluid to minimize interactions of the rock with the drilling fluid, which allows better preservation of samples for further analyses. The drilling was coupled with an industrial vacuum device with particle filtering to minimize dust. The seasonal changes in relative humidity (RH) and temperatures in 2011 were measured by Hedan *et al.* (2014). The temperature at the Tournemire URL in late May and early June 2011 was about 12°C, while the RH was approximately 70–80%. The temperature at the URL in December 2011 was between 11 and 12°C, while the RH was approximately 45–75%.

A series of 1 m-long cores were obtained using a single-tube drill bit (76 mm outer diameter) coupled with a core lifter (61.7 mm inner diameter). Core logging (lithology, palaeontology and structures) was carried out right after core extraction. Each core was then cut into 20–25 cm-long sections, which were wrapped in aluminium foil, vacuumed then sealed. Finally, each sample was bubble-wrapped and all samples were carefully emplaced within a rigid box, which was sent to the CANMET laboratory in Ottawa, Canada, by plane freight.

Rock samples were received at the laboratory as sent (i.e. sealed to preserve original moisture) less than 1 week after their departure from France. The samples were immediately transferred and stored inside the CANMET Rock Mechanics Laboratory environmental chamber at ambient temperature (22°C) with 85% constant RH. Test specimens were prepared according to the American Society for Testing and Materials (ASTM) standards D3967 (ASTM 2008*a*) and D4543 (ASTM 2008*b*), and then sealed back under vacuum (one test specimen

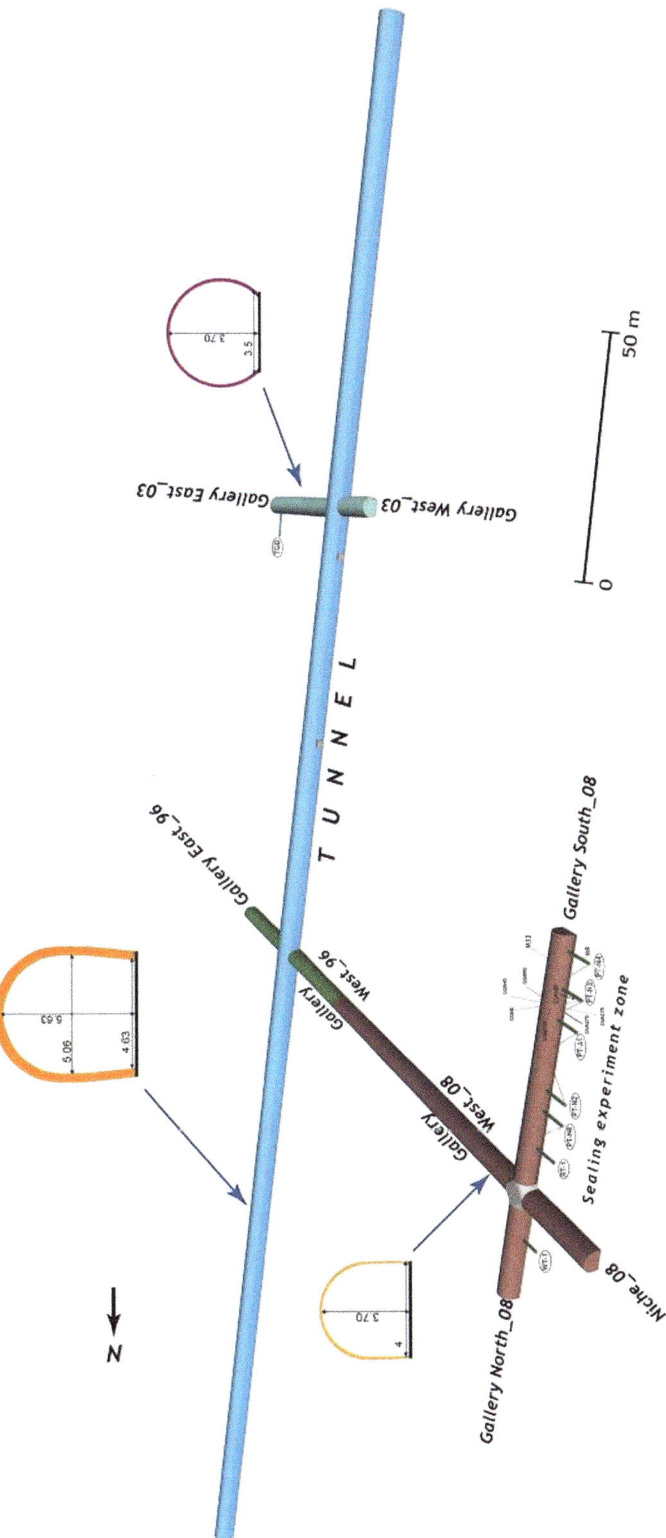

Fig. 1. General layout of the Tournemire URL.

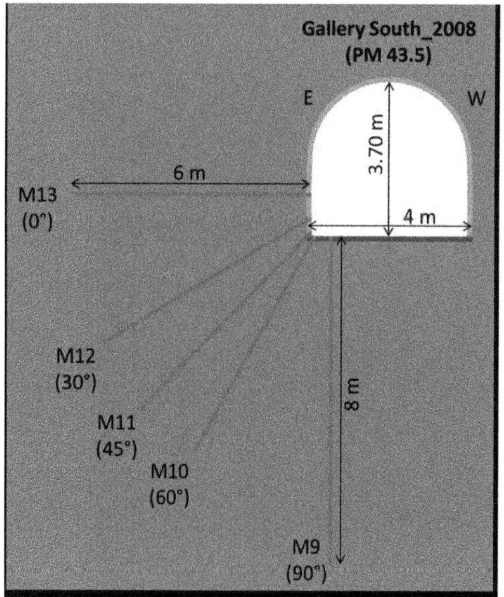

Fig.2. Layout of the sampling boreholes (M9–M13).

per bag) and returned to the environmental chamber. Testing was carried out shortly afterwards according to ASTM standards D3967 (ASTM 2008*a*) and D7012 (ASTM 2007) at the ambient laboratory temperature and humidity. The moisture content and porosity of the Tournemire argillite were measured for four samples before the testing.

The testing specimens were cylindrical samples about 60 mm in diameter and 130 mm in height for uniaxial, triaxial and cyclic tests. Shorter specimens with the same diameter and a nominal length of 40 mm were used for Brazilian tests. The depths of the tested specimens from the gallery boundary are provided in Table 1. The testing equipment consisted of a Mechanical Testing Systems (MTS) rock mechanics system (model 815), with linear transducers to measure the axial displacement, a chain extensometer to measure the circumferential displacement at the centre of the test specimen, transducers to measure porewater pressures and a data acquisition module. A detailed description of the MTS system was reported by Labrie & Conlon (2008). Acoustic emissions were recorded through a measuring system developed by Physical Acoustics Corporation (Princeton Junction, NJ, USA), which was operated simultaneously but independently of the main MTS.

Summary of experimental results

Laboratory investigation of the mechanical behaviour of the Tournemire argillite was performed at the CANMET laboratories in Ottawa, Canada. The experimental programme and the detailed results were presented by Abdi & Evgin (2013) and Abdi *et al.* (2015). Some main facts and experimental results are summarized here for the purpose of numerical characterization of the mechanical behaviour of the Tournemire argillite.

Table 1. *Sample information for the experiments*

Borehole No.	Test type	Confining pressure (KPa)	Sample depth (m)	Borehole orientation (°)
M13	U	0	3.55 ~ 4.55	0
	T	4	3.81 ~ 4.05	
	T	10	4.25 ~ 4.55	
	B	N/A	3.55 ~ 4.25	
M12	U	0	3.55 ~ 4.55	30
	T	4	3.75 ~ 4.04	
	T	10	4.25–4.55	
	B	N/A	3.55–4.05	
M11	U	0	3.75–4.85	45
	T	4	4.10–4.26	
	T	10	4.65–4.85	
	B	N/A	4.05–4.85	
M10	U	0	3.81–4.65	60
	T	4	3.95–4.21	
	T	10	4.35–4.65	
	B	N/A	3.75–4.65	
M8/M9	U	0	5.10–6.85	90
	T	4	5.76–6.18	
	T	10	6.11–6.46	
	B	N/A	4.49–4.95, 7.40	

U, uniaxial test; T, triaxial test; B, Brazillian test.

The test loading pressure was applied at different angles (θ) with respect to the bedding planes (Fig. 3): that is, parallel to the bedding planes ($\theta = 0^\circ$), perpendicular to the bedding planes ($\theta = 90^\circ$) and at three different inclinations ($\theta = 30^\circ$, 45° and 60°). Confining pressures of 0, 4 and 10 MPa were used for uniaxial and triaxial tests based on expected *in situ* stresses of $\sigma_v \approx \sigma_H \approx 4$ MPa. The choice of loading rate was based on the nominal capacity of the test system. A 0.03 mm min^{-1} loading displacement rate was used for all tests.

Basic physical properties of Tournemire argillite

The following properties were determined from laboratory experiments. The moisture content of the rock specimens was 3.86% with a standard deviation (SD) equal to 0.078, the bulk density was 2550 kg m^{-3} (SD = 0.0244), porosity was 9.52% (SD = 0.211) and the anisotropy coefficient of primary (P)-wave velocity was 2.16. The mineralogical composition of the Tournemire argillite was reported by Niandou *et al.* (1997).

Acoustic emission data

The AE technique was used to detect the initiation and propagation of microcracks in rock specimens under mechanical loading. Based on the experimental AE data, new microcracks start to develop at a stress level where the crack initiation (σ_{ci}) is approximately 70–75% of the peak strength. At a stress level above σ_{ci}, the development and propagation of microcracks accelerate with the increase in stress level. In addition, according to the observations made in the experimental data, the crack damage stress level (σ_{cd}) is reached at a stress level where σ_{cd} is approximately 85–90% of the peak strength. The recorded data indicate that the stress levels of damage initiation and growth for the argillite are different to those of hard rocks (Martin 1993).

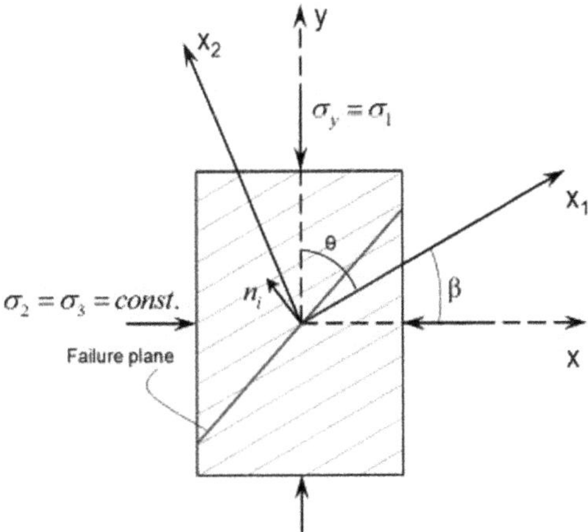

Fig. 3. Loading direction with respect to bedding planes and the geometry of the problem.

Results of Brazilian indirect tensile tests

Brazilian tests were carried out on specimens obtained from different boreholes at depths ranging from 3.55 to 7.4 m (Table 1). The experimental results show that the tensile strength is influenced by the loading orientation angle, θ, and varies between 4.8 MPa ($\theta = 0^\circ$) and 5.9 MPa ($\theta = 90^\circ$). However, the variation of tensile strength with a loading orientation of between 30° and 60° is small (i.e. between 5.2 and 4.9 MPa).

Results of uniaxial and triaxial tests

Uniaxial and triaxial tests were carried out on specimens obtained from depths ranging from 3.55 to 6.85 m (Table 1). Typical stress–strain curves are shown in Figures 4 and 5 for all tested loading orientations and confining pressures. Table 2 presents the experimental results of peak strength, and the axial and volumetric strains at peak strength for different values of loading orientation angles.

The experimental results show that both peak strength and axial strain depend on θ and the confining pressure. The maximum strength was obtained in tests performed at $\theta = 0^\circ$ (i.e. for loading applied parallel to bedding). The strength obtained in tests performed at $30^\circ \leq \theta \leq 60^\circ$ exhibits the lowest values. This indicates that the compressive strength is highly sensitive to the load inclination. The ratio between the principal stress difference at peak strength when $\theta = 0^\circ$ and the principal stress difference at peak strength when $\theta = 90^\circ$ increases slightly with the increase in confining pressure, but its value is nearly equal to unity. This indicates that the strength anisotropy between the two principal axes, $\theta = 0^\circ$ and 90°, is small. The ratio between the maximum and minimum peak strengths for the same confining pressure is nearly equal to 2, and decreases with increase in confining pressure, which indicates that the strength anisotropy becomes smaller for higher confining pressures.

The axial strain at peak strength increases with increase in confining pressure. Small values of θ ($<30^\circ$) have no significant effect on the values of the axial strain at peak strength. However, the axial strain at peak strength increases significantly with larger values of θ ($>30^\circ$). These experimental results indicate that the compressibility of the rock in the direction perpendicular to bedding is higher than the compressibility of the rock along the

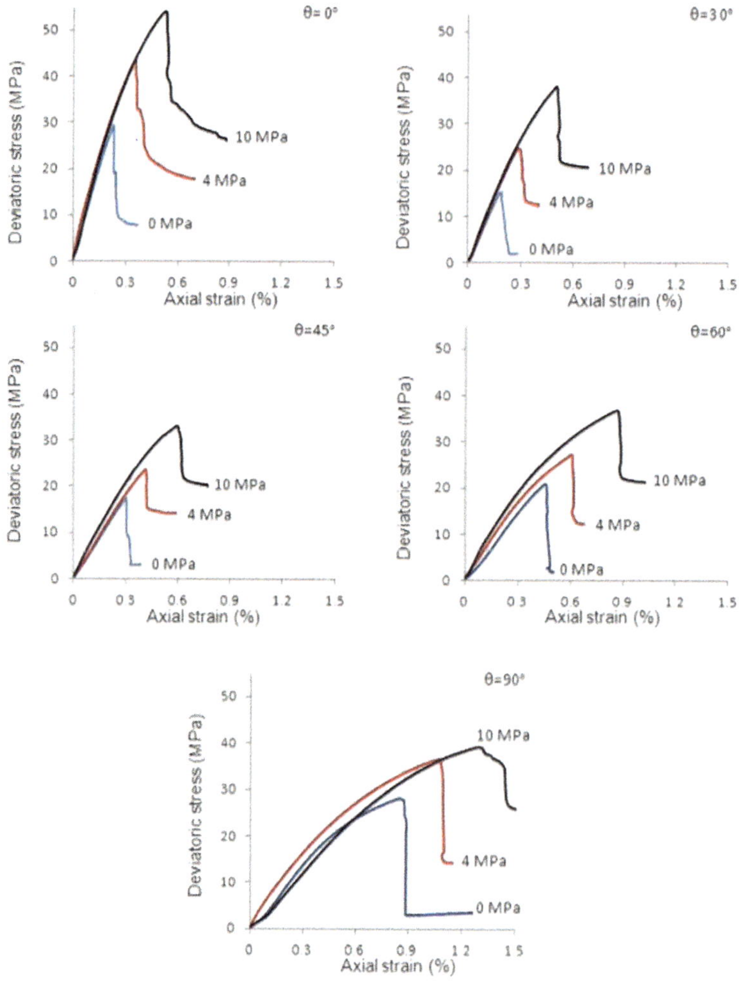

Fig. 4. Deviatoric stress v. axial strain curves at various loading orientations and confining pressures.

bedding. The experimental data also indicate that Young's modulus, E, decreases with increase in θ. The maximum Young's modulus value was obtained at tests performed at $\theta = 0°$; the Young's modulus value was lowest at $\theta = 90°$.

The failure of rock specimens occurs in a brittle manner with a sudden collapse of the material strength after peaking, as shown in Figures 4 and 5. Fractures observed on failed samples indicate that failure occurs in three principal modes: extension (splitting), shearing, and a combination of extension and shearing. The failure mode is a function of the loading orientation, θ, as shown in Figure 6. The range of confining pressures used in this investigation has little effect on the failure mode.

Numerical simulations of triaxial test results

The experimental results of triaxial tests were used for numerical investigation of the anisotropic mechanical behaviour of the Tournemire argillite. The constitutive relationship was developed to describe the stress–strain behaviour of the argillite. In the simulation, we used an elastoplastic framework. In the elastic range, Hooke's law for transversely isotropic materials was used with five parameters obtained from the experimental results to account for the bedding. In the plastic range, a microstructure tensor approach (Pietruszczak & Mroz 2001) was adopted in order to take the anisotropic behaviour of the argillite into account. The

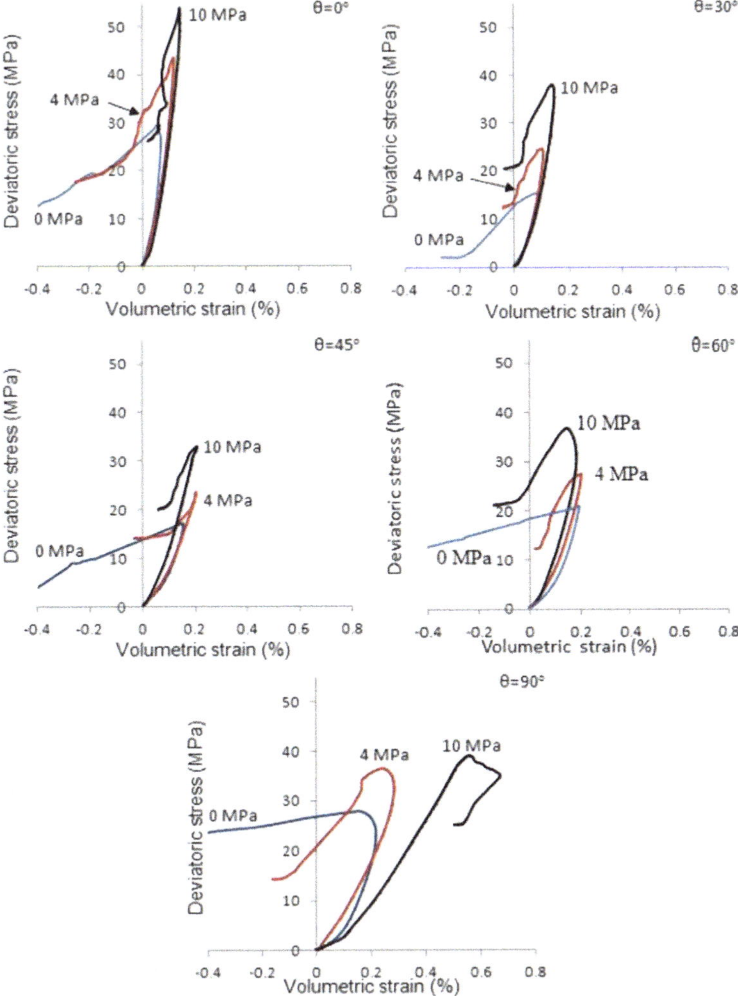

Fig. 5. Deviatoric stress v. volumetric strain curves at various loading orientations and confining pressures.

material functions/parameters and results of the numerical simulation are presented.

Constitutive relationship based on the microstructure tensor

In order to define the anisotropy parameter(s), the formulation employs a generalized loading vector that is defined as:

$$L_i = L_j e_i^{(j)}, L_j = (\sigma_{j1}^2 + \sigma_{j2}^2 + \sigma_{j3}^2)^{1/2} \; (i, j = 1, 2, 3), \tag{1}$$

where $e_i^{(j)}$, $j = 1, 2, 3$, are the base vectors that specify the preferred material axes. Thus, the components of L_i represent the magnitudes of traction vectors on the planes normal to the principal material axes. Note that:

$$L_i^2 = e_k^{(i)} \sigma_{kj} e_l^{(i)} \sigma_{lj} = \mathrm{tr}\left(m_{kp}^{(i)} \sigma_{ql} \sigma_{lk}\right)$$
$$L_k L_k = \sigma_{kj} \sigma_{kj} = \mathrm{tr}(\sigma_{kl} \sigma_{lj}), \tag{2}$$

so that the traction moduli can be expressed as mixed invariants of the stress and microstructure-orientation tensors. The unit vector along L_i is given by:

$$l_i = \frac{L_i}{(L_k L_k)^{1/2}} = \left[\frac{(e_k^{(i)} \sigma_{kj} e_m^{(i)} \sigma_{mj})}{\sigma_{pq} \sigma_{pq}}\right]^{1/2}. \tag{3}$$

A microstructure tensor a_{ij} is now introduced that is a measure of material fabric. While different

Table 2. *Experimental results at the peak strength for different loading orientations*

Loading orientation angle (°)	Confining pressure (MPa)	$(\sigma_1-\sigma_3)$ at the peak (MPa)	Axial strain at peak strength (%)	Volumetric strain at peak strength (%)
0	0	29.17	0.229	0.057
	4	43.53	0.353	0.119
	10	53.90	0.539	0.146
30	0	15.27	0.187	0.075
	4	24.67	0.288	0.105
	10	38.00	0.501	0.138
45	0	17.24	0.301	0.150
	4	23.61	0.415	0.202
	10	33.13	0.592	0.206
60	0	20.89	0.450	0.195
	4	27.40	0.604	0.202
	10	36.80	0.849	0.145
90	0	28.13	0.849	0.152
	4	36.38	1.060	0.246
	10	39.00	1.300	0.558

descriptors may be employed to quantify the fabric, the eigenvectors of this operator are said to be collinear with $e_i^{(a)}$. The projection of the microstructure tensor on l_i becomes:

$$\eta = a_{ij} l_i l_j = \frac{a_{ik}\sigma_{ij}\sigma_{kj}}{\sigma_{pq}\sigma_{pq}}. \qquad (4)$$

Here, the scalar variable η, referred to as anisotropy parameter, specifies the effect of load orientation relative to material axes, and is defined as the ratio of the joint invariant of stress and microstructure tensor $a_{ij}\sigma_{ik}\sigma_{jk}$ to the stress invariant $\sigma_{ij}\sigma_{ij}$. It is a homogeneous function of stress of the degree zero, so that the stress magnitude does not affect its value. Note that equation (4) can be expressed as:

$$\eta = \eta_0(1 + A_{ij} l_i l_j) \quad A_{ij} = \frac{1}{\eta_0} a_{ij} - \delta_{ij} \quad \eta_0 = \frac{1}{3} a_{kk}, \qquad (5)$$

where $A_{ij} = dev(a_{ij})/\eta_0$ is a symmetrical traceless operator. Equation (5) may be generalized by considering higher-order tensors, i.e.:

$$\eta = \eta_0(1 + A_{ij} l_i l_j + A_{ijkl} l_i l_j l_k l_l + \cdots). \qquad (6)$$

The above representation is rather complex in terms of implementation and/or identification. Therefore, it is convenient to use a simplified functional form obtained by replacing the higher-order tensors by dyadic products of second-order tensors, i.e.:

$$\eta = \eta_0\left(1 + A_{ij} l_i l_j + b_1(A_{ij} l_i l_j)^2 + b_2(A_{ij} l_i l_j)^3 + \cdots\right), \qquad (7)$$

where b_s $(s = 1, 2, \ldots)$ are constants.

In view of the consideration above, the failure function can be expressed in the general form:

$$F = F(\sigma_{ij}, a_{ij}) = F(I_1, J_2, J_3, \eta), \qquad (8)$$

where I_1 is the first stress invariant, while J_2, J_3 are the basic invariants of the stress deviator. As mentioned earlier, the parameter η is typically identified with a relevant strength descriptor, the value of which is then assumed to depend on the orientation of the sample relative to the direction of loading. Thus, the existing criteria can be easily extended to an anisotropic regime by assuming that the strength parameters vary according to equation (7). In this work, a linear form of equation (8) has been adopted, which corresponds to the well-known Mohr–Coulomb criterion, i.e.:

$$F = \sqrt{3}\bar{\sigma} - \eta_f g(\theta)(\sigma_m + C) = 0. \qquad (9)$$

Here:

$$\bar{\sigma} = (J_2)^{1/2} \quad \sigma_m = -\frac{1}{3} I_1 \quad \theta = \frac{1}{3}\sin^{-1}\left(\frac{-3\sqrt{3}}{2}\frac{J_3}{\bar{\sigma}^3}\right)$$

and

$$g(\theta) = \frac{3 - \sin\phi}{2\sqrt{3}\cos\theta - 2\sin\theta\sin\phi}$$

$$\eta_f = \frac{6\sin\phi}{3 - \sin\phi} \qquad C = c\cot\phi, \qquad (10)$$

where θ is the Lode's angle, and ϕ and c are the angle of friction and cohesion, respectively.

An extension to the case of inherent anisotropy can be accomplished by assuming that the strength

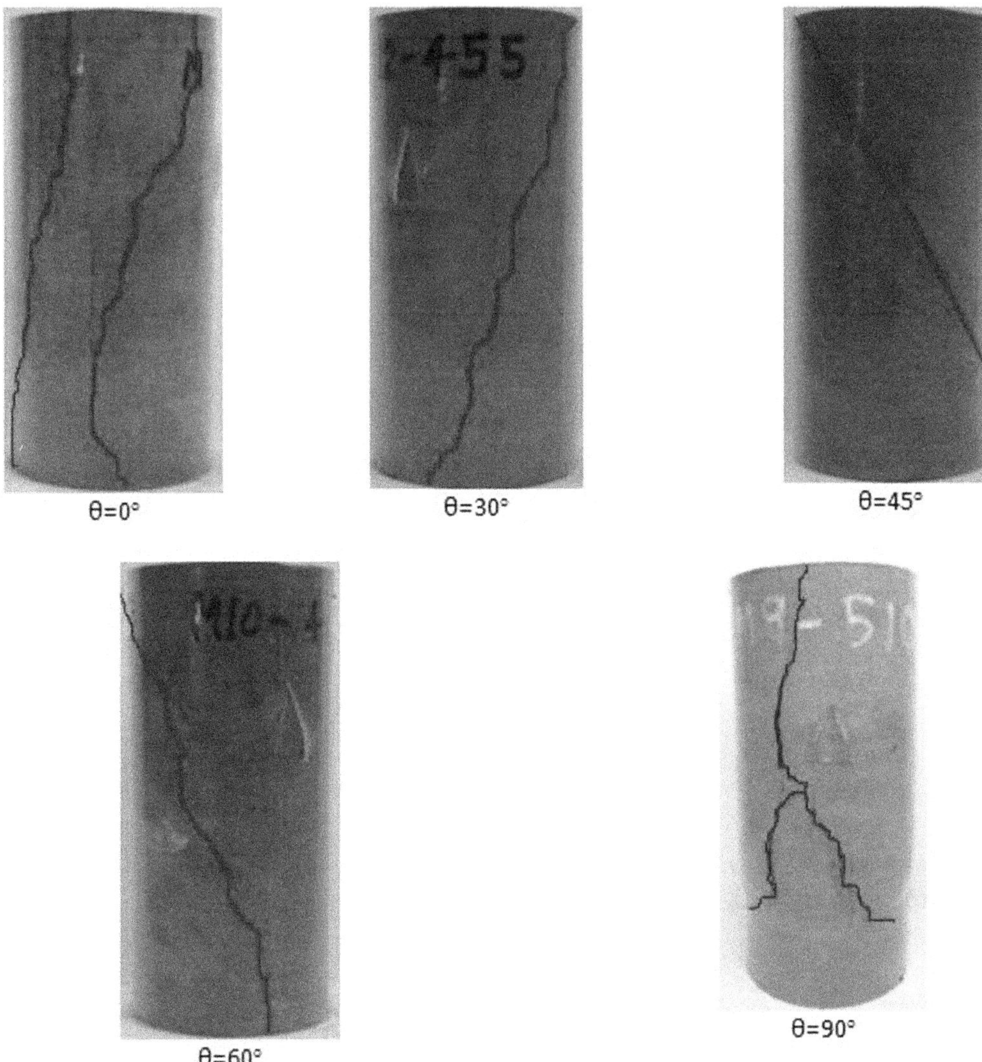

Fig. 6. Failure modes of the Tournemire argillite under different loading directions to the bedding planes.

descriptors, in this case η_f and C, are orientation-dependent and have a representation analogous to that of equation (7). Note that, in the context of the equation (9), the parameter C is associated with a hydrostatic stress state. The latter is, in fact, invariant with respect to the orientation of the sample. Thus, the effects of anisotropy can be primarily attributed to a variation in the strength parameter η_f, i.e.:

$$\eta_f = \hat{\eta}_f\left(1 + A_{ij}l_il_j + b_1(A_{ij}l_il_j)^2 + b_2(A_{ij}l_il_j)^3 + \cdots\right) \quad C = \text{const.} \tag{11}$$

The general plasticity formulation can be derived by assuming that the yield/loading surface in the form is consistent with equation (11), i.e.:

$$f = \sqrt{3}\bar{\sigma} - \vartheta g(\theta)(\sigma_m + C) = 0$$

$$\vartheta = \vartheta(\kappa) = \eta_f \frac{\zeta\kappa}{A + \kappa} \quad d\kappa = \left(\frac{2}{3} de_{ij}^{\mathrm{p}}\, de_{ij}^{\mathrm{p}}\right)^{1/2}, \tag{12}$$

where de_{ij}^{p} is the deviatoric part of the plastic strain increment, A and ζ are the material parameters, and κ is the plastic distortion. According to the hardening rule, for $\kappa \to \infty$, there is $\vartheta \to \zeta\eta_f$, where $\zeta > 1$. The parameter ζ is introduced here in order to define the transition to localized deformation, which is assumed to occur at $\vartheta = \eta_f$. Note that the latter equality implies that $f = F$, so that the conditions at failure are consistent with the Mohr–Coulomb criterion (equation 9). The plastic potential can be chosen as:

$$\psi = \sqrt{3}\bar{\sigma} + \eta_c g(\theta)(\sigma_m + C)\ln\frac{(\sigma_m + C)}{\sigma_m^0} = 0, \quad (13)$$

where $\eta_c \propto \eta_f$, so that $\eta_c = \eta_c(l_i)$. Following now the standard plasticity procedure (Pietruszczak 2010) – i.e. invoking the consistency condition $df = 0$ – the constitutive relationship can be expressed as:

$$d\varepsilon_{ij} = C^e_{ijkl}d\sigma_{kl} + d\varepsilon^p_{ij} = \left(C^e_{ijkl} + H_p^{-1}\frac{\partial f}{\partial\sigma_{ij}}\frac{\partial\psi}{\partial\sigma_{kl}}\right)d\sigma_{kl}$$
$$= C_{ijkl}d\sigma_{kl}, \quad (14)$$

where:

$$H_p = -\sqrt{\frac{2}{3}}\frac{\partial f}{\partial\vartheta}\frac{\partial\vartheta}{\partial\kappa}\left(\text{dev}\frac{\partial\psi}{\partial\sigma_{ij}}\text{dev}\frac{\partial\psi}{\partial\sigma_{ij}}\right)^{1/2}$$

where C^e_{ijkl} is the elastic compliance, the representation of which, once again, depends on the type of material anisotropy.

Procedure for identification of material functions/parameters

The identification process employed here is based on the experimental data of a series of triaxial compression tests presented in the previous sections. The first step involves the specification of the conditions at failure which, in turn, requires the identification of material function η_f and the parameter C, as defined in equation (11). The specification of coefficients appearing in $\eta_f = \eta_f(l_i)$ entails the information on conditions at failure in samples tested at arbitrary orientation β ($\beta = 1 - \theta$) with respect to direction of loading. Consider, for this purpose, the response of the sample under axial compression at a confining pressure p_0. Refer the geometry of the sample to the coordinate system in which $x_i = \{x_1, x_2, x_3\}$ are the material axes (see Fig. 3). In this case,

$$A_{ij}l_il_j = A_1(1 - 3l_2^2) \quad l_2^2 = \frac{p_0^2\sin^2\beta + \sigma_1^2\cos^2\beta}{2p_0^2 + \sigma_1^2}, \quad (15)$$

where β defines the orientation of the bedding planes. Given the representation above, the function $\eta_f = \eta_f(l_i)$ (equation 11) reduces to:

$$\eta_f = \hat{\eta}_f\Big(1 + A_1(1 - 3l_2^2) + b_1A_1^2(1 - 3l_2^2)^2$$
$$+ b_2A_1^3(1 - 3l_2^2)^3 + b_3A_1^4(1 - 3l_2^2)^4 + \cdots\Big). \quad (16)$$

Note that in the absence of confinement, $p_0 = 0$, there is $l_2 = \cos\beta$, so that the anisotropy parameter is an explicit function of the deposition angle β, i.e.:

$$l_2^2 = \cos^2\beta \Rightarrow \eta_f = \hat{\eta}_f\big(1 + A_1(1 - 3\cos^2\beta)$$
$$+ b_1A_1^2(1 - 3\cos^2\beta)^2 + \cdots\big). \quad (17)$$

Given now the stress parameters $\{p, q\}$ at failure, for each specific orientation of the sample, the anisotropy parameter can be determined – i.e. $\eta_f - q/(p + C)$ – together with the corresponding value of l_2 (equation 15). The results can then be plotted in the affine space $\{\eta_f, l_2\}$ to generate a set of data describing equation (15). It should be noted that the above procedure requires an estimate of the value of the constant C, which can also be obtained based on the results of standard triaxial tests.

The specification of strength parameters for Tournemire argillite employs the results plotted in Figures 7–10. Figure 7 shows a linear Mohr–Coulomb approximation of the conditions at failure in samples tested at different orientations of the bedding planes (experimental data points are shown as symbols). Based on these results, the value of C was estimated to be $C = 10.6$ MPa, which represents an orientation-average for tests conducted at $\beta = 0°$, 30°, 45°, 60° and 90°. Given the value of the constant C, the spatial distribution of the anisotropy parameter $\eta_f = \eta_f(l_i)$ was established by following the procedure described above.

Figure 8 shows the experimental results projected in the affine space $\{\eta_f, (1 - 3l_2^2)\}$. Again, the data incorporate the tests conducted at different confining pressures, viz. 0, 4 and 10 MPa, and different orientations of the bedding planes. The best-fit approximation is based on equation (16). In this case, the second-order approximation appears to be adequate and corresponds to the following set of coefficients: $\hat{\eta}_f = 1.0725$, $A_1 = 0.17034$ and $b_1 = 5.4957$.

Figure 9 shows the plot of η_f against the loading angle $\alpha = \cos^{-1} l_2$, which incorporates the parameters given above.

The next step of the identification process pertains to specification of the material parameters and functions that govern the deformation characteristics. The elastic properties were obtained

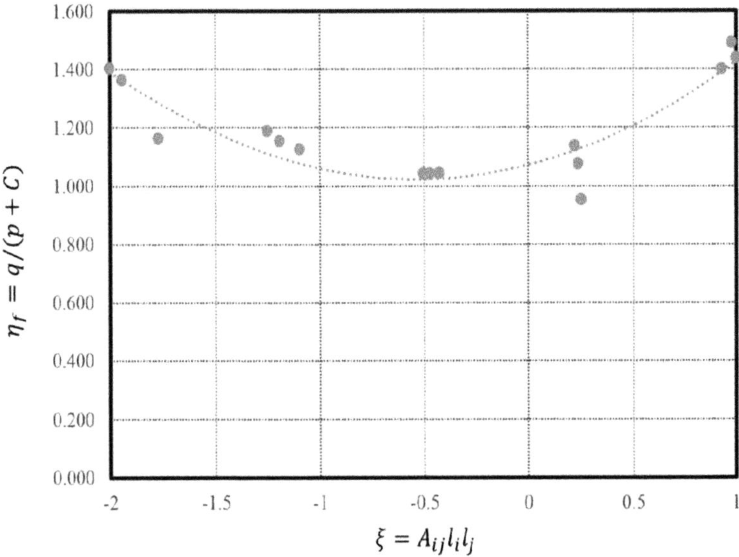

Fig. 7. Approximation of conditions at failure for different sample orientations.

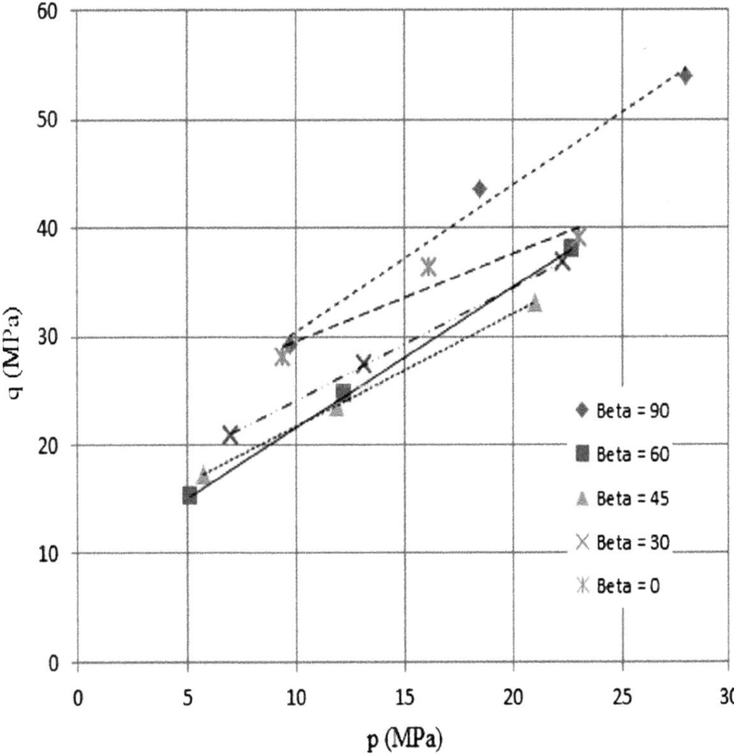

Fig. 8. Variation of the strength parameter η_f with $\xi = A_{ij}l_il_j$.

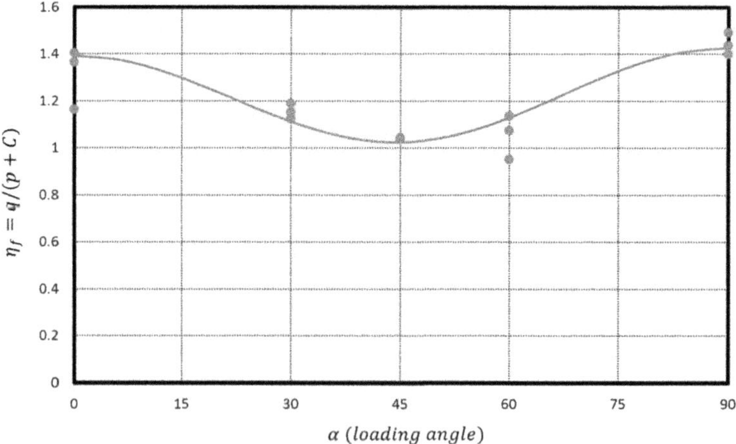

Fig. 9. Variation of the strength parameter η_f with loading angle (α).

based on the results of unconfined cyclic tests as follows: $E_1 = 12.5$ GPa, $E_2 = 21.0$ GPa, $\nu_{21} = 0.16$, $\nu_{13} = 0.08$ and $G_{21} = 4.57$.

Again, these values are referred to the coordinate system associated with principal material axes, as shown in Figure 3.

The evaluation of the plastic properties requires the identification of the potential function, equation (13), which incorporates the parameter η_c. The latter defines the transition from plastic contraction to dilatancy and its value can be assessed by examining the respective volume change characteristics. For the Tournemire argillite, the results of experimental tests as described previously indicate that the transition to dilatancy occurs at deviatoric stress intensities that approach those associated with unstable strain-softening responses. Therefore, it may be adequate to approximate this material function as: $\eta_c(l_i) \approx 0.99\eta_f(l_i)$.

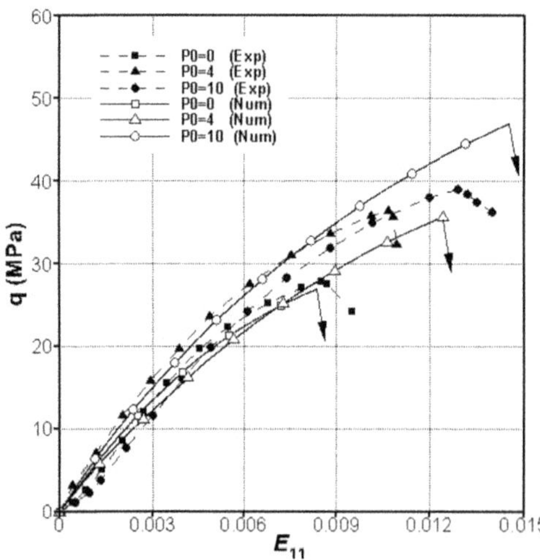

Fig. 10. Trial-and-error assessment of the value of parameter A; the results correspond to $A = 0.0012$.

The specification of the hardening function involves the identification of the material parameter A, namely equation (12). In order to accomplish this, the mechanical characteristics should be replotted in $\{\kappa, \eta = q/(p+C)\}$ space. Note that constructing such characteristics requires the value of the elastic shear modulus so that the plastic deviatoric strain can be determined via the additivity postulate. The deformation properties in the plastic range are likely to be direction-dependent. Thus, the best-fit procedure should be repeated for different orientations of the sample in order to obtain a set of corresponding values of the parameter A. In order to maintain the simplicity of the formulation, it may then be sufficiently accurate to assume that A represents a material constant, the value of which is identified with the orientation average.

It needs to be pointed out here that the experimental results provided are not exactly adequate to follow the identification procedure. This stems from the fact that the deformation measurements comprise only the vertical and volumetric strain, which for samples tested at orientations other than $\beta = 0°$ (i.e. horizontal bedding planes) is not sufficient to uniquely define the value of the plastic distortion κ. Note that at $\beta = 90°$, the lateral strain will not be the same in the radial directions. The deformation in different radial directions is, however, not measured during the laboratory tests. At the same time, for inclined samples, the shear strain will develop under a stress-controlled regime, as the material is anisotropic.

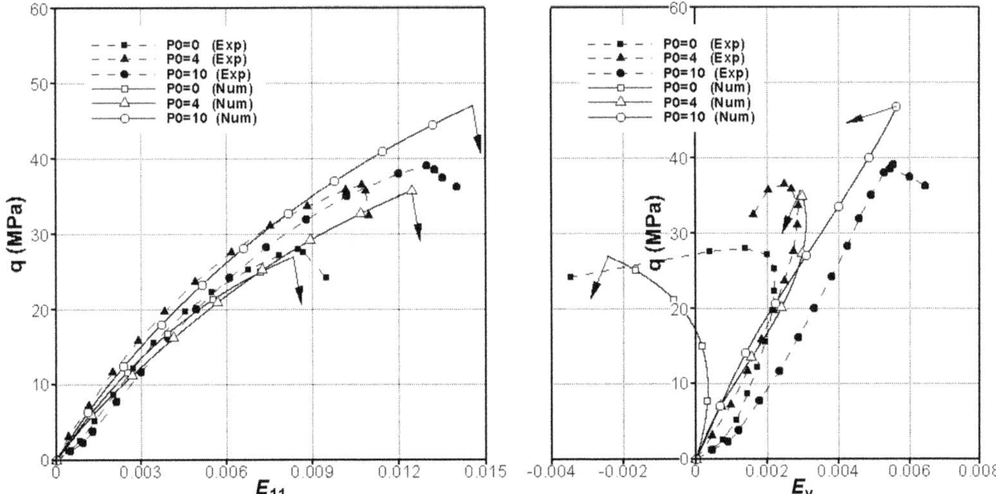

Fig. 11. Numerical simulations of triaxial compression tests: deviatoric stress (**q**) v. axial strain (E_{11}) and volumetric strain (E_v) corresponding to $\beta = 0°$.

Given the restrictions mentioned above, parameter A was determined by a trial-and-error procedure that involved fitting the deviatoric characteristics q v. axial strain for samples with horizontal bedding planes. Figure 10 shows the plots corresponding to the confining pressures of 0, 4 and 10 MPa. It is important to point out that the unstable strain-softening characteristics reported in experimental tests are associated with localized deformation: that is, the formation of macrocracks. Therefore, they do not represent a material response and should be analysed within the context of a boundary-value problem. Thus, the characteristics reported in Figure 10 are restricted to a stable range associated with strain hardening. The experimental response is characterized by an abrupt transition to a localized mode. Within the current mathematical framework, the onset of localized deformation is associated with $\vartheta = \eta_f$ and is defined by means of parameter ζ, which appears in the hardening function (equation 12). In general, the constraint $\zeta > 1$ ensures that the transition occurs along the ascending branch

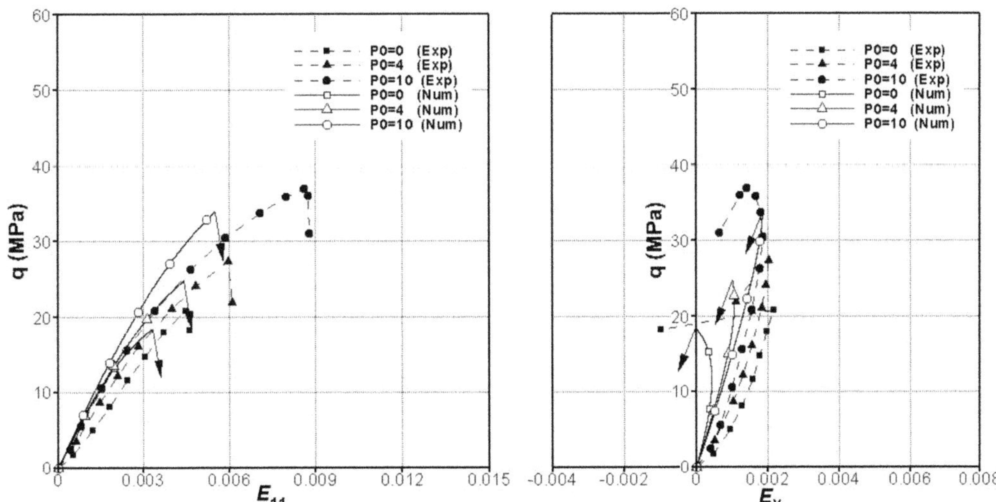

Fig. 12. Numerical simulations of triaxial compression tests: deviatoric stress (**q**) v. axial strain (E_{11}) and volumetric strain (E_v) corresponding to $\beta = 30°$.

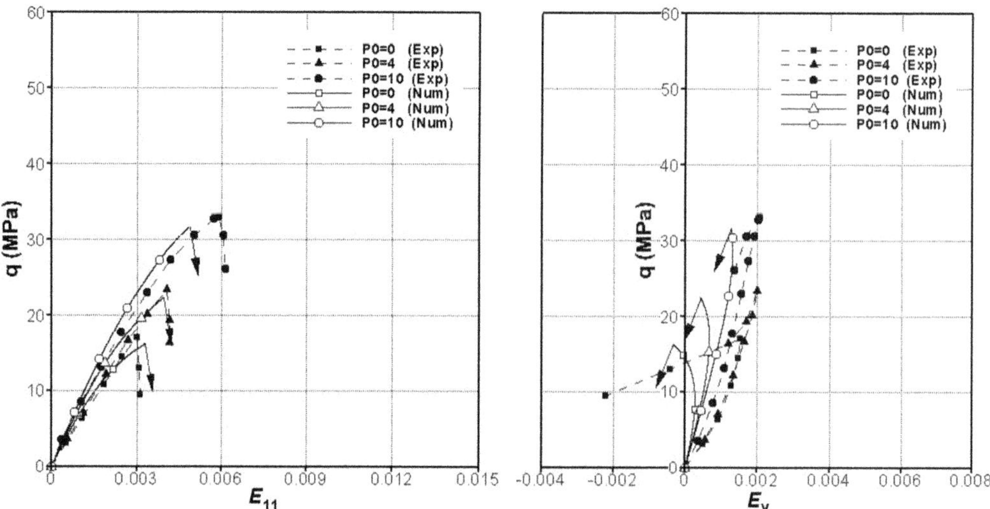

Fig. 13. Numerical simulations of triaxial compression tests: deviatoric stress (**q**) v. axial strain (E_{11}) and volumetric strain (E_v) corresponding to $\beta = 45°$.

of the deviatoric characteristic. The simulations reported in Figure 10 were conducted assuming $\zeta = 1.25$, and resulted in the value of $A = 0.0012$. Thus, the material parameters governing the plastic response of the Tournemire argillite were taken as: $\eta_c \approx 0.99\eta_f(l_i)$, $\zeta = 1.25$ and $A = 0.0012$.

Finally, it should be pointed out that the results of the triaxial tests for inclined samples need to be taken with caution. This stems from the fact that the samples will have tendency to distort under the applied load: however, in a triaxial set-up, such distortion is kinematically constrained by the presence of the loading platens. The latter will lead to a non-uniform stress distribution within the specimen, so that the results are representative of a boundary-value problem rather than a material test. This is, in fact, the main argument why, at the present stage, a simplified procedure for the identification of material parameters in the plastic range has been adopted.

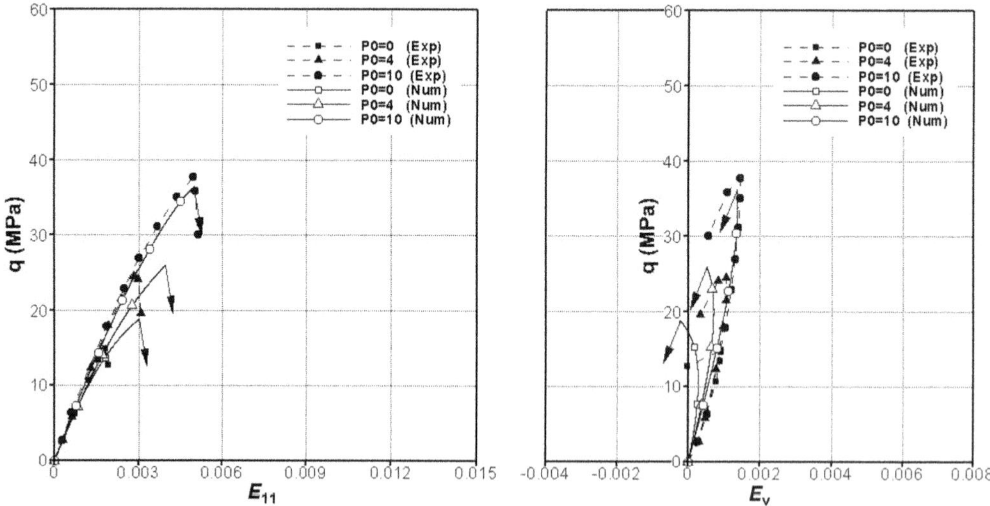

Fig. 14. Numerical simulations of triaxial compression tests: deviatoric stress (**q**) v. axial strain (E_{11}) and volumetric strain (E_v) corresponding to $\beta = 60°$.

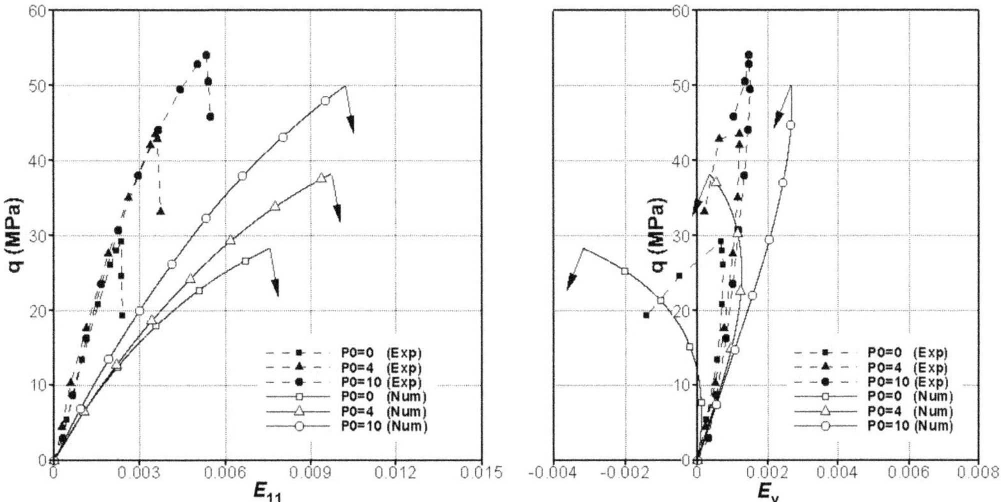

Fig. 15. Numerical simulations of triaxial compression tests: deviatoric stress (**q**) v. axial strain (E_{11}) and volumetric strain (E_v) corresponding to $\beta = 90°$.

Numerical simulation

The results of numerical simulations with the finite-element code Abaqus are shown in Figures 11–15. The figures present both the deviatoric characteristics q v. axial strain (E_{11}), as well as the corresponding evolution of volumetric strain (E_v). It is again noted that the simulations were terminated at the instant the criterion for the onset of localization, defined as $\vartheta = \eta_f$, was satisfied. As mentioned earlier, the subsequent unstable response (indicated by an arrow pointing downwards) constitutes a boundary-value problem and cannot be examined within the context of the point-integration scheme. It appears that, in spite of simplifications embedded in the identification procedure, the results of simulations are quite reasonable in terms of depicting the basic trends. This is certainly the case given the doubts concerning the validity of experimental data for samples tested at orientations other than those involving the horizontal bedding planes.

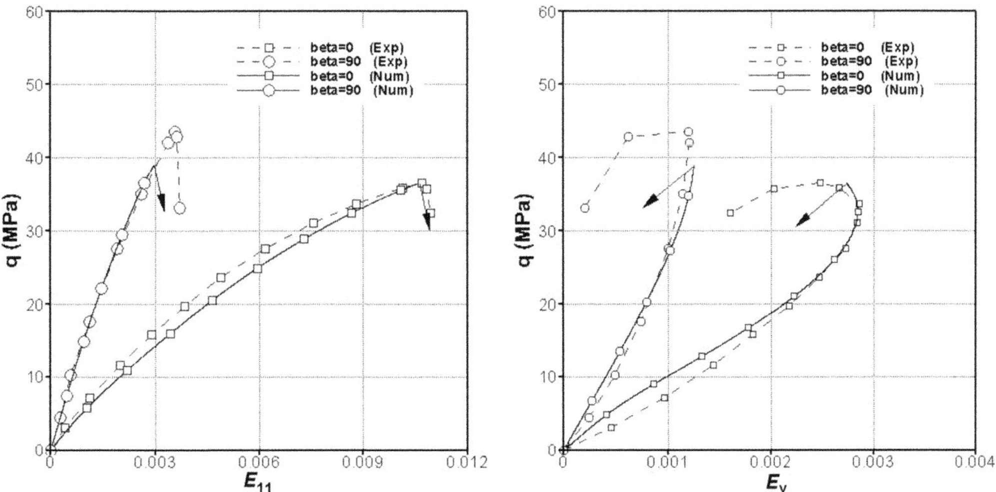

Fig. 16. Numerical simulations for $\beta = 0°$ and $\beta = 90°$, respectively, corresponding to $A = 0.0002$ (confining pressure of 4 MPa).

The most significant discrepancy between the experimental data and the numerical simulations occurs at $\beta = 90^\circ$ (Fig. 15). Here, the conditions at failure are predicted quite accurately. However, the stiffness characteristics are markedly different. The reason for this discrepancy is probably because the strain-hardening effect before failure for $\beta = 90^\circ$ (i.e. the axial loading is parallel to the bedding plane) is stronger than that for other β angles, as shown in the experimental results. Based on the hardening function represented by equation (12), a smaller value of A would reflect a stronger hardening effect, which means that parameter A might also be orientation-dependent. However, this hypothesis would need more experimental and numerical work to support it. Therefore, the predictions can be improved by accounting for the fact that the hardening characteristics are likely to be influenced by the orientation of the sample. This is illustrated in Figure 16, which shows the results for the simulations corresponding to a value of $A = 0.0002$. It is evident that a significant improvement in terms of predictive abilities is achieved.

Conclusion

The Tournemire argillite is an anisotropic material, and its strength and deformation behaviour are direction-dependent. Stress–strain relationships are non-linear, and failure occurs in a brittle manner with a sudden collapse of the material. The failure mode exhibits localization along distinct failure planes and also depends on the loading orientation.

The identification procedure for material functions and parameters was outlined, and the model was applied to simulate the set of triaxial tests performed at different levels of confining pressure and orientation of the bedding planes. It is demonstrated that the model adequately reproduces the anisotropy, the pre-peak stress–strain response, and the onset of material collapse in those tests.

References

Abdi, H. & Evgin, E. 2013. *Laboratory Characterization, Modeling, and Numerical Simulation of an Excavation Damaged Zone Around Deep Geological Repositories in Sedimentary Rocks*. Research Report RSP-0287. Canadian Nuclear Safety Commission, Ottawa, Canada.

Abdi, H., Labrie, D. *et al.* 2015. Laboratory investigation on the mechanical behaviour of Tournemire argillite. *Canadian Geotechnical Journal*, **52**, 268–282.

Amadei, B. 1983. *Rock Anisotropy and the Theory of Stress Measurements*. Springer, Berlin.

Arthur, J.R.F. & Menzies, B.K. 1972. Inherent anisotropy in sand. *Géotechnique*, **22**, 115–128.

ASTM 2007. *Standard Test Method for Compressive Strength and Elastic Moduli of Intact Rock Core Specimens Under Varying States of Stress and Temperatures*. ASTM Standard D7012-07. ASTM International, West Conshohocken, PA, USA.

ASTM 2008*a*. *Standard Test Method for Splitting Tensile Strength of Intact Rock Core Specimens*. ASTM Standard D3967-08. ASTM International, West Conshohocken, PA, USA.

ASTM 2008*b*. *Standard Practices for Preparing Rock Core as Cylindrical Test Specimens and Verifying Conformance to Dimensional and Shape Tolerances*. ASTM Standard D4543-08. ASTM International, West Conshohocken, PA, USA.

Atwell, P.B. & Sandford, M.A. 1974. Intrinsic shear strength of a brittle anisotropic rock – I: experimental and mechanical interpretation. *International Journal of Rock Mechanics and Mining Sciences & Geomechanics Abstracts*, **11**, 423–430.

Blümling, P., Bernier, F., Lebon, P. & Martin, C.D. 2007. The excavation damaged zone in clay formations time-dependent behaviour and influence on performance assessment. *Physics and Chemistry of the Earth, Parts A/B/C*, **32**, 588–599.

Boehler, J.P. & Sawczuk, A. 1977. On yielding of oriented solids. *Acta Mechanica*, **27**, 185–204, https://doi.org/10.1007/BF01180085

Boisson, J.Y., Bertrand, L., Heitz, J.F. & Moreau-Le Golvan, Y. 2001. In situ and laboratory investigations of fluid flow through an argillaceous formation at different scales of space and time, Tournemire tunnel, southern France. *Hydrogeology Journal*, **9**, 108–123.

Casagrande, A. & Carillo, N. 1944. Shear failure of anisotropic materials. *Journal of the Boston Society of Civil Engineers*, **31**, 74–87.

Donath, F.A. 1961. Experimental study of shear failure in anisotropic rocks. *Geological Society of America Bulletin*, **72**, 985–990.

Duveau, G., Shao, J.F. & Henry, J.P. 1998. Assessment of some failure criteria for strongly anisotropic materials. *Mechanics of Cohesive Frictional Materials*, **3**, 1–26.

Gao, Z., Zhao, J. & Yao, Y. 2010. A generalized anisotropic failure criterion for geomaterials. *International Journal of Solids and Structures*, **47**, 3166–3185.

Hedan, S., Fauchille, A.L., Valle, V., Cabrera, J. & Cosenza, P. 2014. One-year monitoring of dessication cracks in Tournemire argillite using digital image correlation. *International Journal of Rock Mechanics & Mining Sciences*, **68**, 22–35.

Hill, R. 1950. *The Mathematical Theory of Plasticity*. Oxford University Press, Oxford.

Hoek, E. 1968. Brittle failure of rock. *In*: Stagg, G. & Zienkiewicz, O.C. (eds) *Rock Mechanics in Engineering Practice*. Wiley, London, 99–124.

Hoek, E. 1983. Strength of jointed rock masses. *Géotechnique*, **33**, 187–205.

Kwasniewski, M.A. 1993. Mechanical behaviour of anisotropic rocks. *In*: Hudson, J.A. (ed.) *Comprehensive Rock Engineering*, **1**. Pergamon, Oxford, 285–312.

Labiouse, V. & Vietor, T. 2014. Laboratory and *in situ* simulation tests of the excavation damaged zone around galleries in Opalinus Clay. *Rock Mechanics and Rock Engineering*, **47**, 57–70.

LABRIE, D. & CONLON, B. 2008 Hydraulic and poroelastic properties of porous rocks and concrete materials. Paper presented at the The 42nd US Rock Mechanics Symposium (USRMS), 29 June–2 July, San Francisco, California. American Rock Mechanics Association, Alexandria, VA, USA.

LADE, P.V. 2007. Modeling failure in cross-anisotropic frictional materials. *International Journal of Solids and Structures*, **44**, 5146–5162, https://doi.org/10.1016/j.ijsolstr.2006.12.027

LERAU, J., SAINT LEU, C. & SIRIEYS, P. 1981. Anisotrophie de la dilatance des roches schisteuses. *Rock Mechanics and Rock Engineering*, **13**, 185–196.

LYDZBA, D., PIETRUSZCZAK, S. & SHAO, J.F. 2003. On anisotropy of stratified rocks: homogenization and fabric tensor approach. *Computers and Geotechnics*, **30**, 289–302, https://doi.org/10.1016/S0266-352X(03)00004-1

MARTIN, C.D. 1993. *The Strength of Massive Lac du Bonnet Granite Around Underground Openings*. PhD Thesis, University of Manitoba, Winnipeg, Manitoba.

MCLAMORE, R. & GRAY, K.E. 1967. The mechanical behaviour of anisotropic sedimentary rocks. *Journal of Manufacturing Science and Engineering, ASME*, **89**, 62–73.

MILLARD, A., MABMANN, J., REJEB, A. & UEHARA, S. 2009. Study of the initiation and propagation of excavation damaged zones around openings in argillaceous rock. *Environmental Geology*, **57**, 1325–1335.

NIANDOU, H., SHAO, J., HENRY, J. & FOURMAINTRAUX, D. 1997. Laboratory investigation of the mechanical behaviour of Tournemire argillite. *International Journal of Rock Mechanics and Mining Sciences & Geomechanics Abstracts*, **34**, 3–16.

NOVA, R. 1980. The failure of transversely isotropic rocks in triaxial compression. *International Journal of Rock Mechanics and Mining Sciences & Geomechanics Abstracts*, **17**, 325–332, https://doi.org/10.1016/0148-9062(80)90515-X

ODA, M., KOISHIKAWA, I. & HIGUCHI, T. 1978. Experimental study of anisotropic shear strength of sand by plane strain test. *Soils & Foundations*, **18**, 25–38.

PARISEAU, W.G. 1968. Plasticity theory for anisotropic rocks and soil. *In*: *Proceedings of the 10th US Symposium on Rock Mechanics*. American Rock Mechanics Association, Alexandria, VA, USA, 267–295.

PIETRUSZCZAK, S. 2010. *Fundamentals of Plasticity in Geomechanics*. CRC Press.

PIETRUSZCZAK, S. & MROZ, Z. 2000. Formulation of anisotropic failure criteria incorporating a microstructure tensor. *Computers and Geotechnics*, **26**, 105–112. https://doi.org/10.1016/S0266-352X(99)00034-8

PIETRUSZCZAK, S. & MROZ, Z. 2001. On failure criteria for anisotropic cohesive-frictional materials. *International Journal for Numerical and Analytical Methods in Geomechanics*, **25**, 509–524, https://doi.org/10.1002/nag.141

PIETRUSZCZAK, S., LYDZBA, D. & SHAO, J. 2002. Modelling of inherent anisotropy in sedimentary rocks. *International Journal of Solids and Structures*, **39**, 637–648, https://doi.org/10.1016/S0020-7683(01)00110-X

RAMAMURTHY, T. 1993. Strength and modulus responses of anisotropic rocks. *In*: HUDSON, J.A. (ed.) *Comprehensive Rock Engineering*, **1**. Pergamon, Oxford, 319–329.

Convergence analysis of an unsupported micro-tunnel at the Meuse/Haute-Marne Underground Research Laboratory

L.-M. GUAYACÁN-CARRILLO[1,2], J. SULEM[2*], D. M. SEYEDI[1], S. GHABEZLOO[2], A. NOIRET[3] & G. ARMAND[3]

[1]*ANDRA R&D, 92298 Chatenay-Malabry, France*

[2]*Université Paris-Est, Laboratoire Navier/CERMES, Ecole des Ponts ParisTech, IFSTTAR, CNRS, 77455 Marne la Vallée, France*

[3]*ANDRA R&D, Meuse/Haute-Marne Underground Research Laboratory, 55290 Bure, France*

**Correspondence: jean.sulem@enpc.fr*

Abstract: Convergence measurements recorded for one representative micro-tunnel (diameter *c.* 0.7 m) in Callovo-Oxfordian claystone were analysed. The micro-tunnel was excavated in the direction of the horizontal principal major stress. *In situ* observations showed anisotropic convergence with the maximum and minimum values in the horizontal and vertical directions, respectively. The horizontal closure of walls was fitted on the basis of a semi-empirical convergence law. This law is a predictive model reflecting the global response of the ground to excavation works. As the convergence measurements were performed after the end of excavation, their evolution in time can only be related to the time-dependent behaviour of the ground and the effect of the face advance cannot be captured. It is shown that some parameters of the semi-empirical law did not change along the micro-tunnel. An easy and efficient method is thus proposed for the long-term prediction of wall closure by the fitting of a single parameter on recorded data. Comparison with a drift (diameter *c.* 5 m) highlighted the influence of the support installation and the rate of excavation on the variation in the parameter values of the semi-empirical law. The vertical closure of the micro-tunnel walls, which showed a very weak evolution over time, was analysed based on the rate of convergence.

The French National Radioactive Waste Management Agency (ANDRA) began constructing the Meuse/Haute-Marne Underground Research Laboratory (M/HM URL) in the year 2000 with the main goal of demonstrating the feasibility of a geological repository in Callovo-Oxfordian claystone. A network of experimental drifts (with diameters ranging between 4 and 9 m) has been excavated in the directions of the horizontal principal stresses. Micro-tunnels (diameter *c.* 0.7 m) were excavated from these experimental drifts to test the feasibility of disposal cells for packages of high-level, long-lived radioactive waste and their impact on the surrounding rock (Renaud & Morel 2010). As reported by Wileveau *et al.* (2007), the major principal horizontal stress (σ_H) at the main level of the M/HM URL (-490 m) is *c.* 16.0 MPa orientated in the direction N 155° E $\pm$ 10°; the minor principal horizontal total stress is very close to the vertical stress ($\sigma_h \approx \sigma_v \approx$ 12.0 MPa). Therefore the initial stress state is quasi-isotropic (in the plane of their section) for the drifts following the direction of the major horizontal principal stress.

Different sections in each drift and in each micro-tunnel have been instrumented to monitor the evolution of convergence and rock deformation. Continuous monitoring of the excavated zone around the drifts and micro-tunnels revealed the development of a fractured zone (extensional and shear fractures) induced by excavation. The fracture distribution depends on the drift orientation with respect to the *in situ* stress field (Armand *et al.* 2014) and has an important influence on the deformation of the drifts (Armand *et al.* 2013). Fractured zones present anisotropic extents in both directions and therefore the convergence measurements show an anisotropic closure for both directions and the ratio of anisotropy depends on the orientation of the drifts. The anisotropic closure is more pronounced for drifts excavated along the minor horizontal principal stress with a ratio of vertical to horizontal convergence of 4.0, whereas for drifts excavated along the major horizontal principal stress this ratio is about 0.5 (Armand *et al.* 2013; Guayacán-Carrillo *et al.* 2016). Most of the micro-tunnels have been excavated in the direction of the

From: Norris, S., Bruno, J., Van Geet, M. & Verhoef, E. (eds) 2017. *Radioactive Waste Confinement: Clays in Natural and Engineered Barriers*. Geological Society, London, Special Publications, **443**, 115–125.
First published online December 20, 2016, https://doi.org/10.1144/SP443.24

major horizontal principal stress. Thus, as for the parallel drifts, the horizontal convergence is significantly higher than the vertical convergence even if the stress state is quasi-isotropic in the plane of their section.

Recent studies have been carried out to analyse the thermo-mechanical behaviour of the micro-tunnels and the thermo-hydro-mechanical impact on the surrounding rock, based on data recorded for a full-scale disposal cell demonstrator developed in the M/HM URL (Morel *et al.* 2013). The short- and long-term behaviours of the damage zone around micro-tunnels have been described based on an isotropic elastic–viscoplastic model (Renaud & Morel 2010). Similar studies have been performed for higher diameter excavations (e.g. Blanco Martín *et al.* 2011*a*; Souley *et al.* 2011; Plassart *et al.* 2013). Pardoen *et al.* (2015*a*, *b*) have recently published a numerical study of the progressive strain localization phenomena to reproduce the observed fracture network around excavations, taking into account the anisotropy of the initial stress state and the influence of the cross-anisotropy of the rock formation. However, modelling of the evolution of anisotropic closure over time has yet to be achieved. This work presents a different approach based on the semi-empirical law proposed by Sulem *et al.* (1987), which describes the evolution of tunnel closure by a direct analysis of measurements of convergence.

It has been observed that the ratio between horizontal closure and vertical closure for micro-tunnels is >5 and increases with time. This paper presents and analyses the convergence measurements in one micro-tunnel following the direction of the major principal horizontal stress. The evolution of horizontal convergence was fitted using the law proposed by Sulem *et al.* (1987). The values obtained for the parameters of the convergence law were compared with those obtained for parallel drifts with a diameter of 5 m reported by Guayacán-Carrillo *et al.* (2016). The vertical convergence, which presents a very weak evolution in time, was analysed independently on the basis of the closure rate.

Micro-tunnel instrumentation and *in situ* observations

Horizontal micro-tunnels *c.* 0.7 m in diameter have been excavated to study the behaviour of disposal cells for packages of high-level, long-lived radioactive waste and their impact on the surrounding rock (Morel *et al.* 2013). The micro-tunnels were excavated with a micro-tunnel boring machine, which has a drilling head that can be adapted to excavate diameters of 0.70–0.75 m (Fig. 1). The machine is laser-guided, which allows control of the micro-tunnel trajectory with a precision greater than ±2 cm. The drilling machine limits the induced damage in the rock by using air drilling and by removing the drilling tool at the end of the excavation by rotation in the opposite direction.

Micro-tunnels were excavated with lengths varying between 20 and 40 m. The excavation could be achieved with or without casing. The casing was made of segments of metal tubes 2 m long, 70 cm in diameter and 2 cm thick. The 2 m long elements were welded or socketed to each other as the excavation advanced (Morel *et al.* 2013). The unsupported micro-tunnels allowed us to follow the formation of breakouts (Fig. 2) and the state of the micro-tunnel walls and their evolution with time. Convergence measurements were carried out using a specially designed system allowing measurements with an accuracy of 0.1 mm (Gay *et al.* 2010). The system was installed in four convergence measurement sections placed 4–8 m from each other depending on the length of the micro-tunnel (Fig. 3). It allowed the long-term monitoring of wall closure and the identification of breakout zones.

We focus here on the *in situ* convergence measurements of the micro-tunnel referenced as ALC1603, which was excavated following the direction of the major principal stress σ_H. This micro-tunnel is unsupported, which allows a good study of the behaviour of the ground around the excavation. It has a diameter of 0.74 m and a length of 20 m. The excavation rate was around 0.6–0.7 m h^{-1}

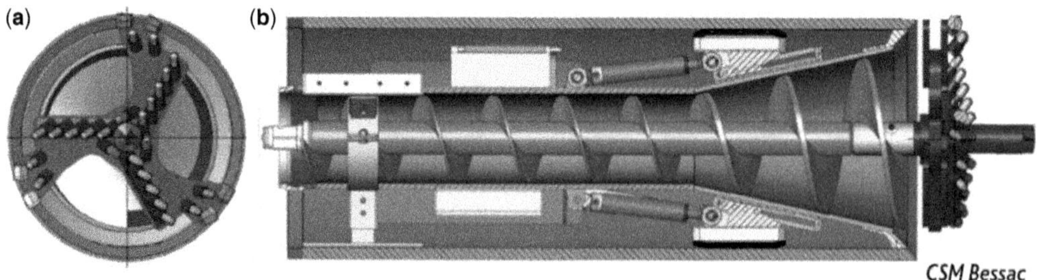

Fig. 1. Schematic diagram of the micro-tunnel boring machine. (**a**) Retractable cutting tool at the machine head; and (**b**) the back of the machine showing the rotating helical screw.

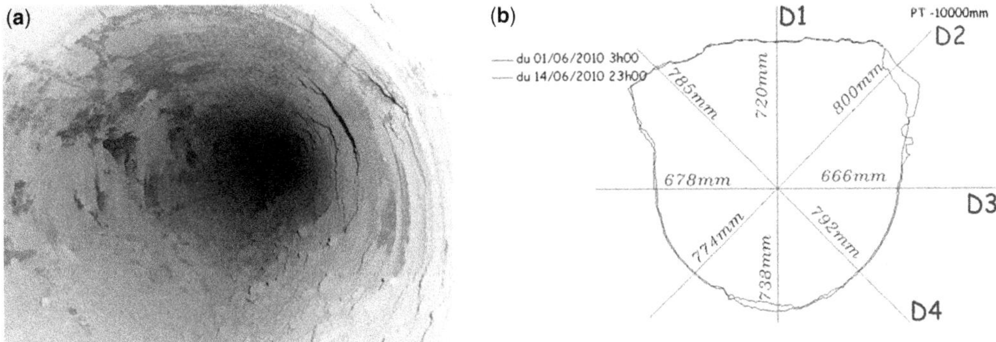

Fig. 2. State of the ALC1603 micro-tunnel wall (adapted from Morel *et al.* 2013). (**a**) Photograph after the end of excavation; and (**b**) typical profile of the micro-tunnel wall after excavation.

and the excavation was completed in three days. As a result of a hydraulic failure, the excavation was stopped for 31 hours at PM 13 (the PM number identifies the metric position from the entrance of the micro-tunnel). The convergence measurement instrumentation was installed three weeks after the end of the excavation. Four measurement sections were installed at PM 4, PM 8, PM 12 and PM 16. Measurements were taken in the vertical, horizontal and transverse directions (Fig. 3).

The convergence was at a maximum in the horizontal direction and at a minimum in the vertical direction. This behaviour was observed for all the sections along the micro-tunnel and was similar to observations performed in parallel drifts (diameter *c.* 5 m). Figure 4 shows the data recorded in the vertical, horizontal and transverse directions for the section at PM 12 in the ALC1603 micro-tunnel. The vertical and horizontal convergence data for the four sections along the micro-tunnel are presented in Figure 5. Eighty per cent of the total horizontal convergence occurred in the first six months of monitoring. The deformation rate then decreased with time and was <1 mm per month after six months of monitoring. The horizontal convergence was larger for the sections deeper in the rock mass. The convergence measurement section at PM 4 showed the lowest amount of wall closure, whereas the monitoring section at PM 16 showed the highest amount of wall closure. The reason for this is not clearly understood, but it can be attributed to the fact that the micro-tunnel was excavated in the lateral wall of a supported drift with a diameter of *c.* 5 m so that the stress field around the micro-tunnel was influenced by this drift in its vicinity (e.g. mainly in the section at PM4). Similar observations have been made in parallel micro-tunnels. The discrepancy in the evolution of closure between the sections at PM8, PM12 and PM16 can be related to the time elapsed between the excavation of the different sections. As the excavation was stopped for 31 hours at PM 13, after the opening of sections

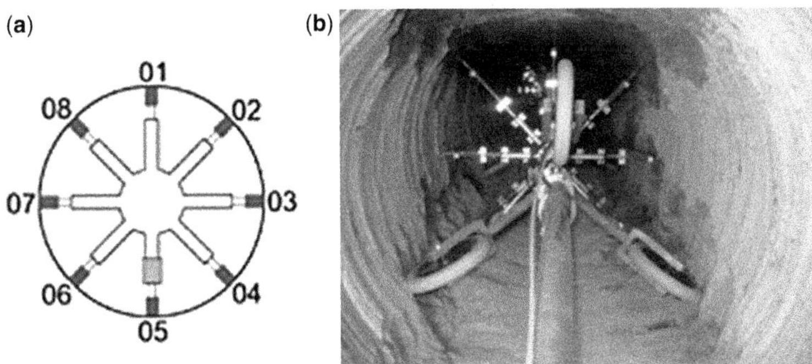

Fig. 3. Instrumentation of a convergence measurement section in the ALC1603 micro-tunnel. (**a**) Schematic diagram of the convergence measurement section. (**b**) Photograph of the installation of the convergence measurement section. The convergence measurements were performed in the vertical (01-05), horizontal (03-07) and transverse (02-06 and 04-08) directions.

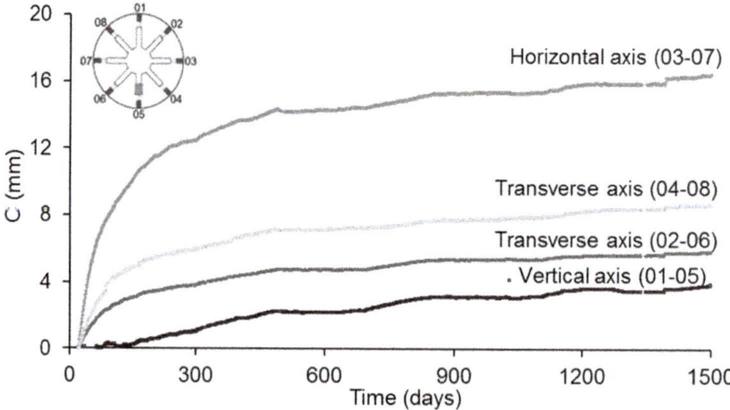

Fig. 4. Evolution of convergence in the ALC1063 micro-tunnel recorded in the convergence measurement section at PM 12 from the entrance of the drift.

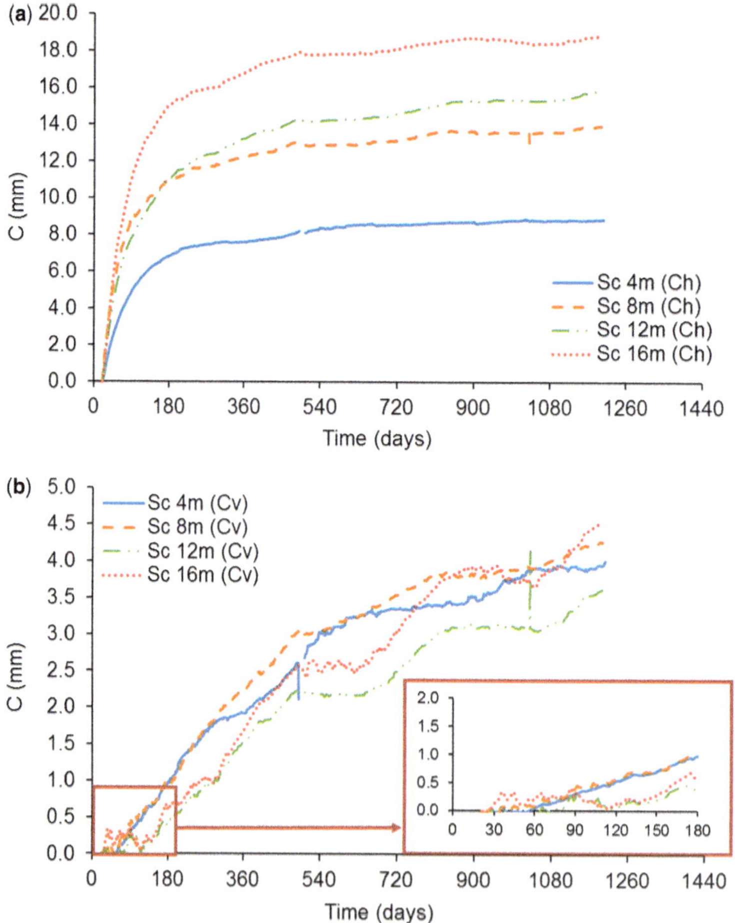

Fig. 5. Evolution of convergence in the ALC1603 micro-tunnel. Evolution of (**a**) horizontal convergence (Ch) and (**b**) vertical convergence (Cv). Sc denotes the measurement section at PM 4, PM 8, PM 12 and PM 16.

PM8 and PM12, part of the convergence of these sections was not recorded. Guayacán-Carrillo *et al.* (2016) reported that the maximum closure rate occurs in the first few days after excavation, so if the initial measurement for a section is delayed, a significant part of the total expected closure is lost. The vertical convergence was almost zero during the first two months of monitoring and then slowly increased with time. The data recorded for vertical convergence are presented and analysed in a later section.

Analysis of the horizontal convergence of a micro-tunnel following the direction of σ_H

As explained by Sulem *et al.* (1987), the convergence is the variation of the distance between two opposite points of the drift wall. It is a function of two variables: x, the distance to the front face; and t, the time after passage of the front face. Thus the evolution of convergence for a circular tunnel can be defined as $C(x, t) = D - D_0$, where D_0 is the initial diameter and D is the final diameter of the tunnel section (for a given distance x and time t after the passage of the tunnel face).

The evolution of convergence was fitted using the semi-empirical law proposed by Sulem *et al.* (1987) and given in equation (1). This convergence law depends on five parameters: T, a parameter related to the time-dependent properties of the system (rock mass formation and support); X, the length related to the distance of influence of the face and to the extension of the decompressed zone around the drift; $C_{\infty x}$, the instantaneous convergence obtained in the case of an infinite rate of face advance (no time-dependent effect); m, a parameter related to the ratio between the time-dependent convergence and the instantaneous convergence; and n, an exponent that describes the form of the fitted curve. Note that these parameters also depend on the excavation process, in particular the rate of excavation. The total convergence in the long term is given by equation (2).

$$C(x, t) = C_{\infty x}\left[1 - \left(\frac{X}{x + X}\right)^2\right] \times \left\{1 + m\left[1 - \left(\frac{T}{t + T}\right)^n\right]\right\} \quad (1)$$

$$C_{\text{Total}} = C_{\infty x}(1 + m) \quad (2)$$

It should be noted that this law has been proposed to describe the evolution of tunnel closure by the direct analysis of convergence measurements. Tran-Manh *et al.* (2016) related the parameters of this law to the ground parameters for the Saint-Martin-la-Porte access adit in France within the framework of a constitutive model with time-dependent degradation of the rock mass.

Evaluation of the convergence law along the horizontal direction

The convergence measurement devices were installed three weeks after the end of excavation and the recorded data therefore only describe the time-dependent behaviour of the rock mass. The face of the micro-tunnel was far enough from the convergence measurement sections to have no influence on the evolution of convergence. Thus the parameter X, which is directly related to the distance of influence of the face, cannot be estimated from these data. The recorded data will therefore be analysed taking into account only that part of the convergence law related to the time-dependent behaviour as given in equation (3).

$$C(t) = C_{\infty x}\left\{1 + m\left[1 - \left(\frac{T}{t + T}\right)^n\right]\right\} \quad (3)$$

The interpretation of the measured convergence must take into account the displacement that has occurred between the opening of the section and the installation of the convergence targets (the so-called lost convergence). The recorded convergence is presented in equation (4), where t_0 is the time elapsed for the first recorded reading since the face crossed the considered section; and t_i is the time elapsed after the passage of the face. By fitting the mathematical function (equation 4) to the measured data, the optimum values of the four parameters (T, $C_{\infty x}$, m and n) can be obtained.

$$\Delta C(t_i) = C(t_i) - C(t_0) \quad (4)$$

The obtained results show similar values of the parameters T, m and n for all sections. Thus an average constant value of these parameters can be assumed along the micro-tunnel. Table 1 presents the results for the four sections in the ALC1603 micro-tunnel. Only parameter $C_{\infty x}$ is fitted independently in each cross-section. This parameter reflects the impact of the induced fractured zone on wall closure and depends on the local conditions of each section. As shown in Figure 6, the evolution of the horizontal convergence is well reproduced as a function of time for all convergence measurement sections along the micro-tunnel, although the instantaneous convergence induced by the excavation advance is lost in this analysis.

Table 1. *Numerical values of convergence law parameters in the ALC1603 micro-tunnel*

Section (PM)	T (days)	m	n	Horizontal convergence	
				$C_{\infty x}$ (mm)	C_{Total} (mm)
4.0	0.2	4.0	0.45	22.6	113.2
8.0				35.3	176.3
12.0				38.6	193.3
16.0				45.7	228.3

Comparison with convergence predictions in drifts

The micro-tunnels and drifts orientated along σ_H both exhibit anisotropic deformation, with the horizontal closure stronger than the vertical closure. The data recorded for parallel drifts of *c.* 5 m diameter has been presented by Armand *et al.* (2013) and Guayacán-Carrillo *et al.* (2016). The origin of this anisotropic response can be related to factors such as the shape of the fracture zone induced by the excavation and to the anisotropic behaviour of the rock formation (Morel *et al.* 2013). Blümling *et al.* (2007) and Armand *et al.* (2014) reported that the development of the fracture network around excavations is related to the anisotropy of the material and its direction with respect to the *in situ* stress field. *In situ* observations show that the extent of the fractured zone is more developed in the horizontal direction for drifts and micro-tunnels following the direction of σ_H. It has been observed that the lateral extent of shear fractures is equal to about one diameter for different sizes of excavation (a drift of diameter 5 m, a borehole of diameter 5 cm and around a micro-tunnel of diameter 0.7 m). *In situ* observations showed that the closure anisotropy was stronger in the micro-tunnels than in the drifts. This can be related to the fact that the closure of the micro-tunnels was very weak in the vertical direction and evolved only slowly over time. Some rock blocks had fallen in the ALC1603 micro-tunnel and breakouts were observed during the course of the excavation, mainly in the vertical direction of the cross-section. This can be linked to the absence of rock bolts. A detailed description is presented in the analysis along the vertical axis of deformation.

In an attempt to compare the evolution of horizontal convergence between the micro-tunnels and the parallel drifts, an analysis was made based on a comparison of the convergence law parameters. A comparison of the evolution of convergence between the ALC1603 micro-tunnel and the GCS drift (5 m diameter) is presented.

An analysis of the evolution of convergence of this drift, based on the same semi-empirical law, has been presented by Guayacán-Carrillo *et al.* (2016). This drift was excavated following the same direction as the micro-tunnel and has a flexible support consisting of radial bolts, 21 cm of fibre shotcrete, yieldable concrete wedges and welded mesh. These wedges can accommodate the deformation of the rock mass and control the ground deconfinement. More information about this support can be found in Bonnet-Eymard *et al.* (2011). It was

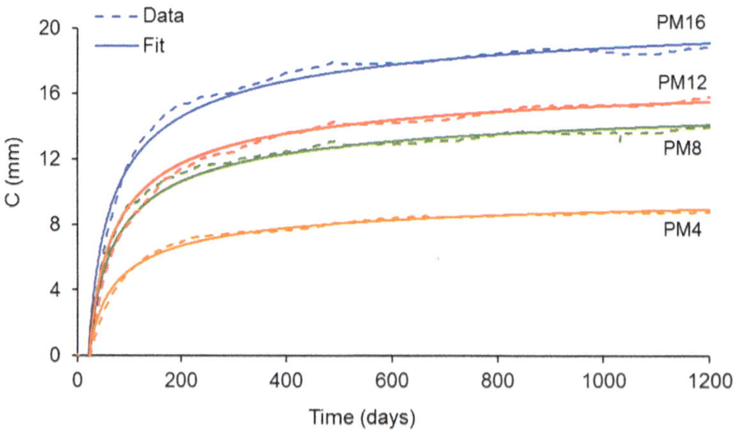

Fig. 6. Evolution of horizontal convergence in the ALC1603 micro-tunnel. The convergence data are represented by dashed lines and the convergence predicted by the convergence law (equations 3 and 4) is represented by continuous lines (time-dependent response).

observed that the two main axes of deformation of the cross-section were in the horizontal and vertical directions, respectively, as observed for the micro-tunnels. It was also observed that constant values of the parameters T, X, m and n can be assumed for the two axes of deformation and along the drift and that only the parameter $C_{\infty x}$ changes in relation to local conditions. Convergence monitoring in the drifts started immediately after the passage of the front at each convergence measurement section, whereas the convergence monitoring for the micro-tunnels started some days after the end of excavation. Thus the analysis of the evolution of convergence for the micro-tunnels differs from the analysis of the drifts. However, a comparison can be performed based on the time-dependent response of the ground, which is described by the parameters T, m and n of the convergence law. Table 2 summarizes the values of these three parameters for the ALC1603 micro-tunnel and the GCS drift.

As the micro-tunnel is unsupported and the rate of excavation was higher than that of the parallel drift (*c.* 2 m per week for the GCS drift), it was expected to have a higher convergence rate during the first few days after excavation and a faster stabilization of convergence over time. Therefore lower values of T and m were both expected and observed. A higher value of the exponent n captured the stronger deconfinement of the ground. The evolution of closure was higher in the micro-tunnel than in the drift, which was directly reflected in the variation of parameter $C_{\infty x}$. The ratio of $C_{\infty x}$ to the diameter of the excavation was higher for the micro-tunnel than for the GCS drift (Table 2). This higher rate of closure of the walls can be attributed to the fact that the drift's support was installed after each step of excavation, whereas the micro-tunnel was unsupported.

Table 2. *Comparison of the time-dependent parameters of the convergence law*

	T (days)	m	n	Horizontal convergence $C_{\infty x}$/diameter (%)
ALC1603 micro-tunnel	0.2	4.0	0.45	*c.* 5
GCS drift	6.0	5.7	0.3	*c.* 0.22

The evolution over time of the convergence (equation 5) obtained with the parameters of the ALC1603 micro-tunnel showed a higher initial rate that stabilized faster over time than the convergence of the GCS drift (Fig. 7). This highlights the influence of the installation of a support after each step of excavation and the influence of the rate of excavation on the evolution of convergence.

$$f(t) = 1 - \left(\frac{T}{t+T}\right)^n \quad (5)$$

Evolution of convergence along the vertical axis of deformation

The vertical convergence showed a weak evolution over time. Figure 5 shows that after six months of monitoring the vertical convergence was about 1 mm, although it was *c.* 10 times higher in the horizontal direction. This differential response in the

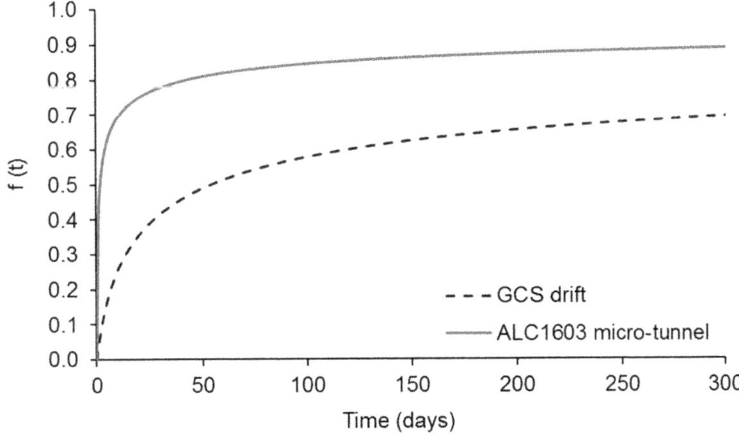

Fig. 7. Comparison of the time function evolution of the ALC1603 micro-tunnel and the GCS drift based on the convergence predictions of Guayacán-Carrillo *et al.* (2016).

horizontal and vertical directions may be related to the different extents of the fracture network, which was mainly developed in the horizontal direction. Figure 8 shows the distribution of fractures around the ALC3005 micro-tunnel excavated parallel to ALC1603. Marschall *et al.* (2006) and Blümling *et al.* (2007) explained that the breakouts occurred mainly at locations where the bedding planes were orientated tangential to the tunnel circumference and, as a consequence, the breakouts mainly developed in the vertical direction of the cross-section. Some rock block falls and breakouts were observed in the upper part of the micro-tunnel during the excavation. The occurrence of breakouts may also be related to the absence of rock bolts. Chappell (1989) and Blanco Martín *et al.* (2011*b*) reported that rock bolts are very efficient in the stabilization of the rock mass and in improving its stiffness.

It is likely that, in the first days after installation, the measurement system was not yet in contact with the rock formation in the zones of breakouts or was disturbed by the rock block falls. This may explain why the vertical convergence measurements were almost zero during the first two months and then started to slowly increase. It was not possible to analyse this response with our model as this was performed in the horizontal direction and therefore an analysis based on the convergence rate is proposed.

The evolution of convergence in drifts of diameter 5 m was higher in the few first days and then stabilized over time. The vertical and horizontal convergence then both followed a similar evolution for drifts parallel to the direction of the micro-tunnels (Guayacán-Carrillo *et al.* 2016). As the fracture network around the excavation was similar and the main axes of deformation were in the horizontal and vertical directions for both the micro-tunnels and drifts, the deferred vertical convergence was expected to present a similar response.

Taking into account the convergence recorded over more than three years, Figure 9 compares the vertical and horizontal rates of convergence. The convergence rates for the vertical and horizontal directions were comparable after about six months of monitoring.

An analysis of the vertical convergence was also performed for another micro-tunnel, ALC3005. This unsupported micro-tunnel was excavated in the face of a drift about 5 m in diameter. This micro-tunnel was excavated in the face of a drift to give a less disturbed initial state of stress. The excavation followed the direction of σ_H over a length of 20 m. Figure 10 shows that the vertical displacement rate was comparable with that observed in the parallel micro-tunnel ALC1603 (over the same time interval of monitoring) and *in situ* observations showed that all the micro-tunnels following this direction had a similar evolution of closure. Thus because the deferred evolution of convergence was similar in the horizontal and vertical directions, the ratio of anisotropic closure could be directly related to the instantaneous response of the ground.

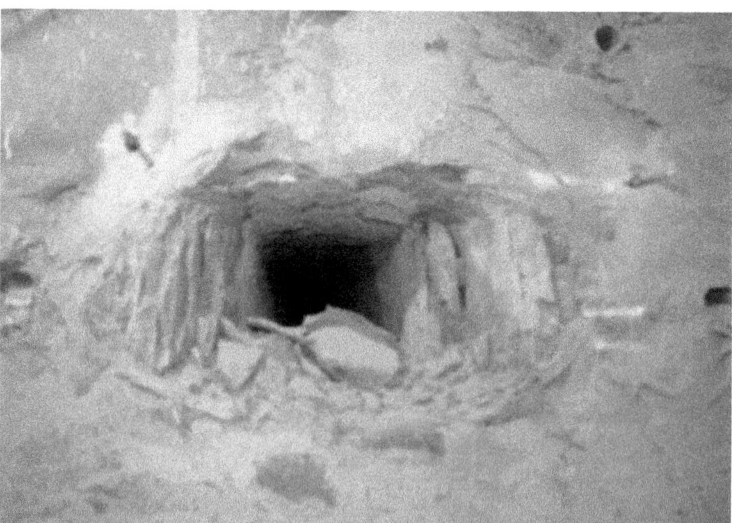

Fig. 8. Induced fracture distribution around micro-tunnel ALC3005 showing the elliptical distribution of fractures around the opening, with a larger extension in the horizontal direction. The breakouts induced by excavation are observed in the ascending vertical direction.

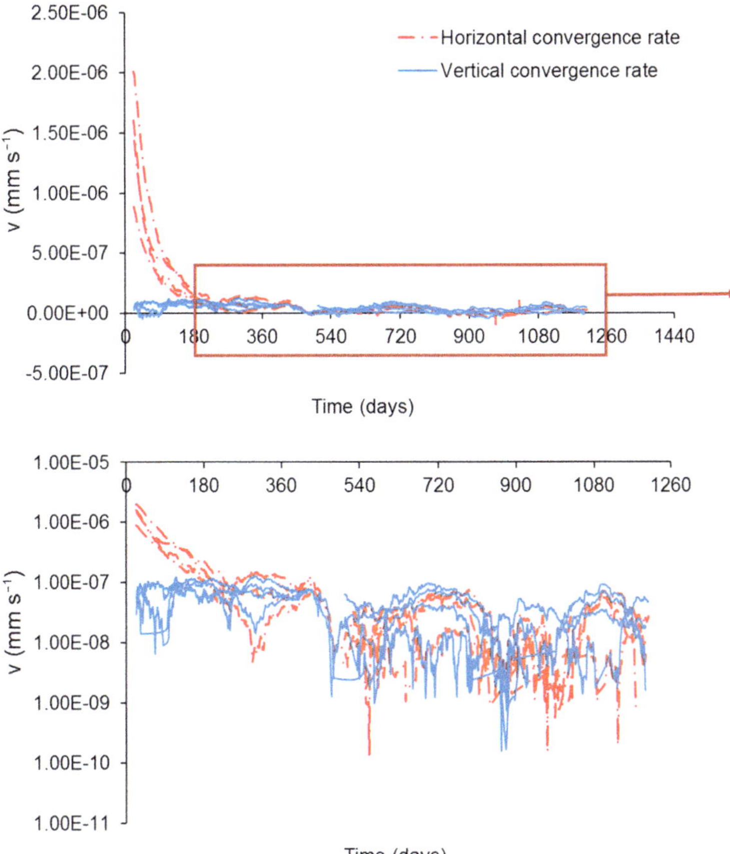

Fig. 9. Evolution of rate of vertical and horizontal convergence in the ALC1603 micro-tunnel.

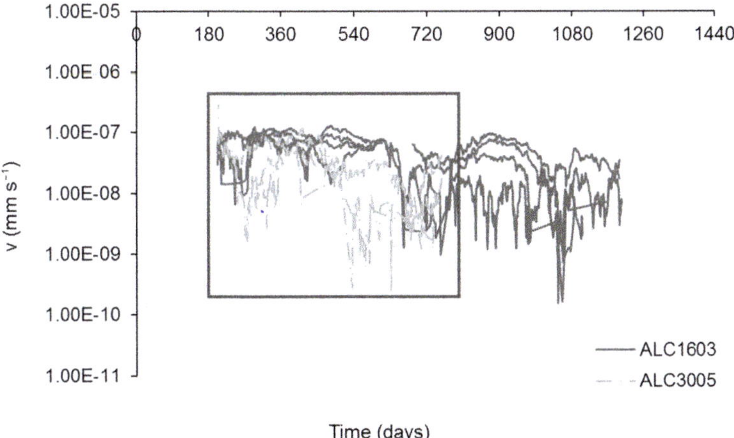

Fig. 10. Evolution of rate of vertical convergence for the ALC1603 and ALC3005 micro-tunnels.

Conclusions

We analysed convergence measurements in a micro-tunnel in Callovo-Oxfordian claystone, with the axis following the direction of the major initial principal horizontal stress. The *in situ* measurements showed an anisotropic closure of the walls that evolved over time. The horizontal convergence showed a stronger evolution about 10 times higher than the vertical convergence after six months of monitoring. The ratio between the horizontal and the vertical convergence also evolved over time because the evolution of the vertical convergence was very weak. The first convergence measurement was made several days after the end of excavation and thus the analysis of the recorded data mainly focused on the time-dependent behaviour of the rock formation.

The evolution of horizontal convergence was analysed based on a proposed convergence law. The four model parameters, which were related to the time-dependent behaviour (T, related to the time-dependent properties of the system rock mass – support; m, related to the ratio between the time-dependent convergence and the instantaneous convergence; $C_{\infty x}$, the instantaneous convergence; and n, an exponent describing the form of the fitted curve) were evaluated for different sections in the micro-tunnel. The results showed similar values for the parameters T, m and n, which can be assumed to be constant along the micro-tunnel. A very good reproduction of convergence was obtained by fitting a single parameter $C_{\infty x}$. This parameter represented the dispersion observed in the different sections of the micro-tunnel. A comparison of the results from the analysis of the convergence measurements in the ALC1603 micro-tunnel with those obtained from the analysis of the convergence in *c.* 5 m diameter drifts showed that the installation of a support after each step of excavation and changes in the rate of excavation played key parts in the time-dependent evolution of ground convergence.

A comparison was made between the rate of convergence in the horizontal and vertical directions to analyse the evolution of vertical convergence. The rate of vertical displacement was comparable with the horizontal rate after about six months of monitoring. The disturbance induced by excavation produced breakouts. Thus monitoring of the vertical convergence may be disrupted if the measurement system is not in good contact with the rock formation in the breakout zones or if it is disturbed by the fall of rock blocks. However, it was interesting to observe that, as for the parallel drifts, the deferred evolution of convergence in the horizontal and vertical directions was similar. Therefore the ratio of anisotropic closure can be directly related to the instantaneous response.

These results were obtained for data recorded in a representative micro-tunnel (0.7 m diameter) and a representative drift (5.2 m of diameter) following the direction of the major initial principal horizontal stress. Ongoing monitoring of the closure of the different drifts is being carried out at M/HM URL to better assess the effect of size, the influence of the sequence of excavation and the time elapsed for the installation of supports on the evolution of the convergence of drifts excavated in Callovo-Oxfordian claystone.

References

Armand, G., Noiret, A., Zghondi, J. & Seyedi, D. 2013. Short- and long-term behaviors of drifts in the Callovo-Oxfordian claystone at the Meuse/Haute-Marne Underground Research Laboratory. *Journal of Rock Mechanics and Geotechnical Engineering*, **5**, 221–230.

Armand, G., Leveau, F. et al. 2014. Geometry and properties of the excavation induced fractures at the Meuse/Haute-Marne URL drifts. *Rock Mechanics and Rock Engineering*, **47**, 21–41, https://doi.org/10.1007/s00603-012-0339-6

Blanco Martín, L., Hadj-Hassen, F., Tijani, M. & Armand, G. 2011*a*. New numerical modelling of the mechanical long-term behaviour of the GMR gallery in ANDRA's Underground Research Laboratory. *Physics and Chemistry of the Earth*, **36**, 1872–1877, https://doi.org/10.1016/j.pce.2011.07.027

Blanco Martín, L., Hadj-Hassen, F., Tijani, M. & Noiret, A. 2011*b*. A new experimental and analytical study of fully grouted rockbolts. *In*: *45th US Rock Mechanics/Geomechanics Symposium*, 26–29 June 2011, San Francisco, CA, USA, ARMA 11–242.

Blümling, P., Bernier, F., Lebon, P. & Derek Martin, C. 2007. The excavation damaged zone in clay formations: time-dependent behaviour and influence on performance assessment. *Physics and Chemistry of the Earth*, **32**, 588–599, https://doi.org/10.1016/j.pce.2006.04.034

Bonnet-Eymard, T., Thiriat, F. et al. 2011. Soutènement souple pour le creusement de galeries dans les argilites du CMHM. *In*: *Proceedings of the International Congress, AFTES, Underground Spaces for Tomorrow*, 17–19 October 2011, Lyon, France, 1–7.

Chappell, B.A. 1989. Rock bolts and shear stiffness in jointed rock masses. *Journal of Geotechnical Engineering*, **115**, 179–197.

Gay, O., Allagnat, D., Morel, J. & Armand, G. 2010. HA cells monitoring at the Underground Research Laboratory (URL) in the CMHM (Andra). *Tunnels et Espace Souterrain*, **221**, 371–382.

Guayacán-Carrillo, L.M., Sulem, J., Seyedi, D.M., Ghabezloo, S., Noiret, A. & Armand, G. 2016. Analysis of long-term anisotropic convergence in drifts excavated in Callovo-Oxfordian claystone. *Rock Mechanics and Rock Engineering*, **49**, 97–114, https://doi.org/10.1007/s00603-015-0737-7

Marschall, P., Distinguin, M., Shao, H., Bossart, P., Enachescu, C. & Trick, T. 2006. Creation and evolution of damage zones around a microtunnel in a claystone formation of the Swiss Jura mountains. *In*: *SPE International Symposium and Exhibition on Formation Damage Control*, 15–17 February 2006, Lafayette, LA, USA, 10.

MOREL, J., BUMBIELER, F., CONIL, N. & ARMAND, G. 2013. Feasibility and behavior of a full scale disposal cell in a deep clay layer. *In*: KWAŚNIEWSKI, M. & ŁYDŻBA, D. (eds) *Rock Mechanics for Resources, Energy and Environment*. CRC Press, Boca Raton, FL, 621–626.

PARDOEN, B., LEVASSEUR, S. & COLLIN, F. 2015*a*. Using local second gradient model and shear strain localisation to model the excavation damaged zone in unsaturated claystone. *Rock Mechanics and Rock Engineering*, **48**, 691–714, https://doi.org/10.1007/s00603-014-0580-2

PARDOEN, B., SEYEDI, D.M. & COLLIN, F. 2015*b*. Shear banding modelling in cross-anisotropic rocks. *International Journal of Solids and Structures*, **72**, 63–87, https://doi.org/10.1016/j.ijsolstr.2015.07.012

PLASSART, R., FERNANDES, R., GIRAUD, A., HOXHA, D. & LAIGLE, F. 2013. Hydromechanical modelling of an excavation in an underground research laboratory with an elastoviscoplastic behaviour law and regularization by second gradient of dilation. *International Journal of Rock Mechanics and Mining Sciences*, **58**, 23–33, https://doi.org/10.1016/j.ijrmms.2012.08.011

RENAUD, V. & MOREL, J. 2010. Calculs 3D prédictifs pour la conception d'alvéoles de stockage de déchets radioactifs HAVL. *Tunnels et Espace Souterrain*, **221**, 383–390.

SOULEY, M., ARMAND, G., SU, K. & GHOREYCHI, M. 2011. Modeling the viscoplastic and damage behavior in deep argillaceous rocks. *Physics and Chemistry of the Earth*, **36**, 1949–1959, https://doi.org/10.1016/j.pce.2011.10.012

SULEM, J., PANET, M. & GUENOT, A. 1987. Closure analysis in deep tunnels. *International Journal of Rock Mechanics and Mining Sciences & Geomechanics Abstracts*, **24**, 145–154, https://doi.org/10.1016/0148-9062(87)90522-5

TRAN-MANH, H., SULEM, J. & SUBRIN, D. 2016. Progressive degradation of rock properties and time-dependent behavior of deep tunnels. *Acta Geotechnica*, **11**, 693–711, https://doi.org/10.1007/s11440-016-0444-x

WILEVEAU, Y., CORNET, F.H., DESROCHES, J. & BLUMLING, P. 2007. Complete in situ stress determination in an argillite sedimentary formation. *Physics and Chemistry of the Earth*, **32**, 866–878, https://doi.org/10.1016/j.pce.2006.03.018

A pragmatic approach to abstract the excavation damaged zone around tunnels of a geological radioactive waste repository: application to the HG-A experiment in Mont Terri

ANDRÉS ALCOLEA[1]*, ULI KUHLMANN[1], PAUL MARSCHALL[2], ANDREA LISJAK[3], GIOVANNI GRASSELLI[3], OMID MAHABADI[3], RÉMI DE LA VAISSIÈRE[4], HELEN LEUNG[5] & HUA SHAO[6]

[1]*TK Consult AG, Hallenstrasse 10, 8008 Zürich, Switzerland*

[2]*National Cooperative for the Disposal of Radioactive Waste (NAGRA), Hardstrasse 73, 5430 Wettingen, Switzerland*

[3]*Geomechanica Inc., 90 Adelaide Street West, Toronto, Ontario M5H 3V9, Canada*

[4]*French National Agency for Radioactive Waste Management (ANDRA), RD 960, 55290 Bure, France*

[5]*Nuclear Waste Management Organization, 22 St Clair Avenue East, Toronto, Ontario M4T 2S3, Canada*

[6]*Federal Institute for Geosciences and Natural Resources, Stilleweg 2, 30655 Hannover, Germany*

**Correspondence: andres.alcolea@tkconsult.ch*

Abstract: The excavation damaged zone (EDZ) around the backfilled tunnels of a geological repository represents a possible release path for radionuclides, corrosion and degradation gases that needs to be adequately addressed by safety assessment (SA) modelling tools. The hydromechanical phenomena associated with the creation and temporal evolution of the EDZ are of high complexity, precluding detailed representations of the EDZ in conventional SA. Thus, simplified EDZ models mimicking the safety-relevant features of the EDZ are required. In this context, a heuristic modelling approach has been developed to represent the creation and evolution of the EDZ in an abstracted and simplified manner. The key features addressed are the stochastic character of the excavation-induced fracture network and the self-sealing processes associated with the re-saturation after backfilling of the tunnels. The approach has been applied to a range of generic repository settings to investigate the impact of repository depth and *in situ* conditions on the hydraulic significance of the EDZ after repository closure. The model has been benchmarked with a dataset from a self-sealing experiment at the Mont Terri underground rock laboratory (URL), demonstrating the ability of the approach to mimic the evolution of the hydraulic significance of the EDZ during the re-saturation phase.

At the Cluster Conference in Luxembourg (November 2003) dedicated to the 'Impact of the Excavation Disturbed or Damaged zone (EDZ) on the Performance of Radioactive Waste Geological Repositories', a general definition of the EDZ around underground excavations was agreed upon by the scientific community:

> The Excavation Damaged Zone (EDZ) is a zone with significant irreversible processes and significant changes in flow and transport properties. These changes, for example, can include one or more orders of magnitude increase in flow permeability. (Tsang *et al.* 2005)

In recent years, the EDZ has been a focus of continuous international research activities, covering the key processes, phenomena and features associated with the creation and evolution of the EDZ for a variety of host-rock formations, including crystalline rock, rock salt, soft clays and indurated clays (Alheid *et al.* 2007; Aranyossy *et al.* 2008). Soft clays and indurated clay formations were identified as a class of sedimentary rocks with a distinctive deformation behaviour, displaying transitional features between ductile yielding and brittle failure in response to the excavation process. In the context of radioactive waste disposal, special emphasis has

From: Norris, S., Bruno, J., Van Geet, M. & Verhoef, E. (eds) 2017. *Radioactive Waste Confinement: Clays in Natural and Engineered Barriers*. Geological Society, London, Special Publications, **443**, 127–147.
First published online June 24, 2016, https://doi.org/10.1144/SP443.8

been given to the favourable capacity of EDZ fractures in clay formations to self-sealing after closure of the backfilled repository structures (Bernier *et al.* 2007; Blümling *et al.* 2007; Bock *et al.* 2010).

The Opalinus Clay, an indurated clay formation in the Molasse Basin of northern Switzerland, has been identified as a candidate host rock for a geological repository for radioactive waste in Switzerland. The EDZ around excavated openings in the Opalinus Clay Formation is predominantly a zone of brittle failure, as confirmed by multiple pieces of evidence gained during construction of motorway and railway tunnels (e.g. Steiner *et al.* 2010). In the long term, the fractures forming the EDZ are expected to re-seal in response to the re-saturation process and the corresponding pore-pressure recovery.

Quantitative analyses in support of the assessment of long-term radiological safety of geological repositories are often based on simplified representations of the EDZ, assuming a configuration of concentric shells (single shell or double shell, respectively) with enhanced hydraulic conductivity. The radius of the shell representing the EDZ typically depends on the tunnel radius and on the orientation of the tunnel axis with respect to the principal horizontal stress direction (e.g. NAGRA 2002). Several numerical EDZ abstraction approaches have been proposed in recent years, from complex EDZ fracture networks to simplified representations with equivalent flow and transport characteristics. Lanyon & Senger (2011) developed a sequential modelling approach for indurated clay formations, based on a stochastic representation of the EDZ in terms of discrete fracture network (DFN) models. The underlying hydraulic DFN models with stochastic distributions of fracture frequency, orientation and transmissivity were converted into equivalent-porous-medium (EPM) models with stochastic permeability distribution using the approach suggested by Jackson *et al.* (2000). Combined water and gas flow along stochastic realizations of the EDZ was simulated with the EPM models. Eventually, effective properties suitable for performance assessment were derived by inverse modelling of the results of stochastic modelling, with a simple single-shell representation of the EDZ. Hawkins *et al.* (2011) presented a similar approach with applications to the Meuse/Haute Marne underground rock laboratory (URL) in the Callovo-Oxfordian Formation in the NE Paris Basin.

The aforementioned approaches for the simplification of the EDZ draw on stochastic representations of the EDZ fracture network derived from structural mapping in tunnels. This work presents a novel heuristic modelling approach based on EDZ fracture networks derived with a fracture-mechanics-based numerical model. Thus, the impact of stress conditions, rock strength and tunnel geometry on fracture patterns can be assessed in a systematic manner using parametric sensitivity analyses. The approach is able to mimic the safety-relevant features of the EDZ in an abstracted and simplified manner. This paper is organized as follows. First, the conceptual framework for the hydraulic significance of the EDZ is outlined. Second, the methodology is presented in detail. Third, the results of a sensitivity analysis of the hydraulic significance of the EDZ to various repository settings are explored. Finally, the application and benchmarking of the abstraction approach to the HG-A experiment ('Gas path through host rock and along seal sections') is presented.

Conceptual framework

Empirical and experimental evidence

When a geological repository is constructed in an indurated clay formation, the host rock around the newly created openings is expected to respond in a brittle manner. Part of the irreversible deformation already occurs as pre-deformation ahead of the tunnel face (Martin 1997; Yong 2008), whereas the development of the system of excavation-induced fractures continues during the excavation process and even after completion of the permanent tunnel support system (Blümling *et al.* 2007). The shape and extension of the EDZ around the openings are controlled by many factors, including the general geological setting (e.g. rock mass strength, stress and pore-pressure distribution, sedimentary and tectonic features, and *in situ* stress field), engineering design (e.g. configuration of the underground structures, excavation technique and support system/lining) and the impacts related to the operational phase (e.g. ventilation and waste emplacement). Rock deformation is complicated by clay-specific hydromechanical couplings, giving rise to long-lasting time-dependent processes in the EDZ even after closure of the backfilled repository structures. A typical inventory of discrete EDZ features consists of extensional fractures (Fig. 1a), caused directly by the short-term excavation-induced unloading (undrained elasto-plastic response), spalling and buckling of the bedding planes (Fig. 1b, c), interactions with the existing tectonic features (Fig. 1b), and bounding shear bands. These features delimit the EDZ and the intact rock zones (Blümling *et al.* 2007). Further EDZ features can be attributed to the operational phase, such as desiccation cracks in response to tunnel ventilation and swelling-induced disaggregation of the rock matrix, when subjected to water-based fluids (e.g. cement water during construction of the invert).

The characterization of structural and hydromechanical features and processes of the EDZ has been

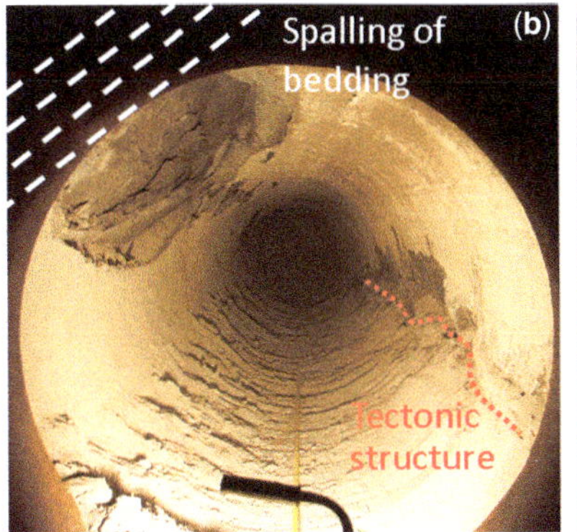

Fig. 1. Excavation-induced brittle features in Opalinus Clay (Mont Terri URL): (**a**) extensional fractures created during the excavation of Gallery 98 (NAGRA 2002); (**b**) bedding-related spalling processes in the crown of the HG-A microtunnel and the interaction of stress-controlled failure at the lower-right side wall with a tectonic structure (Marschall *et al.* 2006); and (**c**) buckling of the bedding planes around a small borehole (Blümling *et al.* 2007).

a key research topic in the scientific programme of the Mont Terri URL in Switzerland. Early work by Martin & Lanyon (2003) and Corkum & Martin (2007) focused on the EDZ around tunnel sections normal to bedding strike, drawing on previous work from Bossart *et al.* (2002). More recently, emphasis was given to excavations parallel to bedding strike because these configurations are more relevant for a future deep repository (Marschall *et al.* 2006; Lanyon *et al.* 2014; Lisjak *et al.* 2015). The impact of pre-existing tectonic features on the development of the EDZ was highlighted by Yong (2008) and Nussbaum *et al.* (2011). Yong (2008) reported a comparison of fracture density in boreholes around tunnels and niches at Mont Terri, examining data from over 100 boreholes (Fig. 2a). All structures, other than those clearly related to tectonic features or drill core handing, were counted over 0.5 m intervals. Almost 75% of the boreholes were drilled into the sidewalls and data were normalized per borehole to reduce the bias. Fracture count in the first 0.5 m can be as high as 10 (with a mean spacing 5 cm) around some excavations, but typically reduces to less than two at 3 m from the tunnel wall.

At the Mont Terri URL, small-scale *in situ* permeability measurements in the EDZ were carried out in boreholes equipped with multi-packer systems orientated radial to the drift, isolating measurement intervals of 10–50 cm at various distances from the wall. Hydraulic pulse and constant head tests were carried out in the saturated part of the EDZ, whereas simple pneumatic tests were undertaken in the unsaturated rock in the immediate vicinity of the tunnel surface to investigate the local effective permeability (Bossart & Thury 2008). Figure 2b shows that permeability increases by several orders of magnitude (up to six or seven) within 20–40 cm from the tunnel. This zone coincides with that exhibiting high fracture density.

Ferrari *et al.* (2013) and Romero & Gómez (2013) performed permeability tests in oedometric cells and triaxial cells on core samples from Mont Terri and from other locations in northern Switzerland to investigate the stress dependence of permeability and porosity of intact Opalinus Clay. A consistent trend in porosity and permeability reduction with increasing effective stress was observed, although each sample showed the influence of preconsolidation and mineralogy. Hydraulic conductivity was typically below 10^{-13} m s^{-1} for effective stresses greater than 20 MPa and reduced further to around 10^{-14} at 100 MPa (Fig. 3a). Clay content, porosity and specific surface data obtained from the laboratory analysis of Opalinus Clay samples were used to infer matrix hydraulic conductivity based on the Kozeny–Carman relationship (Alcolea *et al.* 2014).

The dependence of fracture transmissivity on normal effective stress has been the issue of laboratory studies and *in situ* tests (Blümling *et al.* 2007). As part of the GS ('Gasfrac Self-Healing') experiment at Mont Terri, hydraulic tests were performed in a borehole before and after a hydrofrac experiment. Testing was conducted in a test interval containing a hydrofrac with a well-defined fracture geometry. Interval transmissivity increased by five or six orders of magnitude when the injection pressure exceeded the normal stress on the fracture plane (Fig. 3b). For low injection pressures, however, the interval transmissivity was close to that of the intact rock ($T \leq 10^{-12}$ m^2 s^{-1}). These findings can be explained by the dependence of fracture transmissivity (or hydraulic conductivity) on effective

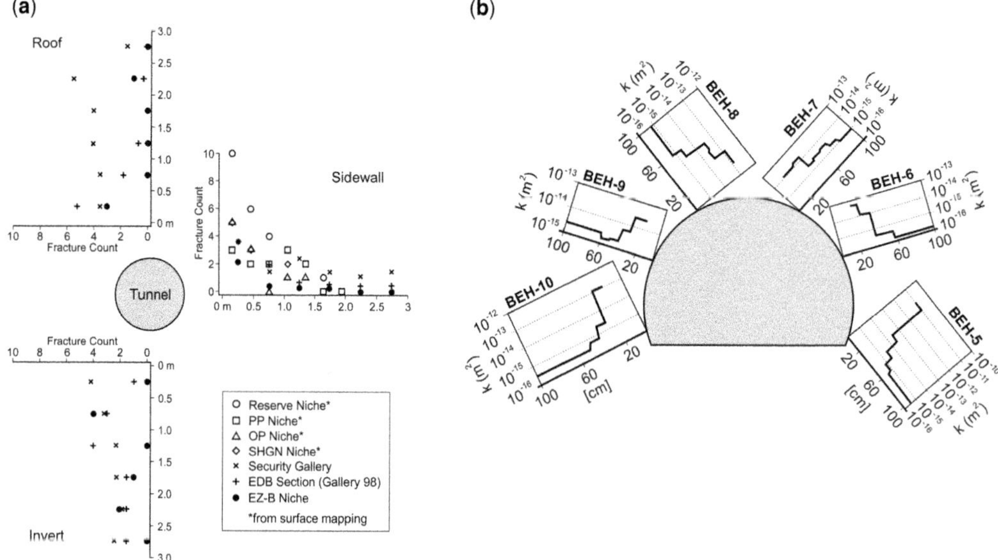

Fig. 2. Structural and hydraulic characterization of the EDZ features around excavations in the Mont Terri URL: (**a**) fracture counts in 0.5 m borehole intervals from around the EZ-B Niche and previous subparallel excavations (modified from Yong 2008); and (**b**) *in situ* permeability measurements conducted in Mont Terri experimental drift (from Bossart & Thury 2008).

normal stress, which follows a hyperbolic fracture closure law (see NAGRA 2004). An effective mechanical self-sealing of the artificial fracture was already observed at moderate effective normal stresses of the order of 1–2 MPa (Fig. 3b).

Further experimental evidence of the self-sealing capacity of excavation-induced fractures has been gained by *in situ* experiments at the Mont Terri URL, namely the SF experiment (Bernier *et al.* 2007) and the HG-A experiment (Lanyon *et al.* 2014). Notably, the long-term self-sealing tests in the HG-A microtunnel (Fig. 1b) served as a benchmark case for the modelling approach suggested in this work.

Conceptualization of fracture self-sealing

Drawing on the presented empirical and experimental evidence for the hydraulic significance of the EDZ during tunnel construction and its evolution during the operational phase and after backfilling of the underground structures, the functional

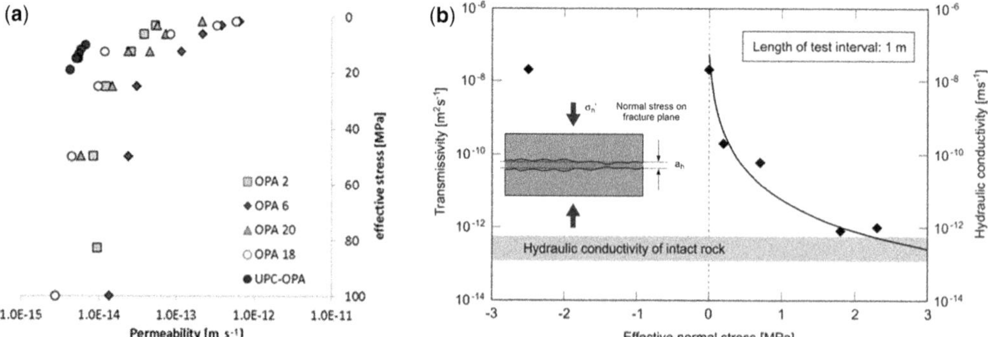

Fig. 3. Empirical relationships between stress and hydraulic properties of the host rock: (**a**) the stress–permeability relationship of the intact rock matrix; and (**b**) hyperbolic transmissivity–stress relationship between fracture transmissivity and normal effective stress (results of the GS experiment, from NAGRA 2004).

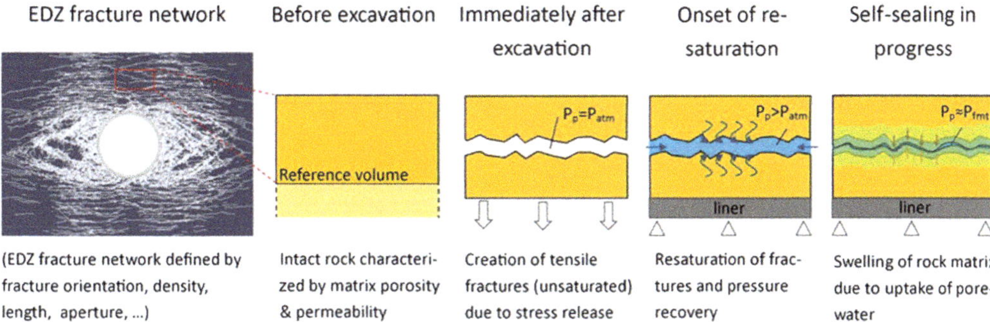

Fig. 4. Schematic sketch of the conceptual framework for EDZ fracture closure in Opalinus Clay, covering the key phenomena and features from the early post-excavation phase until static formation pressure recovery. In the insets, P_p denotes pore pressure, P_{atm} denotes atmospheric pressure and P_{fmt} denotes formation pressure.

requirements associated with a quantitative EDZ closure model can be formulated as follows (Fig. 4):

- The creation of the EDZ is a brittle process (i.e. the increase of the void volume in the damaged rock zone around the excavation is solely attributed to fracture opening, whereas the porosity of non-fractured rock remains essentially unchanged during the early times after excavation).
- Initially, the newly created EDZ fractures are unsaturated and exposed to atmospheric pressure, whereas the non-fractured rock remains saturated and exhibits high matrix suction as a consequence of the high gas entry pressure of the Opalinus Clay (Ferrari *et al.* 2014). The initial transmissivity of the unsaturated EDZ fractures is controlled by the fracture aperture and can be very high. The matrix conductivity remains essentially unchanged (i.e. it is the same as the conductivity of the intact rock).
- With time, the matric suction in the rock matrix decreases due to the uptake of porewater from outer rock zones, and the fractures start to resaturate.
- Porewater uptake of the non-fractured rock zones is associated with swelling and, consequently, with an increase in matrix porosity. The porosity increase of the rock matrix occurs at the expense of fracture aperture (i.e. the fractures start to close and fracture transmissivity reduces drastically, whereas the hydraulic conductivity of the non-fractured matrix zones increases slightly as a consequence of the porosity increase).
- This increase in matrix porosity and hydraulic conductivity is associated with the decrease in the fracture aperture and transmissivity, and progresses until the equilibrium of effective stresses is reached. This is essentially the case, when pore pressure reaches the static formation pressure.

A numerical approach for modelling the hydraulic significance of the EDZ is developed in the next section with the following goals: (i) to cope with a wide range of geological, hydrogeological and geomechanical conditions relevant to the candidate siting regions for a geological repository in northern Switzerland; (ii) to cover the relevant hydraulic and geomechanical phenomena and processes associated with the creation of the EDZ during construction, and its evolution during operational times and after backfilling of the underground structures; and (iii) to consistently capture the experimental evidence of the EDZ self-sealing, which has been gained as part of *in situ* experiments in URLs (e.g. the HG-A experiment at Mont Terri).

The methodology presented herein consists of three sequential main steps (Fig. 5). First, a hybrid finite-discrete element method (FDEM) (Geomechanica 2013, 2014; Lisjak *et al.* 2015, 2016) is used to simulate the geometry and geomechanical conditions of the DFNs forming the EDZ (Fig. 5a). The FDEM simulations are purely mechanical and mimic the excavation and emplacement processes. Secondly, the geometrical properties simulated by the FDEM are mapped onto a finite element mesh (Fig. 5b), which allows the fluid motion equations in the excavation near field to be solved. A salient feature of the suggested methodology is that hydraulic parameters of both fracture and matrix evolve with time as a response to re-saturation of the tunnel surroundings. Thirdly, an abstraction of the complex model is made based on the late-time behaviour (after full re-saturation) of the system (Fig. 5c). The main outputs of the model are the spatio-temporal distributions of hydraulic parameters and the corresponding specific fluxes towards the tunnel, with special emphasis on the late-time behaviour. Finally, the abstraction of the EDZ is implemented by defining a piecewise homogeneous 'shell-like' model with hydraulic behaviour identical to that of the complex model.

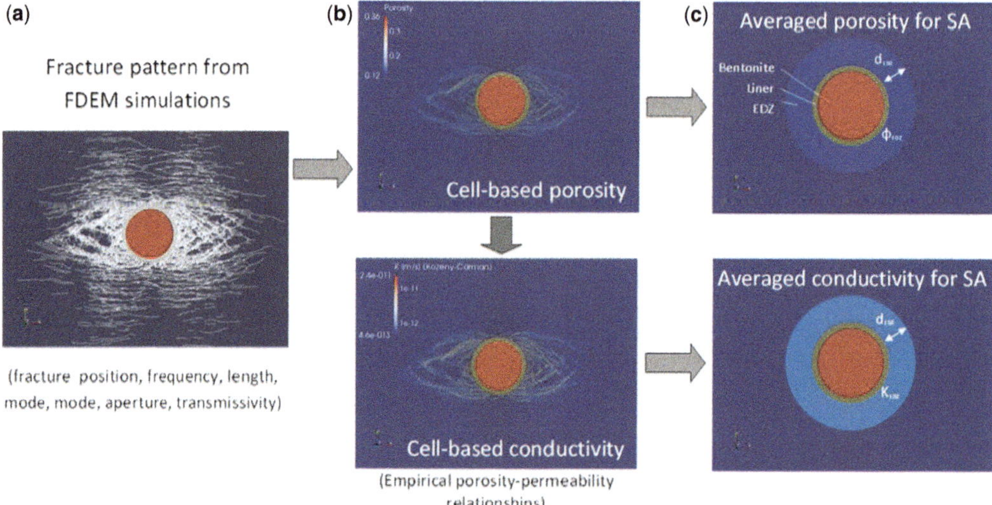

Fig. 5. Concept of the EDZ abstraction process for safety assessment (SA) applicable to a circular tunnel: (**a**) representative fracture patterns are simulated for relevant repository configurations with a discrete element model (FDEM); (**b**) the discrete fracture patterns are converted into heterogeneous porosity and conductivity distributions; and (**c**) in a final abstraction process, the heterogeneous porosity–conductivity distributions are converted in a shell defined by a radius and homogeneous porosity–conductivity.

The present study focuses on the temporal evolution of hydraulic parameters of the EDZ around the tunnel. For illustration purposes, the methodology is applied to the HG-A experiment at the Mont Terri URL.

Modelling approach

Following Lanyon & Senger (2011), an EPM representing a cross-section orthogonal to the axis of the underground structure is used to model the re-saturation of the EDZ. Re-saturation leads to the self-sealing of the EDZ by virtue of mechanical closure of fractures and swelling of the clay-rich minerals of the rock matrix. The modelling approach consists of three basic tools (Fig. 6): (1) mapping of the FDEM geometrical properties; (2) the re-saturation model, which calculates the pressure state of the system at a certain re-saturation stage and updates the FDEM parameters accordingly; and (3) EDZ abstraction. These tools are described in detail below and are applied to an illustrative example.

Mapping of the FDEM to an equivalent-continuum model

The fluid flow caused by the re-saturation of the EDZ is predominantly radial towards the tunnel owing to the shape of the EDZ and of the underground structure. Therefore, a radial finite element mesh is used, thus helping to alleviate the numerical artefacts caused by structured meshes while solving radial flow and reducing the computational time. A zonation of hydraulic parameters is also adopted to assign different properties to different materials (Table 1). The mesh is finer at the zone encompassing the EDZ. Cell size increases according to a geometric progression. Hydraulic properties at the cells are calculated from geometrical properties of the FDEM. To this end, the properties of fractures intersecting each cell (if any) are upscaled.

Cell porosity is calculated as:

$$\phi = \frac{V_{\rm p}}{V_{\rm tot}} = \frac{V_{\rm p,f}(t) + (1 - (V_{\rm p,f}(t)/V_{\rm tot}))V_{\rm p,m}(t)}{V_{\rm tot}} = \phi_{\rm f}(t)(1 - \phi_{\rm m}(t)) + \phi_{\rm m}(t) \quad (1)$$

where ϕ [$-$] is total porosity, $V_{\rm p}$ and $V_{\rm tot}$ [$\rm L^3$] are the void volume and total volume of the cell, respectively, and t [T] is time. The numerical model is based on a unit length in the axial direction of the tunnel. Therefore, cell volumes are equivalent to cell areas. Subscripts 'f' and 'm' refer to fracture and matrix, respectively. Constant deterministic values of matrix porosity for the different materials defining the model zonation are summarized in Table 1.

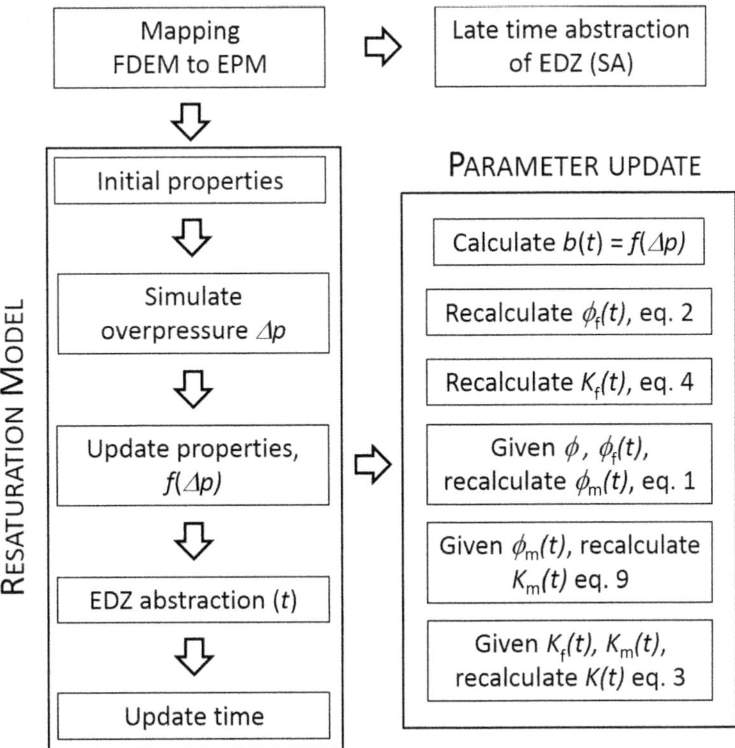

Fig. 6. Suggested modelling approach. The starting point is a mechanical simulation of the EDZ after excavation. Geometrical properties of the fractures delineating the EDZ are mapped to a continuum model. After mapping, and assuming that all fractures are closed (i.e. attributed with a residual aperture/irreducible transmissivity), the late-time behaviour of the EDZ ($t > 20\,000$ years) is analysed and an abstraction consisting of a piecewise homogeneous model is proposed. Such abstraction is amenable to SA analysis. In addition, the temporal evolution of the EDZ during re-saturation can also be addressed by the continuum model. A decoupled hydromechanical model is posed that updates fracture aperture at each time step as a function of overpressure (i.e. above atmospheric pressure). Hydraulic properties of the EDZ vary in time in response to re-saturation. The EDZ abstraction can therefore be made at intermediate times of re-saturation. In the insets, t denotes time, b is aperture, K is the hydraulic conductivity and ϕ is porosity. Subscripts f and m denote fracture and matrix, respectively.

Fracture porosity of a given cell is calculated as an average of the geometrical parameters defining the fractures intersecting the cell:

$$\phi^j_{\mathrm{f}}(t) = \frac{\sum_i b_i(t) L^j_i}{V_j} \tag{2}$$

where b_i [L] and L^j_i [L] are the aperture and trace length of fracture i intersecting cell j. V_j is the cell volume (area).

Hydraulic conductivity of a cell K [LT^{-1}] is calculated as:

$$K(t) = K_{\mathrm{m}}(t) + K_{\mathrm{f}}(t) \tag{3}$$

where K_{m} [LT^{-1}] denotes hydraulic conductivity of the matrix (Table 1) and K_{f} [LT^{-1}] is fracture conductivity of cells intersected by fractures (0 otherwise). K_{f} is calculated from fracture trace length and transmissivity, calculated from fracture aperture using a cubic law approximation:

$$K^j_{\mathrm{f}} = \frac{\sum_i T_i(t) L^j_i}{V_j}; \quad T_i = \frac{\rho g b(t)^3_i}{12\mu} \tag{4}$$

where T_i [$\mathrm{L^2T^{-1}}$] is the transmissivity of fracture i intersecting cell j, g is gravity [$\mathrm{LT^{-2}}$], and ρ [$\mathrm{ML^{-3}}$] and μ [$\mathrm{ML^{-1}T^{-1}}$] are the fluid density and dynamic viscosity, respectively.

Within the modelled cross-section, the flow occurs mainly in the fracture plane (i.e. along its longest edge). Thus, anisotropic hydraulic conductivities are calculated from the maximum

Table 1. *Model zonation and deterministic values of hydraulic properties*

Zone	Approximate radial extent (m)	Porosity	In-plane conductivity	
			K_{xx} (m s^{-1})	K_{zz} (m s^{-1})
Canister	0–0.5	NA	NA	NA
Bentonite (buffer zone)	0–1.25	0.35	10^{-13}	10^{-13}
Steel liner	1.25–1.5	0.20	10^{-10}	10^{-10}
EDZ	Variable	Variable	Variable	Variable
Intact Opalinus Clay (far field)	$R_{\text{model}} = 120$	0.12	10^{-13}	2×10^{-14}

The canister, emplaced at the centre of the model, is not simulated. Instead, a no-flow boundary condition is set there. R_{model} denotes the radius of the outer boundary of the model.

conductivity by standard projection:

$$K_{f,xx}(t) = K_f(t)\cos\theta$$
$$K_{f,zz}(t) = K_f(t)\sin\theta \quad (5)$$

where θ is the inclination of the fracture (measured counter-clockwise from the positive x-axis). x and z are local Cartesian axes in the plane of the cross-section. The positive y-axis runs along the tunnel towards its end.

Hydraulic conductivities in the x- and z-directions are then calculated at each cell:

$$K_{xx}(t) = K_{f,xx}(t) + K_{m,xx}(t)$$
$$K_{zz}(t) = K_{f,zz}(t) + K_{m,zz}(t)\,. \quad (6)$$

Finally, cell conductivity, K_{yy}, orthogonal to the cross-section (i.e. parallel to the axis of the opening) is calculated as:

$$K_{yy}(t) = \sqrt{K_{xx}(t)\,K_{zz}(t)}. \quad (7)$$

In equation (7), it is assumed that the cell size is sufficiently small and that the anisotropy of hydraulic conductivity is not significant. Note also that equations (1) and (3) involve time-dependent hydraulic parameters. The way in which these parameters vary with time is described in the next subsection.

FDEM parameters at a certain re-saturation stage

The re-saturation of the system has two implications. On the one hand, an increase in pressure inside the fractures (from atmospheric to hydrostatic conditions) causes a reduction in effective normal stresses. This reduction is assumed to be insufficient to either stimulate existing fractures (i.e. fracture trace length is constant) or to generate new fractures. Yet, the reduction of effective stresses ($\sigma > 0$ means compression) leads to mechanical closure of fractures. Fracture closure leads to a reduction in fracture porosity and hydraulic conductivity, according to equations (2) and (4). On the other hand, the high percentage of clay minerals causes the matrix to swell at low-pressure variations (<2 MPa). Swelling causes both matrix porosity and matrix hydraulic conductivity to increase with time.

Based on previous analyses of the results of the GS experiment (Fig. 3b) (see also NAGRA 2004), fracture closure is accommodated in the model through a function relating fracture aperture and overpressure, Δp [$ML^{-1}T^{-2}$] (i.e. pressure above atmospheric):

$$b(t) = b_0\left(1 - \frac{1}{(b_0 K_{n0}/\Delta p(t))^{1-\alpha} + 1}\right); \quad \alpha < 1 \quad (8)$$

where b_0 is the initial fracture aperture (here, before the re-saturation starts, immediately after excavation and emplacement) and K_{n0} is the fracture normal stiffness ($ML^{-2}T^{-2}$: *c.* 920 MPa m^{-1}) (Yong *et al.* 2010). The exponent α controls the velocity of fracture closure and is referred to as the closure rate. Note that at early times ($\Delta p \rightarrow 0$), the aperture tends to the initial one, in the absence of pressure perturbations. Low α values lead to smaller apertures for a given value of overpressure (alternatively, for a given value of re-saturation time). Correspondingly, the time required for a full re-saturation of the system is smaller for small α values. α is estimated by assuming that, after full re-saturation, all fractures are closed and attributed with a residual aperture of $b_r = 8.9 \times 10^{-8}$ m, corresponding to an irreducible transmissivity of $T_r = 5.7 \times 10^{-16}$ m^2 s^{-1}, according to the cubic law in equation (4). The residual aperture is selected in such a way that, after full re-saturation (i.e. after a differential pressure $\Delta p \rightarrow 7.8$ MPa, from atmospheric to hydrostatic conditions), the fracture hydraulic conductivity of the most conductive cell is small and equivalent to the matrix conductivity of the intact Opalinus Clay. This low residual transmissivity is in agreement with the results of the GS experiment at Mont Terri (NAGRA 2004) (Fig. 3b). Plugging Δp, b_r and a mean initial aperture $b_0 = 6.7 \cdot 10^{-4}$ m in equation (8) yields a α value of approximately -3.4.

Matrix porosity, $\phi_m(t)$, is calculated from $\phi_f(t)$ by assuming that the total porosity, ϕ, is constant during the re-saturation process (equation 1). This assumption can be justified by the fact that fracture porosity diminishes with time owing to mechanical closure caused by a reduction in effective stresses, whereas matrix porosity increases with time due to the swelling of the rock matrix. The rates of porosity reduction/increase of $\phi_f(t)$ and $\phi_m(t)$ are unknown and, therefore, it is assumed for simplicity that the two effects cancel each other out. Matrix porosity and hydraulic conductivity are related through the well-known Kozeny–Carman equation:

$$K_m(t) = \frac{\phi_m(t)^3}{(1 - \phi_m(t))^2} \frac{\rho g}{\mu} \frac{d_{10}^2}{180} \tag{9}$$

where d_{10} [L] is the particle size for which 10% of the soil is finer. d_{10} is calculated using the pair of porosity–hydraulic conductivity values of the intact Opalinus Clay given in Table 1, giving $d_{10} = 1.94 \times 10^{-8}$ m.

The re-saturation process is simulated using a numerical model with the zonation and deterministic parameters given in Table 1, and with the following boundary conditions: zero drawdown at the external boundary (i.e. the far field is not perturbed by the presence of the underground structure due to the low hydraulic conductivity of the intact Opalinus Clay); and zero flux at the inner zone representing the canister, which, for simplicity, is assumed to be impervious. The initial pressure condition is hydrostatic at cells not intersected by fractures. Otherwise, atmospheric pressure is imposed. At a given time, t, the overpressure distribution is calculated and fracture apertures are recalculated according to equation (8). Fracture transmissivities are then calculated according to the cubic law in equation (4). Then, fracture porosity and hydraulic conductivity are calculated according to equations (2) and (4).

Figures 7 and 8 display the temporal evolution of pressure and total hydraulic conductivity at selected times. Early stages of the re-saturation process are controlled by the high transmissivity of the many fractures defining the EDZ. This process is initially very fast, but very slow at mid and late times, when a certain degree of re-saturation has already been achieved. In fact, the buffer zone (backfilled with

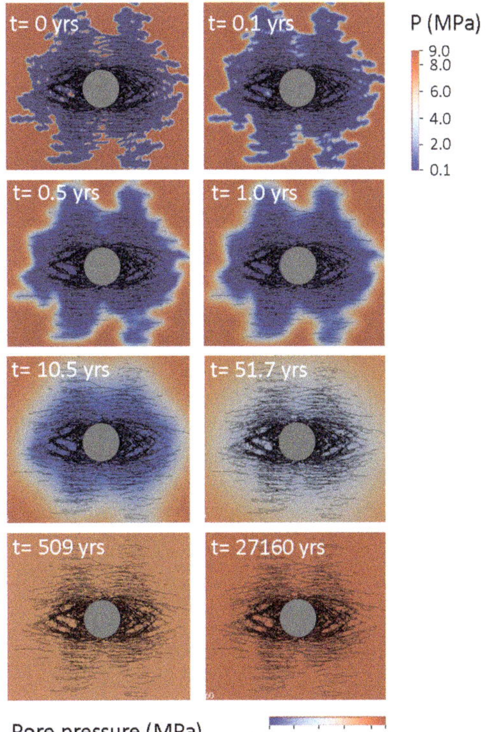

Fig. 7. Temporal evolution of pressure at selected time steps. Fractures defining the FDEM are superimposed (in black).

bentonite) and some fractures are initially unsaturated. A set of bi-phase scoping simulations was carried out to address the impact of the initial unsaturated state. It was observed that the saturation of the initially unsaturated pores takes less than 10 years. This time frame is negligible compared to the total time required for the full pressure recovery of the system (>20 000 years). Therefore, the initial unsaturated state was ignored in order to decrease the overall computational time. The time at which full re-saturation occurs is uncertain and largely depends on the closure rate, α. This issue is discussed later in this paper.

The temporal evolution of the system can be observed in Figure 9, which displays the number of closed fractures (i.e. those having residual aperture values) as a function of time. Most fractures are closed at early re-saturation times. This number steadily decreases with time and reaches a plateau after approximately 1000 years. The number of fractures defining the EDZ is >43 000. However, the number of closed fractures at the end of the simulation period is approximately 34 000. This means that full re-saturation of the EDZ was not achieved.

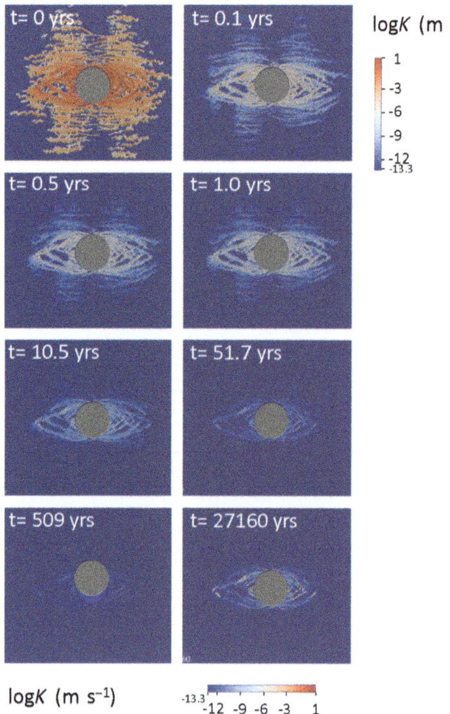

Fig. 8. Temporal evolution of axial hydraulic conductivity, K_{yy}, at selected time steps.

The fractures that remain open are clustered around the tunnel and are those with the largest initial aperture. The fast hydraulic response of the system is also observed in this plot.

Abstraction of the EDZ

This subsection describes the methodology for the abstraction of the EDZ. The abstraction consists of a simple piecewise homogeneous shell model that includes the liner, buffer zone, an equivalent EDZ and the intact Opalinus Clay (i.e. the far field). Homogeneous hydraulic properties are assigned to each zone. Hydraulic properties of the EDZ are averaged over its extent. Such abstraction is amenable to SA and frees the modeller from the burden of having to deal with complex, heterogeneous, EDZ models. Other geometrical conceptual models can be considered at this stage (e.g. elliptical zonation of the EDZ). The main advantage of the shell model proposed in this work, besides its simplicity, is that it allows the use of very simple analytical solutions or one-dimensional (1D) models for quick and simple calculations.

We consider the axial flow (i.e. parallel to the axis of the tunnel) through the modelled section, calculated as:

$$Q_y(r_i) = \sum_{j \in \text{shell } i} K_y^{i,j} A_{i,j} \qquad (10)$$

where r_i [L] denotes the radius of a shell i, j is a counter of cells defining that shell, $A_{i,j}$ is the cell area and $K_y^{i,j}$ denotes the axial hydraulic conductivity (equation 7). $Q_y(r_i)$ [L^3T^{-1}] is the axial flow rate under unit hydraulic head gradient between the extremes of the tunnel (i.e. the hydraulic conductance). To eliminate spurious mesh-related effects, Q_y is normalized by the area of the shell, thus becoming a specific flux, q_y [LT^{-1}]. Under unit hydraulic head gradient conditions, the specific

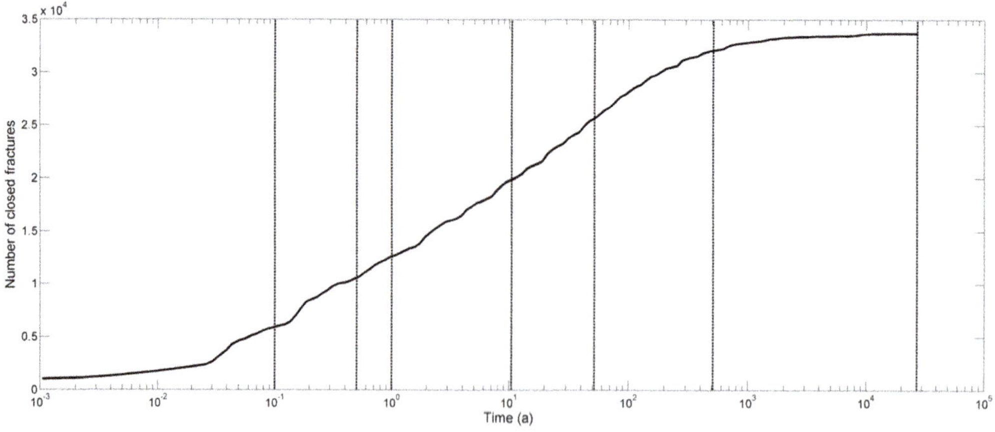

Fig. 9. Number of closed fractures v. time. The FDEM is made of 43 000 fractures. Vertical dashed lines depict the time steps in Figures 7 and 8.

flux at a certain shell is the average of the axial hydraulic conductivity.

Figure 10 depicts the temporal evolution of the equivalent EDZ hydraulic conductivity and the flow rate across the abstracted EDZ, Q_y. The equivalent hydraulic conductivity of the EDZ (or, equivalently, the total circulating flow rate) drops by approximately 11 orders of magnitude due to the mechanical closure of fractures and the swelling of the clay-rich materials of the rock matrix.

Notwithstanding the radial fluxes from/to the tunnel caused by regional (i.e. large-scale) porewater flow conditions in the host-rock formation, the axial fluxes and porosities at late times (i.e. after full re-saturation of the system, when all fractures are closed) are most relevant to SA (Bengtsson *et al.* 1991; Poller *et al.* 2014). Several observations can be made based on Figure 11:

- The specific flux, q_y, across fractures drops by almost 14 orders of magnitude after full re-saturation of the system (Fig. 11a) and by 13 orders of magnitude when the matrix is considered (Fig. 11b). Note that the total axial flux is already the residual one (*c.* 10^{-13} m s^{-1}) approximately 6 m far away from the centre of the opening, corresponding to the equivalent radius of the EDZ.
- At early times (Fig. 11c), fracture porosity decreases several orders of magnitude along a very small radial distance of approximately 6 m (i.e. the equivalent radius of the EDZ). In contrast, matrix porosity remains unchanged because swelling of the rock matrix has not yet taken place. At late times (Fig. 11d), fracture porosity is reduced by approximately four orders of magnitude due to mechanical closure. Instead, matrix porosity increases and coincides with total porosity as all fractures are closed. In fact, there is no distinction between the red and black lines in Figure 11d.

The late-time spatial distributions of effective porosity, $\phi(r)$, and hydraulic conductivity, $K(r)$, are the subject of further analysis, given its relevance to SA. First, averaged profiles $\langle\phi(r)\rangle$ and $\langle K(r)\rangle$ are obtained at each shell of cells defining the mesh of the EPM. The arithmetic mean is adopted for the porosity profiles, whereas the geometric mean is used to estimate the averaged conductivity profiles. Second, the average values of void volume $\langle V_p\rangle$ and total axial flux $\langle Q\rangle$ are calculated by integration:

$$\langle V_p\rangle = \int_{r_0}^{R_{\text{model}}} [\phi(r) - \phi_m]\, r\, dr$$
$$\langle Q\rangle = \int_{r_0}^{R_{\text{model}}} [K(r) - K_m]\, r\, dr \qquad (11)$$

where r_0 [L] and R_{model} [L] are the radius of the unlined tunnel and of the modelled section, respectively. Third, $\langle V_p\rangle$ and $\langle Q\rangle$ are used to define the homogenized properties of the abstracted EDZ in

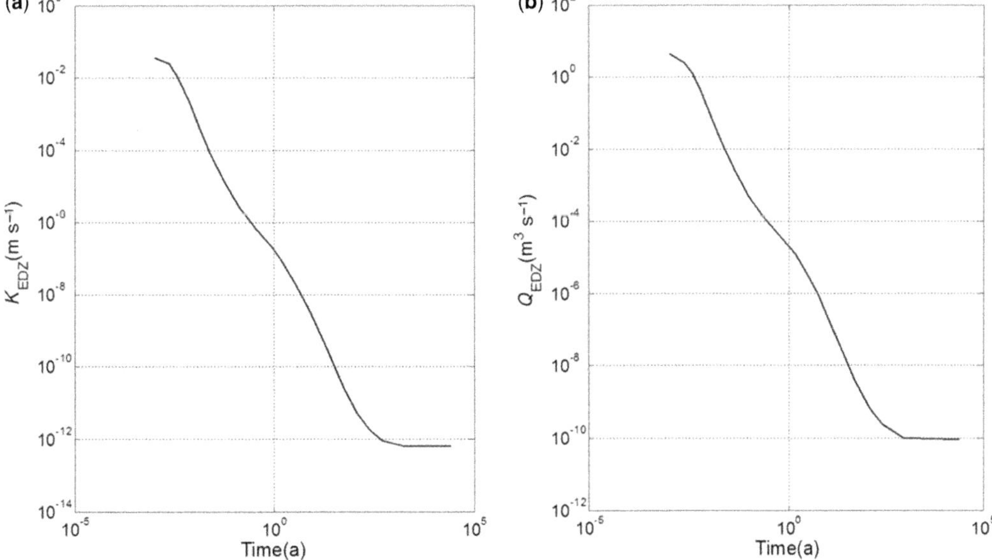

Fig. 10. Temporal evolution of (**a**) the equivalent axial hydraulic conductivity of the EDZ and (**b**) the axial flow rate.

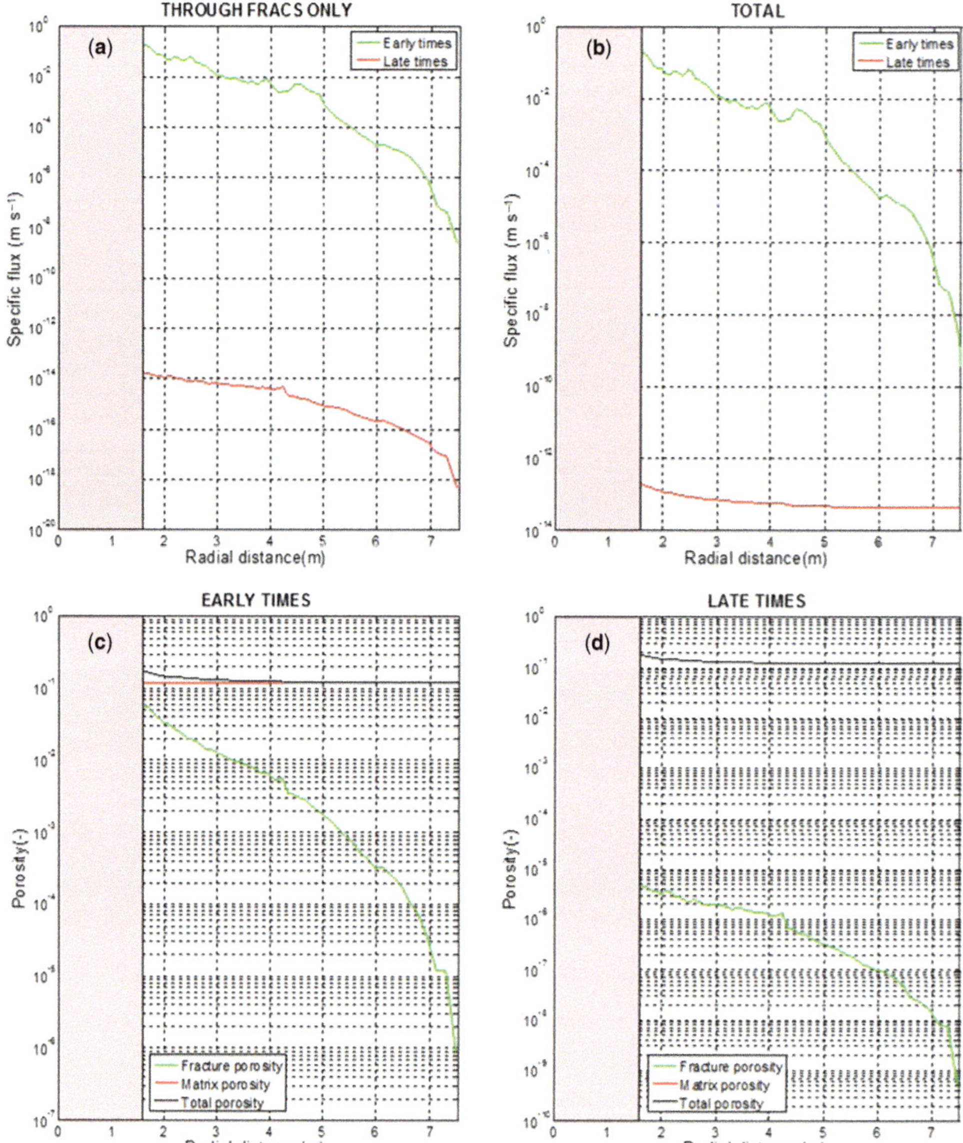

Fig. 11. Spatial distribution of: (**a**) the specific axial flux, q_y, through EDZ fractures only at early and late times; (**b**) the total specific axial flux at early and late times; (**c**) porosities at early times; and (**d**) porosities at late times ($t > 20\,000$ years).

terms of a relationship between the EDZ equivalent porosity, ϕ_{equiv}, hydraulic conductivity, K_{equiv}, and radius, r_{equiv}:

$$\phi_{\mathrm{equiv}} = \phi_{\mathrm{m}} + \frac{\langle V_{\mathrm{p}} \rangle}{\pi(r_{\mathrm{equiv}}^2 - r_0^2)}$$
$$K_{\mathrm{equiv}} = K_{\mathrm{m}} + \frac{\langle Q \rangle}{\pi(r_{\mathrm{equiv}}^2 - r_0^2)}. \tag{12}$$

Sensitivity analysis

This section illustrates the sensitivity analysis of the EDZ significance to various repository settings, relevant to a deep geological repository in the proposed siting regions in northern Switzerland. Excavation-induced fracture patterns around the repository structures were simulated for repository depths of between 450 and 800 m below ground, horizontal to vertical stress ratios, $K_{\mathrm{o}} = S_{\mathrm{h}}/S_{\mathrm{v}}$, of between

0.8 and 1.3, and rock mass strengths varying by a factor 5 around the reference values representative for the intact Opalinus Clay (Table 2) (Geomechanica 2013). Further simulations included stiff and soft tunnel support systems, and hypothetical repository configurations such as an intersecting fault system. A 3.0 m-diameter circular opening for the high-level waste SF-HLW emplacement tunnel (HAA hereinafter) is analysed in this study. Figure 12 displays simulated fracture patterns around the HAA cross-section for different repository depths (650 and 800 m below ground), different stress ratios ($K_o = 0.8$, 1.0 and 1.3), different rock strengths (low, reference, high) and different stiffnesses of the tunnel support (rigid, reference, soft). Notably, the fracture patterns exhibit marked differences in terms of size, shape and fracture intensity.

Alcolea *et al.* (2014) analysed the hydraulic behaviour of the EDZ at HAA cross-sections and also considered larger openings, such as shafts. The cross-sections of the HAA models were assumed perpendicular to the strike of flat-lying bedding planes (i.e. HAA models are vertical sections). The effect of the tectonic imprint on the EDZ response was investigated using five variants of the HAA model, in which three deterministic fractures dipping at 80° intersect the back and invert of the tunnel. These deterministic fractures have a thickness of 10 cm and are spaced at 0.5 m. Table 2 summarizes the main features of the simulations being analysed.

Figure 13a displays the radial distribution of fracture density, defined as the number of fractures within a given shell of mesh cells normalized by the area of the shell. The number of fractures close to the liner depends greatly on stress conditions and strength parameters under which the FDEM simulation was generated. However, a characteristic common to all FDEM simulations is that fracture density drops dramatically to almost negligible values at radial distances greater than about 4 m (radial distances are always measured with respect to the centre of the tunnel). The total extension of the EDZ (i.e. zero fracture density) is nearly the same for all FDEM models: approximately 8 m.

Table 2. *Summary of HAA model simulations*

Model	S_v (MPa)	S_h (MPa)	Strength properties	Elastic modulus of support (MPa)	Core-softening ratio γ	Number of EDZ fractures	Average aperture (mm)	Average initial transmissivity ($\log10T$ ($m^2 s^{-1}$))
Sensitivity to *in situ* stress conditions and core-softening ratio								
HAA-01	19.6	15.7	OPA × 2	32	0.01	43058	0.40	−4.28
HAA-02	19.6	19.6	OPA × 2	32	0.01	46366	0.40	−4.28
HAA-03	19.6	25.5	OPA × 2	32	0.01	55543	0.43	−4.19
HAA-04	19.6	15.7	OPA × 2	32	0.05	22419	0.30	−4.66
HAA-05	19.6	19.6	OPA × 2	32	0.05	25237	0.31	−4.62
HAA-06	19.6	25.5	OPA × 2	32	0.05	34891	0.36	−4.43
HAA-07	19.6	19.6	OPA × 2	32	0.008	50411	0.41	−4.26
HAA-08	15.9	20.7	OPA × 2	32	0.008	42331	0.41	−4.24
HAA-09	15.9	15.9	OPA × 2	32	0.003	36658	0.45	−4.11
Sensitivity to strength parameters								
HAA-10	15.9	20.7	OPA × 1.5	32	0.008	62413	0.37	−4.38
HAA-11	15.9	20.7	OPA × 3	32	0.008	18492	0.51	−3.96
HAA-12	15.9	20.7	OPA × 4	32	0.008	6697	0.68	−3.59
HAA-13	15.9	20.7	OPA × 5	32	0.008	3363	0.84	−3.32
Sensitivity to the presence of faults								
HAA-14	19.6	19.6	OPA × 2	32	0.01	56057	0.40	−4.28
HAA-15	19.6	19.6	OPA × 1.5	32	0.008	85324	0.38	−4.34
HAA-16	19.6	19.6	OPA × 3	32	0.008	25671	0.43	−4.20
HAA-17	19.6	19.6	OPA × 4	32	0.008	15037	0.45	−4.12
HAA-18	19.6	19.6	OPA × 5	32	0.008	9724	0.41	−4.26
Sensitivity to shotcrete stiffness								
HAA-19	15.9	20.7	OPA × 2	16	0.008	42376	0.41	−4.24
HAA-20	15.9	20.7	OPA × 2	3.2	0.008	43260	0.45	−4.13
HAA-21	15.9	20.7	OPA × 1.5	3.2	0.008	65376	0.40	−4.27
HAA-22	15.9	20.7	OPA × 3	3.2	0.008	19994	0.55	−3.86
HAA-23	15.9	20.7	OPA × 4	3.2	0.008	7378	0.71	−3.54
HAA-24	15.9	20.7	OPA × 5	3.2	0.008	3487	0.85	−3.30

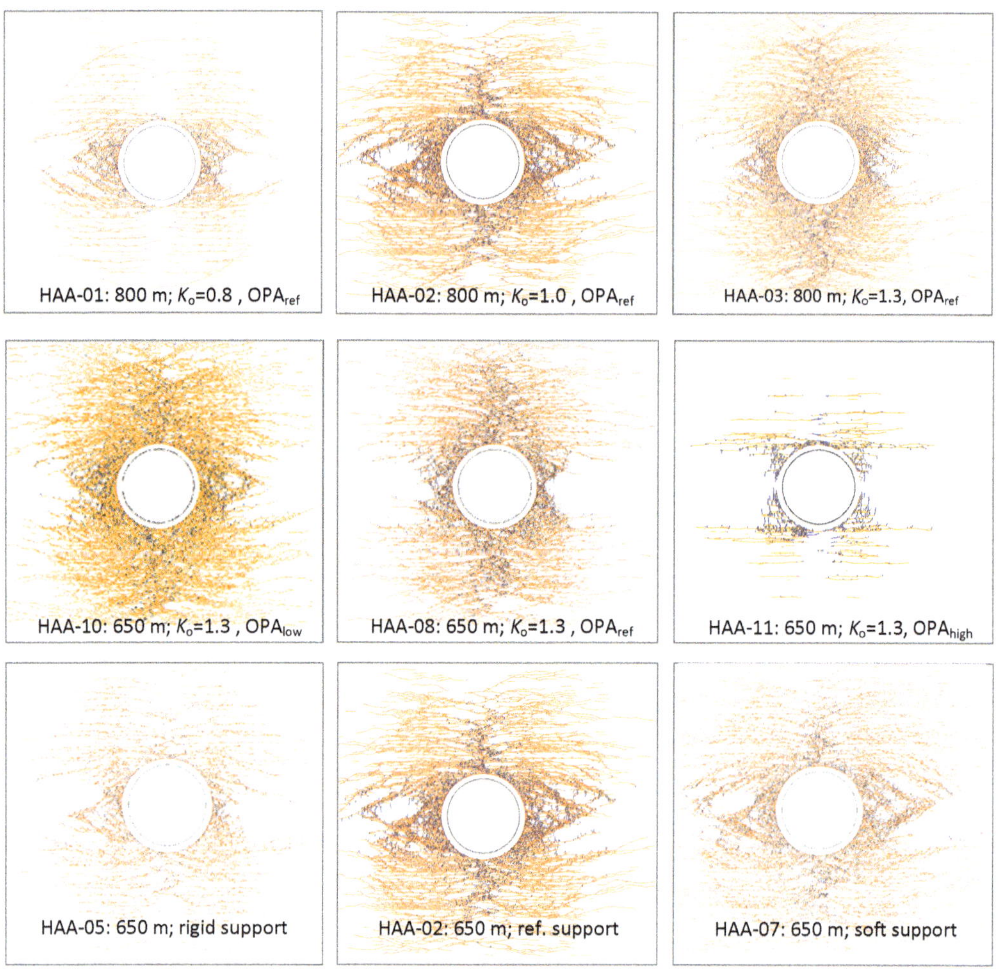

Fig. 12. EDZ fracture patterns around a circular SF-HLW emplacement tunnel ('HAA cross-section') simulated for various repository settings. The sensitivity cases comprise the variation of repository depth, the stress ratio $K_o = S_h/S_v$, the rock strength ($OPA_{low} = 0.5 \times$ reference strength, OPA_{ref}; $OPA_{low} = 3 \times OPA_{ref}$) and the stiffness of the tunnel support (stiff lining: core-softening ratio $\gamma = 0.003$; reference support: $\gamma = 0.01$; soft lining $\gamma = 0.05$). OPA denotes Opalinus Clay. Examples are taken from Geomechanica (2013).

Figure 13b displays the spatial distribution of matrix porosity after full re-saturation of the system. Swelling of the clay-rich materials forming the matrix increases matrix porosity from its initial value (0.12: Table 1) to values greater than 0.2. As observed, the sensitivity of the matrix porosity to different stress–strength conditions is large close to the liner. Yet, the overall sensitivity of matrix porosity is low at a radial distance of 2 m, even if fracture densities are still high.

Figure 13c, d depicts the fracture porosity at early and late times. Initial fracture porosities are high, up to 0.12, but drop drastically (more than five orders of magnitude) after full re-saturation of the system. The sensitivity of fracture porosity to stress–strength conditions and other factors defining the repository settings is also low. The same conclusion becomes apparent from Figure 13e, f, reporting on total specific fluxes. The very high initial transmissivity of the EDZ generates axial fluxes of up to 1 m s^{-1} (assuming unit gradient conditions). These drop substantially by more than 12 orders of magnitude after re-saturation of the system due to the mechanical closure of the fractures and the swelling of the clayey matrix.

Overall, the sensitivity of the hydraulic behaviour of the EDZ to repository depth, *in situ* stress conditions, rock mass strength and tunnel support system is low. The same conclusion can be drawn from Figure 14, which displays the homogenized

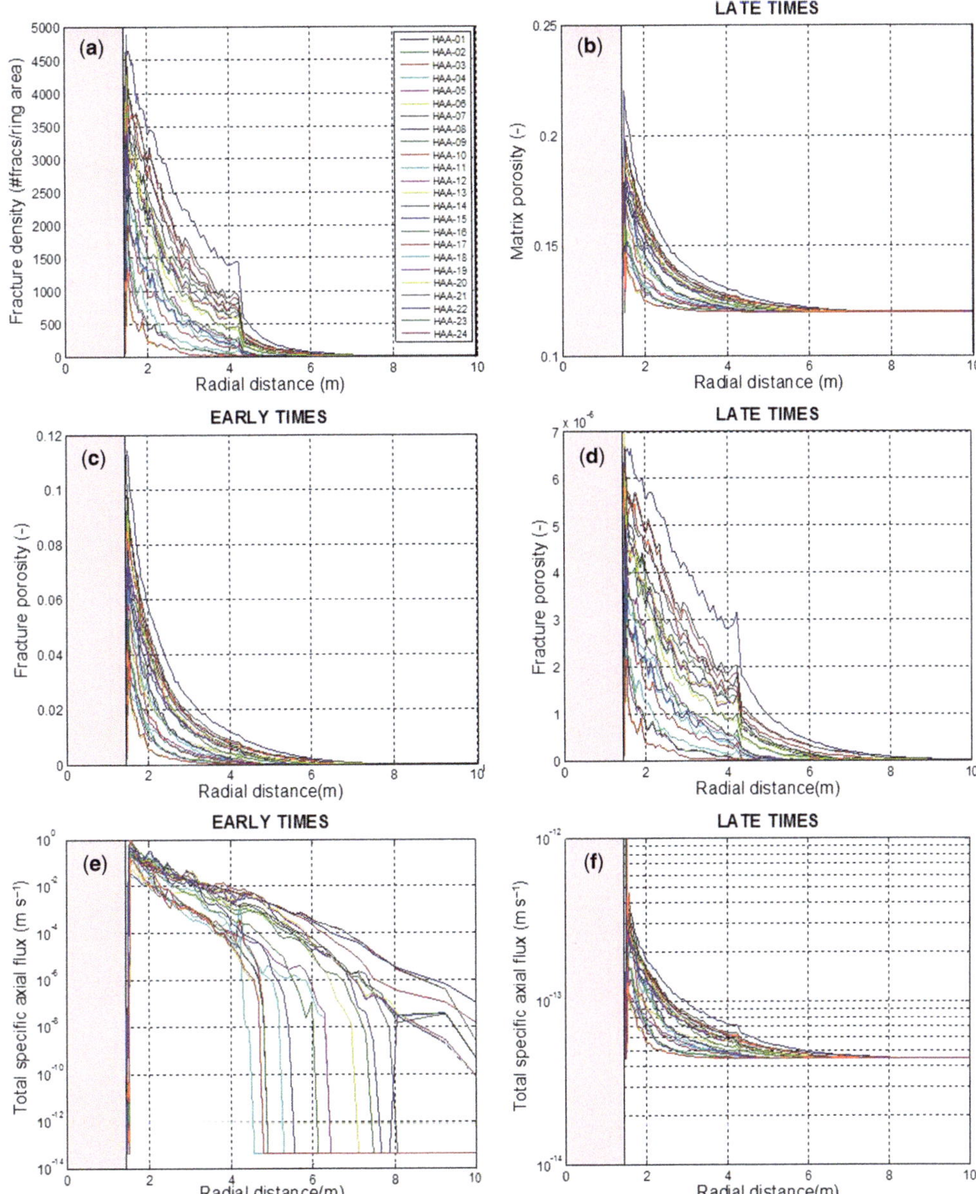

Fig. 13. Radial distribution of: (**a**) fracture density; (**b**) matrix porosity at late times, after full re-saturation of the system; (**c**) & (**d**) fracture porosity at early times after excavation and at late times, respectively; and (**e**) & (**f**) total specific axial fluxes at early and late times. The pink insets depict the buffer and liner zone.

equivalent properties of the EDZ. As observed, the radius of the EDZ is 5–6 m, regardless of repository settings. The choice of the EDZ radius and corresponding EDZ parameters (equation 12) is driven by the pragmatic considerations of the end-users' applications, such as probabilistic safety assessments. The similarity of results in terms of EDZ radius, despite the broad range of tested repository settings, is ascribed to the role of the extensile fractures, forming onion-like fracture patterns with significant fracture apertures in the immediate vicinity of the tunnel surface. The far-reaching shear fractures are characterized by smaller fracture apertures and emerge with a greater diversity of fracture patterns in response to different *in situ* stress conditions or to the presence of tectonic imprint. This is reflected (1) in the geometry and fracture density of the EDZ and (2) in the hydraulic behaviour at early times, when all fractures are open. At late times, self-sealing of fractures takes place. This leads to a certain homogenization of the system because the contribution of the shear fractures to the total conductance of the EDZ reduces when compared to that of the extensile fractures in the immediate vicinity of the tunnel surface.

Application to the HG-A experiment

The suggested approach is benchmarked with a dataset from the HG-A experiment: an *in situ* self-sealing experiment at the Mont Terri URL in Switzerland. A 1 m-diameter, 13 m-long microtunnel was excavated during February 2005 using a steel auger from a niche in Gallery 04 of the URL. The microtunnel was excavated parallel to the strike of bedding planes dipping at about 50°. The first 6 m of the microtunnel was lined with a steel casing immediately after excavation in order to stabilize

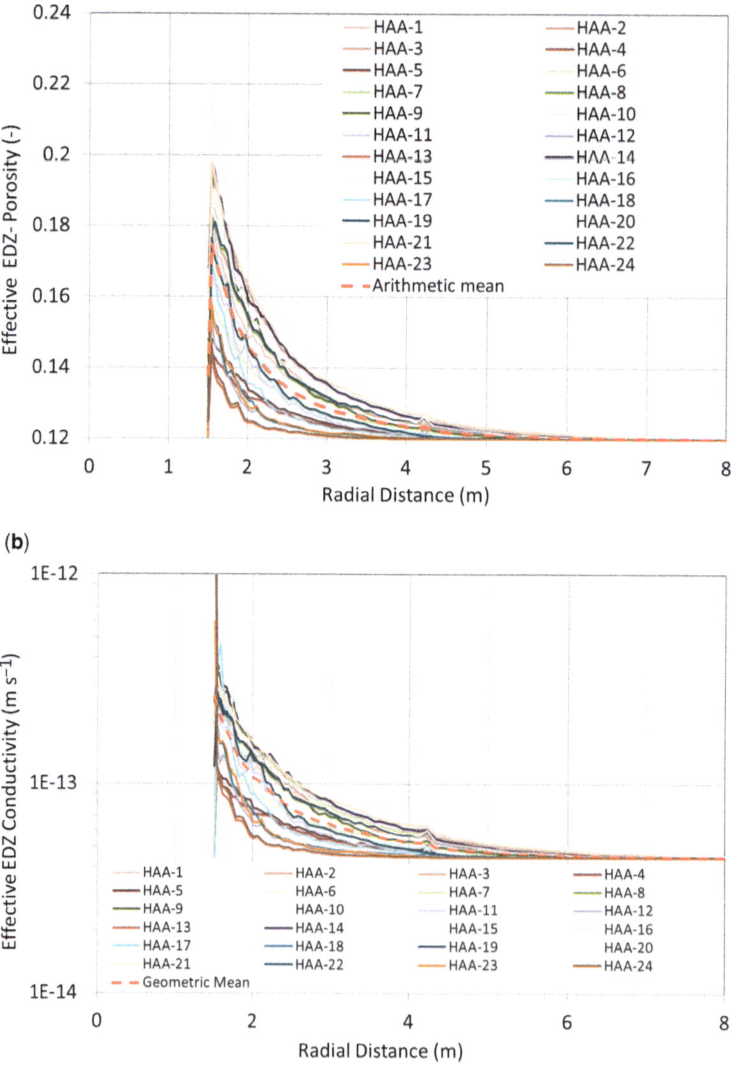

Fig. 14. Spatial distribution of equivalent porosity, ϕ_{equiv}, and equivalent hydraulic conductivity, K_{equiv}, for a variety of cases involving different repository depths, *in situ* stress conditions, rock mass strengths and tunnel support systems.

the opening. The gap behind the liner was then cement-grouted, but not sealed. A purpose-built hydraulic megapacker (diameter of 940 mm and sealing section length of 3000 mm) was installed in 2006. The sealing section was located at 6–9 m with a 1 m grouted zone containing the non-sealing part of the packer and a retaining wall from 9 to 10 m. The final 3 m of the microtunnel (from 10 to 13 m) formed the test section, which was instrumented and backfilled prior to megapacker emplacement (Fig. 15a) (see Lanyon *et al.* 2014 for further details).

Following emplacement of the megapacker, the 1 year-long saturation period was followed by extensive multi-rate injection testing over 2 years. Water was injected into the test section at the end of the micro-tunnel. During the multi-rate test, the injection rate was reduced from an initial

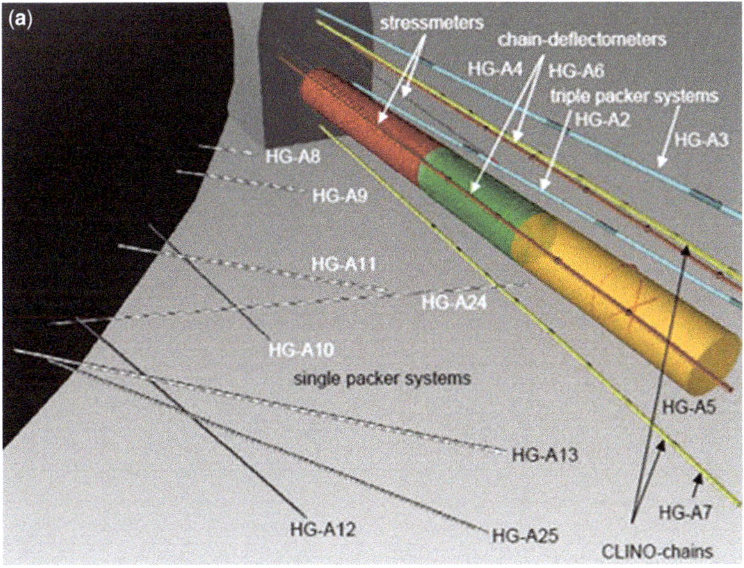

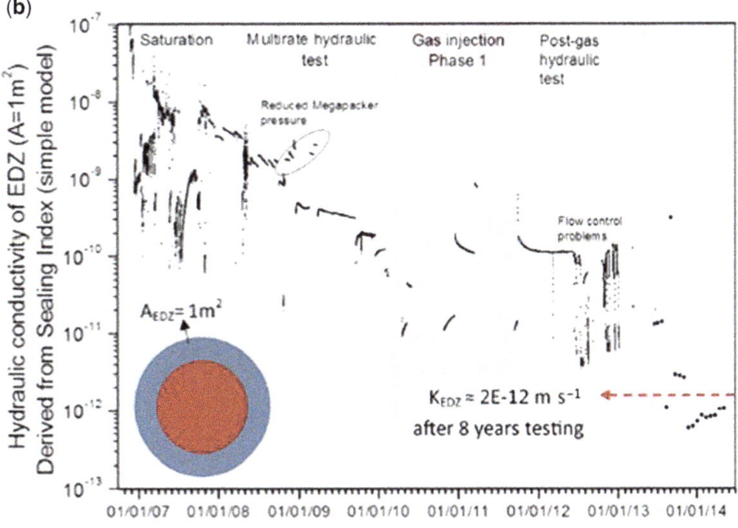

Fig. 15. (**a**) Layout and borehole instrumentation (colour coding: red, the steel liner: green, the seal section; orange, the backfilled test section). (**b**) Sealing index observed during the HG-A experiment. A significant reduction in the ability to inject water into the test section caused by the reduction of the EDZ conductive features is observed. At the end of the observation period, the effective hydraulic conductivity of an equivalent double-shell EDZ model with a cross-sectional area of 1 m^2 is around 2×10^{-12} m s^{-1}.

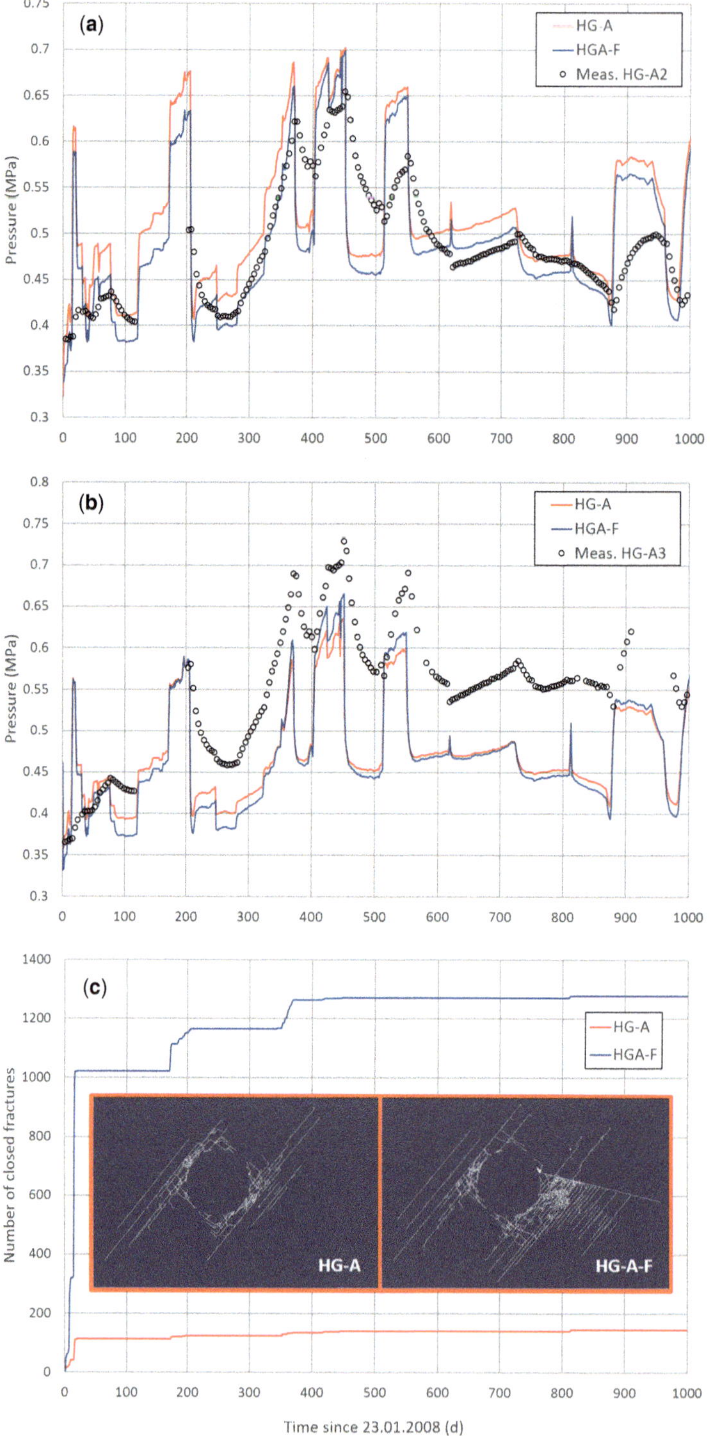

Fig. 16. (**a**) Pressure fits at observation borehole HG-A2 for simulations HG-A and HG-A-F; (**b**) pressure fits at observation borehole HG-A3; and (**c**) the number of closed fractures v. time. In the inset, the discrete fracture networks.

10 ml min^{-1} to less than 0.1 ml min^{-1} as the permeability of the EDZ dropped. The sealing index (SI), a simple measure of the hydraulic conductance of the EDZ, was calculated from the applied injection rate and the test section pressure, and is shown in Figure 15b. Assuming linear flow and a flow-path length of 3 m (seal section), the sealing index can be converted into an EDZ conductance (m^3 s^{-1}) by multiplying it by a factor of 5×10^{-7} (Lanyon *et al.* 2009, 2014).

The model applied for the benchmarking is slightly different to that presented in the previous section. In this case, we modelled a vertical cross-section at the seal section, and the canister and buffer zones were replaced by the megapacker. Available pressure measurements during the multi-rate injection test at the packered sections of observation boreholes HG-A2 and HG-A3 (Fig. 15a) were used in the calibration. Two parameters are calibrated by least-squares optimization: that is, the closure rate, α, and a dampening factor, β. The latter multiplies the prescribed pressure at the inner boundary condition (i.e. that caused by megapacker inflation/deflation) and mimics (a) a skin effect around the microtunnel and (b) the transfer of mechanical energy to fluid and solid phases in the vicinity of the megapacker external membrane. Two different FDEM models of the EDZ were considered: without and with tectonic fractures (Fig. 16b: models HG-A and HG-A-F, respectively). The model HG-A is initially not too transmissive compared to HG-A-F owing to its relatively low fracture density.

Figure 16a, b displays the pressure fits at observation boreholes HG-A2 and HG-A3 for both models. Note that the suggested modelling methodology is not aimed at obtaining good fits of available measurements for several reasons: (1) the model is 2D, whereas the reaction of the system to the re-saturation of the test section and to the inflation/deflation of the megapacker is clearly 3D (Lanyon *et al.* 2014); and (2) the hydraulic and mechanical responses are fully decoupled (actually, the elasto-plastic stress–strain equations are not solved). Instead, we seek a simple model, well suited to SA and capable of capturing the main physical processes caused by EDZ re-saturation. Such a simplistic model is shown to achieve relatively good fits of available measurements (see, for instance, the time period 300–400 days: Fig. 16a) and captures the main trends of the reaction to the forcing term measured at boreholes.

Best-fitting parameters are $\alpha = 1$ and $\beta = 1.05$ for HG-A, and $\alpha = 1$ and $\beta = 1.25$ for HG-A-F. The fact that a similar closure rate is attained by the two models (i.e. in the absence and presence of tectonic imprint) reveals that the calculated responses at observation points are not very sensitive to the geometry of the EDZ. However, the large difference between the best-fitting β factors shows the great sensitivity of the model to the hydraulic significance of the EDZ (Fig. 16c). In the absence of a tectonic imprint, fracture density is smaller and the forcing term, calibrated through the dampening factor β, is smaller. Instead, the higher EDZ transmissivities in the presence of tectonic imprint require a stronger forcing term to achieve the same goodness of fit.

Conclusions

The hydraulic significance of the EDZ around the backfilled underground structures of a geological repository has been identified as an important aspect in safety assessments (SAs). Indurated clay formations are characterized by brittle failure and complex coupled hydromechanical phenomena in response to the excavation process, precluding a detailed representation of the EDZ in conventional SA modelling tools. Instead, simplified models – still able to mimic the safety-relevant functional features of the EDZ – are required. This work presents a heuristic sequential modelling approach that is capable of representing adequately the key features and processes of the EDZ in indurated clays. First, brittle failure during the excavation process is simulated with a hybrid finite–discrete element code, providing realistic fracture patterns and fracture statistics that are in broad accordance with *in situ* observations. Secondly, EDZ self-sealing processes in response to pore-pressure recovery are modelled by the interplay of hydromechanical fracture closure mechanisms and swelling of the intact clay matrix. Thirdly, an abstraction towards a simplified shell-like representation of the EDZ is made. Such a simple model is hydraulically equivalent to the original discrete fracture network, but is amenable to SA.

A sensitivity analysis has been carried out to assess the hydraulic significance of the EDZ for various repository settings, relevant to a deep geological repository in the proposed siting regions in northern Switzerland. The sensitivity analysis included variations in repository depth, stress ratio, rock mass strength and design of the support system for a wide range of parameters. As a result, fracture patterns exhibiting marked differences in terms of size, shape and fracture density were obtained. The diversity of fracture patterns is reflected in the hydraulic simulations, which yield significantly different (several orders of magnitude) hydraulic conductances at early times of re-saturation. With time, the self-sealing of fractures takes place, causing fracture porosity and hydraulic conductivity to drop several orders of magnitude, and, conversely, matrix porosity and hydraulic conductivity to increase. At late times (typically of the order of 20 000 years after repository closure), the EDZ is restricted to a radial zone with a radius of 5–6 m (equivalent to a thickness of less than two tunnel diameters) for all simulation cases. A significant enhancement of hydraulic conductivity is observed only in a zone with a thickness of less than half the tunnel diameter. There, the enhancement of effective hydraulic conductivity is less than one order of magnitude with respect to that of the intact rock matrix. Furthermore, the corresponding increase in porosity in this zone is less than 20% of the porosity of the intact rock matrix.

The methodology was then benchmarked with a dataset from an *in situ* self-sealing experiment at the Mont Terri URL, demonstrating the capacity of the modelling approach to mimic the evolution of hydraulic conductance of the EDZ around a backfilled tunnel section during the re-saturation phase. The fact that a simple model, which captures the main physical phenomena behind the long-term evolution of the EDZ, reproduces well the trends observed in available measurement records should be viewed as a hopeful step in the direction of abstracting simple models, amenable to safety assessment, from complex representations of the EDZ.

This work has been supported by the Swiss National Cooperative for the Disposal of Radioactive Waste (NAGRA) and the Mont Terri Consortium, representing the partners of the HG-A experiment ANDRA (France), BGR (Germany), NWMO (Canada), NAGRA and Swisstopo (Switzerland).

References

ALCOLEA, A., KUHLMANN, U., LANYON, G.W. & MARSCHALL, P. 2014. *Hydraulic Conductance of the EDZ Around Underground Structures of a Geological Repository for Radioactive Waste – A Sensitivity Study for the Candidate Host Rocks in the Proposed Siting Regions in Northern Switzerland*. NAGRA Arbeitsbericht NAB 13-94. NAGRA, Wettingen, Switzerland.

ALHEID, H.J., ARANYOSSY, J.F., BLÜMLING, P., HOTEIT, N. & VAN GEET, M. 2007. *EDZ Development and Evolution – State of the Art*. NF-PRO. Contract Number F16W-CT-2003-02389. European Commission, Luxembourg.

ARANYOSSY, J.F., MAYOR, J.C. ET AL. 2008. *EDZ Development and Evolution (RTDC 4) – Final Synthesis Report (D. 4.5.3)*. FP6-EURATOM/NF-PRO, F16W-CT-2003-02389. European Commission, Luxembourg.

BENGTSSON, A., GRUNDFELT, B., MARKSTROM, A. & RASMUSON, A. 1991. *Impact from the Disturbed Zone on Nuclide Migration – A Radioactive Waste Repository Study*. SKB Technical Report 91-11. SKB, Stockholm.

BERNIER, F., LI, X.L. ET AL. 2007. *Fractures and Self-healing within the Excavation Disturbed Zone in Clays (SELFRAC)*. Final Report, EURATOM, Fifth Framework Programme, Contract No. FIKW-CT2001-00182. European Commission, Luxembourg.

BLÜMLING, P., BERNIER, F., LEBON, P. & MARTIN, C.D. 2007. The excavation damaged zone in clay formations time-dependent behaviour and influence on performance assessment. *Physics and Chemistry of the Earth*, **32**, 588–599.

BOCK, H., DEHANDSCHUTTER, B., MARTIN, C.D., MAZUREK, M., DE HALLER, A., SKOCZYLAS, F. & DAVY, C. 2010. *Self-Sealing of Fractures in Argillaceous Formations in the Context of Geological Disposal of Radioactive Waste Review and Synthesis Report*. Radioactive Waste Management. OECD NEA 6184. NEA OECD, Paris.

BOSSART, P. & THURY, M. (eds). 2008. *Mont Terri Rock Laboratory Project, Programme 1996 to 2007 and Results*. Reports of the Swiss Geological Survey, **3**. Swiss Geological Survey, Wabern.

BOSSART, P., MEIER, P.M., MOERI, A., TRICK, T. & MAYOR, J.C. 2002. Geological and hydraulic Characterization of the excavation disturbed zone in the Opalinus Clay of the Mont Terri Rock Laboratory. *Engineering Geology*, **66**, 19–38.

CORKUM, A.G. & MARTIN, C.D. 2007. Modelling a Mine-By test at the Mont Terri rock laboratory, Switzerland. *International Journal of Rock Mechanics and Mining Sciences*, **44**, 846–859, https://doi.org/10.1016/j.ijrmms.2006.12.003

FERRARI, A., FAVERO, V., MANCA, D. & LALOUI, L. 2013. *Geotechnical Characterization of Core Samples from the Geothermal Well Schlattingen SLA-1*. NAGRA Arbeitsbericht NAB 12-50. NAGRA, Wettingen, Switzerland.

FERRARI, A., FAVERO, V., MARSCHALL, P. & LALOUI, L. 2014. Experimental analysis of the water retention behavior of shales. *International Journal of Rock Mechanics and Mining Sciences*, **72**, 61–70.

GEOMECHANICA 2013. *Extent and Shape of the EDZ around Underground Structures of a Geological Repository for Radioactive Waste – A Sensitivity Study for the Candidate Host Rocks in the Proposed Siting Regions in Northern Switzerland*. NAGRA Arbeitsbericht NAB 13-78. NAGRA, Wettingen, Switzerland.

GEOMECHANICA 2014. *HGA Experiment: Numerical Simulation of the EDZ Formation and Mechanical Re-compaction Process Using a FDEM Approach*. Mont Terri Project Technical Note TN 2014-89. Mont Terri Project, St-Ursanne, Switzerland.

HAWKINS, I.R., SWIFT, B.T., HOCH, A.R. & WENDLING, J. 2011. Comparing flows to a tunnel for single porosity, double porosity and discrete fracture representations of the EDZ. *Physics and Chemistry of the Earth*, **36**, 1990–2002.

JACKSON, C.P., HOCH, A.R. & TODMAN, S. 2000. Self-consistency of a heterogeneous continuum porous medium representation of a fractured medium. *Water Resources Research*, **36**, 189.

LANYON, G.W. & SENGER, R. 2011. A structured approach to the derivation of effective properties for combined water and gas flow in the EDZ. *Transport in Porous Media*, **90**, 95–112.

LANYON, G.W., MARSCHALL, P., TRICK, T., DE LA VAISSIÈRE, R., SHAO, H. & LEUNG, H. 2009.

Hydromechanical evolution and self-sealing of damage zones around a microtunnel in a claystone formation of the Swiss Jura Mountains. *In*: *43rd US Rock Mechanics Symposium & 4th US–Canada Rock Mechanics Symposium*, June 28–July 1, 2009, Asheville, North Carolina. American Rock Mechanics Association, Alexandria, VA, 652–663 (paper 09-333).

Lanyon, G.W., Marschall, P., Trick, T., De La Vaissiere, R., Shao, H. & Leung, H. 2014. Self-sealing experiments and gas injection tests in a backfilled microtunnel of the Mont Terri URL. *In*: Norris, S., Bruno, J. *et al.* (eds) *Clays in Natural and Engineered Barriers for Radioactive Waste Confinement.* Geological Society, London, Special Publications, **400**, 93–106, https://doi.org/10.1144/SP400.8

Lisjak, A., Garitte, B., Grasselli, G., Müller, H. & Vietor, T. 2015. The excavation of a circular tunnel in a bedded argillaceous rock (Opalinus Clay): short-term rock mass response and FDEM numerical analysis. *Tunnelling and Underground Space Technology*, **45**, 227–248.

Lisjak, A., Tatone, B.S.A. *et al.* 2016. Hybrid finite-discrete element simulation of the EDZ formation and mechanical sealing process around a microtunnel in Opalinus Clay. *Rock Mechanics and Rock Engineering*, **49**, 1849–1873, https://doi.org/10.1007/s00603-015-0847-2

Marschall, P., Distinguin, M., Shao, H., Bossart, P., Enachescu, C. & Trick, T. 2006. Creation and evolution of damage zones around a microtunnel in a claystone formation of the Swiss Jura Mountains. Paper SPE-98537 presented at the SPE International Symposium and Exhibition on Formation Damage Control, 15–17 February 2006, Lafayette, LA, USA, https://doi.org/10.2118/98537-MS

Martin, C.D. 1997. Seventeenth Canadian Geotechnical Colloquium: the effect of cohesion loss and stress path on brittle rock strength. *Canadian Geotechnical Journal*, **34**, 698–725.

Martin, C.D. & Lanyon, G.W. With Contributions from P. Bossart & P. Blümling 2003. *EDZ in Clay Shale: Mont Terri.* Mont Terri Technical Report 2001. GeoScience Ltd, Falmouth.

NAGRA 2002. *Projekt Opalinuston – Synthese der geowissenschaftlichen Untersuchungs-ergebnisse. Entsorgungsnachweis für abgebrannte Brennelemente, verglaste hochaktive sowie langlebige mittelaktive Abfälle.* NAGRA Technical Report NTB 02-03. NAGRA (National Cooperative for the Disposal of Radioactive Waste), Wettingen, Switzerland.

NAGRA 2004. *Effects of Post-Disposal Gas Generation in a Repository for Spent Fuel, High-Level Waste and Long-Lived Intermediate Level Waste Sited in Opalinus Clay.* NAGRA Technical Report NTB 04-06. NAGRA (National Cooperative for the Disposal of Radioactive Waste), Wettingen, Switzerland.

Nussbaum, C., Bossart, P., Amann, F. & Aubourg, C. 2011. Analysis of tectonic structures and excavation induced fractures in the Opalinus Clay, Mont Terri underground rock laboratory (Switzerland). *Swiss Journal of Geosciences*, **104**, 187–210, https://doi.org/10.1007/s00015-011-0070-4

Poller, A., Smith, P., Mayer, G. & Hayek, M. 2014. *Modelling of Radionuclide Transport along the Underground Access Structures of Deep Geological Repositories.* NAGRA Technischer Bericht NTB 14-10. NAGRA (National Cooperative for the Disposal of Radioactive Waste), Wettingen, Switzerland.

Romero, E. & Gómez, R. 2013. *Water and Air Permeability Tests on Deep Core Samples from Schlattingen SLA-1 Borehole.* NAGRA Arbeitsbericht NAB 13-51. NAGRA (National Cooperative for the Disposal of Radioactive Waste), Wettingen, Switzerland.

Steiner, W., Kaiser, P.K. & Spaun, G. 2010. Role of brittle fracture on swelling behaviour of weak rock tunnels: hypothesis and qualitative evidence. *Geomechanics and Tunnelling*, **3**, 583–596.

Tsang, C.F., Bernier, F. & Davies, C. 2005. Geohydromechanical processes in the Excavation Damaged Zone in crystalline rock, rock salt, and indurated and plastic clays – in the context of radioactive waste disposal. *International Journal of Rock Mechanics and Mining Sciences*, **42**, 109–125.

Yong, S. 2008. *A three-dimensional analysis of excavation-induced perturbations in the Opalinus Clay at the Mont Terri Rock Laboratory.* PhD thesis, ETH Zurich.

Yong, S., Kaiser, P.K. & Loew, S. 2010. Influence of tectonic shears on tunnel-induced fracturing. *International Journal of Rock Mechanics and Mining Sciences*, **47**, 984–907.

The influence of different supports on the properties of the excavation damaged zone along the FE tunnel in the Mont Terri Underground Rock Laboratory

HUA SHAO[1]*, BENJAMIN PAUL[1], XUERUI WANG[1], JÜRGEN HESSER[1], JENS BECKER[2], BENOIT GARITTE[2] & HERWIG MÜLLER[2]

[1]*Federal Institute for Geosciences and Natural Resources (BGR), Hanover, Germany*

[2]*National Cooperative for the Disposal of Radioactive Waste (NAGRA), Wettingen, Switzerland*

**Correspondence: shao@bgr.de*

Abstract: Permeability and its spatial distribution around an underground opening in a geological formation are important for the interpretation of thermal, hydraulic and mechanical findings from an *in situ* demonstration experiment. Within the site characterization programme of the Full-scale Emplacement (FE) experiment, permeability measurements with nitrogen gas have been conducted from six short boreholes. Four of them were located in a section without shotcrete support and two in a section with a three-layer-shotcrete lining. As expected, the extension of the zone with an increased permeability was larger (up to 2 m) in the area without shotcrete support than that in the section with a shotcrete lining (less than 1.5 m).

The water content in the sections with or without shotcrete linings also showed different behaviour over long-term monitoring. The water content in the deep borehole section in the area with a shotcrete lining stayed almost constant, while the water content in the deep borehole section in the area without shotcrete tended to continuously decrease. In general, the water content close to the tunnel is influenced by the seasonal change in the temperature and relative humidity within the tunnel, especially in the section without a shotcrete lining.

Analysis of the abovementioned observations/findings was done by performing FEM (finite-element method) calculations with OpenGeoSys (OGS) software using a coupled hydromechanical model. Owing to the high stiffness of shotcrete, the displacement in the section with a shotcrete lining was smaller. This, in turn, results in a smaller extension in the excavation damaged zone (EDZ). However, shotcrete has a relatively high suction capacity and high initial water content: thus, the interface between the shotcrete and the Opalinus Clay becomes more saturated. Therefore, the excavation-induced fractures in the Opalinus Clay close to the shotcrete can be sealed by swelling. The water content decreases continuously, as a result of desaturation occurring during the operational phase and the associated change in porewater pressure.

To better understand the coupled thermohydromechanical (THM) processes of heated Opalinus Clay, many *in situ* experiments have been performed at the Mont Terri Rock Laboratory located in northern Switzerland. Based on the HE-D Experiment, a borehole-scale (diameter 30 cm) experiment without backfill material (Wileveau 2005), and HE-E Experiment, a micro-tunnel (diameter 1 m) with two heaters filled with backfill materials (MX-80 bentonite and sand–bentonite mixture) (Gaus *et al.* 2014), a full-scale *in situ* heating experiment (FE) is under construction and in operation (Fig. 1) (Müller *et al.* 2012). The FE tunnel was excavated in 2012 as an extension of the so-called mine-by (MB) tunnel, it has a length of about 50 m and a diameter of about 3 m. Different tunnel support systems (without shotcrete, two-layer shotcrete and three-layer shotcrete) were emplaced to investigate the influence of the technical supporting system on the deformation behaviour (Fig. 2). Three heaters with a power of 1500 W, each surrounded by backfill materials, simulate the heat generated by a real high-level waste repository following the Swiss design concept. Studies will focus on the long-term coupled THM behaviour of the host rock and the buffer material, as well as their interaction.

Within the FE Experiment, a site characterization project was initiated by NAGRA (Müller *et al.* 2015). BGR, as an FE project partner, is involved in the near-field investigation programme (Shao *et al.* 2008). The site-specific know-how about the permeability distribution around the FE tunnel is of particular importance for the design and numerical interpretation of long-term monitoring. Boreholes have been drilled into two gallery sections: one is placed in a section with a three-layer shotcrete

From: Norris, S., Bruno, J., Van Geet, M. & Verhoef, E. (eds) 2017. *Radioactive Waste Confinement: Clays in Natural and Engineered Barriers*. Geological Society, London, Special Publications, **443**, 149–157.
First published online September 26, 2016, https://doi.org/10.1144/SP443.18

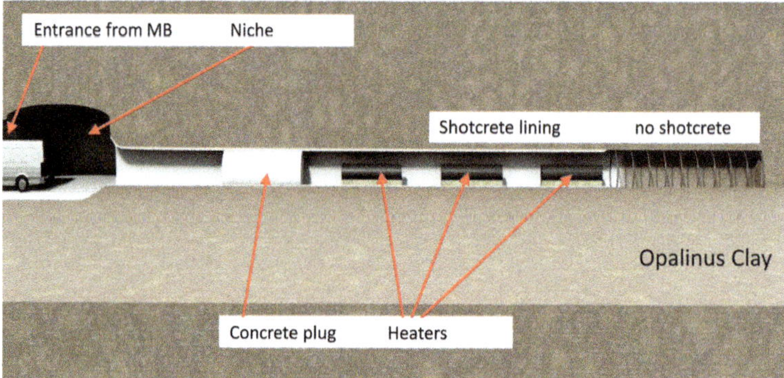

Fig. 1. Schema of the Full-scale Emplacement (FE) experiment (Müller *et al.* 2012).

lining (14.7 m into the tunnel), the other in a section without shotcrete support (38–42 m into the tunnel) (Fig. 2). Geo-electrical and ultra-sonic measurements, as well as permeability measurements, have been conducted in all boreholes.

For the permeability tests, nitrogen gas is injected into short intervals of about 10 cm along the boreholes. In the immediate near-field (up to 1 m away from the tunnel surface), high permeabilities of 10^{-13} m^2 have been measured. Because of different extensions of the zone with increased permeability, numerical simulations that took into account the monitored temperature and humidity during the excavation were applied in order to analyse the mechanisms of fracture generation and development.

This paper focuses on the permeability measurements within the FE site characterization programme and the related numerical interpretation using the fully coupled THM code OGS.

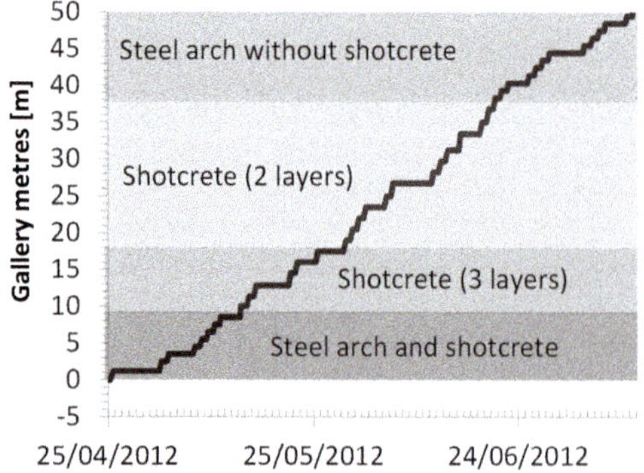

Fig. 2. Excavation sequence and supporting systems emplaced in the FE tunnel (Garitte *et al.* 2014). Gallery metres is the distance into the tunnel.

In situ measurements

Near-field permeability tests

In total, six boreholes in two sections were drilled for geophysical and permeability measurements. These boreholes were orientated in different directions with respect to the orientation of the tunnel and the bedding plane in order to find the bedding dependency of the excavation damaged zone (EDZ). Four of these were orientated perpendicular to the bedding plane because breakouts in such an orientation are less likely. A borehole camera scan was conducted before testing in order to identify suitable packer positions for the nitrogen injection tests. Breakouts were found in some boreholes, especially in the horizontal borehole BFEE004 (Fig. 3). A local ‘wetspot’ was also found in one borehole. The predefined test locations were verified using information about anomalies obtained from seismic measurements.

The permeability measurements were obtained using pulsed injections of nitrogen gas. In total, 107 injections were carried out in the six boreholes. The monitored gas pressure during the recovery phases was used for a 3D numerical evaluation that was able to simulate the entire experimental procedure including, for example, the spatial gas flow around the packer. Therefore, more precise values of the permeability of the rock could be obtained (Kunz *et al.* 2013). The uncertainty of the estimated permeability was lower than half an order of magnitude.

Measured permeability distributions in both sections with and without shotcrete support fit well with the near-field assumptions of the permeability. The highest permeability was around 10^{-13} m^2 in the first 1 m in the tunnel wall and decreased with the depth of the borehole. The minimum permeability of Opalinus Clay was lower than 10^{-20} m^2, which may be suggested as the permeability of the

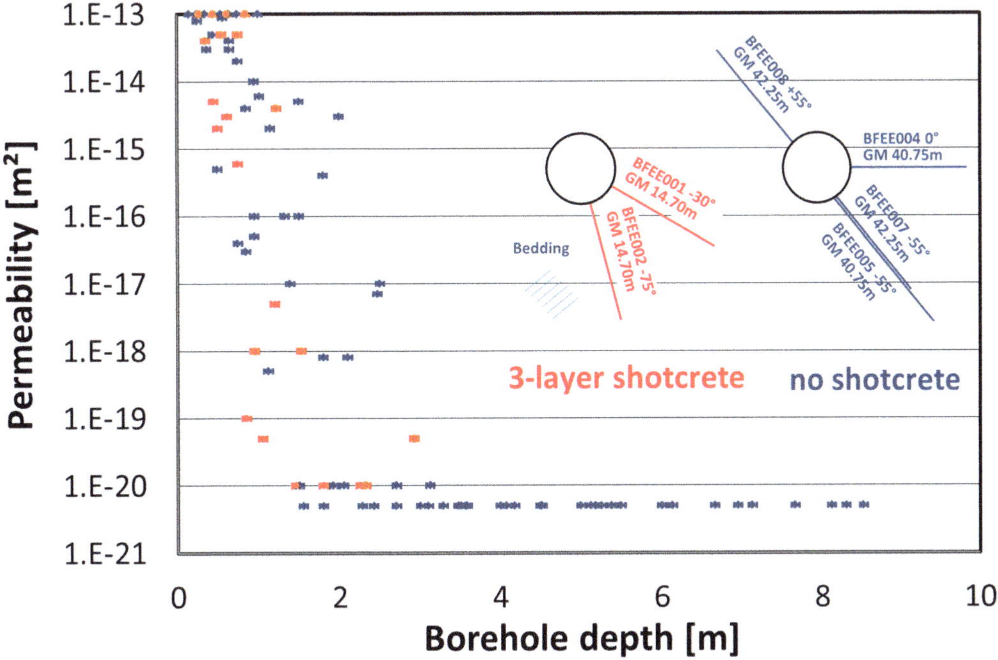

Fig. 3. Permeability distribution in the area without shotcrete lining (40–42 m inside the tunnel: blue lines) and in the section with a three-layer shotcrete lining (14.7 m inside the tunnel: red lines).

undisturbed clay rock. The extent of the zone with an increased permeability is generally, without taking into consideration the orientation of the boreholes, larger (up to 2 m) in the section without shotcrete support than that in the section with a shotcrete lining, which is less than 1.5 m (Fig. 3).

Long-term climate data in the laboratory

The hydromechanical (HM) properties of Opalinus Clay are strongly subject to the thermal, hydraulic and mechanical conditions acting on it. To interpret the *in situ* observation, data from the long-term monitoring performed in the entrance niche of the FE tunnel were analysed. Data on the relative humidity and temperature were provided by the site management (Swisstopo, Switzerland). The database covered the period before, during and after the excavation phase of the FE tunnel; the seasonal effects on the measured data were evident (Fig. 4) and, in addition, site activities during the excavation were clearly reflected in the temperature patterns.

As the fluctuation in temperature was limited (the difference between the maximum and minimum temperature was only 7°C), the relative humidity changed significantly with time. A difference of more than 30% in the relative humidity between winter and summer can be observed. Both temperature and relative humidity changes have an critical impact on the suction pressure acting on the tunnel surface. The suction pressure (p^c) can be calculated according to Kelvin's equation (equation 1):

$$p^c = -\rho^w \frac{RT}{M_w} \ln(\mathrm{RH}) \quad (1)$$

where R is the gas constant (8.314 J mol^{-1} K^{-1}), M_w is the water molar mass, ρ^w is the water density, T is the temperature in the tunnel and RH is the relative humidity.

Differences of almost one order of magnitude in the suction pressure can be related to changes in relative humidity and temperature (Fig. 5). In the winter period, it is dry and cold, and the suction pressure is therefore high, which indicates low saturation based on the relationship between the capillary pressure and the water saturation (equation 2). Low water saturation may induce shrinking and therewith deformation of the clay rock. During the summer time, processes are reversed due to high temperatures and relative humidity.

Measurements during the tunnel excavation

The 50 m-long FE tunnel was constructed using an excavator equipped with a spade chisel, and a rotary drum cutter for profiling (Fig. 2). The excavation and support of the FE tunnel were performed at an

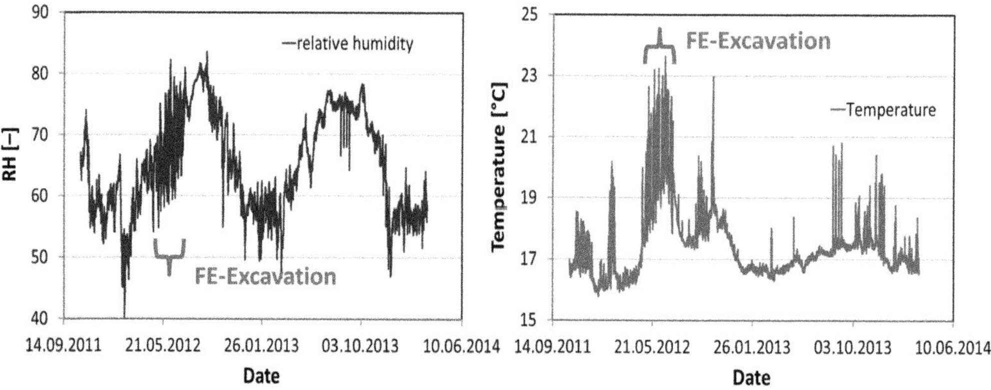

Fig. 4. Measured relative humidity (RH) and temperature after the excavation (Swisstopo 2014).

average rate of approximately 1 m per day. For the first 0–38 m of the tunnel, the support in the tunnel consisted of mesh-reinforced shotcrete. Between 9 and 38 m, the shotcrete was applied in two layers, with a total thickness of at least 16 cm; from 9 to 18 m, a third additional 10 cm-thick layer of shotcrete with a second layer of wire mesh was installed. The far end of the FE tunnel (from 38 to 50 m) was constructed using only steel arches and reinforcement mesh, but with no shotcrete, for rock support.

Two inclinometers above the tunnel were installed beforehand to monitor the settlement during the excavation. Borehole BFEA011 was located in a zone where a large amount of deformation could occur due to mechanical anisotropy. It was clear that the largest amount of settlement was that for ICL_40 at the far end of the FE tunnel where only steel arches were available for support. The smallest amount of settlement was for ICL_10, which was located in the area with a three-layer shotcrete support.

Only the settlement for ICL_01 represents a positive value, which may be explained as having been influenced by the pre-excavated niche between the MB and the FE tunnels.

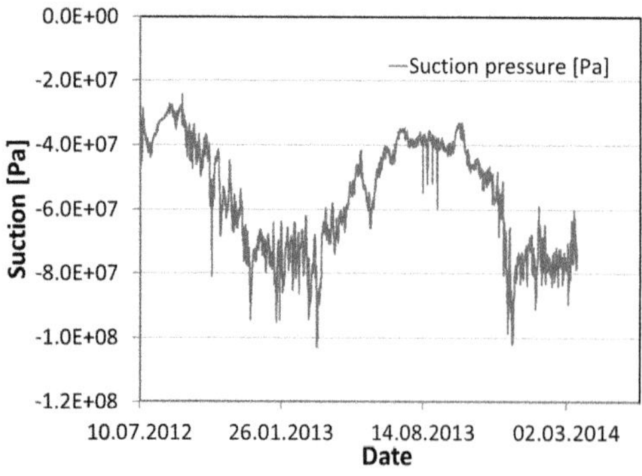

Fig. 5. Calculated suction pressure evolution after excavation, after equation (1).

Borehole water content

Three types of sensors (relative humidity sensor, equitensiometer and time-domain reflectometer (TDR)) were located in the boreholes in the FE tunnel to obtain measurements of water content. The sensors were installed at different depths in the borehole to obtain a profile along it (Gräfe *et al.* 2013).

The water content measured from two boreholes – one located in the deep tunnel section without shotcrete (e.g. BFEC006) and another located in the section with three-shotcrete layers (e.g. BFEB050) – was analysed. The water content in borehole BFEC006, especially in the near field (at a depth of 0.08 m from the tunnel wall), reacted seasonally, while the water content in borehole BFEB050 showed a decreasing trend. Shortly after applying the shotcrete, an increased water content at all intervals (BFEB050) was measured, this decreased, however, after some hours. In the deeper intervals, the water content in both boreholes remained high, at a value of 0.17, which corresponds to the initial water content of the Opalinus Clay.

Hydromechanical properties of shotcrete

Both hydraulic and mechanical properties of shotcrete were considered in the interpretation of the results. In the unsaturated flow system, the suction pressure in the retention curve was the driving force for flow. Based on laboratory data for concrete and mortar (Schneider *et al.* 2012), parameters in the Van Genuchten function (equation 2) have been determined for the shotcrete material:

$$p^{c} = p^{0}\left[S_{\text{eff}}^{-1/m} - 1\right]^{(1-m)} \tag{2}$$

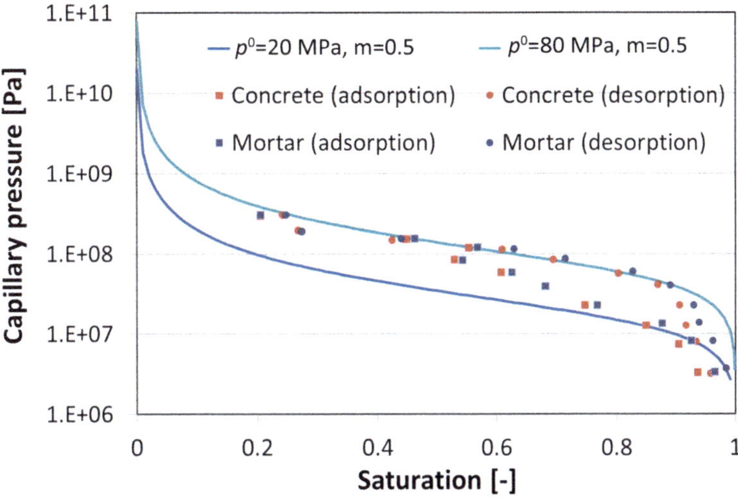

Fig. 6. Water-retention curves for shotcrete adopted for coupled HM calculations (symbols are the laboratory data from Schneider *et al.* 2012).

where p^0 is the air entry pressure, m is the index for pore-size distribution and S_{eff} is the effective saturation. The parameters p^0 and m can be determined by fitting to the laboratory data for concrete and mortar (Fig. 6).

The permeability of emplaced shotcrete was assumed to be 10^{-19} m^2 (Table 1).

A time-dependent Young's (or elastic) modulus for shotcrete was adopted according to the measured data in the laboratory, based on the fact that the stiffness increases with time after emplacement. A 10-day hardening process was simulated using a linear function (Fig. 7).

Modelling

Numerical background and parameters

Numerical calculations were performed in both 2D and 3D models using the finite-element code OGS (Kolditz *et al.* 2014) to interpret the measured data and to understand the experimental findings. These models combine an unsaturated flow model (Richards' flow) and an elastic model, taking into account anisotropic deformation behaviour. The effective stress (σ^{eff}) is defined by using the water saturation as a weighting function for the water pressure (equation 3) (Xu *et al.* 2013):

$$\sigma^{\mathrm{eff}} = \sigma^{\mathrm{tot}} + \alpha (S^{\mathrm{w}})^{\chi} p^{\mathrm{w}} \tag{3}$$

where σ^{tot} is the total stress, α is Biot's coefficient, S^{w} is the degree of water saturation, χ is Bishop's coefficient and p^{w} is the water pressure.

A suction curve (Fig. 5) was applied as a boundary condition on the tunnel wall in both the 2D and 3D models. The parameters used are summarized in Table 1.

2D models

Three 2D coupled HM models were used to analyse the different extents of the increased permeability zone. All three models had a similar mesh structure

Table 1. *Parameters used in the model*

Parameter	Unit	Rock(‖/⊥)	Shotcrete
Elastic modulus (E)	MPa	8000/4000	$f(t)$ (Fig. 7)
Poisson ratio (ν)	–	0.35	$f(t)$
Porosity (φ)	%	13	13
Permeability (K)	m^2	$10^{-19}/10^{-18}$	8×10^{-19}
Storage coefficient (S_s)	m^{-1}	3×10^{-6}	3×10^{-6}
Biot's coefficient (α)	–	0.6	0.6
Bishop's coefficient (χ)	–	1.4	1.4

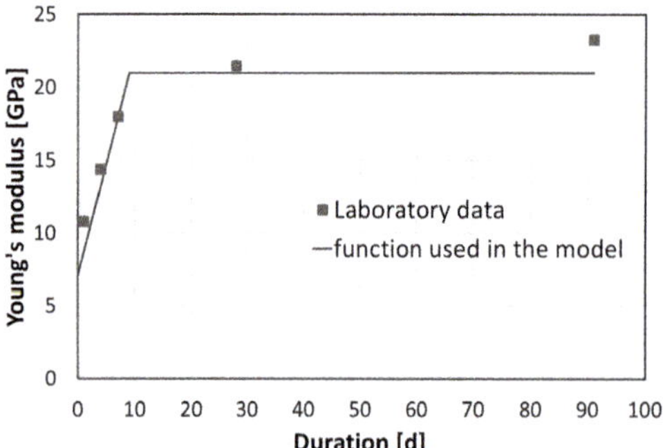

Fig. 7. Time-dependent Young's modulus for shotcrete (laboratory data from NAGRA).

with or without shotcrete layers and an EDZ zone with predefined EDZ properties that were not included in the models (Fig. 8). As initial conditions, measured principal stress (6.5, 4 and 2 MPa) and initial porewater pressure (20 bars) were used. No displacement and a constant pore pressure of 20 bars were assigned as the outer boundary conditions. After the excavation, free deformation and suction pressure curve (Fig. 5) were used in the tunnel wall. As expected, a relatively small deformation was modelled because of the high stiffness of the shotcrete in the case of the three-layer shotcrete model (Fig. 9). The extent of the zone with considerable deformation perpendicular to the bedding plane 1 year after the gallery construction was calculated to 2 m in the case of no shotcrete, while it was only 1 m in the case of the three-layer shotcrete. These results agree quantitatively well with the measured extent of the increased permeability zone described in the earlier subsection on 'Near-field permeability tests' (Fig. 3).

A more reasonable distribution of water content along borehole BFEC006 can be achieved using the 2D model without shotcrete (Fig. 10, left). The calculated distribution of water content along borehole BFEB050, however, is underestimated (Fig. 10, right). The reason for this could be the neglect of the initial water content in the shotcrete layers using the current model, on the one hand, and, on the other hand, an overestimation of the permeability of the shotcrete.

Sensitivity studies were performed by the variation of parameters: for example, the permeability of the shotcrete lining (Jeoung *et al.* 2008), and the specific storage of Opalinus Clay and shotcrete. Special attention was paid to the effect of the retention curves of Opalinus Clay and shotcrete.

3D numerical modelling of the excavation process

To better understand the excavation-induced processes, a 3D model was used to simulate the excavation of the FE tunnel (Fig. 11). Based on the calibrated parameters from the previous study on the MB Experiment, development of the measured deformation and pore pressure during the FE excavation can be well interpreted. In this model, a certain area representing the EDZ was considered. However, a constant thickness of 19 cm (the mean value) for the different shotcrete layers in the entire tunnel was used to determine the influence of the shotcrete on vertical deformation.

The measured data can be pretty well reproduced in trend using the suggested model. Particularly shortly after the passage of the excavation front, the agreement between the measured and calculated data is excellent both in trend and magnitude, except for the measured point ICL_01 of both boreholes (Fig. 12) that may have been influenced by the pre-excavated niche between the MB and FE tunnels, and so was not included in the current model. As expected, the low calculated settlements, especially for sensors ICL_26 and ICL_40, are definitively reduced to the application of a constant value. In fact, only two shotcrete layers with a thickness of about 16 cm in the section where ICL_26 is located and no shotcrete layer in the last section where ICL_40 is located were emplaced.

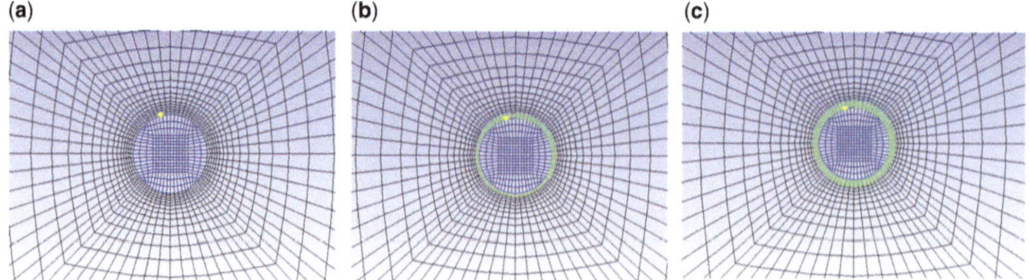

Fig. 8. 2D models: without shotcrete layer (**a**); with two-layer shotcrete (**b**); and with three-layer shotcrete (**c**).

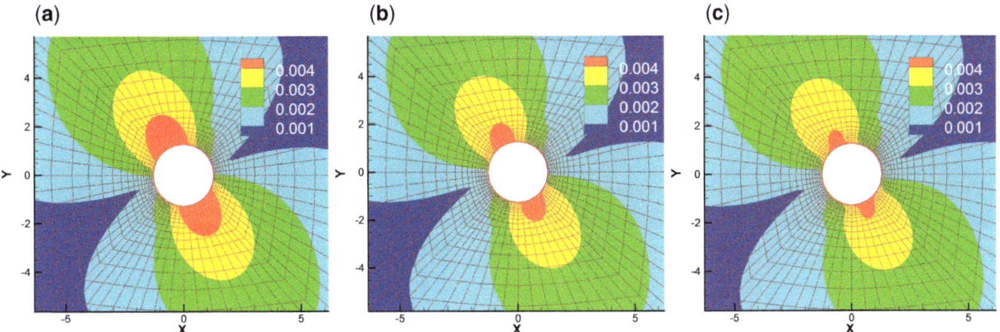

Fig. 9. Calculated displacement in three models: without shotcrete layer (**a**); with two-layer shotcrete (**b**); and with three-layer shotcrete (**c**).

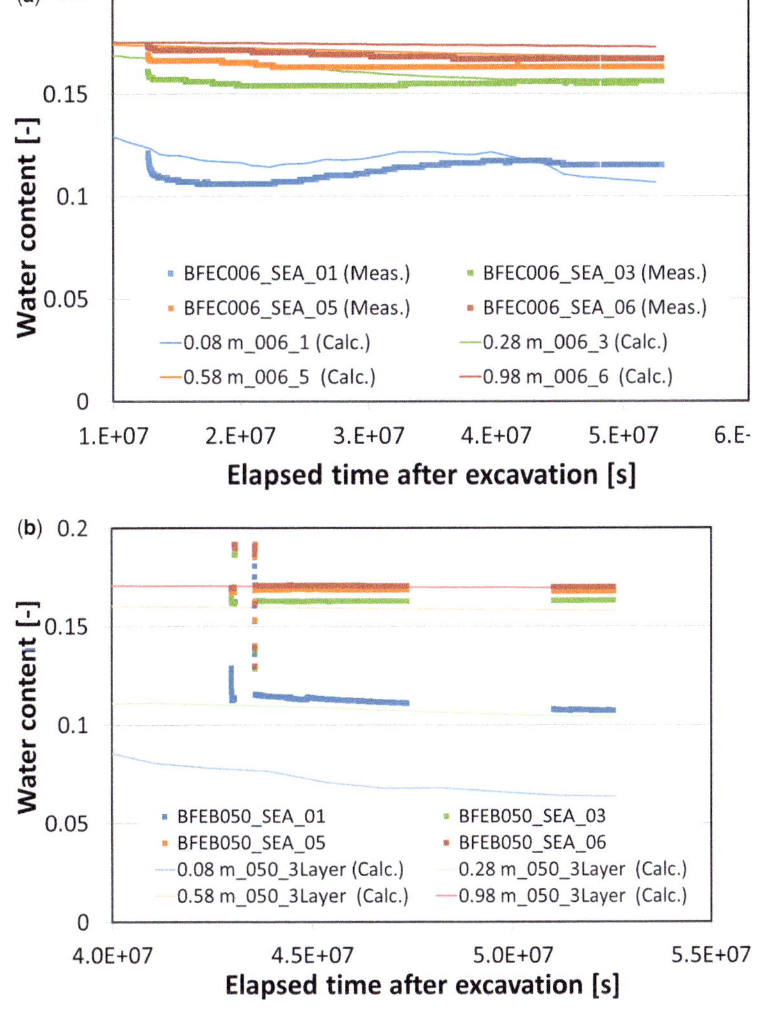

Fig. 10. Measured and modelled water content in the section without shotcrete (46 m inside the tunnel) (**a**) and in the section with three-layer shotcrete (**b**).

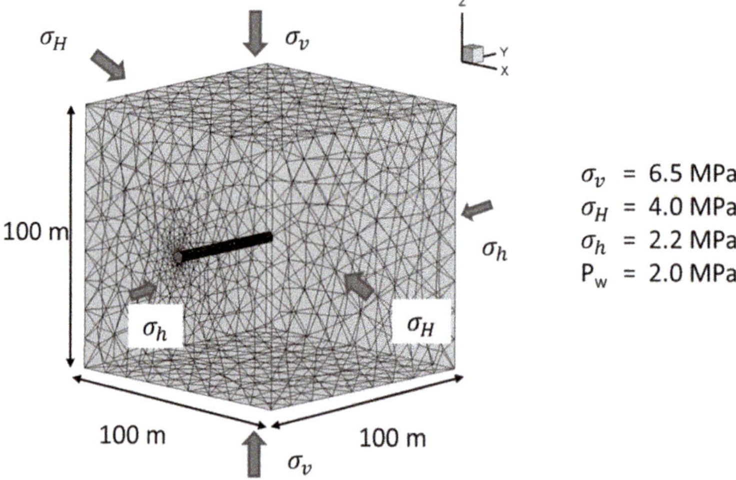

Fig. 11. 3D numerical model for the simulation of the FE tunnel excavation (σ_v is the vertical principal stress, σ_H is the maximum horizontal principal stress and σ_h is the minimum horizontal principal stress; P_w is the pore water pressure).

Results and conclusions

Within the full-scale emplacement (FE) project at the Mont Terri Underground Rock Laboratory, a site characterization programme was initiated by NAGRA. *In situ* measurements and numerical interpretations were conducted in order to better understand the generation and development of the excavation damaged zone.

The measured permeability distribution may serve as a complement to the database for the conceptual and numerical modelling of long-term THM monitoring. The permeability of undisturbed Opalinus Clay in the FE tunnel can be suggested to be as lower as 10^{-20} m^2 (Fig. 3), which is comparable to the assumed laboratory value. In the immediate near field (<1 m), permeability can be much higher (>10^{-14} m^2) than in the undisturbed rock. The extension of the increased permeability zone may be defined up to 2 m in the area without shotcrete lining support and about 1.5 m in the area with a three-layer shotcrete construction. Unfortunately, no measurement was carried out in the area with a two-layer shotcrete construction, which is

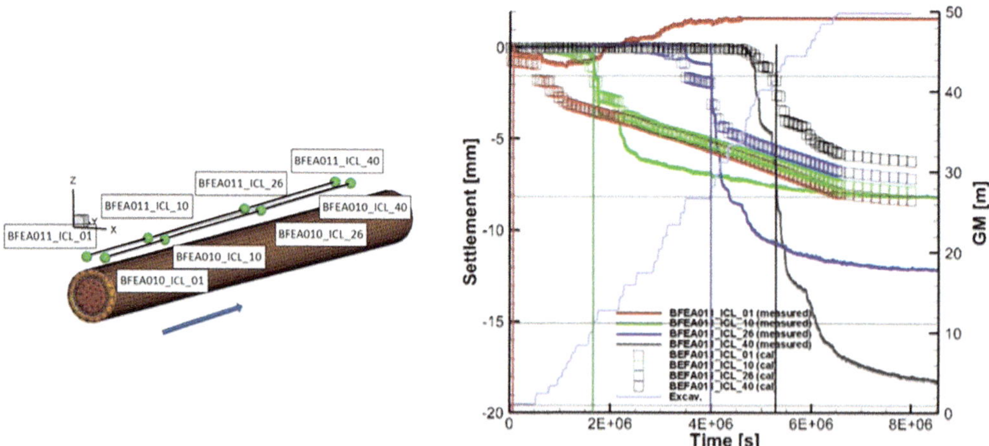

Fig. 12. Displacement of one inclinometer borehole (BFEA011) along the tunnel axis. GM, gallery metres (the distance into the tunnel).

where the heaters were located. Reference values from the zone with the three-layer shotcrete support may be suggested for use, based on the numerical analysis (Fig. 10, middle).

From both 2D and 3D model calculations, it is obvious that the shotcrete lining may play an important role as support for the clay rock in any underground opening. Mechanically, owing to the high stiffness of shotcrete, the displacement in the section with a shotcrete lining is smaller than in the section without shotcrete. This leads to a smaller EDZ extension in the section with a shotcrete lining. This is more obvious in the direction perpendicular to the bedding, where the elastic modulus (E) value for clay is only half that of the clay parallel to the bedding. Hydraulically, owing to the low permeability of shotcrete, the shotcrete lining may essentially prevent a desaturation process induced by seasonal change of temperature and relative humidity.

The shotcrete has a relatively high suction capacity (Schneider *et al.* 2012) and a high initial water content, and so the interface between the shotcrete and Opalinus Clay will become more saturated after the emplacement of the shotcrete construction. Thus, the excavation-induced fractures near the interface of the shotcrete and Opalinus Clay can be sealed as a result of swelling deformation. The water content decreases continuously, as a result of desaturation occurring during the operational phase and the associated change in porewater pressure.

This work was supported by BMWi (Bundesministerium für Wirtschaft und Energie, Berlin). The authors would like to thank all BGR technicians (D. Nowotny, S. Sanchez and S. Börner) involved in the field work, and Swisstopo (Ch. Nussbaum and D. Jäggi) for the field supporting and data delivery.

References

GARITTE, B., MÜLLER, H., VOGT, T., VIETOR, T., THATCHER, K. & SENGER, R. 2014. Scoping computations for the full-scale emplacement (FE) experiment at the Mont Terri underground research laboratory. Paper presented at the International Conference on the Performance of Engineered Barriers, 6–7 February 2014, Hanover, Germany.

GAUS, I., JOHNSON, L. ET AL. 2014. EBS performance at temperatures above 100°C. Paper presented at the International Conference on the Performance of Engineered Barriers, 6–7 February 2014, Hanover, Germany.

GRÄFE, K., VRZBA, M. & RÖSLI, U. 2013. *FE-B Experiment: Installation of Water Content and Suction Sensors for Gallery Near Field Instrumentation during Ventilation Phase*. Technical Note 2013-30. Mont Terri Project, Switzerland.

JEOUNG, J.Y., LEE, S. & LEET, G. 2008. Evaluation of hydraulic conductivity of shotcrete lining samples. Paper presented at the World Tunnel Congress 2008 – Underground Facilities for Better Environment and Safety, September 2008, New Delhi, India.

KOLDITZ, O., SHAO, H., WANG, W.-Q. & BAUER, B. 2014. Thermo-Hydro-Mechanical-Chemical Processes in Fractured Porous Media – Modelling and Benchmarking. Springer, Heidelberg.

KUNZ, H., SHAO, H., ZARETZKI, B., ZHAO, H.G., SU, R. & WANG, J. 2013. 3D Numerical model evaluation of packer tests for hydraulic characterization of fractured rock. *In*: *Proceedings of SINOROCK 2013 – Shanghai*, 18–20 June 2013, China, Taylor & Francis, 479–484.

MÜLLER, H.R., WEBER, H.P., KÖHLER, S., VOGT, T. & VIETOR, T. 2012. The Full-scale Emplacement (FE) Experiment at the Mont Terri URL. Paper presented at the 5th International Meeting on Clays in Natural and Engineered Barriers for Radioactive Waste Confinement, October 2012, Montpellier, France.

MÜLLER, H.R., VOGT, T., GARITTE, B., KÖHLER, S., SAKAKI, T., WEBER, H.P. & VIETOR, T. 2015. The Full-Scale Emplacement Experiment. Implementation of a Multiple Heater Experiment at the Mont Terri URL. Paper presented at the 6th International Meeting on Clays in Natural and Engineered Barriers for Radioactive Waste Confinement, March 2015, Brussels, Belgium.

SCHNEIDER, S., MALLANTS, D. & JACQUES, D. 2012. Determining hydraulic properties of concrete and mortar by inverse modelling. *Materials Research Society Proceedings*, **1475**, mrc11-1475-nw35-p60, https://doi.org/10.1557/opl.2012.601

SHAO, H., SCHUSTER, K., SÖNNKE, J. & BRÄUER, V. 2008. EDZ development in indurated clay formations – In-situ borehole measurements and coupled HM-modelling. *Physics and Chemistry of the Earth*, **33**(Suppl. 1), S388–S395, https://doi.org/10.1016/j.pce.2008.10.031

WILEVEAU, Y. 2005. *THM Behaviour of Host Rock: (HE-D Experiment)*. Technical Report TR 2005-03. Mont Terri Project, Switzerland.

XU, W.J., SHAO, H., MARSCHALL, P., HESSER, J. & KOLDITZ, O. 2013. Analysis of flow path around the sealing section HG-A experiment in the Mont Terri Rock Laboratory. *Environmental Earth Sciences*, **70**, 3363–3380, https://doi.org/10.1007/s12665-013-2403-2

Characterization of transport and water retention properties of damaged Callovo-Oxfordian claystone

S. M'JAHAD[1]*, C. A. DAVY[2], F. SKOCZYLAS[2] & J. TALANDIER[3]

[1]*Institute of Mechanical and Industrial Applications, UMR EDF-CNRS-CEA-ENSTA ParisTech 9219, EDF Lab Paris-Saclay, 7 boulevard Gaspard Monge, 91120 Palaiseau, France*

[2]*Ecole Centrale de Lille and LML UMR CNRS 8107, 59650 Villeneuve d'Ascq, France*

[3]*ANDRA, 1–7 rue Jean Monnet, 92298 Châtenay-Malabry Cedex, France*

**Correspondence: sofia.m-jahad@edf.fr*

Abstract: In the context of the underground storage of radioactive waste, the aim of this experimental study is to characterize the effect of damage on transport and water retention properties of Callovo-Oxfordian (COx) argillite. The originality of the study is to simultaneously investigate the pore-size distribution, water retention, the dry, effective and relative gas permeability, and the gas breakthrough pressure (GBP) of damaged COx argillite. These different properties are all relevant to characterizing the fluid transport ability of COx argillite.

Results show that the damage has a significant impact on the properties of the COx argillite. It induces a decrease in its water retention capacity and GBP, and it increases its gas permeability and apparent porosity available to water owing to the creation of micro-cracks.

Another objective is to show which of these properties is the most suitable to detect early damage states in COx argillite, with a potential use being to identify them *in situ*. GBP appears to be the best 'detector' of damage because of its sensitivity to damage even under high confinement pressures. Gas permeability could be a good indicator of damage, as it increases significantly (one or several orders of magnitude) after the damage. Finally, the water permeability curve is a poor indicator of COx argillite damage.

The French National Radioactive Waste Management Agency (ANDRA) has designed an underground repository for high-level and long-lived nuclear waste. This repository will be located in the east of France (Meuse/Haute-Marne) and will be built within an argillaceous formation (Callovo-Oxfordian claystone – COx) 500 m underground. This claystone has favourable properties, such as low permeability and seismic stability.

In this context, ANDRA has constructed an underground research laboratory (URL) that includes excavated tunnels and galleries, and where many *in situ* experiments are being conducted. The excavation of the repository induces local damage around excavated tunnels. Following the excavation stage, an instantaneous decrease in containment is observed in the argillite, which corresponds to an increase in the deviatoric stress around the structure (ANDRA 2005). Consequently, this induces the creation of an excavated damaged zone (EDZ). This EDZ can be schematically divided into two concentric zones with a gradual transition: 'the fractured zone', which is close to the surface of the tunnel and has a dense network of connected fractures; and 'the micro-cracked zone' which is further away, and contains isolated fractures and diffusive damage (ANDRA 2012). In both cases, the EDZ is potentially the preferential pathway for fluid transport and radionuclide migration. For performance and safety assessments, it is important to characterize the impact of such damage on the behaviour of the host rock.

The long-term efficiency of the deep disposal of radioactive waste depends on the limitation of transfer (of gas mainly, but also of liquids) to the geological environment. Indeed, the corrosion of metallic elements (e.g. waste packages) can induce a significant production of hydrogen (H_2), and the gas pressure can rise up to 7 MPa in some storage facilities. It is admitted that the increase in gas pressure through the storage materials implies the migration of gas through (and out of) the facilities: hence, the need to take gas entry pressure into consideration in the storage safety. The aim of this study is to investigate the influence of damage on the hydraulic properties of partially or completely saturated COx claystone. The tested properties are water retention properties, gas permeability and gas breakthrough pressure (GBP).

Several studies (ANDRA 2005, 2012; Escoffier *et al.* 2005; Pham 2006; Davy *et al.* 2007; Distinguin & Lavanchy 2007; Boulin *et al.* 2008; Zhang *et al.*

From: NORRIS, S., BRUNO, J., VAN GEET, M. & VERHOEF, E. (eds) 2017. *Radioactive Waste Confinement: Clays in Natural and Engineered Barriers*. Geological Society, London, Special Publications, **443**, 159–177.
First published online December 22, 2016, https://doi.org/10.1144/SP443.23

2010; Skoczylas 2011; Cariou *et al.* 2012; M'Jahad 2012) have been conducted by ANDRA and the scientific community to investigate the behaviour of COx argillite under varied thermal, hydric, mechanical and chemical conditions, corresponding to deep underground storage conditions, in terms of water retention properties (Pham 2006; Distinguin & Lavanchy 2007; Cariou *et al.* 2012) and gas transport ability (Escoffier *et al.* 2005; Davy *et al.* 2007; Boulin *et al.* 2008), as well as the possible coupling between water retention and gas permeability (Zhang *et al.* 2010). However, very few studies have simultaneously investigated the pore-size distribution, water retention, and relative gas permeability and gas migration properties of damaged COx argillite that are afforded in this study. These different properties are all relevant to characterize the fluid transport ability of COx argillite. Another objective is to show which of these properties is the most suitable for detecting early damage states in COx argillite and with the potential to identify them *in situ*. We assess, experimentally, water retention (i.e. suction curves), relative gas permeability and gas breakthrough pressure for samples from different COx cores and with different orientations (from the bedding plane).

Experimental methods

Sample preparation and induced damage protocol

The samples were prepared from different cores of COx argillite obtained from drilling with given *in situ* orientation: vertical, horizontal and sloped, as shown in Figure 1. Cylindrical samples, 37.7 mm in diameter and 10 mm in height, were used for water retention isotherms, and gas permeability tests and gas breakthrough pressure tests. Cylinders of 20 mm diameter and 40 mm height were only used for water retention isotherms. Previous studies (Zhang *et al.* 2010; Skoczylas 2011; M'Jahad 2012) showed that the sample size does not impact on the water retention properties and permeability results. A damage procedure, which consists of one imbibition (with relative humidity (RH) at 100%)–drying (60°C) cycle, was conducted on the 'pre-damaged' (D) argillite samples in order to induce a small amount of diffusive damage, which is representative of the EDZ zone. However, this damage procedure produces similar (micro-cracking) damage to the COx as that produced when COx is subjected to ambient humidity and temperature conditions, under free volume and no mechanical loading. The main challenge of the study was to prepare 'intact' samples because a number of samples were initially damaged or were damaged during the preparation and testing phases (see Fig. 2). Therefore, the number of 'intact' argillite samples was very low compared to the number of samples initially prepared. In addition, it was difficult to develop a repeatable cracking methodology on argillite because it was difficult to obtain 'intact' samples. Therefore, a sorting methodology was developed, instead, in order to compare gas permeability results with those of previous Laboratoire de Mécanique de Lille (LML) studies (Skoczylas 2011).

Water retention experiment

For water retention experiments, samples were placed in hermetic chambers, above salt-saturated solutions, each at a fixed RH value of 100%, or 98, 92, 85, 75, 70, 59, 43 or 11% at 20°C. The evolution mass changes were recorded continuously: first, under constant RH until mass stabilization, and then after placing the samples in an oven at 65°C until dry mass stabilization was reached. The mass changes recorded at RH = 100% correspond to the saturated sample mass. The water saturation level, S_w, was calculated at each RH value using the following definition, where m_{RH} is the stabilized mass at given RH, m_{dry} is the sample dry

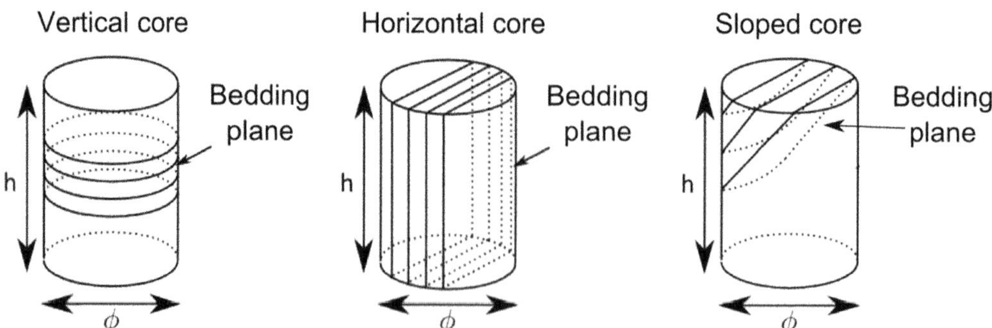

Fig. 1. Schematic orientation of the COx core samples and the position of the bedding plane.

Fig. 2. Photograph of cracked COx argillite showing pyrite inclusions, and the development of iron oxide and cracks in their area.

mass and $m_{saturated}$ is water-saturated mass (i.e. at RH = 100%):

$$S_w = \frac{m_{RH} - m_{dry}}{m_{saturated} - m_{dry}}. \quad (1)$$

Water retention curves were represented by plots of S_w (RH). In order to provide a single suction curve for numerical simulations, experimental data were fitted (in the least-squares sense) using the van Genuchten law (van Genuchten 1980), which is written as follows:

$$S_w = \left(1 + \left(\frac{P_{cap}}{P_r}\right)^n\right)^m \quad (2)$$

where P_{cap} is capillary pressure (see the following subsection on 'Pore-size distribution'), and P_r and n are the only two independent model parameters, m being linked to n through $m = 1 - (1/n)$. The n parameter is often called the 'pore-size distribution index' and could be related to the measure of the pore-size distribution (large values of m or n are associated with narrower distributions).The Pr parameter is usually interpreted as a gas entry pressure through the porous medium. At constant temperature, T, capillary pressure, P_{cap}, within a porous material is given by Kelvin's law (Gregg & Sing 1982):

$$P_{cap} = -\frac{\rho_{water} RT}{M_{water}} \ln(\mathrm{RH}) \quad (3)$$

where ρ_{water} is the density of water (taken as $1000\ \mathrm{kg\ m^{-3}}$ at 20°C), R is the gas constant ($8.314\ \mathrm{J\ mol^{-1}\ K^{-1}}$), T is the temperature in K, M_{water} (taken as $18.015\ \mathrm{g\ mol^{-1}}$) is the water molar mass and RH is the relative humidity.

Pore-size distribution

Despite the complexity of the argillite microstructure and in order to quantify its pore-size distribution, it is generally assumed that the porous material is basically described as a set of cylindrical pores of variable diameter, placed in parallel. At a given RH, an assessment of the largest water-saturated pore size is provided by Laplace's law by assuming the previous simplified model of the pore network. Using Laplace's law, capillary pressure is also related to the drained pore diameter, as follows:

$$d_{drained} = \frac{4\gamma}{P_{cap}} \quad (4)$$

where γ is the water–air surface tension (equal to $72.86 \times 10^{-3}\ \mathrm{N\ m^{-1}}$ at 20°C). In this case, all of the pores with diameters larger than $d_{drained}$ are dry. The adsorption/desorption isotherms can be used to estimate the pore-size distribution of a material. Rougelot *et al.* (2009) proposed a pore-distribution curve that uses experimental sorption isotherms to evaluate the pore-size distribution of the material. As for the mercury intrusion porosimetry (MIP) method, this model gives a 'pore distribution' that expresses the pore distribution, r, as a function of the drained pore diameter using the relationship below:

$$r(\mathrm{RH}^{i+1}) = \frac{S_w(\mathrm{RH}^i) - S_w(\mathrm{RH}^{i+1})}{\log(d_i) - \log(d_{i+1})} \quad (5)$$

where r is the pore distribution at the relative humidity RH^i, and d_i and d_{i+1} are the drained pore diameters corresponding, respectively, to the relative humidifies RH^i and RH^{i+1}.

With this model, pore diameters as small as 1 nm (corresponding to RH = 11% at 20°C using the Kelvin–Laplace law) can be reached, instead of 6 nm for MIP. However, very low pore diameters need to be considered with caution because of the limit of validity of Laplace's law (not valid for RH < 50%). Nevertheless, as pointed out by Silva & Grifoll (2007), capillary pressure and Kelvin's equation can be applied in soil water retention curves in the range of low water content.

For RH = 100%, Laplace's law does not allow the drained diameter to be estimated (theoretically, it is equal to infinity). It is assumed, arbitrarily, that the biggest pore drained, at RH = 100%, has a diameter of 500 nm (Brue *et al.* 2012). This value is usually used for concrete, as mentioned in Rougelot *et al.* (2009). For COx argillite, the authors do not know the size of the largest pore drained of argillite. This can be greater than 500 nm, but the authors chose the value arbitrarily. The authors assumed that the relative error of using this value is low. In fact, the results are similar when $d_{max} = 500$ nm and $d_{max} = 10$ μm as shown in Figure 3a, b.

Gas transport experiment

The gas permeability, K_{gas} (the gas used in the experimental tests in this study was argon), was measured for each sample at a given RH (or S_w) after mass stabilization and under a fixed confinement pressure (P_c) of 6 or 12 MPa. Gas relative permeability is defined as:

$$K_{rg} = \frac{K_{gas}(RH)}{K_{dry}} = \frac{K_{gas}(S_w)}{K_{dry}} \tag{6}$$

where K_{dry} is the dry gas permeability (after oven-drying at 65°C) of the same sample as that tested at a given RH. To be rigorous, absolute gas permeability, measured only when gas fills the porosity, should be used instead of K_{dry}. Nevertheless, the authors assumed that both previous permeabilities were similar. The drying at 65°C is assumed to remove enough water to consider that the argillite is 'dry' without damaging the material. This drying temperature is the temperature reference in the Laboratoire de Mécanique de Lille and was validated by ANDRA. All permeability tests were performed with the quasi-static technique (e.g. see Loosveldt *et al.* 2002 for details). Relative permeability is preferred to effective permeability, $K_{gas}(RH)$, in order to limit the effect of sample variability.

Following the description of suction curves, the relative gas permeability, K_{rg}, is represented by the van Genuchten–Mualem (VGM) model (Mualem 1976) as follows:

$$K_{rg} = (1 - S_w)^{\eta}(1 - S_w^{1/m})^{2m} \tag{7}$$

where m is determined by fitting the previous water retention curve, and η is determined by fitting equation (7) to the measured K_{rg} data and using equation (6).

Gas breakthrough pressure experiment

When gas (here, argon) is injected into one side of a saturated porous medium, its migration is progressive, as shown in Figure 4 (see Hildenbrand *et al.* 2002; Marschall *et al.* 2005). When gas pressure reaches the gas entry pressure, the gas begins to pass through the material and pushes a small quantity of water out of the material on the downstream sample side only. Experimentally, it is difficult

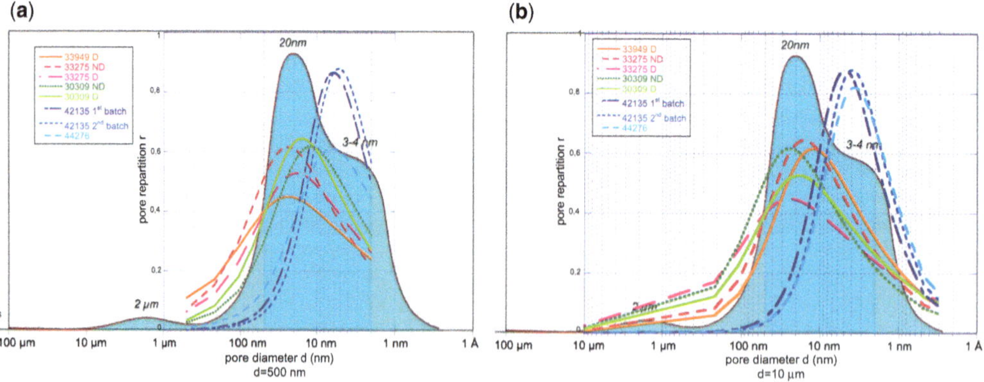

Fig. 3. Comparison of the pore-size distribution of COx argillite using isotherms obtained using the van Genuchten model for different initial hydric states – red ($S_w = 30\%$), green ($S_w = 54\%$) and blue ($S_w > 70\%$) – with that of reference for ANDRA (ANDRA 2005).

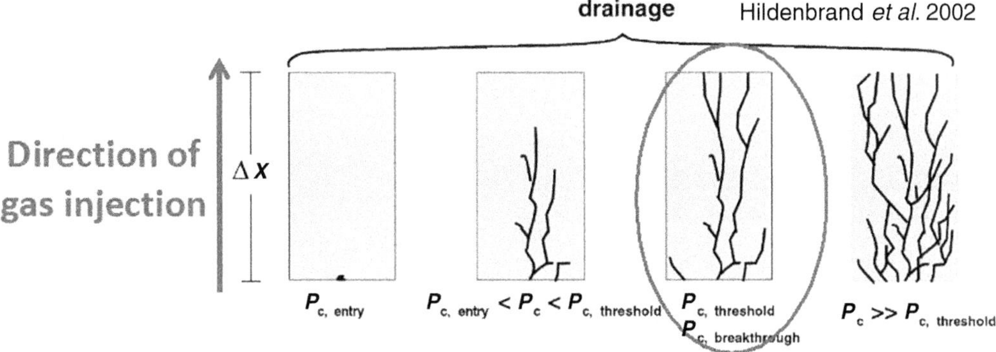

Fig. 4. The various steps in the gas penetration process in a porous material initially fully water-saturated (taken from Hildenbrand *et al.* 2002).

to measure accurately the gas entry pressure, in particular when its value is so low that it could be mistaken for thermally induced pressure variations. This may occur even in thermally regulated experimental rooms. Preferably, GBP is measured, which is defined as the gas injection pressure at which gas is expelled on the downstream side of the sample. Figure 5 shows the experimental device used to measure the GBP.

The recording method for gas breakthrough pressure is as follows.

- Prior to the GBP test, the sample is fully water saturated in a hydrostatic cell at a confinement pressure of 5, 6 or 12 MPa, by applying a 4 MPa water pressure gradient until the liquid permeability reaches 1×10^{-20}–1×10^{-21} m^2 range.
- Next, the upstream and downstream pipes are emptied of water and the sample kept at a constant confining pressure of $P_c = 5$, 6 or 12 MPa. The 2 cl (by volume) downstream chamber is then closed with a dedicated valve and its pressure is recorded with a pressure transducer.
- The argon gas pressure is increased very slowly on the upstream side (at a rate of 1–10 days between two $DP_{gas} = 0.5$ MPa steps; DP_{gas} is the pressure gradient, i.e. the pressure difference between the upstream and downstream sides of the sample), until the presence of gas is detected in the downstream chamber. Gas detection is performed using both the downstream pressure transducer (± 100 Pa) and a dedicated argon gas-flow detector (± 0.1 $\mu l\ s^{-1}$). Fluid movement towards the sample downstream side is detected via the downstream pressure increase (it can be related to water and/or gas transport).
- Every 24 h, the valve connected to the downstream chamber is opened and connected to the gas-flow detector. Whenever it is actually present, argon is expelled and identified using

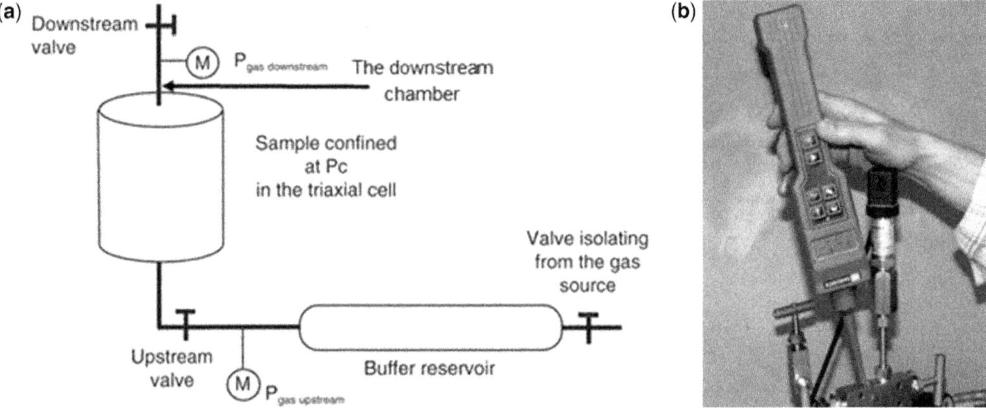

Fig. 5. (**a**) The GBP experimental device and (**b**) the gas detector.

the detector. GBP corresponds to the first pressure at which argon is identified in the downstream chamber (using the argon detector).

Results and discussion

Table 1 summarizes all the argillite cores used, their drilling orientation and the type of results obtained.

After the tests, almost all of the samples of our study were damaged, either initially or as a result of the laboratory protocol. The results of dry permeability show that the damage procedure increases the average gas permeability by a factor 5 (cf. Table 2). However, the permeability values vary considerably from one sample to another, probably due to the non-reproducibility of the damage protocol. In addition, the COx samples have different levels of damage within the same core. As most samples were already damaged, and as the damage procedure is not reproducible, a sorting methodology was developed for the samples. The damage can actually occur at different times before permeability is measured (see M'Jahad 2012), such as:

- during the *in situ* coring (in the URL);
- during the transport (in the URL) and during the preparation of the ANDRA T1 cell;
- at the opening of the T1 cell (in the laboratory) and during sample preparation (e.g. coring);
- in constant-RH hermetic chambers (in the laboratory).

Water retention experiment

Effect of damage at RH = 100%. Plots given in Figures 6–9 represent the evolution of relative mass variations in four batches of COx argillite – during stabilization at constant RH, drying at 60°C and under saturation at RH = 100% in Figures 6, 7 and 9; and during stabilization at constant RH, drying at 11% RH, saturation at RH = 100% and drying at 60°C in Figure 8:

- from horizontal drilling: Est 30309 D pre-damaged and Est 30309 ND not pre-damaged batches;
- from vertical drilling: Est 42135 ND and Est 44276 ND, both not pre-damaged.

Except for the Est 44276 samples (for the Est 44276 samples, the drying under different RHs was not long enough to reach a conclusion but other tests in the LML laboratory demonstrate the following for almost cores of Cox argillite – this core was the latest to be tested at the laboratory), for a RH of between 11 and 98%, the relative mass variation for all of the COx argillite samples stabilizes relatively fast, after only a few days. This confirms the quick-drying kinetics under constant temperature of argillite observed by Skoczylas (2011). Similarly, sample mass is stable after only a few days when drying at 60°C. Because of the large number of samples tested and the fact that the experimental equipment was shared, the drying phase at 60°C lasted for more than 1 week (and up to 100 days). This over-drying does not affect the water stability of the material. However, it may damage the argillite samples.

Unfortunately, the mass evolution at RH = 100% does not stabilize quickly. In fact, under this hydric condition, the claystone continues its water uptake over very long periods (*c.* 100 days), as shown in Figures 6–9. This is also observed by Zhang *et al.* (2010), as shown in a continuous water uptake reported for COx claystone and Opalinus clay. This water uptake results from the coupling of hydration and clay mineral swelling, which generates damage (cracking) under free-volume boundary conditions (Saiyouri *et al.* 2000). According to Hildenbrand *et al.* 2002), hydration induces swelling in clay minerals, mainly smectites, which causes cracking damage. This damage increases the apparent porosity, which is accessible to water. This process is self-powered, with a very slow evolution. Consequently, the apparent porosity increases from 14 to 19% for intact samples, to 20–25% after swelling (cf. Fig. 10). This is a signature for damage caused by hydration. This also has a significant influence on the water retention properties of COx argillite, as shown in Figure 11. The coupling between the damage and the swelling reduces the water retention capacity of the COx argillite (cf. Fig. 11) through the development of micro-cracks. Thus, for the same relative humidity, the sample is less water saturated because the damage induces a larger space available to water (larger drained pore diameter).

The van Genuchten model (see the 'Experimental methods' section) was applied to the earlier experimental isotherms of the COx argillite, depending on the damage (duration of the saturation step at RH = 100%). The model parameters (n and P_r) were optimized using the least-squares method and are given in Table 3. The damage/swelling during the imbibition phase at RH = 100% decreases significantly parameter P_r, which is interpreted as a gas entry pressure, from 13.5 to 6.8 MPa. Using the Kelvin–Laplace law, this corresponds to an increase in the drained pore diameter from 21.6 to 43 nm. This means that the damage enlarges the pore size of the material by opening interparticle pores or by creating micro-cracks. This is also consistent with the porosity increase due to the damage/swelling explained earlier. Higher porosities or a larger pore radius ease the fluid flow (in particular, gas flow) through the porous media. These results

Table 1. *Summary of all batches of COx argillite used in the study*

Core number	Cracking state	Drilling orientation	Initial saturation (%)	Results obtained
Est 30309 ND	No damage procedure	Horizontal	54	Relative mass variation(RH), S_w (RH), P_{cap} (S_w), K_{rg} (RH)
Est 30309 D	Damage procedure applied	Horizontal	54	Relative mass variation(RH), S_w (RH), P_{cap} (S_w), K_{rg} (RH)
Est 33275 ND	No damage procedure	Sloped	30	Relative mass variation(RH), S_w (RH), P_{cap} (S_w)
Est 33275 D	Damage procedure applied	Sloped	30	Relative mass variation(RH), S_w (RH), P_{cap} (S_w)
Est 33949 D	Damage procedure applied	Horizontal	30	Relative mass variation(RH), S_w (RH), P_{cap} (S_w)
Est 34450 ND	No damage procedure	Horizontal		Relative mass variation(RH), GBP
Est 34381 ND	No damage procedure	Horizontal		Relative mass variation(RH)
Est 42135 ND	No damage procedure	Vertical	75	Relative mass variation(RH), S_w (RH), P_{cap} (S_w), K_{rg} (RH)
Est 44276 ND	No damage procedure	Vertical	70	Relative mass variation(RH), S_w (RH), P_{cap} (S_w)
Est 34394 ND	No damage procedure	Horizontal		GBP
Est 33271 ND	No damage procedure	Sloped	30	GBP

ND, no pre-damage; D, pre-damage.

Table 2. *Effect of the damage procedure on the average gas permeability of the COx argillite in the dry state (Est 30309 from horizontal drilling, seven samples per batch)*

Argillite (Est 30309)	P_c (MPa)	Average K_{dry} value (m^2)	Permeability range (m^2)
'Intact' batch ND	6	1.55×10^{-17}	1.35×10^{-19}–6.42×10^{-17}
	12	6.61×10^{-18}	9.44×10^{-19}–2.42×10^{-17}
Damaged 'batch' D	6	7.72×10^{-17}	6.6×10^{-19}–3.36×10^{-16}
	12	3.51×10^{-17}	3.43×10^{-19}–1.6×10^{-16}

confirm that argillite is damaged during stabilization at RH = 100%.

Effect of the initial hydric state. As argillite may be damaged at RH = 100%, the samples were not pre-saturated before stabilizing at a constant relative humidity. However, the COx cores have different water saturation states when received in the laboratory (see Table 1). This can significantly influence its water retention properties because of the hysteresis between the adsorption and desorption isotherms. The objective here is to analyse and quantify the effect of the initial water saturation and hysteresis on the argillite adsorption/desorption isotherms. (Almost samples (i.e. 'intact' samples without pre-damage) were dried at different RHs from their initial saturation states. The samples have been classified as located in the adsorption or desorption branches, while the post-treatment of experimental results depend on their initial degree of water saturation.) Figure 12 plots sorption isotherms of COx argillite obtained for COx from different initial hydric states, characterized here by the initial water saturation, and ranging from 30 to 75%.

Figure 12 shows that there is a variation in water saturation of more than 20% for the same level of RH. However, the water retention capacity of COx argillite decreases significantly when the initial saturation is low. Figure 13 reports the distinction between the effects of initial water saturation (cf. Fig. 12), and the effects of the hysteresis between adsorption and desorption (cf. Fig. 13). Even though the variability in the results is high (*c.* 20%), there is no significant effect of the hysteresis between adsorption and desorption on sorption isotherms (cf. Fig. 13). This means that only the initial hydric state of COx argillite is responsible for the loss of its retention capacity. It also means that samples with a low initial saturation are already damaged as their water retention capacity is low (cf. Fig. 12). Samples with a high initial

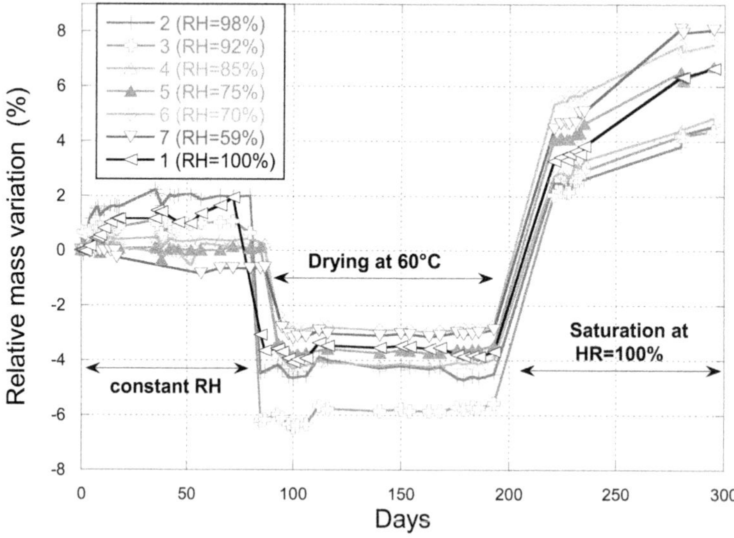

Fig. 6. Evolution of the relative mass variation in COx samples from Est 30309 ND (horizontal core), non-pre-damaged, under conditions of: constant RH, then drying at 60°C and, finally, at RH = 100%.

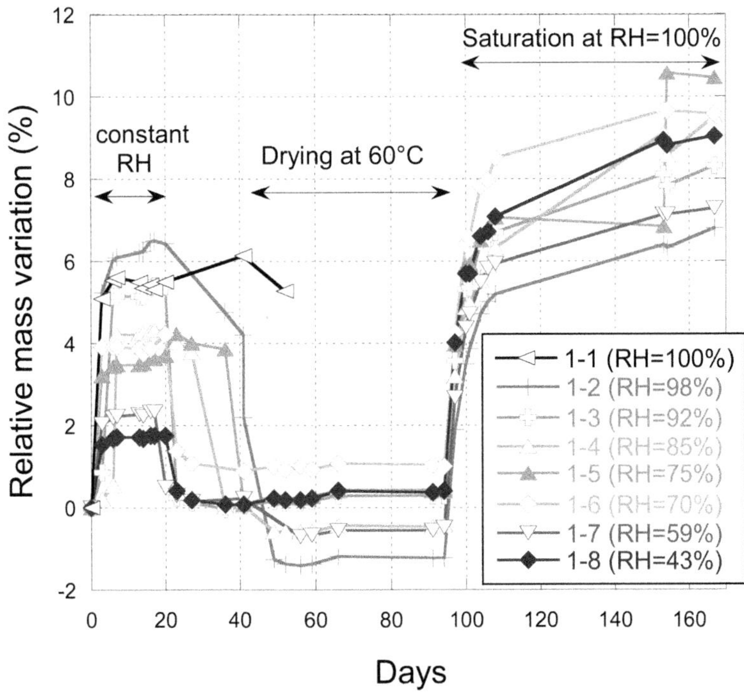

Fig. 7. Evolution of the relative mass variation in pre-damaged COx samples from Est 30309 D (horizontal core), under conditions of: constant RH, then drying at 60°C and, finally, at RH = 100%.

saturation have a better water retention capacity, which suggests that their level of damage is lower (cf. Fig. 12).

As done above, the van Genuchten model parameters (n and P_r) were optimized using the least-squares method in order to fit experimental sorption

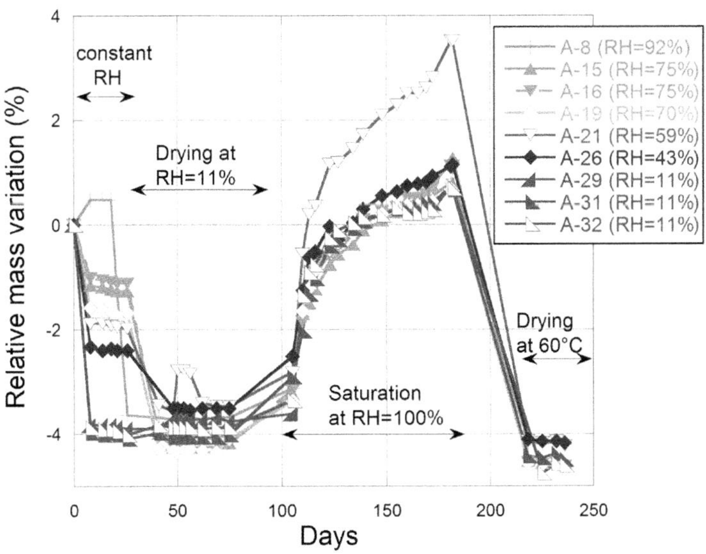

Fig. 8. Evolution of the relative mass variation in COx samples from Est 42135N ND (vertical core), non-pre-damaged, under conditions of: constant RH, at RH = 11%, at RH = 100% and, finally, drying at 60°C.

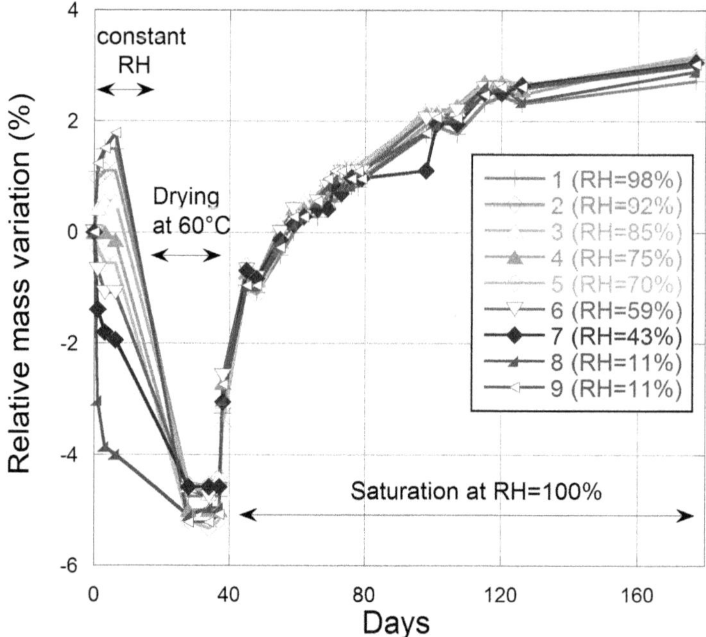

Fig. 9. Evolution of the relative mass variation in COx samples from Est 44276 ND (vertical core), non-pre-damaged, under conditions of: constant RH, then drying at 60°C and, finally, at RH = 100%.

curves (Table 4). On the one hand, the parameter n controls the curvature of the water adsorption isotherm, particularly for high RH, and the pore-size distribution (see the subsection on 'Pore-size distribution' later). This parameter has low values (1.38, 1.48 and 1.6) for batches Est 33949 and Est 33275, mainly representing the initially low saturated samples. However, n is higher (1.59, 1.97 and 1.99) for batches Est 44276 and Est 42135, mainly representing initially high saturated samples. On

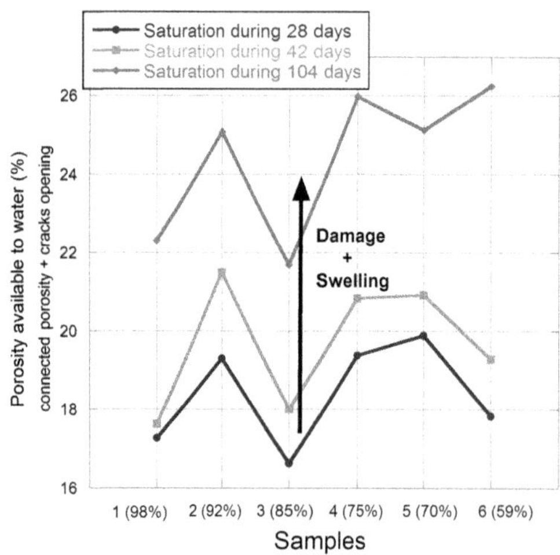

Fig. 10. Evolution of apparent porosity available to water (initial porosity + cracks due to damage and swelling) in COx samples from Est 30309 ND for different saturation durations at RH = 100%.

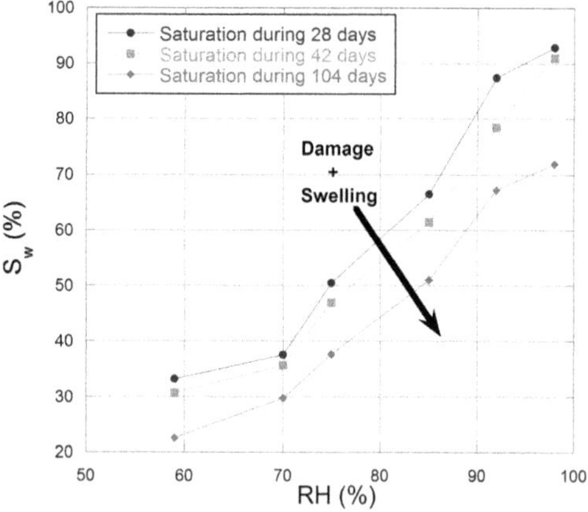

Fig. 11. Evolution of the sorption isotherm for COx samples from Est 30309 ND core, with an initial water saturation of 54%,for different saturation durations at RH = 100%: samples stabilized at RH = 59, 70 and 75% were in a desorption path, while the samples stabilized at RH = 85, 92 and 98% were in an adsorption path.

Table 3. *Effect of the damage at RH = 100% on the van Genuchten parameters*

COx argillite batch	Duration of the saturation (days)	n parameter of the VG model	P_r parameter of the VG model (MPa)
Est 30309 ND	28	1.6	13.5
(horizontal)	42	1.6	10.9
	104	1.6	6.8

the other hand, the parameter P_r varies between 5.1 and 7.8 MPa for initially highly desaturated argillite (Est 33949 and Est 33275: $S_w = 30\%$) (see Table 4). Using the Kelvin–Laplace law, this corresponds to drained pore diameters ranging between 36 and 56 nm. As these samples are in the adsorption path, we can deduce that adsorption in argillite is controlled by pores with diameters ranging between 36 and 56 nm, which mainly corresponds to interparticle pores. Furthermore, for initially highly saturated argillite (Est 42135 and Est 44276: $S_w > 70\%$), P_r varies between 25 and 40 MPa. Using the Kelvin–Laplace law, this corresponds to drained pore diameter ranging between 7 and 12 nm. However, as most of these samples are in the desorption path, these results mean that the pores diameters driving desorption are between 7 and 12 nm, which also represents interparticle pores. The higher the initial water saturation of argillite, the smaller the drained pore diameter. This confirms that the lower the initial water saturation of argillite, the more damaged the argillite will be. This can be attributed to the enlargement of the pores or the creation of micro-cracks due to swelling.

Effect of the drilling orientation. For initially highly saturated argillite ($S_w > 70\%$), which is assumed to be less damaged, water adsorption isotherms were compared with those from Skoczylas (2011) (see Fig. 14). Figure 14 shows a good consistency between the results of both studies, with a dispersion of about 10%. This means that there is no significant

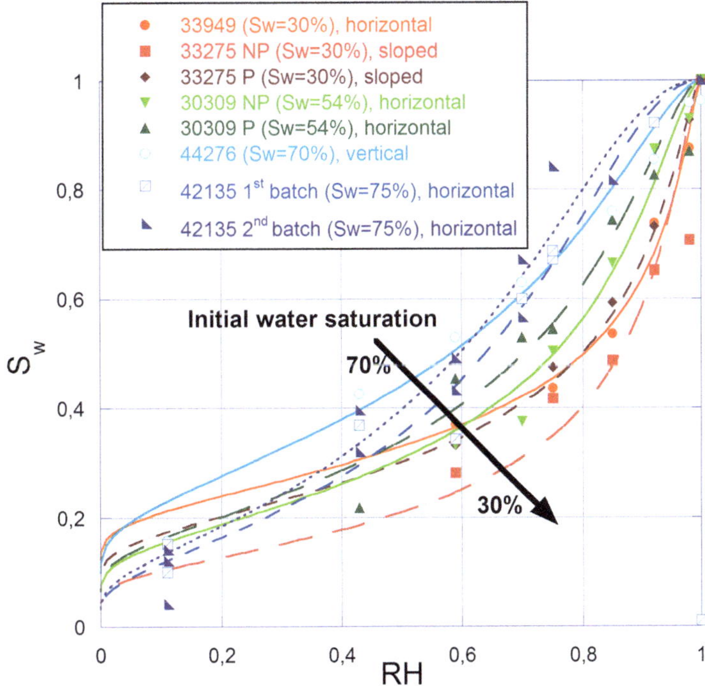

Fig. 12 Comparison of water adsorption isotherms of COx argillite for different initial hydric states (initial water saturation) using the following colour code: red for a low initial water saturation (30%), green for a middle initial saturation (54%) and blue for a high initial saturation (>70%).

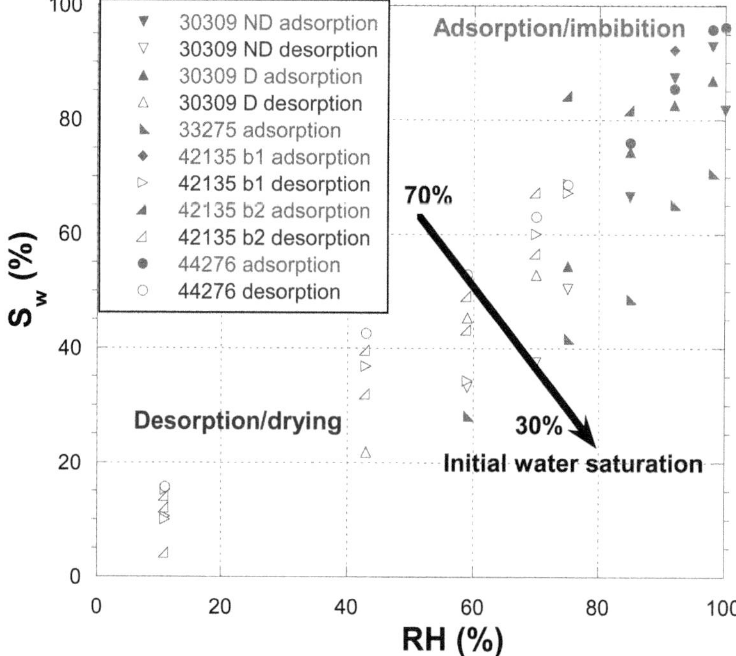

Fig. 13. Comparison of water adsorption isotherms of COx argillite for different initial hydric states: adsorption path (dark symbols) and desorption path (light-coloured symbols).

effect of the drilling orientation on the water retention properties of the COx argillite. In addition, these results confirm previous conclusions: the dispersion is mainly a consequence of the damage, even if low hysteresis is reported between adsorption and desorption. For an equivalent initial degree of water saturation, sorption isotherms show a very low dispersion (<10%) for argillite. In this case, we have one fit of the sorption isotherm for 'intact argillite' using the van Genuchten model: $n = 1.58$ and $P_r = 21.9$ MPa as shown in Figure 14.

Pore-size distribution

The water adsorption isotherms of the present work and the pore-distribution curve given by equation (5) (Rougelot *et al.* 2009) were used to determine the pore-size distribution of COx argillite. The pore distribution vs. pore diameter is plotted for all batches of the study using experimental isotherms (Fig. 15a) and isotherms fitted by the van Genuchten model (see Fig. 15b). The objective is to quantify the pore size of pores or their grooves, which drive the water (and gas) transfer.

Table 4. *Effect of initial saturation (hydric state) on the van Genuchten parameters for COx argillite*

COx argillite batch	Drilling orientation	Initial saturation (%)	n parameter of the VG model	P_r parameter of the VG model (MPa)
Est 33949 D	Horizontal	30	1.38	5.1
Est 33275 ND	Sloped	30	1.6	7
Est 33275 D	Sloped	30	1.48	7.8
Est 30309 ND	Horizontal	54	1.6	13.5
Est 30309 D	Horizontal	54	1.64	18
Est 44276 ND	vertical	70	1.59	25.1
Est 42135 – first batch ND	Horizontal	75	1.97	34.1
Est 42135 – second batch ND	Horizontal	75	1.99	40.1

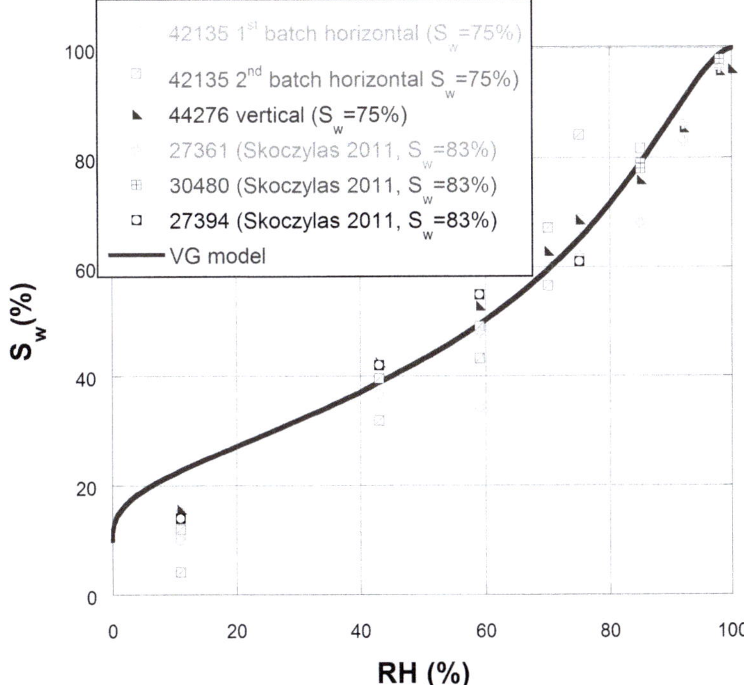

Fig. 14. Comparison of sorption isotherms of COx argillite in the present study (horizontal and vertical drilling) with those of Skoczylas (2011) (vertical drilling) for a high initial hydric state ($S_w > 70\%$) – a single fit is given by the van Genuchten model: $n = 1.58$ and $P_r = 21.9$ MPa.

The pore-size distributions of argillite Est 33949 and Est 33275, in an adsorption path, reveal three distinct peaks (see Fig. 15a): a first peak centred around a diameter of 106 nm, and a second and third located between 6 and 13 nm, respectively. These pores correspond to spaces between the clay particles (ANDRA 2005). However, as most of the samples are damaged, a part of this porosity may

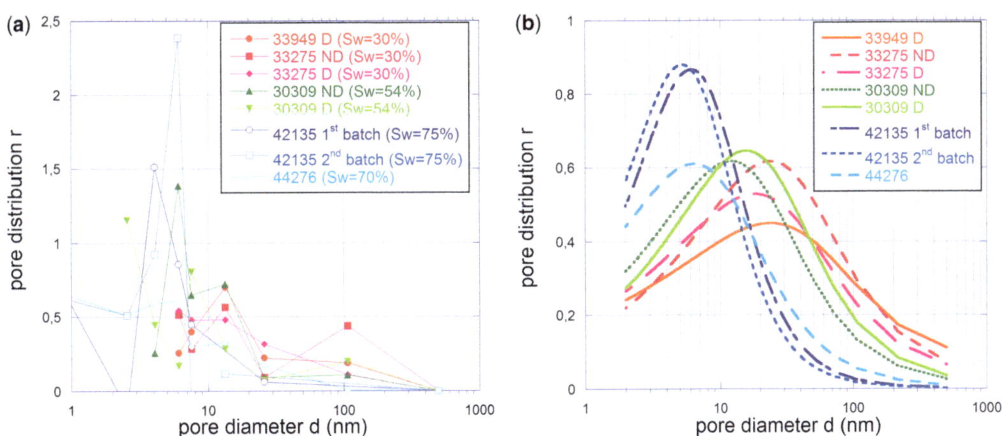

Fig. 15. Comparison of the pore-size distribution of COx argillite for different initial hydric states: red ($S_w = 30\%$), green ($S_w = 54\%$) and blue ($S_w > 70\%$): (**a**) using experimental isotherms for different initial hydric states; and (**b**) using isotherms obtained using the van Genuchten model and equation (5).

be representative of micro-cracks. Furthermore, the pore-size distribution of argillite Est 42135 and 44276 Est (Fig. 15a) shows different peaks in the desorption paths, for which diameters range from 4 to 6 nm. This means that, during desorption, the water is transferred through pores with entrance diameters of between 4 and 6 nm. Results on argillite Est 30309 confirm the previous analysis. In fact, the samples in the adsorption path show peaks around 13–106 nm, while those in the desorption path highlight a peak at around 6 nm.

When isotherms fitted to the van Genuchten model (Fig. 15b) are used and compared with those of ANDRA reference values (Fig. 3), the pore-size distributions of all batches highlight three groups of argillite: a first group (No. 1) containing argillite with an initial water saturation of <30% (Est 33949 and Est 33275); a second group (No. 2) containing argillite with an initial water saturation of ≥70% (Est 42135 and 44276); and a third group (No. 3) with an intermediate initial saturation (*c.* 54%). The pore-size distribution of the third group is between the distributions of the first two groups. The first group shows a peak centred around 20–30 nm, confirming the values given by the P_r parameter, while the second group has peaks centred around 5–6 nm. This also confirms the range of pore diameters calculated from the variability in P_r. According to Robinet (2008), interlayer pores of the clay minerals have a maximum size of 0.5 or 1 nm, while interparticle pores vary between 1 nm and 1 µm. Therefore, argillite batches used in the present study highlight interparticle pores and/or micro-cracks resulting from to damage.

Otherwise, as explained above, the parameter n could be related to the measure of the pore-size distribution. Thus, high values of n (1.59, 1.97 and 1.99: see Table 4) for batches Est 44276 and Est 42135 are associated with narrower distributions and smaller pore diameters. Conversely, the curves of batches Est 33949 and Est 33275 have lower values of n and larger distributions. This confirms the enlargement of pore diameters as a result of damage in the samples that swell more.

The pore-size distributions determined from the van Genuchten retention curves (cf. Fig. 15b) were compared to those of ANDRA reference values (Andra 2005). This comparison is shown in Figure 3. A good consistency is observed. However, even if both pore-size distributions are similar in term of the pore locations, the results of this study are less representative of the ANDRA reference values in terms of their distribution (see Fig. 3).

Gas transport experiment

As mentioned earlier, it is difficult to obtain undamaged samples for COx argillite because of the damage that occurs during the sampling experimental procedures. Moreover, we were only able to test two cores of COx argillite in order to measure the gas permeability: samples Est 30309 and Est 42135 from an horizontal and vertical drilling, respectively (cf. Table 1). In addition, the first experimental campaign conducted on argillite Est 30309 showed that damage to the argillite, initial and/or induced by the damage protocol, was not reproducible. For these reasons, in this subsection we propose developing a sorting methodology to distinguish 'intact' samples from damaged samples, and to identify gas transfer properties for each family.

Dry and effective gas permeability at the dry state. The gas permeability in the dry state ($T = 65°C$) is recorded at two confining pressures: 6 and 12 MPa. The results of COx argillite Est 30309 ND (not pre-damaged) and Est 30309 D (pre-damaged) are plotted in Figure 16. For the non-pre-damaged batch, for $P_c = 6$ MPa, the dry gas permeability results range between 1.35×10^{-19} and 6.42×10^{-17} m^2. Similarly, values for the pre-damaged samples are spread between 6.6×10^{-19} and 3.36×10^{-16} m^2 for $P_c = 6$ MPa. These results show a variability of three orders of magnitude throughout samples for both batches. Nevertheless, they show that the average value of both batches in the dry state is multiplied by a factor of 5 after the damage protocol – this factor of 5 is obtained by comparing results of the dry gas permeability for both batches and could be attributed to the damage induced by the damage protocol. This result applies to both confining pressures. For the argillite Est 42135 batch, a similar difference of three orders of magnitude (see Fig. 17) was observed. This may be considered as an identifier for the damage. Because permeability measurements for the samples were too dispersed, a sorting methodology was developed based on the following three criteria:

- the value of the dry gas permeability;
- the value of the effective gas permeability $K_{gas}(S_w)$;
- the correlation between the effective gas permeability and the degree of water saturation.

Using these criteria, three families of COx argillite samples were identified:

- The 'highly cracked' argillite – samples of this family: (1) have very high permeabilities in the dry state ($K_{dry} > 1 \times 10^{-17}$ m^2 for $P_c = 6$ MPa); (2) their effective gas permeabilities are $>1 \times 10^{-18}$ m^2 for $P_c = 12$ MPa; and (3) they do not present any correlation between effective permeability and water saturation. This behaviour is attributed to cracking in argillite, which

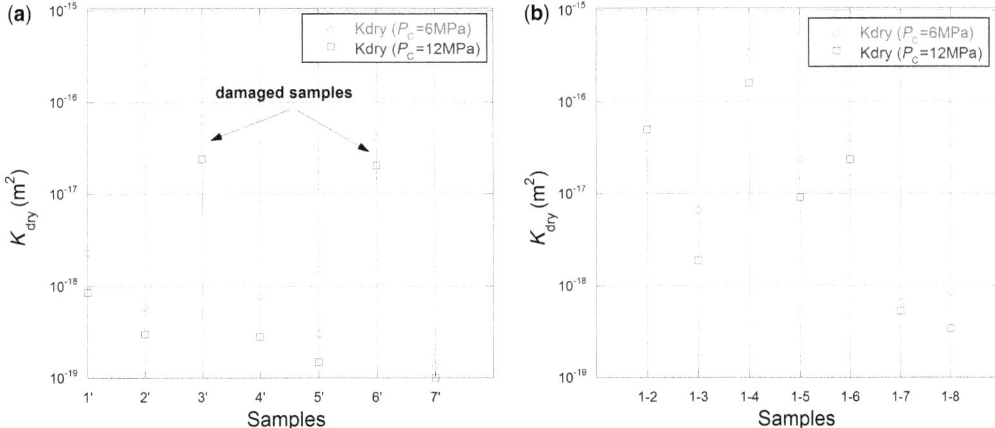

Fig. 16. Gas permeability of COx argillite (horizontal drilling) in the dry state ($T = 60°C$): (**a**) for Est 30309 ND non-pre-damaged and (**b**) Est 30309 D pre-damaged, for two confinement pressures: $P_c = 6$ and 12 MPa.

generates new paths for gas migration, independent of the saturation state of the material. No clogging at high S_w is observed for this family. This family is not plotted in Figure 18, but some samples are shown in Figure 16.

- The supposed 'intact' argillite (or low damaged) – samples of this family: (1) present both dry and effective permeability values of $<1 \times 10^{-18}$ m^2 for $P_c = 6$ MPa; and (2) a significant correlation between the effective permeability and saturation (see Fig. 18). This family mainly constitutes samples from Skoczylas (2011) and few samples of argillite Est 30309 (see Fig. 18).
- The 'micro-cracked' argillite – samples of this family: (1) have a permeability in the dry state of $K_{dry} < 1 \times 10^{-17}$ m^2 for $P_c = 6$ MPa; (2) $K_{gas}(S_w) < 1 \times 10^{-18}$ m^2 for $P_c = 12$ MPa; and (3) a low correlation between effective permeability and the degree of water saturation. These samples differ from 'intact' samples by an effective permeability that is more than one order of magnitude higher (see Fig. 18). Almost all samples from argillite Est 42135 belong to this family. Therefore, the permeability measurements of these samples are consistent with those of Zhang *et al.* (2010), when gas injection is parallel to the bedding plane.

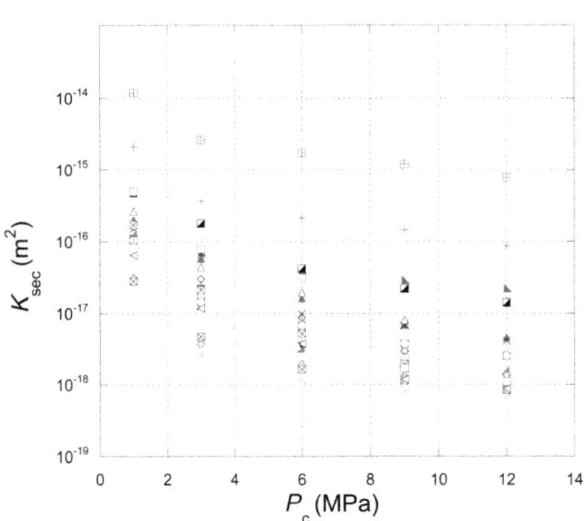

Fig. 17. Gas permeability of COx argillite Est 42135 (horizontal drilling) in the dry state (RH = 11%, $T = 20°C$) for different confinement pressures.

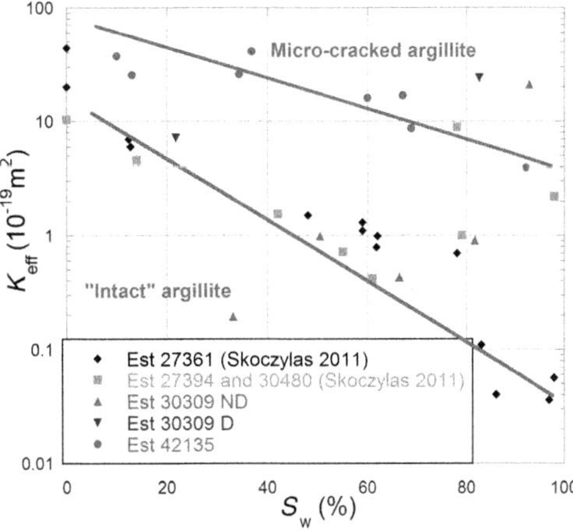

Fig. 18. Gas effective permeability of COx argillite in this study (Est 30309 and Est 42135 from horizontal drilling) compared with that from Skoczylas (2011) (Est 27361, Est 27394 and Est 30480).

Relative gas permeability. Using the previous sorting method, the relative gas permeability v. water saturation of the COx argillite (Est 30309 and Est 42135) was compared to the results of Skoczylas (2011) (Est 27361, Est 27394 and Est 30480) for $P_c = 6$ and 12 MPa. For a given confinement pressure, each Cox argillite family has its characteristic $K_{rg}(S_w)$ curve: the curve of the 'intact' family is shown in Figure 19, while the curve of the 'micro-cracked' argillite is presented in Figure 20. The comparison of these curves suggests that the gas transfer behaviour depends on the degree of damage to the material. 'Intact' argillite displays lower permeability, and a good correlation between relative gas permeability and saturation. Conversely, the micro-cracked argillite is more permeable to gas, and shows very limited correlation between K_{rg} and water saturation. Furthermore, for $P_c =$ 6 MPa, the relative gas permeability is insensitive to saturation except at high values of saturation, which is a sign of damage.

For stronger confinement (12 MPa), the 'micro-cracked' samples (of the Est 42135 batch) have relative gas permeability values similar to those of the intact argillite recorded at $P_c = 6$ MPa. This means that a stronger confinement (12 MPa) contributes to a mechanical contraction of the micro-cracks. By contrast, 'highly cracked' samples from Est 30309, especially those from the pre-damaged batch, have high values of relative permeability even with strong confinement. Indeed, they are so damaged that a confinement of 12 MPa is not enough to close the cracks.

The VGM model (equation 7) was fitted to the experimental relative gas permeability curves for each family of COx argillite (Table 5).The parameter P_r is equal to 21 MPa for the 'intact' argillite. Using the Kelvin–Laplace law, this corresponds to a drained pore diameter of 14 nm. However, the parameter P_r for 'micro-cracked' argillite is 50% lower than that of the 'intact' argillite. This means that the drained pore diameter is much larger (*c.* 30 nm) for the 'micro-cracked' argillite, which is an indicator of damage. The size of this drained pore is characteristic of interparticular pores

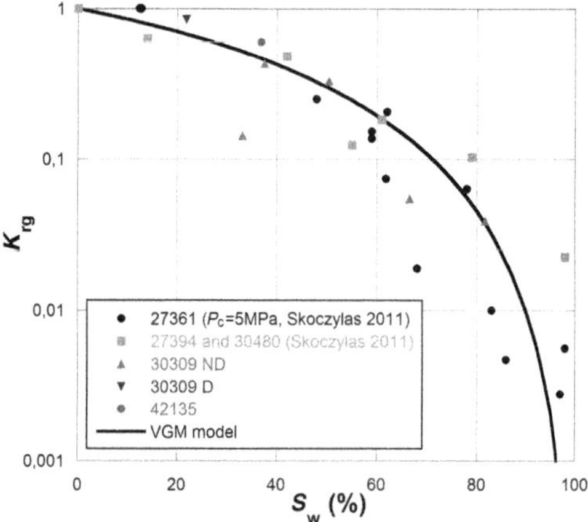

Fig. 19. Relative gas permeability of 'intact' argillite from samples in this study (Est 30309 and Est 42135 from horizontal drilling) compared with that from Skoczylas (2011) (Est 27361, Est 27394 and Est 30480).

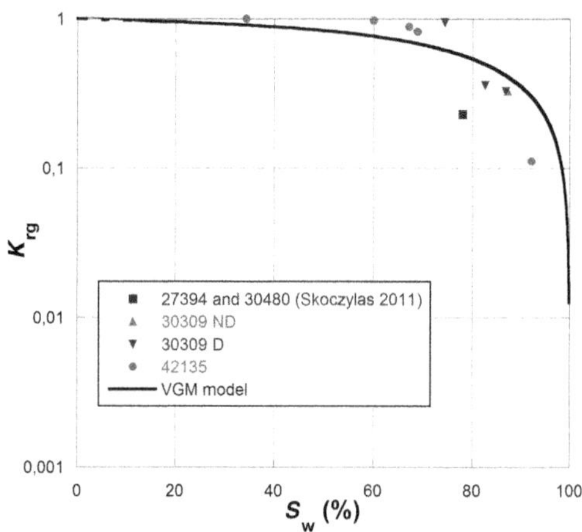

Fig. 20. Relative gas permeability of 'micro-cracked' argillite from samples in this study (Est 30309 and Est 42135 from horizontal drilling) with that from Skoczylas (2011) (Est 27361, Est 27394 and Est 30480).

Table 5. *Parameters of the VGM model for each COx argillite family*

Argillite batch	n	P_r (MPa)	η	
			$P_c = 6$ MPa	$P_c = 12$ MPa
'Intact' argillite	1.63	21	1.525	1.693
'Micro-cracked' argillite	1.36	10	0.195	0.147
ANDRA (ANDRA 2005)	1.49	15	0.5	

Table 6. *Water permeability and gas breakthrough pressure (GBP) of COx argillite*

Sample	Drilling orientation	P_c (MPa)	Water injection duration (days)	K_{water} (m^2)	GBP (MPa)
EST 34394-6	Horizontal	6	6	2.3×10^{-20}	0.4
EST 34394-7	Horizontal	6	4	6×10^{-19}	0.3
		12			1.4
EST 34394-4	Horizontal	12	19	1.5×10^{-20}	0.6
EST 33271	Sloped	11	8	4.2×10^{-21}	1.4
EST 34450	Horizontal	11–12	13	9×19^{-21}	0.4

and micro-cracks, especially for 'micro-cracked' argillite.

However, the parameter η of the Mualem model is usually associated with the tortuosity of the pore network. Its default value is equal to 0.5 for soils. For both confinement pressures, $\eta > 1$ for 'intact' argillite, which means that its porous network is tortuous and complex. By contrast, for 'micro-cracked' argillite $\eta < 0.5$, which means that the porosity is less tortuous. In fact, the cracks increase the connectivity of pores and, consequently, make the passage of gas through argillite easier.

Gas breakthrough pressure experiment

Gas breakthrough pressure was measured for five samples of COx argillite. These samples were cylinders 37 mm in diameter and 10 mm in height, and were from horizontal (Est 34394 and Est 34450) and from sloped drillings (Est 33271). Consequently, the GBP was measured in the direction parallel to the bedding plane for the Est 34394 and Est 34450 samples (Fig. 1). First, water permeability of the fully saturated sample was recorded (Table 6). The water permeability values varied between $4.2 \times 10^{-21}$and 1×10^{-19} m^2. These values are within the variability range of water permeability found in the literature (1×10^{-21} $m^2 < K_w < 1 \times 10^{-19}$ m^2) (see Escoffier *et al.* 2005; Davy *et al.* 2007; Boulin *et al.* 2008; Zhang *et al.* 2010; Skoczylas 2011). The high values of water permeability emphasize a state of damage in some samples (10^{-19} m^2).

Second, the GBP was measured at two confining pressures 6 and 12 MPa. For $P_c = 6$ MPa, GBP values are about 0.4 MPa (Est 34394-6 and Est 34394-7). These values are lower than the GBP measured by Skoczylas (2011) on macro-cracked argillite for a similar confinement (*c.* 1.23–1.6 MPa). These low GBP values are due mainly to damage (Fig. 21) and partly to the limited duration of water saturation (4–6 days). In fact, one sample (34394-4) that was saturated for 19 days has a similar GBP (*c.* 0.6 MPa), meaning that the damage has a significant effect on the GBP. In addition, argillite Est 34394 contains a number of pyrite inclusions, which contribute to the macro-cracking (see Fig. 2).

GBP was also measured under a higher confinement pressure ($P_c = 11$–12 MPa) for four samples from horizontal and sloped drillings (Est 34394-4, Est 34394-7, Est 33271 and Est 34450). The

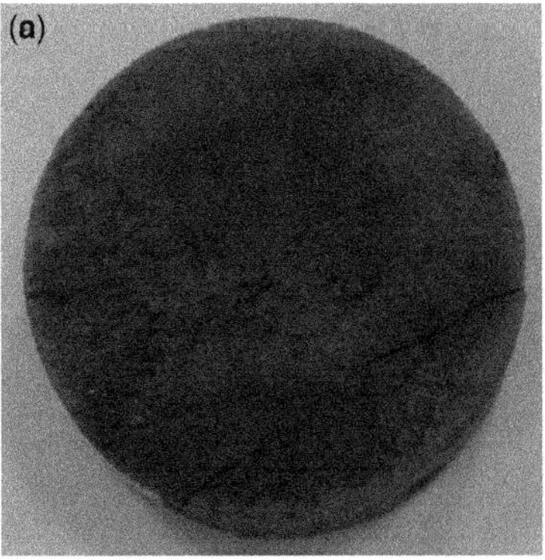

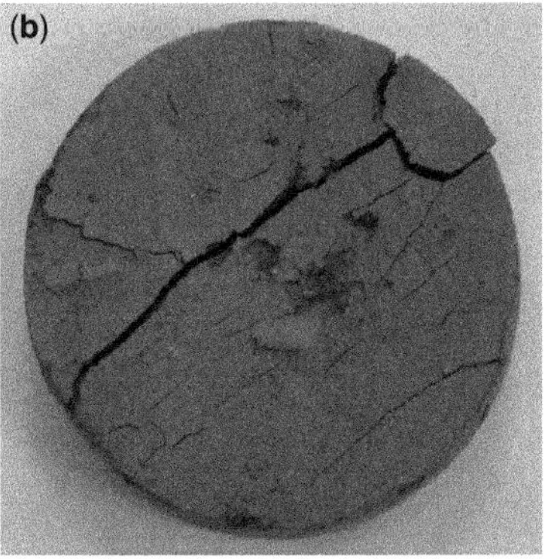

Fig. 21. Photographs of argillite samples: (**a**) Est 34394-6 and (**b**) 34394-7 after measurement of GBP and after drying under free volume.

GBP ranged between 0.4 and 1.4 MPa. These values are very low compared to those of 'intact' argillite (*c.* 5.5 MPa) measured by Harrington *et al.* (2014). Nevertheless, the GBP values are of the same order of magnitude as micro-cracked samples from Skoczylas (2011) and Davy *et al.* (2012). In fact, the GBP is equal to 1.8 MPa for Est 27361 (Skoczylas 2011), and ranges between 1.53 and 2.25 for Est 34386 (Davy *et al.* 2012). The low values of GBP are representative of the cracked argillite, meaning that they were initially damaged.

Finally, the GBP representing the 'intact' argillite was measured by Harrington *et al.* (2014) on samples with a much greater height ($h = 54$ mm for Est 27367 and $h = 86$ mm for Est 30341). Consequently, the gas passes easily through a thicker sample (10–30 mm) because the sample is more easily damaged. In particular, if the sample contains inclusions, its damage during drying increases the connectivity of the porosity. This partly explains why the larger samples of Harrington *et al.* (2014) are less damaged.

Conclusion

The mass stabilization of COx argillite was achieved quickly (in a few weeks) under drying conditions at 60°C, and at a RH of between 11 and 98%, as observed by Skoczylas (2011), while the mass stabilization at RH = 100% required more than 100 days. The associated long duration of water uptake is explained as the results of the coupling between the hydration, the clay mineral swelling and the damage under free-volume boundary conditions. This mechanism led to an increase in the apparent porosity of the sample, to a decrease in the water retention capacity and, subsequently, to damage, as observed by the appearance of a network of (micro-)cracks. This conclusion is supported by the work of Hildenbrand *et al.* (2002) and Zhang *et al.* (2010).

Based on several input-damaged COx samples, experiments also showed that:

- The lower the initial water saturation of the argillite, the lower the water retention capacity and the higher the damage of the argillite.
- The higher the initial water saturation of the argillite, the smaller the drained pore diameter.

The sorption isotherms and the model of Rougelot *et al.* (2009) were used to determine the pore-size distribution of the COx argillite. The obtained pore-size distributions highlighted three groups of argillite: a first argillite group characterized by drained pores with a diameter of 20–30 nm; a second group with pore diameters of between 4 and 6 nm; and a third group with an intermediate pore diameter distribution. The pore-size distributions showed good consistency with the previously published results from ANDRA (2005), especially in terms of the location of pores.

Since almost all input samples were already damaged, and the damage procedure was not sufficiently reproducible, the samples were sorted based on their dry and effective gas permeability, and on the correlation between effective gas permeability and the degree of water saturation. Using these criteria, three families of COx argillite were identified: the 'highly cracked' argillite family; the 'intact' (or low damaged) argillite family; and the 'micro-cracked' argillite family.

The GBP was measured on COx argillite from horizontal or sloped drilling. The GBP measurements obtained in the present work ranged between 0.3 and 1.4 MPa for $P_c = 12$ MPa, which is significantly below the 'intact' argillite GBP measurements of Harrington *et al.* (2014) that were around 5–6 MPa. Nevertheless, the GBP values were of the same order of magnitude as micro-cracked samples from Skoczylas (2011) and Davy *et al.* (2012), meaning that the samples were initially damaged.

In conclusion, the damage has a significant impact of the following properties of COx argillite. It induces:

- a decrease in its water retention capacity;
- a decrease in its GBP;
- an increase in its apparent porosity available to water due to creation of micro-cracks;
- an increase in its gas permeability (and probably the pore network + micro-crack connectivity).

Finally, as pointed in the introduction to this paper, one of the objectives of this work was to identify the most suitable properties for detecting early damage states in COx argillite, in order to use them in the laboratory or for *in situ* COx identification. These properties can be classified in descending order of damage detection relevance:

- Gas breakthrough pressure (GBP) appears to be the best 'detector' of damage in the studied properties. In fact, it is very sensitive to damage, and allows the damage state (intact and damaged) of samples to be determined with its close connection to water permeability values. This is also possible under high confinement pressures.
- Gas permeability (dry, effective and relative permeability) is also a good indicator of damage. Damaged argillite has a higher gas permeability value than that of intact argillite, and this increases from about one up to several orders of magnitude.
- Water permeability is a poor indicator of damage in COx argillite. In fact, argillite presents similar values of water permeability for both intact and damaged samples, which is attributed to the COx argillite's capacity for self-sealing.

The authors gratefully acknowledge the financial support provided by ANDRA (Agence Nationale pour la Gestion des Déchets Radioactifs).

References

ANDRA 2005. Projet HAVL – Dossier 2005. *Référentiel site Meuse/Haute-Marne.* Rapport ANDRA C.RP.ADS. 04.0022. ANDRA, Châtenay-Malabry, France.

ANDRA 2012. *Referentiel du comportement THM des formations sur le site de Meuse/Haute-Marne Centre de Meuse/Haute-Marne.* Rapport ANDRA D.RP.AMFS. 1 2.0024. ANDRA, Châtenay-Malabry, France.

BOULIN, P.F., ANGULO-JARAMILLO, R., DAIAN, J.-F., TALANDIER, J. & BERNE, P. 2008. Experiments to estimate gas intrusion in Callovo-oxfordian argillites. *Physics and Chemistry of the Earth, Parts A/B/C*, **33**, S225–S230.

BRUE, F., DAVY, C.A., SKOCZYLAS, F., BURLION, N. & BOURBON, X. 2012. Effect of temperature on the water retention properties of two high performance concretes. *Cement and Concrete Research*, **42**, 384–396.

CARIOU, S., SKOCZYLAS, F. & DORMIEUX, L. 2012. Experimental measurements and water transfer models for the drying of argillite. *International Journal of Rock Mechanics & Mining Sciences*, **54**, 56–69.

DAVY, C.A., SKOCZYLAS, F., BARNICHON, J.-D. & LEBON, P. 2007. Permeability of macro-cracked argillite under confinement: gas and water testing. *Physics and Chemistry of the Earth, Parts A/B/C*, **32**, 667–680.

DAVY, C.A., M'JAHAD, S., SKOCYLAS, F., TALANDIER, J. & GHAYAZA, M. 2012. Evidence of discontinuous and continuous gas migration through undisturbed and self-sealed COx claystone. Oral presentation at a Plenary Session of the 5th International Meeting on Clays in Natural and Engineered Barriers for Radioactive Waste Management, 22–25 October 2012, Montpellier, France.

DISTINGUIN, M. & LAVANCHY, J.M. 2007. Determination of the hydraulic properties of the Callovo-Oxfordian Argillites at the Bure Site: synthesis of the results obtained in deep boreholes using several in-situ investigation techniques. *Physics and Chemistry of the Earth, Parts A/B/C*, **32**, 379–392.

ESCOFFIER, S., HOMAND, F., GIRAUD, A., HOTEIT, N. & SU, K. 2005. Under stress permeability determination of the Meuse/Haute-Marne mudstone. *Engineeering Geology*, **81**, 329–340.

GREGG, S.J. & SING, S.W. 1982. *Adsorption, Surface Area and Porosity.* 2nd edn. AcademicPress, New York.

HARRINGTON, J.F., CUSS, R.J., NOY, D.J. & TALANDIER, J. 2014. Processes governing advective gas flow in the Callovo-Oxfordian Claystone (COx). Paper presented at the 4th EAGE Shale Workshop 'Shales: What do they have in Common?', 6–9 April 2014, Porto, Portugal.

HILDENBRAND, A., SCHÖMER, S. & KROOSS, B.M. 2002. Gas breakthrough experiments on fine-grained sedimentary rocks. *Geofluids*, **2**, 3–23.

LOOSVELDT, H., LAFHAJ, Z. & SKOCZYLAS, F. 2002. Experimental study of gas and liquid permeability of a mortar. *Cement and Concrete Research*, **32**, 1357–1363.

MARSCHALL, P., HORSEMAN, S. & GIMMI, T. 2005. Characterisation of gas transport properties of the Opalinus Clay, a potential host rock formation for radioactive waste disposal. *Oil & Gas Science and Technology – Revue d'IFP Energies nouvelles*, **60**, 121–139.

M'JAHAD, S. 2012. *Impact de la fissuration sur les propriétés de rétention d'eau et de transport des géomatériaux. Application au stockage des déchets radioactifs.* Thèse de doctorat, Ecole Centrale de Lille, Lille, France.

MUALEM, Y. 1976. A new model for predicting the hydraulic conductivity of unsaturated porous media. *Water Resources Research*, **12**, 513–522.

PHAM, Q.T. 2006. *Effet de la désaturation re-saturation sur l'argilite dans les ouvrages souterrains.* Thèse de doctorat, Ecole Polytechnique, Palaiseau, France.

ROBINET, J.C. 2008. *Minéralogie, porosité et diffusion des solutés dans l'argilite du Callovo-Oxfordien de Bure (Meuse/Haute Marne, France) de l'échelle centimétrique à micrométrique.* Thèse de doctorat, University of Poitiers, Poitier, France.

ROUGELOT, T., SKOCZYLAS, F. & BURLION, N. 2009. Water desorption and shrinkage in mortars and cement pastes: experimental study and poromechanical model. *Cement and Concrete Research*, **39**, 36–44.

SAIYOURI, N., HICHER, P.Y. & TESSIER, D. 2000. Microstructural approach and transfer water modelling in highly compacted unsaturated swelling clays. *Mechanics of Cohesive-Frictional Materials*, **5**, 41–60.

SILVA, O. & GRIFOLL, J. 2007. A soil-water retention function that includes the hyper-dry region through the BET adsorption isotherm. *Water Resources Research*, **43**, W11420, https://doi.org/10.1029/2006WR005325

SKOCZYLAS, F. 2011. *Rapport final pour l'étude AS2 du GL Gaz, Mesure des perméabilités intrinsèques et relatives au gaz, Effet de la pression d'injection et de la nature du gaz.* Note technique Andra n° C NT FMFS 11 0002, 2011. Laboratoire de Mécanique de Lille (LML), France.

VAN GENUCHTEN, M.T. 1980. A closed-form equation for predicting the hydraulic conductivity of unsaturated soils. *Soil Science Society of America Journal*, **44**, 1892–1898.

ZHANG, C.L., CZAIKOWSKI, O. & ROTHFUCHS, T. 2010. *Thermo-Hydro-Mechanical Behaviour of the Callovo-Oxfordian Clay Rock.* Final Report of the BURE-HAUPT/EC-TIMODAZ Project, GRS-266. Gesellschaft für Anlagenund Reaktorsicherheit (GRS), Cologne, Germany.

Vertical distribution of helium and $^{40}Ar/^{36}Ar$ in porewaters of the Eastern Paris Basin (Bure/Haute-Marne): constraints on transport processes through the sedimentary sequence

P. JEAN-BAPTISTE[1]*, B. LAVIELLE[2], E. FOURRE[1], T. SMITH[2] & M. PAGEL[3]

[1]*Laboratoire des Sciences du Climat et de l'Environnement (LSCE), CEA-Saclay, 91191 Gif-sur-Yvette cedex, France*

[2]*Centre d'Etudes Nucléaires de Bordeaux-Gradignan, CNRS-Université de Bordeaux, Gradignan, France*

[3]*Géosciences Paris Sud (GEOPS), CNRS-Université Paris-Sud, Orsay, France*

**Correspondence: pjb@lsce.ipsl.fr*

Abstract: As part of its ongoing project on repositories for high-activity, long-lived radioactive waste, a 2000 m deep borehole was drilled by the French Nuclear Waste Agency (ANDRA) in the layered structure of alternating aquifers and aquitards of the Eastern Paris Basin. Among the information retrieved from this borehole, the vertical distribution of chloride in porewaters showed that, in addition to vertical diffusion, lateral advection in the aquifers plays a major part in transporting chlorine away from the study area. Helium concentrations were also measured in porewaters along the borehole. Because the helium input function is different from that of chlorine, it represents an excellent alternative tracer to further constrain transport characteristics. We applied an advection–diffusion model to the helium profiles with the appropriate source term for ^4He based on U–Th measured concentrations of uranium and thorium. $^{40}Ar/^{36}Ar$ data, which were available along the whole sequence, were also simulated. The modelled and measured ^4He profiles were in good agreement, indicating that the transport parameters used for the chlorine simulations were robust. $^{40}Ar/^{36}Ar$ simulations also gave coherent results and confirmed that most of the radiogenic ^{40}Ar remained trapped in the rocks (primarily in clays and feldspars).

The Paris Basin is filled with a thick accumulation of Mesozoic and Cenozoic sediments (up to 3000 m thick in its centre) with contrasting lithologies and permeabilities, creating a multi-layered structure of alternating aquifers and aquitards. Since 1996, the French national radioactive waste management agency (ANDRA) has been investigating the Meuse/Haute-Marne area (Gaucher *et al.* 2004; Delay *et al.* 2007*a*, *b*), located in the eastern part of the Paris Basin (Fig. 1), to assess the feasibility of a repository for high- to intermediate-activity, long-lived radioactive waste in the low permeability Callovo-Oxfordian (COx) Formation located at a depth of about 600 m. This 150 m thick impervious clay layer is in contact with two limestone formations, the underlying Dogger and the overlying Oxfordian formations, which host major aquifers. Therefore understanding the hydrogeological processes controlling solute transport is an important prerequisite for the evaluation of the safety of this project.

A 2000 m deep borehole was drilled in 2008 with the aim of gaining a deeper insight into the hydrogeological processes (Landrein *et al.* 2013). Among the information retrieved from this borehole, the results of Rebeix *et al.* (2014), based on measurements of the vertical distribution of chloride in porewaters along the entire sedimentary pile and computer simulations using a two-dimensional (2D) advection–diffusion model, have shown that, in addition to vertical diffusion, lateral advection in the aquifers plays a major part in transporting chlorine away from the study area.

Helium (^4He) concentrations and $^{40}Ar/^{36}Ar$ ratios were also measured in the porewaters along the borehole. Here, we report helium and argon simulations performed with the same advection–diffusion model as that used for chlorine (Rebeix *et al.* 2014). ^4He and ^{40}Ar excesses over their air-saturated water concentrations are produced by the radioactive decay of the uranium and thorium series and of ^{40}K, respectively, whereas chlorine, apart from the initial chlorine content of the porewaters, which is of marine origin, is supplied by the thick halite-rich layer of the Keuper Formation located at the base of the modelled domain. As a result of these very different origins for the noble gases and chlorine, helium and argon appear to be excellent alternative tracers to test the robustness of the conclusions based on chlorine and to

From: NORRIS, S., BRUNO, J., VAN GEET, M. & VERHOEF, E. (eds) 2017. *Radioactive Waste Confinement: Clays in Natural and Engineered Barriers*. Geological Society, London, Special Publications, **443**, 179–192.
First published online December 21, 2016, https://doi.org/10.1144/SP443.25

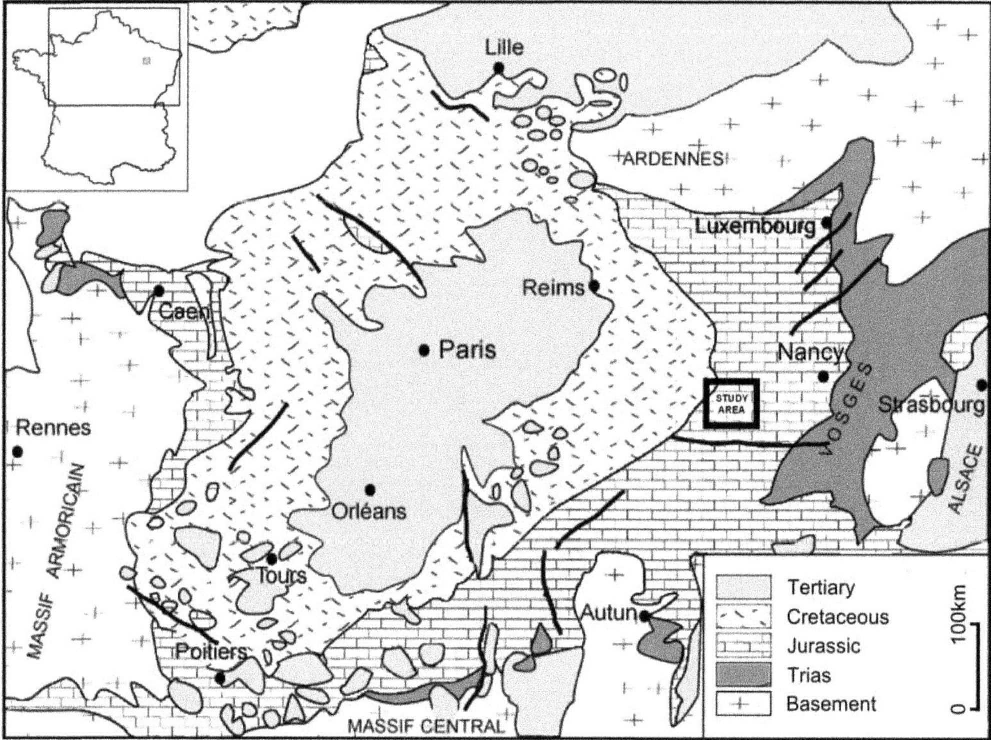

Fig. 1. Simplified map of the Paris Basin showing the location of the study area.

further constrain the transport characteristics of the system.

Geological and hydrogeological framework

The Meuse/Haute-Marne site is located in the eastern part of the Paris Basin (Fig. 1). A simplified sedimentary log of the borehole is shown in Figure 2. The sedimentary cover, which was deposited during the Triassic and Jurassic, is a succession of permeable (limestones and sandstones) and impervious (marls and clays) layers. The Triassic and Jurassic formations are separated by a 200 m thick halite-rich unit belonging to the Keuper Formation (Fig. 2). The clay-rich COx Formation, located between −544 and −691 m, is of particular interest because it is being considered by the French Nuclear Waste Agency (ANDRA) to host a long-term underground nuclear waste repository. The COx layer is in contact with two major (*c.* 300 m thick) limestone aquifers, the Oxfordian Formation at the top and the Dogger Formation at the bottom. Both are fed by meteoric waters infiltrating along the eastern and southeastern outcrops of the Paris Basin, located several tens of kilometres from the drilling sites (Buschaert *et al.* 2007).

The hydrogeology of the area is characterized by the presence of a shallow aquifer in Tithonian limestones (between −100 m and the surface) and a deeper aquifer in the Buntsandstein sandstones (between −1859 and −1980 m). Based on the observation of a well-defined chlorine minimum at the Keuper–Lias boundary, Rebeix *et al.* (2014) also suggested the occurrence of horizontal advection in the *c.* 20 m thick Rhetian sandstones and overlying porous bioclastic limestone layer. Details of the geology and hydrogeology of the area have been reported by Landrein *et al.* (2013).

Noble gas data

The sampling procedures and noble gas measurements used here have been described in detail in Battani *et al.* (2011). The noble gases in each sample porewater were extracted by molecular diffusion into a pre-evacuated helium-tight stainless-steel container over a two to six month period. The noble gas extraction efficiency was checked by performing a second gas extraction after an additional storage period of several months. For helium, the extraction efficiency was close to unity. For argon, the efficiency was not as good as for helium,

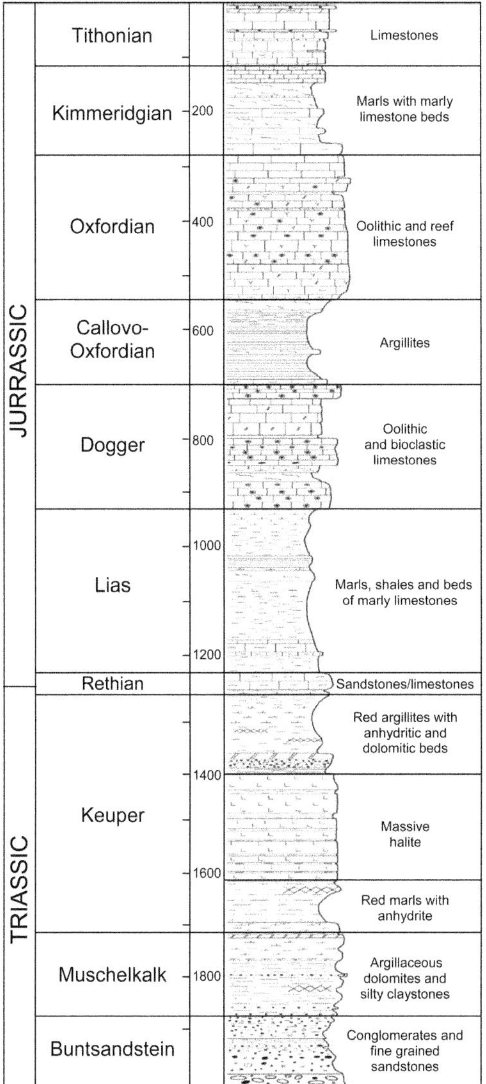

Fig. 2. Stratigraphy of the sedimentary pile at borehole location. Details of the stratigraphy can be found in Landrein *et al.* (2013).

probably because of the lower mobility of argon. For that reason, we preferred to use $^{40}Ar/^{36}Ar$ ratios instead of ^{40}Ar concentrations. As there is no source or loss of ^{36}Ar, the ^{36}Ar concentration is equal to the solubility equilibrium with atmospheric air everywhere in the system. Therefore the $^{40}Ar/^{36}Ar$ increase faithfully reflects the addition of radiogenic ^{40}Ar to the system. Helium concentrations and $^{40}Ar/^{36}Ar$ ratios were measured at the University of Bordeaux (Centre d'Etudes Nucléaires de Bordeaux-Gradignan, France) using a VG 1200 mass spectrometer (Smith 2010). All data were corrected for possible air contamination. Another set of helium samples was measured at the University of Bern with a QMS200 quadrupole mass spectrometer using a similar sampling methodology (Waber & Vinsot 2010). The helium data are available in Battani *et al.* (2011). The helium concentrations increased strongly with depth down to the Keuper–Muschelkalk boundary and then decreased again, approaching those of the groundwater sampled in the Buntsandstein aquifer (Fig. 3a). The agreement between the two datasets was satisfactory, except below the Lias–Keuper boundary, where large discrepancies were observed. There is no obvious explanation for this disagreement. However, as the sampling and helium extraction methods used by both laboratories were almost identical, we favour some analytical problems at high helium concentrations, possibly in relation to non-linearity problems in the response of one mass spectrometer. The $^{40}Ar/^{36}Ar$ vertical profile displayed similar trends (Fig. 3b, Table 1); however, the $^{40}Ar/^{36}Ar$ values remained low compared with what should have been observed if the ^{40}Ar production was entirely transferred to the porewaters, indicating that the vast majority of radiogenic argon remained trapped in the rock, in agreement with the $^{40}Ar/^{36}Ar$ ratios reported by Tolstikhin *et al.* (1996) for a sedimentary sequence from northern Switzerland.

Advection–diffusion model

Previous work on the same borehole using a one-dimensional (1D) geometry (Battani *et al.* 2011) has shown that the measured helium concentrations are well below the theoretical curve describing 1D helium vertical diffusion. This tracer deficit points to the role of aquifers in transporting species laterally away from the study zone and to the need to use a 2D diffusion–advection model. Such a 2D approach was adopted by Rebeix *et al.* (2014) to successfully simulate the chloride concentration profile based on initial modelling work reported in Fourré *et al.* (2011). Here, we applied the same model to helium and argon, with the appropriate source terms and boundary conditions for ^{4}He and ^{40}Ar.

Modelled domain

The massive salt layer of the Keuper Formation is characterized by a very low porosity (0.5 vol.%) and water content (0.3 wt.%). As shown in Battani *et al.* (2011), the pores are completely disconnected. The diffusion coefficients for helium in salt rocks are extremely low at 3.2×10^{-25} m^2 s^{-1} at 40°C (Gentner & Trendelenburg 1954). Measurements

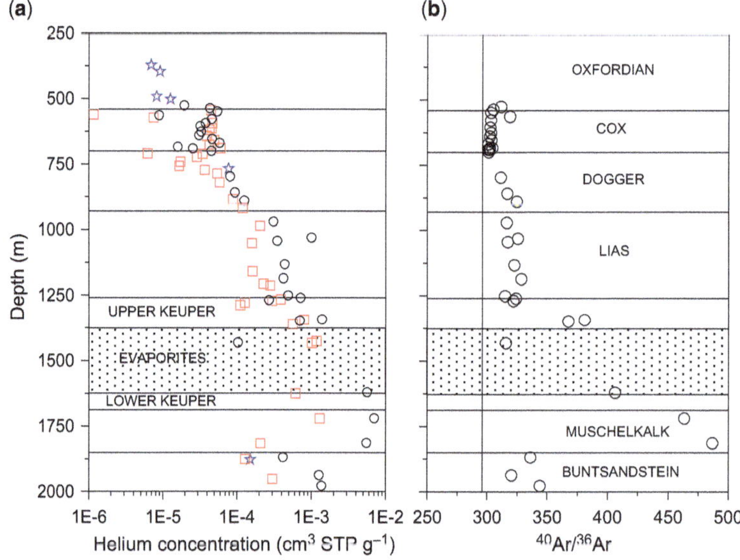

Fig. 3. Depth profiles of (**a**) helium and (**b**) $^{40}Ar/^{36}Ar$ in porewaters. Open circles refer to the University of Bordeaux dataset (Smith 2010; Battani *et al.* 2011) and red squares to the University of Bern dataset (Waber & Vinsot 2010). Blue stars represent helium concentrations in aquifer free water samples reported by Fourré *et al.* (2011). COX, Callovo-Oxfordian.

performed by ANDRA on similar salt samples using the pulse decay method in triaxial cells have given permeability values of $K < 10^{-21}$ m^2 for helium and argon (Chemin 1990). Therefore the 200 m thick halite layer of the Keuper Formation between 1400 and 1600 m depth acts as a natural barrier between the Jurassic and the deeper Triassic geological units. Considering the discrepancies between our two helium datasets for the Triassic formations and the scarcity of information available about these deep layers compared with those of the Jurassic, we focused our modelling study on the sedimentary formations above the halite layer, taking advantage of the fact that this layer imposes a convenient zero-flux boundary condition at the base of the domain. The same choice was made by Rebeix *et al.* (2014) for modelling chlorine, the Keuper halite formation representing in that case a quasi-infinite chloride reservoir for the sedimentary pile.

Model description

The geometry of the modelled domain is shown in Figure 4. As for chlorine (Rebeix *et al.* 2014), it was assumed that transport along the vertical axis was purely diffusive. In addition to vertical diffusion, lateral advection in the aquifer layers was taken into account explicitly in the model, with a groundwater flow velocity v in each aquifer defined as the distance to recharge divided by the transit time of the groundwater. It follows that helium and argon distributions are described by the classic diffusive–advective equation:

$$\frac{\omega_t\,\partial C}{\partial t} = \frac{D_e\,\partial^2 C}{\partial z^2} + \frac{\omega_a v\,\partial C}{\partial x} + \rho_r(1-\omega_t)P \qquad (1)$$

where C is the ^4He or ^{40}Ar concentration in porewater, D_e is the effective diffusion coefficient, ω_t is the total accessible water porosity, ω_a is the free water porosity (i.e. the porosity open to horizontal advection), P is the ^4He or ^{40}Ar radiogenic production per kilogram of rock per second and ρ_r is the rock density. The $^{40}Ar/^{36}Ar$ ratio is calculated using a constant value for ^{36}Ar, equal to the argon concentration at the solubility equilibrium with atmospheric air at 10°C (Weiss 1970) divided by the atmospheric $^{40}Ar/^{36}Ar$ ratio ($^{40}Ar/^{36}Ar = 295.5$).

The diffusion–advection equation (1) is solved numerically with vertical and horizontal steps $\Delta z = 10$ m and $\Delta x = 5$ km, respectively, using time steps of 5 ka.

Diffusion coefficients

In porous media, the effective diffusion coefficient D_e of a species through porewater is classically

Table 1. *$^{40}Ar/^{36}Ar$ data (Smith 2010)*

Depth (m)	$^{40}Ar/^{36}Ar$
526.6	311.39
537.0	304.89
549.2	303.11
564.1	318.91
578.2	302.94
605.7	302.50
626.5	303.19
639.8	302.11
654.2	303.33
669.1	300.74
684.1	304.20
684.1	302.10
690.4	301.96
690.4	300.85
700.2	301.06
797.0	311.30
859.3	316.69
890.1	324.67
970.7	316.28
1031.4	325.55
1044.1	317.09
1132.3	322.60
1186.1	328.48
1251.2	314.85
1260.5	323.89
1269.5	321.77
1343.0	381.04
1348.0	367.34
1431.7	315.60
1621.1	406.19
1719.5	463.37
1813.8	486.77
1868.5	336.25
1937.7	320.31
1978.6	344.11

related to its diffusion coefficient in free water D_0 by $D_e = D_0/F$, where the formation factor F is a geometric parameter accounting for the tortuosity/constrictivity of the medium. We plotted the available literature data for the effective diffusion coefficient of tritium (HTO) as a function of porosity in the various formations of the Bures site and in similar geological environments (Fig. 5). Most of the data were bracketed by the empirical relationship $D_e = D_0 \times \omega_t^n$ (Archie's law) with exponents between 2 and 3.5, in agreement with previous findings in similar environments (Mazurek *et al.* 2011). As diffusion coefficient data are only available at a few discrete levels, Archie's law formulation is a convenient interpolator that allows us to define the formation factor F at every point of the model grid. For our simulations, we adopted the same value $n = 2.5$ (intermediate diffusion case) as Rebeix *et al.* (2014) to remain consistent with the chlorine simulation. However, the high ($n = 2$) and low ($n = 3$) diffusion cases were also explored for helium. The selected D_0 values for helium and argon are 8×10^{-9} and 3×10^{-9} m^2 s^{-1}, respectively (Wise & Houghton 1966).

Porosity vertical distribution

The *in situ* total water accessible porosity ω_t and the free water porosity ω_a (accessible to advective transport) were measured with a nuclear magnetic resonance wire-logging probe (Combinable Magnetic Resonance, trademarked by Schlumberger) down to −1995 m (Fig. 6a, Table 2). Observations made by ANDRA on 16 boreholes down to the Oxfordian Formation and 11 boreholes down to the Dogger Formation since 1994 suggest that this vertical

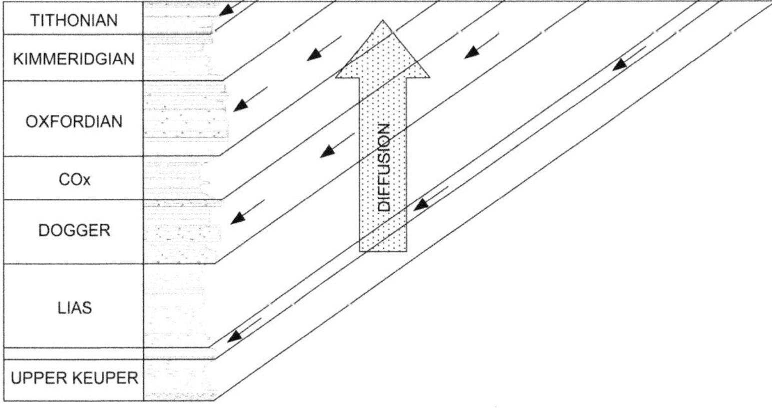

Fig. 4. Geometry of the 2D advection–diffusion model. COx, Callovo-Oxfordian.

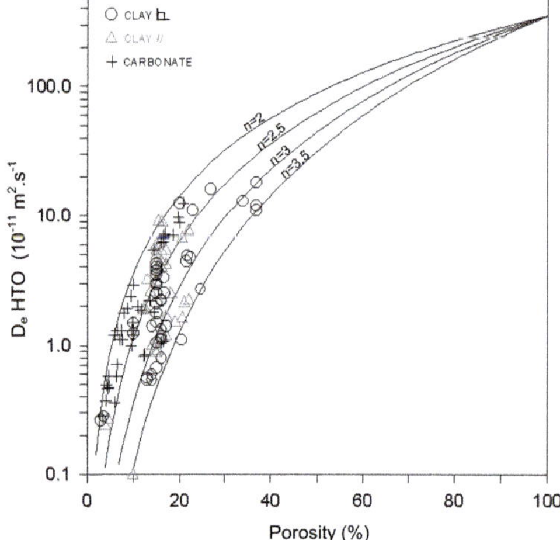

Fig. 5. Data compilation of effective diffusion coefficient for tritium (HTO) in clay/marl and limestone formations as a function of porosity (Put & De Cannière 1994; Palut *et al.* 2003; Van Loon *et al.* 2003*a*, *b*, 2004*a*, *b*; Melkior *et al.* 2004, 2007; Tevissen *et al.* 2004; Wersin *et al.* 2004, 2008; Garcia-Gutiérrez *et al.* 2006, 2008; Anthony *et al.* 2007; Cormenzana *et al.* 2008; Descostes *et al.* 2008; Samper *et al.* 2008; Appelo et al. 2010; Tachi *et al.* 2011). Diffusion coefficient in free water is taken from Mills (1973).

distribution is typical of the study area (100 km^2). Therefore, given the lack of data outside the zone, we assumed that there was no major lateral change in the vertical structure of porosity along the *x*-axis of the model.

Helium and argon radiogenic production

Uranium, thorium and potassium radioactive isotopes and their daughter products emit gamma radiation according to specific energy spectra (e.g. Hearst *et al.* 2000). Gamma ray spectra were acquired every 15 cm in the COx Formation and the underlying rock units using a specific tool (the Hostile Environment Natural Gamma Ray Spectral tool, trademarked by Schlumberger) composed of two bismuth germanate crystal detectors. At each depth, the U–Th–K concentrations were retrieved by deconvolution of the gamma ray spectrum based on the characteristic emission energies of potassium, uranium and thorium, with an accuracy of $\pm 0.5\%$ for potassium, $\pm 2\%$ for thorium and $\pm 2\%$ for uranium (Fig. 6b, Table 2). In the first 500 m, where no such data is available, the U–Th–K concentrations measured on a representative set of rock samples were used instead (Ahmadi 1997; ANDRA 2005). The *in situ* radiogenic helium and argon production rates $P(^4He)$

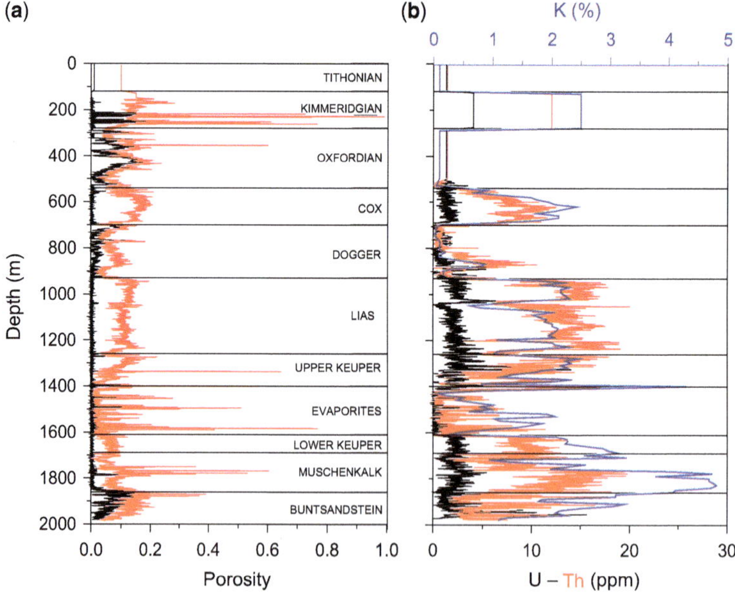

Fig. 6. (**a**) Total water porosity (accessible to diffusion) in red and free water porosity (accessible to horizontal advection) in black, obtained by *in situ* nuclear magnetic resonance spectrometry (Linard *et al.* 2011). (**b**) Potassium (%), thorium (ppm) and uranium (ppm) concentrations. COX, Callovo-Oxfordian.

Table 2. *Radionuclide concentrations and porosity used in the model (see Fig. 6)*

Formation	Depth (m)	^{238}U (ppm)	^{232}Th	^{40}K (%)	Free porosity (%)	Total porosity (%)
Barrois	5	1.50	0.70	0.10	1.00	10.00
Barrois	15	1.50	0.70	0.10	1.00	10.00
Barrois	25	1.50	0.70	0.10	1.00	10.00
Barrois	35	1.50	0.70	0.10	1.00	10.00
Barrois	45	1.50	0.70	0.10	1.00	10.00
Barrois	55	1.50	0.70	0.10	1.00	10.00
Barrois	65	1.50	0.70	0.10	1.00	10.00
Barrois	75	1.50	0.70	0.10	1.00	10.00
Barrois	85	1.50	0.70	0.10	1.00	10.00
Barrois	95	1.50	0.70	0.10	1.00	10.00
Barrois	105	1.50	0.70	0.10	1.00	10.00
Barrois	115	1.50	0.70	0.10	1.00	10.00
Kimmeridgian	125	1.50	5.00	2.00	0.10	15.00
Kimmeridgian	135	1.50	5.00	2.00	0.10	15.00
Kimmeridgian	145	1.50	5.00	2.00	0.10	15.00
Kimmeridgian	155	1.50	5.00	2.00	0.09	15.99
Kimmeridgian	165	1.50	5.00	2.00	0.34	21.76
Kimmeridgian	175	1.50	5.00	2.00	0.20	13.68
Kimmeridgian	185	1.50	5.00	2.00	0.13	15.08
Kimmeridgian	195	1.50	5.00	2.00	0.17	13.19
Kimmeridgian	205	1.50	5.00	2.00	0.40	13.53
Kimmeridgian	215	1.50	5.00	2.00	5.56	21.15
Kimmeridgian	225	1.50	5.00	2.00	20.50	34.36
Kimmeridgian	235	1.50	5.00	2.00	0.38	13.10
Kimmeridgian	245	1.50	5.00	2.00	4.89	20.24
Kimmeridgian	255	1.50	5.00	2.00	14.89	29.79
Kimmeridgian	265	1.50	5.00	2.00	0.36	10.27
Kimmeridgian	275	1.50	5.00	2.00	0.39	6.76
Oxfordian	285	1.50	0.70	0.10	0.23	6.18
Oxfordian	295	1.50	0.70	0.10	7.80	12.59
Oxfordian	305	1.50	0.70	0.10	3.57	8.75
Oxfordian	315	1.50	0.70	0.10	0.23	5.67
Oxfordian	325	1.50	0.70	0.10	9.03	12.88
Oxfordian	335	1.50	0.70	0.10	5.88	11.46
Oxfordian	345	1.50	0.70	0.10	8.76	16.78
Oxfordian	355	1.50	0.70	0.10	11.12	15.04
Oxfordian	365	1.50	0.70	0.10	10.81	15.82
Oxfordian	375	1.50	0.70	0.10	4.01	9.38
Oxfordian	385	1.50	0.70	0.10	4.40	9.27
Oxfordian	395	1.50	0.70	0.10	6.03	10.52
Oxfordian	405	1.50	0.70	0.10	8.99	12.28
Oxfordian	415	1.50	0.70	0.10	16.14	17.30
Oxfordian	425	1.50	0.70	0.10	15.31	16.93
Oxfordian	435	1.50	0.70	0.10	13.15	16.59
Oxfordian	445	1.50	0.70	0.10	8.75	12.75
Oxfordian	455	1.50	0.70	0.10	9.39	12.55
Oxfordian	465	1.50	0.70	0.10	5.42	8.46
Oxfordian	475	1.50	0.70	0.10	2.61	6.54
Oxfordian	485	1.50	0.70	0.10	2.58	6.70
Oxfordian	495	1.50	0.70	0.10	0.61	3.72
Oxfordian	505	1.28	0.79	0.07	2.79	6.53
Oxfordian	515	1.46	0.68	0.06	2.82	8.18
Oxfordian	525	1.42	0.52	0.04	4.84	9.73
Oxfordian	535	1.42	2.12	0.44	0.22	9.42
Callovo-Oxfordian	545	1.17	4.21	0.91	0.37	11.82
Callovo-Oxfordian	555	1.08	4.25	0.69	1.18	11.15
Callovo-Oxfordian	565	1.15	5.56	1.12	0.83	12.13

(*Continued*)

Table 2. *Radionuclide concentrations and porosity used in the model (see Fig. 6)* (*Continued*)

Formation	Depth (m)	^{238}U (ppm)	^{232}Th	^{40}K (%)	Free porosity (%)	Total porosity (%)
Callovo-Oxfordian	575	1.22	7.01	1.65	0.35	15.28
Callovo-Oxfordian	585	1.14	7.54	1.75	0.28	16.00
Callovo-Oxfordian	595	1.54	7.93	1.94	0.28	17.05
Callovo-Oxfordian	605	1.47	8.46	2.06	0.15	17.56
Callovo-Oxfordian	615	1.21	10.01	2.44	0.18	17.44
Callovo-Oxfordian	625	1.23	10.31	2.23	0.17	16.57
Callovo-Oxfordian	635	1.56	8.85	2.05	0.14	15.95
Callovo-Oxfordian	645	1.74	7.37	1.75	0.15	12.82
Callovo-Oxfordian	655	1.74	9.17	2.13	0.20	14.81
Callovo-Oxfordian	665	1.55	9.37	2.13	0.11	14.44
Callovo-Oxfordian	675	2.04	7.75	1.76	0.17	14.34
Callovo-Oxfordian	685	1.42	6.58	1.09	0.17	10.47
Callovo-Oxfordian	695	0.79	1.87	0.13	1.37	5.10
Dogger	705	0.22	0.55	0.03	9.28	11.35
Dogger	715	0.36	0.67	0.04	7.56	9.72
Dogger	725	0.68	1.06	0.05	3.03	5.98
Dogger	735	0.60	0.54	0.02	2.40	5.00
Dogger	745	0.96	0.67	0.03	1.47	4.34
Dogger	755	1.00	0.43	0.06	0.56	3.39
Dogger	765	0.86	0.66	0.10	8.58	10.49
Dogger	775	1.04	0.59	0.11	2.64	6.07
Dogger	785	0.94	0.77	0.10	1.41	5.78
Dogger	795	1.06	1.30	0.10	1.28	5.90
Dogger	805	0.23	0.90	0.13	1.88	7.14
Dogger	815	0.41	2.51	0.20	0.88	6.85
Dogger	825	0.14	3.22	0.10	2.02	9.14
Dogger	835	0.44	2.85	0.10	2.40	8.71
Dogger	845	0.62	4.74	0.35	1.28	8.24
Dogger	855	1.02	5.69	0.60	1.36	9.49
Dogger	865	1.11	6.91	0.87	1.17	10.13
Dogger	875	1.41	5.15	0.77	0.63	7.99
Dogger	885	1.89	1.76	0.31	0.85	4.96
Dogger	895	2.25	1.48	0.16	1.11	3.61
Dogger	905	0.38	1.05	0.17	1.00	3.23
Dogger	915	1.89	1.41	0.16	0.81	3.95
Dogger	925	1.46	3.81	0.25	0.97	7.07
Lias	935	1.87	10.29	1.76	0.42	11.74
Lias	945	2.15	14.06	2.21	0.34	13.64
Lias	955	2.52	12.47	1.86	0.30	12.72
Lias	965	2.44	15.67	2.31	0.38	13.96
Lias	975	2.48	14.20	2.16	0.39	12.57
Lias	985	2.20	14.84	2.15	0.33	12.73
Lias	995	2.44	14.84	2.21	0.29	12.69
Lias	1005	2.55	14.51	2.26	0.23	12.96
Lias	1015	2.38	15.06	2.32	0.25	12.96
Lias	1025	4.37	8.48	1.59	0.16	12.54
Lias	1035	3.15	8.37	1.13	0.35	11.28
Lias	1045	0.73	11.37	0.63	0.78	11.56
Lias	1055	1.13	10.83	0.81	0.27	10.23
Lias	1065	1.66	12.98	1.60	0.21	9.80
Lias	1075	2.15	12.85	2.00	0.15	11.58
Lias	1085	1.93	13.18	1.69	0.23	10.48
Lias	1095	2.17	13.04	1.79	0.20	10.89
Lias	1105	1.96	12.07	1.92	0.20	10.41
Lias	1115	1.83	13.51	1.89	0.22	10.19
Lias	1125	1.76	15.22	2.02	0.24	10.78
Lias	1135	2.28	15.91	2.01	0.24	11.02

(*Continued*)

Table 2. (*Continued*)

Formation	Depth (m)	^{238}U (ppm)	^{232}Th	^{40}K (%)	Free porosity (%)	Total porosity (%)
Lias	1145	2.14	14.59	2.12	0.24	11.29
Lias	1155	2.54	13.12	2.08	0.25	11.24
Lias	1165	2.34	13.19	2.02	0.30	11.83
Lias	1175	1.94	12.72	1.95	0.31	11.31
Lias	1185	2.06	13.58	2.05	0.34	11.03
Lias	1195	2.87	7.28	1.07	0.33	9.74
Lias	1205	2.11	16.35	2.01	0.29	10.73
Lias	1215	2.81	15.51	2.07	0.44	12.29
Lias	1225	1.93	16.20	2.30	0.39	13.46
Lias	1235	2.78	11.72	1.58	0.30	10.90
Rethian	1245	2.74	6.81	1.09	0.30	7.81
Rethian	1255	3.22	6.43	1.02	0.26	6.75
Keuper	1265	2.30	13.80	1.77	0.53	8.60
Keuper	1275	2.46	14.28	2.47	0.68	14.51
Keuper	1285	3.08	13.03	2.24	0.34	13.59
Keuper	1295	2.65	9.55	2.34	0.29	11.78
Keuper	1305	3.28	9.20	2.16	0.34	10.43
Keuper	1315	3.58	9.72	2.28	0.29	8.98
Keuper	1325	3.84	9.89	1.88	0.33	6.81
Keuper	1335	1.83	8.85	1.70	3.73	11.67
Keuper	1345	2.12	5.46	1.32	0.26	5.48
Keuper	1355	1.52	5.69	1.14	0.24	6.27
Keuper	1365	2.45	10.03	2.36	0.20	7.23
Keuper	1375	1.95	9.02	1.99	0.22	4.18
Keuper	1385	1.30	2.52	0.81	0.22	3.36
Keuper	1395	3.34	12.67	4.30	2.88	15.08

All reported values are the average of the wireline-logging measurements acquired every 10 m (corresponding to the vertical discretization of the model $\Delta z = 10$ m).

and $P(^{40}Ar)$, in mol g^{-1} rock a^{-1}, were then calculated using the following equations:

$$P(^4He) = 8\lambda_{238}\,^{238}U(t) + 7\lambda_{235}\,^{235}U(t) + 6\lambda_{232}\,^{232}Th(t)$$

$$P(^{40}Ar) = B_R\lambda_{40}\,^{40}K(t)$$

where the U–Th–K concentrations (in mol g^{-1} rock) are those at time t; λ_{238}, λ_{235}, λ_{232} and λ_{40} are the decay constants of ^{238}U, ^{235}U, ^{232}Th and ^{40}K, respectively; and B_R is the branching ratio of ^{40}Ar (Nier 1950; Steiger & Jäger 1977; Berglund & Wieser 2011).

For helium, the release efficiency is close to unity and, for modelling purposes, was set to unity. The values for the COx Formation were 0.96–0.98 (Bigler *et al.* 2005; Smith 2010; Waber 2012); similar values have been reported for the Toarcian–Domerian clayrocks at Tournemire (0.97–0.99, Bensenouci *et al.* 2011) and the Opalinus Clay at Benken and Schlattingen (>0.95 and 0.92–0.99, Lehmann *et al.* 2001). For argon, the high retentivity of the rock is considered to be an adjustable parameter of the model.

Initial and boundary conditions

The model was run for 200 Ma. The main geological units were added sequentially according to their age. At the surface and in recharge areas, the helium and argon concentrations were taken at the solubility equilibrium at temperature $T = 10°C$ (Weiss 1970, 1971).

The onset of present day hydrogeological conditions, with generalized groundwater horizontal advection in the aquifers, is thought to have occurred around −20 Ma, in relation to the cropping out of their host geological formations at their eastern border (Guillocheau *et al.* 2000; Rebeix *et al.* 2014). However, oxygen isotope investigation of calcite cements deposited in the Oxfordian and Dogger limestones during the Lower Cretaceous (−145 to −110 Ma) indicate that they originated from meteoric fluids, hence calling on active circulations of meteoric water during this period of time (Vincent *et al.* 2007; Brigaud *et al.* 2009). This change in palaeohydrological conditions is due to the uplift of the Paris Basin margin (Guillocheau *et al.* 2000). Therefore meteoric water is assumed to have circulated in the aquifer formations between −145 and −110 Ma. The circulation ceased at −110 Ma due to marine transgression. During this remote period (−145 to −110 Ma), the recharge areas were further away from the site, so the transit times of water in the aquifers may have been longer than today, although their values are unknown. Sensitivity tests carried out with various transit times from $t = 10^6$ years to $t = \infty$ (=purely diffusive case) showed that the difference between the helium profiles at −20 Ma was <7%, indicating that the system had almost recovered from the Lower Cretaceous perturbation before −20 Ma. It was no surprise that this did not really make much difference because the transit times for the aquifers (on the order of 1 Ma) were short compared with the modelling time. Therefore the lack of precise knowledge of the palaeohydrological conditions during the Lower Cretaceous has only limited consequences for the initial conditions at the onset of the present day hydrogeological conditions at −20 Ma.

Transit time of the groundwaters

From −20 Ma (representing the onset of present day hydrological conditions) to the present, the transit times of groundwaters in the various aquifers were those adopted by Rebeix *et al.* (2014) for their chlorine simulation. For the Tithonian and Oxfordian aquifers, these values are fully consistent with the tracer age estimates deduced from radiocarbon and ^{36}Cl measurements, respectively (Rebeix *et al.* 2014). No tracer age data was available for the Dogger Formation and Keuper–Lias boundary, so the values adopted by Rebeix *et al.* (2014) were those that produced the best fit between the modelled and measured chlorine profiles.

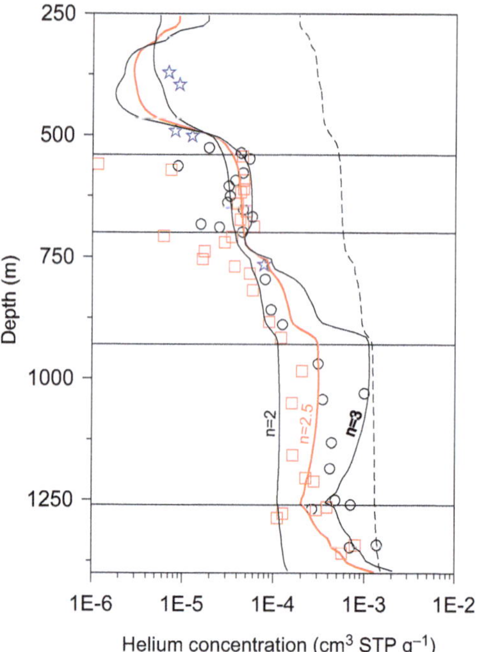

Fig. 7. Simulated helium profile (in red) using the same parameters as for chlorine (see Fig. 8, taken from Rebeix *et al.* 2014). Simulated helium profile for high ($n = 2$) and low ($n = 3$) diffusivities are also shown for comparison purpose (black curves). The dotted curve corresponds to pure diffusion (no advection). Open circles refer to the University of Bordeaux dataset (Smith 2010; Battani *et al.* 2011) and red squares to the University of Bern dataset (Waber & Vinsot 2010). Blue stars represent helium concentrations in aquifer free water samples reported by Fourré *et al.* (2011).

Modelling results and discussion

Helium

The result of the helium simulation is shown in Figure 7. The transport parameters (groundwater residence time in the aquifers and formulation of the diffusion coefficient as a function of porosity) that produced the best results for chlorine (Fig. 8) also did a reasonably good job of simulating the helium data (red curve in Fig. 7). As chlorine and helium are independent tracers with very different sources and boundary conditions, this agreement was certainly not fortuitous. As for chlorine, the intermediate diffusivity case ($n = 2.5$) produced the best result, whereas the high diffusivity ($n = 2$) and low diffusivity ($n = 3$) cases led to less realistic profiles (Fig. 7). This indicates that the transport parameters of the model are robust and that this advection–diffusion model, although very simple, captures reasonably well the main processes responsible for the transport of the dissolved species across the whole sedimentary sequence.

Horizontal advection and vertical diffusion are the two transport processes at work in the system and interfere with each other. When only diffusion was considered, the helium profile (dotted line in Fig. 7) showed a general decreasing trend from the bottom to the surface where the boundary condition for porewaters was the solubility equilibrium with atmospheric air. There was a considerable loss of helium in the 2D configuration compared with the 1D profile. This is because helium can diffuse towards the nearest aquifer, from where it is transported away by the horizontally moving waters, ensuring a persistent concentration gradient between the aquifer level and the surrounding less permeable formations. The higher the advection, the greater the difference in helium concentrations between the groundwaters and porewaters as a result of the

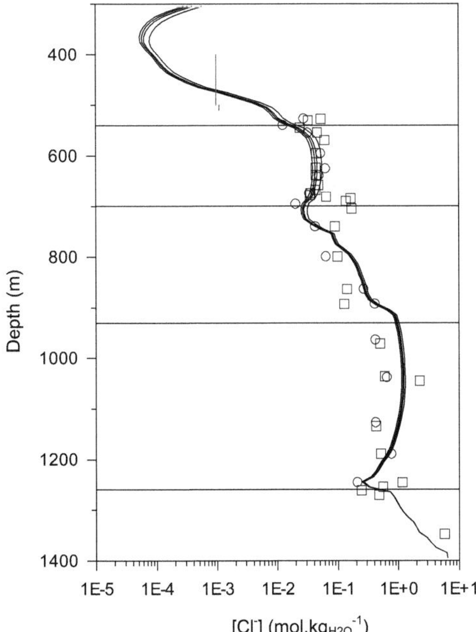

Fig. 8. Plot of the chlorine measurements and the fit of the model results to the data (adapted from figure 9 in Rebeix *et al.* 2014) using the same parameters as for helium (red curve in Fig. 7). Squares and circles represent pore water data from Rebeix *et al.* (2014) and Bensenouci *et al.* (2013), respectively.

dilution effect on diffusive helium by advecting groundwater and, hence, the more efficient the removal of helium. Therefore the aquifers act as drains for helium.

As for chlorine (Fig. 8), the helium profile predicts a marked minimum in the Oxfordian Formation. Unfortunately, there is no corresponding porewater data for the upper part of the sedimentary pile because the non-destructive drilling started at the level of the COx clay formation.

As a whole, the helium modelling confirms the conclusions of Rebeix *et al.* (2014) that lateral advection in the aquifers plays a major part in shaping the vertical distribution of dissolved species in the porewaters. Twenty-five years ago, Torgersen (1989) was among the first to point out that the effective diffusivity of helium is too low to balance helium crustal production and that the transport of helium by fluid movements through the crust is therefore implied. The advective v. diffusive transfer of noble gases in basins is a complex problem and depends on their geological and hydrogeological characteristics (see Ballentine & Burnard 2002; Ballentine *et al.* 2002; and references cited therein). A recent estimate of the helium flux released by large aquifers worldwide suggests that *c.* 53% of the total crustal helium production is transported by horizontal advection (Aggarwal *et al.* 2015). In this study, comparison of helium production in the 1400 m thick sedimentary pile

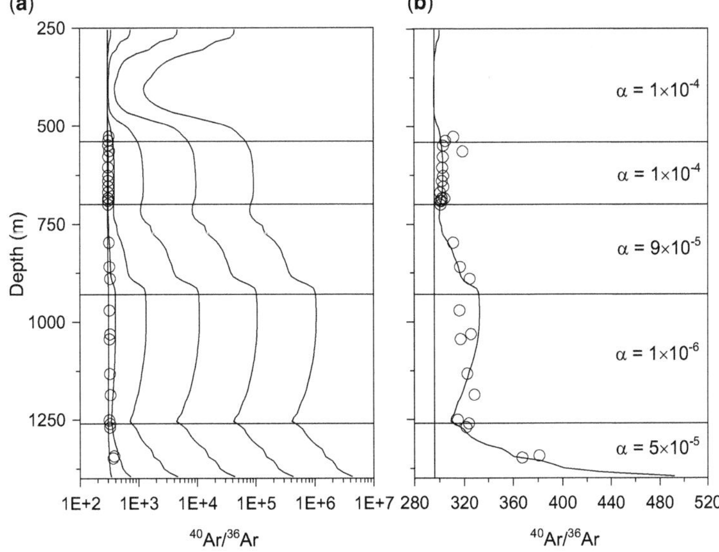

Fig. 9. (**a**) Simulated $^{40}Ar/^{36}Ar$ profile for release efficiencies decreasing by one order of magnitude from 1 to 10^{-5}. (**b**) Simulated $^{40}Ar/^{36}Ar$ profile using the same parameters as for helium (release efficiency α is indicated for each geological layer). Open circles refer to the University of Bordeaux $^{40}Ar/^{36}Ar$ dataset (Smith 2010).

with the vertical flux at the surface indicated that 94% of the helium production was indeed taken away by groundwater (note that this figure corresponds to a rather extreme case because the multi-layered aquifer of the Paris Basin system is highly favourable to transfer from aquitards to neighbour aquifers).

Argon

In contrast with helium, rocks are usually highly retentive with respect to argon (Tolstikhin *et al.* 1996), as evidenced by the high $^{40}Ar/^{36}Ar$ values measured in rocks. As the transport processes for both noble gases (helium and argon) are similar, it seems reasonable to make the hypothesis that the model also works for argon and to use the model to deduce the average argon release efficiency in the various formations. Figure 9a shows the simulated $^{40}Ar/^{36}Ar$ ratios as a function of the release efficiency of argon (the transport parameters are the same as for helium, except D_0 which is different for helium and argon). It shows that almost all of the ^{40}Ar production remains trapped in the rock, with release efficiencies inferred from the model fit as low as 10^{-4}–10^{-5}. This high retentivity is consistent with the thermal evolution of sediments based on several independent geothermometers (Blaise *et al.* 2014), which indicates low burial temperatures (<100°C) throughout the geological history of the Paris Basin.

Figure 9b tentatively shows a set of release efficiencies that produces a reasonable fit of the data. Because of the high, but largely unknown, retentivity of argon in the rocks, argon should be considered as a much less efficient tracer than helium in groundwater and porewater hydrology. However, it offers the possibility of applying K–Ar dating to clays (illite) and feldspar, as evidenced by the agreement between the K–Ar age of illite and the U–Pb age of calcite at −150 Ma (Pisapia *et al.* 2011; Blaise *et al.* 2014).

Conclusions

We have applied to helium and argon the 2D advection–diffusion model of Rebeix *et al.* (2014), designed to simulate the vertical distribution of chlorine in the sedimentary sequences of the Eastern Paris Basin (Bures/Haute-Marne area). The results showed that this simple advection–diffusion model does a reasonably good job of simulating helium data. The model also satisfactorily simulated the $^{40}Ar/^{36}Ar$ data, but, in the case of argon, the very low fraction of radiogenic argon released into the porewaters and groundwaters makes argon a less interesting tracer for the study of groundwater movement.

Many simplifying assumptions have been made due to our limited knowledge of the system.

- The first major assumption was clearly to adopt a 2D description of the system because it is obviously a 3D problem.
- The porosity and U–Th–K concentrations were taken from measurements down a single borehole. Although this was the only option here, extrapolating this across the aquifer is very uncertain.
- Archie's law reduces the calculation of D_e values for the different lithologies to their variability in porosity, disregarding other factors such as tortuosity. In addition, the choice made of calculating of the geometry factor F for a fixed $n = 2.5$ is also an oversimplification. Choosing different values of n for clays and carbonates was clearly an option, but in the light of the scatter in the diffusion coefficient data as a function of porosity, we felt that this choice would have remained arbitrary.
- The data on the past history of the hydraulic system were scarce. We have shown, however, that the lack of a precise knowledge of palaeohydrological conditions during the Early Cretaceous has only limited consequences for the initial conditions at the onset of present day hydrogeological conditions at −20 Ma.

There are many weaknesses and problems inherent with such a model. However, because of the very different origin of the noble gases and chlorine, the coherency between the chlorine and helium simulations demonstrates that the model captures well the main processes responsible for the transport of the dissolved species across the whole sedimentary sequence and that the transport parameters of the model (groundwater residence time in the aquifers and formulation of the diffusion coefficient as a function of porosity) are robust. Therefore it represents a useful tool to assess the long-term environmental consequences of a repository for high- to intermediate-activity, long-lived radioactive waste in the COx Formation.

The authors thank the TAPSS 2000 research programme 'Present and past transfers in a sedimentary aquifer – aquitard system: a 2000 m deep drill-hole in the Mesozoic of the Paris Basin'. This study was funded by CNRS and ANDRA via the GNR FORPRO programme. We are particularly grateful to the people at ANDRA who helped with the acquisition of samples and data, and who provided very useful geological and hydrogeological information.

References

AGGARWAL, P.K., MATSUMOTO, T. *ET AL.* 2015. Continental degassing of ^4He by surficial discharge of deep groundwater. *Nature Geoscience*, **8**, 35–39.

AHMADI, Z.M. 1997. *Sequence stratigraphy using wireline logs from the Upper Jurassic of England*. PhD thesis, Durham University.

ANDRA 2005. *Référentiel du site de Meuse/Haute-Marne*. ANDRA Report **C.RP.ADS.04.0003, 3**.

ANTHONY, C., APPELO, J. & WERSIN, P. 2007. Multicomponent diffusion modeling in clay systems with application to the diffusion of tritium, iodide and sodium in Opalinus clay. *Environmental Science & Technology*, **41**, 5002–5007.

APPELO, C.A.J., VAN LOON, L.R. & WERSIN, P. 2010. Multicomponent diffusion of a suite of tracers (HTO, Cl, Br, I, Na, Sr, Cs) in a single sample of Opalinus Clay. *Geochimica et Cosmochimica Acta*, **74**, 1201–1219.

BALLENTINE, C.J. & BURNARD, P.G. 2002. Production, release and transport of noble gases in the continental crust. *In*: PORCELLI, D., BALLENTINE, C.J. & WIELER, R. (eds) *Noble Gases in Geochemistry and Cosmochemistry*. Reviews in Mineralogy and Geochemistry, **47**. Mineralogical Society of America, Washington DC, 481–538.

BALLENTINE, C.J., BURGESS, R. & MARTY, B. 2002. Tracing fluid origin, transport and interaction in the crust. *In*: PORCELLI, D., BALLENTINE, C.J. & WIELER, R. (eds) *Noble Gases in Geochemistry and Cosmochemistry*. Reviews in Mineralogy and Geochemistry, **47**. Mineralogical Society of America, Washington DC, 539–614.

BATTANI, A., SMITH, T., ROBINET, J.C., BRULHET, J., LAVIELLE, B. & COELHO, D. 2011. Contribution of logging tools to understanding helium porewater data across the Mesozoic sequence of the East of the Paris Basin. *Geochimica et Cosmochimica Acta*, **75**, 7566–7584.

BENSENOUCI, F., MICHELOT, J.L., MATRAY, J.M., SAVOYE, S., LAVIELLE, B., THOMAS, B. & DICK, P. 2011. A profile of helium-4 concentration in pore-water for assessing the transport phenomena through an argillaceous formation (Tournemire, France). *Physics and Chemistry of the Earth*, **36**, 1521–1530.

BENSENOUCI, F., MICHELOT, J.L., MATRAY, J.M., SAVOYE, S., TREMOSA, J. & GABOREAU, S. 2013. Profiles of chloride and stable isotopes in pore-water obtained from a 2000m deep borehole through the Mesozoic sedimentary series in the Eastern Paris Basin. *Physics and Chemistry of the Earth*, **65**, 1–10.

BERGLUND, M. & WIESER, M.E. 2011. Isotopic compositions of the elements 2009. *Pure and Applied Chemistry*, **83**, 397–410.

BIGLER, T., IHLY, B., LEHMANN, B.E. & WABER, H.N. 2005. *Helium Production and Transport in the Low-Permeability Callovo-Oxfordian Shale at the Site Meuse/Haute Marne, France*. Nagra Arbeitsbericht **NAB 05-07**. Nagra, Wettingen.

BLAISE, T., BARBARAND, J. ET AL. 2014. Reconstruction of low temperature (<100°C) burial in sedimentary basins: a comparison of geothermometer in the intracontinental Paris Basin. *Marine and Petroleum Geology*, **53**, 71–87.

BRIGAUD, B., DURLET, C., DECONINCK, J.-F., VINCENT, B., THIERRY, J. & TROUILLER, A. 2009. The origin and timing of multiphase cementation in carbonates: impact of regional scale geodynamic events on the Middle Jurassic Limestones diagenesis (Paris Basin, France). *Sedimentary Geology*, **222**, 161–180.

BUSCHAERT, S., GIANNESINI, S. ET AL. 2007. The contribution of water geochemistry to the understanding of the regional hydrogeological system. *In*: LEBON, P. (ed.) *A Multi-Disciplinary Approach to the Eastern Jurassic Border of the Paris Basin (Meuse/Haute-Marne)*. Mémoires, Societé géologique de France, **178**, 91–114.

CHEMIN, P. 1990. *Etude du rôle des inclusions fluides dans les mécanismes de déformation des roches halitiques. Application aux formations salifères du bassin bressan*. PhD thesis, Ecole Nationale des Ponts et Chaussées.

CORMENZANA, J.L., GARCIA-GUTIERREZ, M., MISSANA, T. & ALONSO, U. 2008. Modeling large-scale laboratory HTO and strontium diffusion experiments in Mont Terri and Bure clay rocks. *Physics and Chemistry of the Earth*, **33**, 949–956.

DELAY, J., VINSOT, A., KRIEGUER, J.M., REBOURS, H. & ARMAND, G. 2007*a*. Making of the underground scientific experimental programme at the Meuse/Haute-Marne underground research laboratory, North Eastern France. *Physics and Chemistry of the Earth*, **32**, 2–18.

DELAY, J., REBOURS, H., VINSOT, A. & ROBIN, P. 2007*b*. Scientific investigation in deep wells for nuclear waste disposal studies at the Meuse/Haute-Marne underground research laboratory, North Eastern France. *Physics and Chemistry of the Earth*, **32**, 42–57.

DESCOSTES, M., BLIN, V. ET AL. 2008. Diffusion of anionic species in Callovo-Oxfordian argillites and Oxfordian limestones (Meuse/Haute–Marne, France). *Applied Geochemistry*, **23**, 655–677.

FOURRÉ, E., JEAN-BAPTISTE, P., DAPOIGNY, A., LAVIELLE, B., SMITH, T., THOMAS, B. & VINSOT, A. 2011. Dissolved helium distribution in the Oxfordian and Dogger deep aquifers of the Meuse/Haute-Marne area. *Physics and Chemistry of the Earth*, **36**, 1511–1520.

GARCIA-GUTIÉRREZ, M., CORMENZANA, J.L., MISSANA, T., MINGARRO, M. & MARTIN, P.L. 2006. Large-scale laboratory diffusion experiments in clay rocks. *Physics and Chemistry of the Earth*, **31**, 523–530.

GARCIA-GUTIÉRREZ, M., CORMENZANA, J.L. ET AL. 2008. Diffusion experiments in Callovo-Oxfordian clay from the Meuse/Haute-Marne URL, France: experimental setup and data analyses. *Physics and Chemistry of the Earth*, **33**, S125–S130.

GAUCHER, E., ROBELIN, C. ET AL. 2004. ANDRA underground research laboratory: interpretation of the mineralogical and geochemical data acquired in the Callovo-Oxfordian formation by investigative drilling. *Physics and Chemistry of the Earth*, **29**, 55–77.

GENTNER, W. & TRENDELENBURG, E.A. 1954. Experimentelle Untersuchung über die Diffusion von Helium in Steinsalzen und Sylvinen. *Geochimica et Cosmochimica Acta*, **6**, 261–267.

GUILLOCHEAU, F., ROBIN, C. ET AL. 2000. Meso-Cenozoic geodynamic evolution of the Paris Basin: 3D stratigraphic constraints. *Geodinamica Acta*, **13**, 189–245.

HEARST, J.R., NELSON, P.H. & PAILLET, F.L. 2000. *Well Logging for Physical Properties*. 2nd edn. Wiley, New York.

LANDREIN, P., VIGNERON, G., DELAY, J., LEBON, P. & PAGEL, M. 2013. Lithologie, hydrodynamisme et thermicité dans le système sédimentaire multicouche recoupé par les forages Andra de Montiers-sur-Saulx

(Meuse). *Bulletin Societé géologique de France*, **184**, 519–543.

LEHMANN, B.E., LOOSLI, H.H., PURTSCHERT, R., TOLSTIKHIN, I., GAUTSCHI, A., KIPFER, R. & AESCHBACH-HERTIG, W. 2001. Helium in a 1000 m-borehole: isotope analyses on sedimentary rocks and related pore- and groundwaters. *Water–Rock Interaction*, **1–2**, 1537–1540.

LINARD, Y., VINSOT, A. *ET AL.* 2011. Water flow in the Oxfordian and Dogger limestone around the Meuse/Haute-Marne underground research laboratory. *Physics and Chemistry of the Earth*, **36**, 1450–1468.

MAZUREK, M., ALT-EPPING, P. *ET AL.* 2011. Natural tracer profiles across argillaceous formations. *Applied Geochemistry*, **26**, 1035–1064.

MELKIOR, T., MOURZAGH, S., YAHIAOUI, S., THOBY, D., ALBERTO, J.C., BROUARD, C. & MICHAU, N. 2004. Diffusion of an alkaline fluid through clayey barriers and its effect on the diffusion properties of some chemical species. *Applied Clay Science*, **26**, 99–107.

MELKIOR, T., YAHIAOUI, S., THOBY, D., MOTELLIER, S. & BARTHÈS, V. 2007. Diffusion coefficients of alkaline cations in Bure mudrock. *Physics and Chemistry of the Earth*, **32**, 453–462.

MILLS, R. 1973. Self-diffusion in normal and heavy water in the range 1-45°. *Journal of Physical Chemistry*, **77**, 685–688.

NIER, A.O. 1950. A redetermination of the relative abundances of the isotopes of carbon, nitrogen, oxygen, argon, and potassium. *Physics Review*, **77**, 789–793.

PALUT, J.M., MONTARNAL, P., GUATSCHI, A., TEVISSEN, E. & MOUCHE, E. 2003. Characterization of HTO diffusion properties by an in situ tracer experiment in Opalinus clay at Mont Terri. *Journal of Contaminant Hydrology*, **61**, 203–218.

PISAPIA, C., DESCHAMPS, P., HAMELIN, B., BATTANI, A., BUSCHAERT, S. & DAVID, J. 2011. U/Pb dating of geodic calcites: a tool for paleohydrological reconstructions. *Mineralogical Magazine*, **75**, 1647.

PUT, M.J. & DE CANNIÈRE, P. 1994. Migration behaviour of ^{14}C labelled bicarbonate, HTO and ^{131}I in Boom clay. *Radiochimica Acta*, **66/67**, 385–388.

REBEIX, R., LE GAL LA SALLE, C. *ET AL.* 2014. Chlorine transport processes through a 2000m aquifer/aquitard system. *Marine and Petroleum Geology*, **53**, 102–116.

SAMPER, J., YANG, Q. *ET AL.* 2008. Numerical modeling of large-scale solid-source diffusion experiments in Callovo-Oxfordian clay. *Physics and Chemistry of the Earth*, **33**, S208–S215.

SMITH, T. 2010. *Transfert vertical des Gaz Rares à l'échelle des différentes formations de la zone de transposition du site de Meuse/Haute-Marne et à l'échelle des eaux porales de l'argilite et du Callovo-Oxfordien.* PhD thesis, Bordeaux-I University.

STEIGER, R.H. & JÄGER, E. 1977. Subcommission on geochronology: convention on the use of decay constants in geo- and cosmochronology. *Earth and Planetary Science Letters*, **36**, 359–362.

TACHI, Y., YOTSUJI, K., SEIDA, Y. & YUI, M. 2011. Diffusion and sorption of Cs^+, I^- and HTO in samples of the argillaceous Wakkanai formation from the Horonobe URL, Japan: clay-based modeling approach. *Geochimica et Cosmochimica Acta*, **75**, 6742–6759.

TEVISSEN, E., SOLER, J.M., MONTARNAL, P., GAUTSCHI, A. & VAN LOON, L.R. 2004. Comparison between in situ and laboratory diffusion studies of HTO and halides in Opalinus clay from Mont Terri. *Radiochimica Acta*, **92**, 781–786.

TOLSTIKHIN, I., LEHMANN, B.E., LOOSLI, H.H. & GAUTSCHI, A. 1996. Helium and argon isotopes in rocks, minerals, and related groundwaters: a case study in northern Switzerland. *Geochimica et Cosmochimica Acta*, **60**, 1497–1514.

TORGERSEN, T. 1989. Terrestrial helium degassing fluxes and the atmospheric helium budget: implications with respect to the degassing processes of continental crust. *Chemical Geology*, **79**, 1–14.

VAN LOON, L.R., SOLER, J.M. & BRADBURY, M.H. 2003*a*. Diffusion of HTO, $^{36}Cl^-$ and $^{125}I^-$ in Opalinus Clay samples from Mont Terri – effect of confining pressure. *Journal of Contaminant Hydrology*, **61**, 73–83.

VAN LOON, L.R., SOLER, J.M., JAKOB, A. & BRADBURY, M.H. 2003*b*. Effect of confining pressure on the diffusion of HTO, $^{36}Cl^-$ and $^{125}I^-$ in a layered argillaceous rock (Opalinus clay): diffusion perpendicular to the fabric. *Applied Geochemistry*, **18**, 1653–1662.

VAN LOON, L.R., WERSIN, P. *ET AL.* 2004*a*. In-situ diffusion of HTO, $^{22}Na^+$, Cs^+ and I^- in Opalinus clay at the Mont Terri underground rock laboratory. *Radiochimica Acta*, **92**, 757–763.

VAN LOON, L.R., SOLER, J.M., MULLER, W. & BRADBURY, M.H. 2004*b*. Anisotropic diffusion in layered argillaceous rocks: a case study with Opalinus clay. *Environmental Science & Technology*, **38**, 5721–5728.

VINCENT, B., EMMANUEL, L., HOUEL, P. & LOREAU, J.-P. 2007. Geodynamic control on carbonate diagenesis: petrographic and isotopic investigation of the Upper Jurassic formations of the Paris Basin (France). *Sedimentary Geology*, **197**, 267–289.

WABER, H.N. 2012. *Laboratoire de Recherche Souterrain Meuse/Haute-Marne – Geochemical Data of Borehole EST433.* Nagra Arbeitsbericht **NAB 09-16**. Nagra, Wettingen.

WABER, H.N. & VINSOT, A. 2010. Unraveling solute fluxes across the aquitard-aquifer sequence in the eastern Paris Basin; pore water tracer data from borehole EST 433. Abstract presented at the 4th International Meeting, Clays in Natural and Engineered Barriers for Radioactive Waste Confinement, 29 March–1 April 2010, Nantes, France.

WEISS, R.F. 1970. Solubility of nitrogen, oxygen and argon in water and seawater. *Deep-Sea Research*, **17**, 721–735.

WEISS, R.F. 1971. Solubility of helium and neon in water and seawater. *Journal of Chemical & Engineering Data*, **16**, 235–241.

WERSIN, P., VAN LOON, L. *ET AL.* 2004. Long-term diffusion experiment at Mont Terri: first results from field and laboratory data. *Applied Clay Science*, **26**, 123–135.

WERSIN, P., SOLER, J.M. *ET AL.* 2008. Diffusion of HTO, Br^-, I^-, Cs^+, $^{85}Sr^{2+}$ and $^{60}Co^{2+}$ in a clay formation: results and modeling from an in situ experiment in Opalinus Clay. *Applied Geochemistry*, **23**, 678–691.

WISE, D.L. & HOUGHTON, G. 1966. The diffusion coefficients of ten slightly soluble gases in water at 10-60°C. *Chemical Engineering Science*, **21**, 999–1010.

Study of ^{85}Sr transport through a column filled with crushed granite in the presence of bentonite colloids

KATEŘINA KOLOMÁ* & RADEK ČERVINKA

Fuel Cycle Chemistry Department, ÚJV Řež, a. s., Hlavní 130, Řež, 250 68 Husinec, Czech Republic

**Correspondence: Katerina.Koloma@ujv.cz*

Abstract: The present work is focused on the study of strontium transport through crushed granite in the presence of bentonite colloids under dynamic arrangement. The aim of the experiments was to investigate the effect of bentonite colloids on strontium migration in crushed granite. The tracer behaviour was studied in a column set-up under aerobic conditions with a continuous inlet of the liquid phase of a constant tracer concentration (activity) and flow rate. Defined volumes of liquid phase were sampled at periodic time intervals at the column outlet for the measurements of tracer concentrations (activity). The transport was described by breakthrough curves. The stepwise approach included these steps: (1) an evaluation of the hydrodynamic column properties by the non-sorbing tracer ^{3}H; (2) a column experiment with bentonite colloids in deionized water was performed; (3) migration of ^{85}Sr solution in two liquid phases (deionized and synthetic granitic water); and (4) the transport of a radiocolloid suspension in deionized water was studied. Results showed different behaviour of bentonite colloids and strontium in the column. Bentonite colloids behaved as a non-sorbing tracer: conversely, strontium showed strong sorption on granitic material. The strontium transport in the presence of bentonite colloids differed from strontium transport itself. The strontium transport in the presence of colloids was faster than transport without the bentonite colloids. The observed retention of strontium on granite suggests a higher affinity of strontium towards granitic rock than towards bentonite colloids, and showed the reversibility of the sorption of strontium on bentonite colloids.

In many countries, including the Czech Republic, the final storage of spent nuclear fuel and high-level radioactive waste is planned in a deep geological repository (DGR) based on the multibarrier system (engineered and natural barriers). One part of the engineered barriers is a bentonite, which is emplaced around the container with the radioactive waste, and as a filling and sealing material within the DGR. The bentonite in contact with groundwater can generate clay colloid particles. The clay colloids are solid particles smaller than 1 μm (by definition) with charge and high surface area. The formation and stability of clay colloids may have a direct impact on DGR safety owing to two main aspects: the generation of colloids may degrade the engineered barrier; and the colloid transport of radionuclides may reduce the efficiency of the natural barrier. The colloids are critical for radionuclide transport provided that they are in a non-negligible concentration, they are mobile and stable in the environment, and that they adsorb some of the radionuclides irreversibly (Missana *et al.* 2004). Under certain conditions, colloid-facilitated transport can be the main transport mechanism of sorbing elements in a geological formation. However, colloid transport in the environment is strongly affected by the tendency of colloid particles to aggregate and deposit as a result of increased ionic strength, the presence of multivalent counter ions and the adsorption of oppositely charged ions to the surface of colloidal particles (Kretzschmar & Schäfer 2005).

The influence of bentonite colloids on the radionuclide behaviour in granite fractures was widely studied both at laboratory scale and in *in situ* experiments. Results showed that the kinetics of sorption/desorption processes and attachment/detachment of colloids play a major role in the transport of colloid-associated radionuclides. Radionuclides bound by bentonite colloids may be released slowly and this dissociation kinetics depends on the chemistry of the radionuclide. Mono- and divalent cations show a rather fast dissociation kinetic, with the exception of Cs, where the geological origin of the clay mineral triggering the frayed edge sites determines the reversibility of sorption. Conversely, tri- and tetravalent radionuclides, such as Th(IV), Pu(IV) and Am(III), show quite slow dissociation (Huber *et al.* 2011). In the presence of bentonite colloids, radionuclides can diffuse into the granite sorbed onto clay particles. The result is that radionuclides that are considered as immobile (e.g. Pu, Eu)

From: Norris, S., Bruno, J., Van Geet, M. & Verhoef, E. (eds) 2017. *Radioactive Waste Confinement: Clays in Natural and Engineered Barriers*. Geological Society, London, Special Publications, **443**, 193–203.
First published online July 8, 2016, https://doi.org/10.1144/SP443.14

may be mobile in the presence of bentonite colloids (Alonso *et al.* 2003; Delos *et al.* 2008; Missana *et al.* 2008). Numerous studies have demonstrated that colloid particles can increase the mobility of Am, Np, Pu and U (Kretzschmar & Schäfer 2005).

In situ transport experiments have been widely performed at the Grimsel Test Site (GTS) (Missana *et al.* 2008). For example, the international Colloid and Radionuclide Retention (CRR) project demonstrated that bentonite colloids influenced in a significant way the *in situ* retardation behaviour of tri- and tetravalent actinides. The migration of Am and Pu was strongly mediated by the bentonite colloids, and, in the presence of colloids, the recoveries of both radionuclides increased in comparison with the system without the bentonite colloids. Caesium showed similar behaviour to Am and Pu: however, strontium showed a reversible sorption to the fracture surface without a significant interaction with colloids (Möri *et al.* 2003; Kretzschmar & Schäfer 2005).

This paper summarizes experiments where we studied the influence of bentonite colloids on strontium transport in crushed granite under dynamic conditions (a simulation of transport in fracture infill). In comparison with a static batch experiment, the dynamic arrangement was closer to the conditions that exist in natural systems (e.g. a low ratio of the solid and liquid phases, and a low flow rate of the liquid phase).

Materials and method

Solid and liquid phase

As solid materials, granite and mica mineral muscovite were used. The granite came from the Rejčkov locality (Melechov Massif, Czech Republic), directly sampled from a bore core originating from a depth of 97.5–98.7 m. The granite was a fine-grained two-mica granite of the Lipnice type. The core was crushed, sieved in defined fractions and washed in deionized water in an ultrasonic bath in order to remove the fine particles. For static and dynamic experiments, the fraction with a grain size of 0.125–0.63 mm was used. X-ray diffraction analysis (XRD) of the crushed granite was performed on a Philips X'Pert System diffractometer using a graphite crystal monochromator and CuKα radiation, at 40 kV and 40 mA, over the range 2°–85° 2θ, with a 0.05° step.

The muscovite was separated manually from the crushed granite. The remaining granite depleted of muscovite was not used for other experiments.

The deionized water (DW) and synthetic granitic water (SGW) were used in the preparation of the tracer solution. The chemical composition of the

Table 1. *Composition and chemical parameters of SGW*

pH: 8.03	
Ionic strength: 3.61×10^{-3} mol l^{-1}	
Ion	Concentration (mg l^{-1})
Na^+	10.6
K^+	1.8
Ca^{2+}	27.0
Mg^{2+}	6.4
Cl^-	42.4
SO_4^{2-}	27.7
NO_3^-	6.3
HCO_3^-	30.4
F^-	0.2

SGW is shown in Table 1 (Havlová *et al.* 2010) and corresponds to granitic waters from Czech granitic massifs from depths of 20–100 m. All chemicals were of analytical grade and were dissolved in deionized water.

Bentonite colloids

The colloids were prepared from Czech bentonite (commercial product Bentonite 75, Keramost a.s.), which was first purified. The purification procedure included several steps: (1) the removal of carbonates using a sodium acetate–acetic acid buffer; (2) the removal of iron using sodium dithionite in an acetic acid buffer and washing in NaCl; (3) separation of the clay fraction <2 μm by sedimentation in a sufficient amount of distilled water; (4) transfer to the homoionic Na^+ form by treatment with NaCl and washing with distilled water; (5) dialysis until the conductivity of the solution was <10 μS cm^{-1}; and (6) freeze-drying of the final product.

Purified bentonite consists mainly of montmorillonite, quartz, kaolinite and anatase (XRD). It has a cation-exchange capacity equal to 52.6 ± 1.8 meq 100 g^{-1}, and an occupancy of cations: Ca^{2+}, 19.24 meq 100 g^{-1}; Mg^{2+}, 6.21 meq 100 g^{-1}; Na^+, 33.97 meq 100 g^{-1}; and K^+, 1.39 meq 100 g^{-1} (Cu(II)-trien method adopted from Meier & Kahr 1999).

From this powder, the bentonite suspension with a particle concentration of 100 mg l^{-1} in distilled water was prepared for experiments. The stability and size of colloids was controlled by photon cross-correlation spectroscopy (PCCS) equipped with a He–Ne laser ($\lambda = 632.8$ nm) and the photomultiplier was located at a scattering angle of 90° (Nanophox instrument, Sympatec, Germany). The bentonite colloids were stable (the size of the colloids did not change with the time) and their initial mean hydrodynamic diameter was 420 nm. The concentration of bentonite particles was estimated from the relationship between the response of the photomultiplier (kcps) and the known concentration of the clay colloids (linear or polynomial calibration curve). The detection limit for bentonite colloids using PCCS is around 1 mg l^{-1}.

Radionuclides

The tritium solution (as HTO) in deionized water was used as a conservative tracer and for describing the hydrodynamic properties of the column. The activity of ^{3}H was measured by liquid scintillation spectrometry with an automatic liquid scintillation counter HIDEX 300 SL (Hidex, Finland). The migration experiments were conducted with ^{85}Sr, which is a chemical analogue of ^{90}Sr, one of the critical radionuclides in the context of DGR. The $SrCl_2$ solution (molar concentration $(c) = 10^{-6}$ mol l^{-1}) in SGW or distilled water spiked with ^{85}Sr was used, the final solution activity of which was 0.7 kBq cm^{-3}. The activity of strontium was measured using a Perkin Elmer 1480 Wizard 3″ Gamma counter (Wallac Oy, Finland).

Radiocolloid suspension

Radiocolloid suspension was prepared by mixing a specific volume of strontium solution in deionized water and the same volume of bentonite colloids. The final strontium concentration in the radiocolloid suspension was 10^{-6} mol l^{-1} and the bentonite colloid concentration was 100 mg l^{-1}. Initially, the mean hydrodynamic size of bentonite particles in suspension was 420 nm using PCCS, and this did not change after the addition of the strontium solution. The contact time between the strontium and the colloid particles before the dynamic experiments was 7 days, to ensure equilibrium. The sorption of strontium on the colloid particles was checked by centrifugation. The radiocolloid suspension, with a volume of 1.5 ml, was centrifuged (for 30 min at a relative centrifugal force (RCF) of 30 000*g*) and the activity was measured in the upper (1 ml) and the bottom part, where the colloid particles settled. Over 90% of strontium activity was present in the bottom part, which confirmed that all the strontium was sorbed (exchanged) onto colloid particles. The stability of the radiocolloids was monitored by PCCS, and the addition of the strontium did not affect the stability of the bentonite colloids.

Static experiment methodology

The sorption of strontium (as $^{85}SrCl_2$) onto the solid material was studied using the static batch method. All batch experiments were conducted under standard laboratory conditions. Reaction mixtures were prepared by mixing the crushed solid material (granite or muscovite with a grain size of 0.125–0.63 mm) with the defined volume of deionized water (g ml^{-1}), the ratio of the solid to the liquid phase was 1:10. The mixtures were shaken for 1 week, which is sufficient for equilibrium to be reached. After that, the samples were centrifugated (10 min at RCF 1500*g*) and the strontium activity in the solution was determined. The blank test without any solid material in solution was conducted following the same procedure. The sorption distribution coefficient, K_d (ml g^{-1}), was calculated as follows:

$$K_d = \frac{(A_0 - A) \cdot V}{A \cdot m} (\text{ml g}^{-1}) \quad (1)$$

where K_d is the distribution coefficient, A_0 is the initial activity of strontium in solution (in counts per min (cpm)), A is the activity of strontium in the liquid phase after sorption (cpm), V is liquid volume (ml) and m is solid mass (g).

Dynamic experiment methodology

The transport of strontium was studied under dynamic conditions in a column arrangement with a continuous inlet of the liquid phase. The crushed granite was placed into 5 cm^3 suitable plastic columns with a 1.2 cm inner diameter and a height of 5.4 cm. The basic hydraulic parameters of the filled column were calculated using experimentally determined values of the mass and volume of the column filled with granite of a given grain size (see Table 2). The experimental solution was added into a suitable reservoir: the concentration (activity) of the tracer and the flow rate were constant during the column experiments. The flow rate of the experimental solution was about 0.03 ml min^{-1}. The aliquot volume (2 ml) samples of the liquid phase were taken in defined time (t) intervals at the column outlet. The measurement of the tracer concentration (activity) (c_t) was performed until the concentration at the outlet reached the concentration (activity) at the column inlet (c_0). This also meant that the column experiments lasted until the tracer (strontium) occupied all the possible sorption sites in the granite. The breakthrough curves (BTC) were constructed (c_t/c_0 v. n_{PV}) from the relative concentrations (c_{rel} ($c_{rel} = c_t/c_0$)) and the number of pore volumes (n_{PV}) of the liquid phase. The obtained sorption BTC had an S-shape with an inflection point (see index i). At the inflection point, $c_{rel} = 0.5c_0$ and the experimental value R is equal to the number of pore volumes ($R = n_{PV,i}$). The resulting value of the sorption distribution coefficient was calculated from the retardation

Table 2. *Parameters of the columns used in the dynamic experiments*

Column code	Grain size (mm)	Mass (g)	Bulk density (g cm^{-1})	Porosity (cm^3 cm^{-3})	Pore volume (ml)	Seepage velocity (cm min^{-1})
B75	0.125–0.63	9.6943	1.491	0.441	3.01	0.066
^{85}Sr in DW	0.125–0.63	9.4602	1.455	0.455	3.10	0.051
^{85}Sr in SGW	0.125–0.63	9.6400	1.506	0.436	3.09	0.093
RC	0.125–0.63	10.091	1.552	0.419	2.86	0.054

Note: B75, bentonite colloids; ^{85}Sr in DW, $SrCl_2$ solution spiked by ^{85}Sr in deionized water (DW); ^{85}Sr in SGW, $SrCl_2$ solution spiked by ^{85}Sr in synthetic granitic water (SGW); RC, radiocolloids.

factor R:

$$R = 1 + \rho \frac{K_d}{\theta} \quad (2)$$

where ρ is the bulk density (g cm^{-3}) and θ is the porosity (cm^3 cm^{-3}).

The experimental BTC were fitted using the CXTFIT code (STANMODE, version 2.08) (Toride *et al.* 1995). For initial estimate experimental values of R, D and v were used. The fitted parameters were flow velocity (v) and retardation factor (R). The value of D was estimated by erfc-type equation, which is a result of the analytical solution of the one-dimensional (1D) advection–dispersion equation (ADE) for boundary conditions of the sorption process:

$$A_{\mathrm{rel,theor}} = 0.5\mathrm{erfc}\left[\frac{R - n_{\mathrm{PV}}}{2(R \cdot n_{\mathrm{PV}}/\mathrm{Pe})^{0.5}}\right] \quad (3)$$

$$A_{\mathrm{rel,theor}} = 1 - \left\{0.5\mathrm{erfc}\left[\frac{-(R - n_{\mathrm{PV}})}{2(R \cdot n_{\mathrm{PV}}/\mathrm{Pe})^{0.5}}\right]\right\} \quad (4)$$

where A_{rel} is relative activity, index 'theor' denotes the theoretical value, R is the retardation factor, n_{PV} is the number of pore volume and Pe is the Péclet number.

It should be noted that equation (3) can be used directly up to $A_{\mathrm{rel}} \leq 0.5$ (i.e. $(R - n_{\mathrm{PV}}) \geq 0$), but for the calculation of $A_{\mathrm{rel}} \geq 0.5$ (i.e. if $(R - n_{\mathrm{PV}}) < 0$) equation (4) should be used.

Table 3. *Mineral composition of granite (in wt%)*

Mineral	Content (wt%)
Quartz	32
Orthoclase	31
Plagioclase	18
Mica	14
Chlorite	5

Results and discussion

The influence of mineralogy on strontium sorption

The main components of the Czech granite samples (from the Melechov Massif) were quartz and feldspars (orthoclase and plagioclase), the minority components were mica and chlorite (see Table 3). It is known that strontium affinity to individual minerals varies, so the strontium sorption may be influenced by the amount of a specific mineral in the granitic matrix. Andersson *et al.* (1983) published the K_d values for strontium in pure minerals, the highest value of K_d was determined for the mica mineral muscovite (see Table 4). Thus, we performed sorption experiments with crushed granite and muscovite to confirm the influence of mica minerals on the strontium sorption.

However, the assumption of high strontium sorption on muscovite was not confirmed. The K_d value of strontium on granite was 30 ml g^{-1}, which corresponds with 60% of the sorption yield. This value is consistent with published data. For example, Carbol & Engkvist (1997) published K_d values for strontium in the range of 0.1–30 ml g^{-1}, where a K_d value of 30 ml g^{-1} corresponds to low saline solutions. Andersson & Allard (1983) mentioned K_d values in the range from 30 ml g^{-1} (low-salinity

Table 4. *The K_d values of strontium on pure minerals (Andersson* et al. *1983)*

Mineral	K_d (ml g^{-1})	pH
Biotite	16	6–8
	24	8–9
Muscovite	71	6–8
	114	8–9
Chlorite	50	8–8.5
Quartz	0	6–8
	1	8–9
Orthoclase	2	6–8
	5	8–9
Plagioclase	4	8–8.5

Table 5. *Sorption of strontium on granite and mica,* c *(Sr)* = 10^{-6} *mol* l^{-1}

Granite		Mica	
K_d (ml g^{-1})	Sorption yield (%)	K_d (ml g^{-1})	Sorption yield (%)
30	60	21	50

c, molar concentration.

solution) to 1 ml g^{-1} (saline solution), and Baik *et al.* (2008) achieved K_d values in the range of 3.5–69 ml g^{-1} for an initial concentration of strontium of 10^{-6}–10^{-5} mol l^{-1}.

However, the K_d value for strontium on muscovite was only 21 ml g^{-1}, which corresponds with a 50% sorption yield. Therefore, the strontium sorption on granite is slightly higher than strontium sorption on mica minerals (see Table 5). The granite contains 14% of the mica mineral muscovite. From experimental sorption data, it follows that 1.2×10^{-8} mol of strontium is sorbed on 1 g of granite, and 1×10^{-8} mol of strontium is sorbed on 1 g of muscovite. If 1 g of granite contains 0.14 g of mica, only 12% of strontium from the experimental solution is sorbed there (see Table 6). The remaining 88% of strontium is sorbed on other minerals in the granitic rock. In the case of the Czech granite, the static sorption experiment showed that the mica mineral muscovite is not a dominant sorbent of strontium. Although there are many sorption data of strontium on crystalline rocks (e.g. granite, diorite, granodiorite, tonalite), the comparison of results is not trivial. The resulting values of K_d differ significantly from each other (Crawford *et al.* 2006) and the solid-phase properties (mineralogical composition, the presence of secondary minerals and the alteration of rock or porosity, and pore structure) are important factors affecting strontium sorption (Andersson *et al.* 1983; Gascoyne *et al.* 1995; Başçetin & Atun 2006).

The influence of the liquid-phase composition on Sr^{2+} migration

The strontium sorption on granitic rock was studied in terms of the composition of the liquid phase. The chemistry of the liquid phase (pH, ionic strength and chemical composition) is one of the crucial parameters influencing sorption of strontium on granite (Andersson & Allard 1983; Wallace *et al.* 2012). In the dynamic experiments, we used two types of liquid phase – deionized water (DW) and synthetic granitic water (SGW) – and studied the influence of the ionic strength of the liquid phase and the influence of competitive ions on the strontium sorption. The results of the column experiments are shown in Figure 1 and Table 7. The difference in strontium transport in the DW and SGW is obvious: the significant differences are also apparent in terms of sorption data and the duration of the experiments. The transport of strontium in SGW was fast, the experiment lasted for about 2 days and strontium sorption on the granitic surface was weak. The K_d value of strontium was 6.5 ml g^{-1}, which corresponds to $R = 24.5$. The transport of strontium through granite in deionized water was completely different. The experiment lasted for about 2 months and strontium performed strong sorption on granite. The K_d value was 210.9 ml g^{-1}, and the R value was 675.7. Such significant differences between the behaviour of strontium are caused by the different composition of the liquid phase. The SGW contains a high concentration of Ca^{2+} (Table 1). Calcium and magnesium have the same chemical charge as strontium, and compete with strontium in exchange reactions within a crystal structure or on mineral surfaces (Ali Khan *et al.* 1995; Liszewski *et al.* 1998). The calcium cations occupy sorption sites on the granitic surface and thus reduce the sorption of strontium.

Table 6. *Sorption of strontium on minerals in granite,* c *(Sr)* = 10^{-6} *mol* l^{-1}

Sorption yield on granite (%)	
Mica	Other minerals
12	88

c, molar concentration.

The ionic strength of the experimental SGW was 3.61 mmol l^{-1} and, from the comparison of strontium sorption in the DW and SGW, it is clear that the strontium sorption decreased with increasing ionic strength of the liquid phase. The reason is that under higher ionic strength the concentration of the competing ions is higher and so causes a reduction in the Sr sorption: the K_d value is reduced by nearly two orders of magnitude. The increasing ionic strength has a major impact on Sr sorption (Kütahyali *et al.* 2012; Wallace *et al.* 2012). The calcium and magnesium are the most important competitive ions; however, sodium and potassium do not influence the sorption of strontium (Liszewski *et al.* 1998; EPA 1999; Cronstrand 2005; Crawford *et al.* 2006).

Column experiments with conservative tracer and bentonite colloids

The column experiment with conservative tracer ^3H was performed for each experimental column. The ^3H migration simulates the transport of deionized water in the filled column and describes the

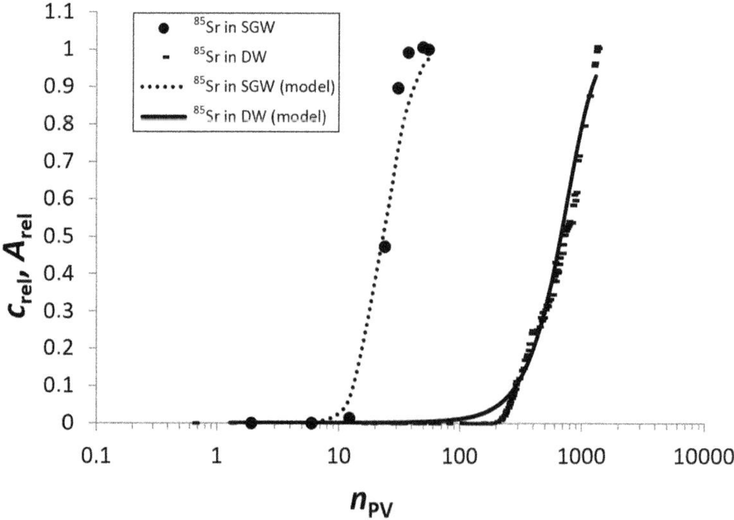

Fig. 1. Experimental (points) and theoretical (lines) BTC of strontium in SGW (•) and in DW (−), c (Sr) = 10^{-6} mol l^{-1}, water flow rate 0.035 ml min^{-1}.

hydrodynamic parameters of columns filled with crushed granite. The ^{3}H transport was identical for all experimental columns and verified the properties of the columns. For illustration purposes, the resulting BTC of ^{3}H is shown in Figure 2. ^{3}H behaved as a non-sorbing, inert tracer and no retention was observed. The retardation factor was 1.02, corresponding to zero sorption, the breakthrough of ^{3}H was observed at approximately 1 pore volume.

Figure 2 also shows the results of the dynamic experiment with bentonite colloids in deionized water without the presence of other tracers (coded B75). The breakthrough of the bentonite colloids was observed at 1.17 pore volume, which means that it was slightly delayed in comparison with the breakthrough of ^{3}H. The result indicates that bentonite colloids migrate almost identically to ^{3}H: however, the colloid transport is slightly slower. Similar results showing the unretarded transport of bentonite colloids have been published widely (Möri *et al.* 2003; Missana *et al.* 2008; Albarran *et al.* 2011).

Table 7. *Retardation factor* (R) *and distribution coefficient* (K_d) *of strontium in the deionized water (DW) and synthetic granitic water (SGW),* c *(Sr)* = 10^{-6} *mol* l^{-1}*, water flow rate 0.035 ml* min^{-1}

Result parameters	^{85}Sr in DW	^{85}Sr in SGW
K_d (ml g^{-1})	210.9	6.5
R	675.7	24.5
t (days)	63.0	2.2

c, molar concentration.

The influence of bentonite colloids on Sr^{2+} *transport*

The dynamic experiment with a radiocolloid suspension was performed in a column denoted as RC. The BTC are displayed in Figure 3. The bentonite colloids in the presence of ^{85}Sr (B75-radiocolloid) passed through crushed granite during the first 24 h and left the strontium (^{85}Sr-radiocolloid) in the column. At the beginning of the experiment, about 90% of strontium was sorbed onto the bentonite colloids. However, the sorption potential of the granite material was higher than the forces of interaction between strontium and the bentonite colloids (ion exchange), and the radiocolloid complex was partitioned into the crushed granite. The colloid particles free of ^{85}Sr passed through the crushed granite quickly, immediately after the injection of the radiocolloids. Bentonite appeared at the column outlet after 1.32 pore volumes.

In comparison with bentonite colloid particles, strontium transport duration was significantly longer. The first strontium activity was measured at the column outlet after 100 pore volumes and the breakthrough of strontium was observed at 209 pore volumes. Strontium transport lasted for about 22 days. The observed retention of strontium on granite can be explained by the higher affinity of strontium towards granitic rock than towards bentonite colloids, and showed the reversibility of the sorption of strontium on bentonite colloids. However, the high sorption of strontium on granite may also be caused by a larger reaction surface of the granite in comparison with the bentonite

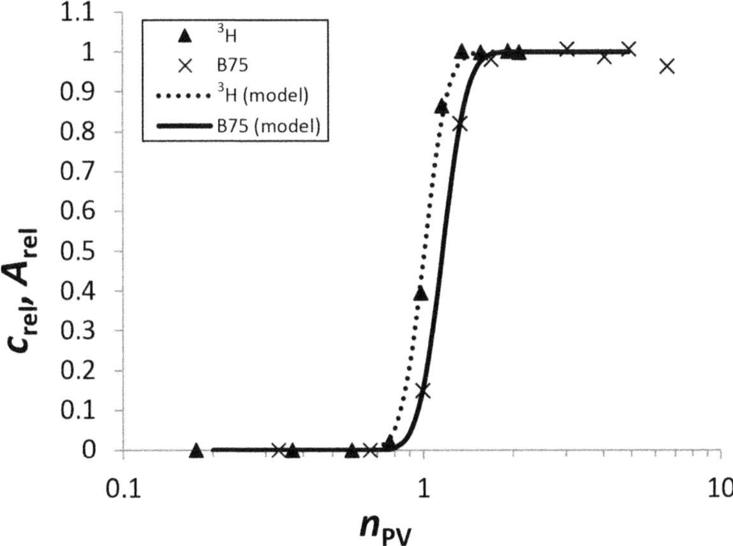

Fig. 2. Experimental (points) and theoretical (lines) BTC of ^{3}H (▲) and bentonite colloids B75 (×), c (B75) = 100 mg l^{-1}, water flow rate 0.035 ml min^{-1}.

colloids. The amount of granite is significantly larger than the amount of bentonite colloids, from which also results a greater number of sorption sites.

The transport parameters R and K_d were calculated from the BTC. Their values are summarized in Table 8. The retardation factors of strontium decrease with an increasing content of the other components in the liquid phase. The highest value of R was observed for strontium in deionized water without the presence of any other tracers or ions. Significantly lower R values were calculated for strontium in the presence of bentonite colloids, and the lowest value of R was observed for strontium in the SGW. Regardless of the presence of bentonite colloids, the observed K_d values are significantly higher in comparison with figures in the literature (see Table 9). The reason is the different composition of the liquid phase and the influence of the competitive ions. The median K_d value for non-saline conditions is about two orders of

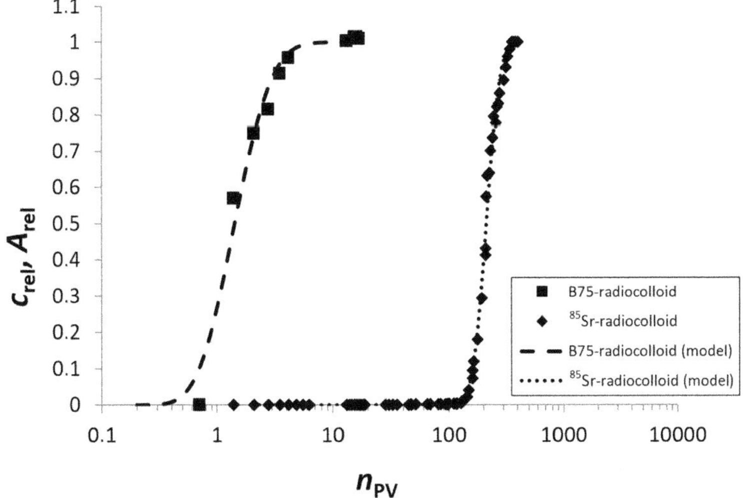

Fig. 3. Experimental and theoretical BTC of ^{85}Sr in the presence of bentonite colloids (■, B75-RC, ◆, ^{85}Sr-RC).

Table 8. *Transport parameters of different tracers through the crushed granite*

Column code	Tracer	R	R_{theor}	K_d (ml g^{-1})
B75	B75	1.17	1	0.05
^{85}Sr in DW	^{85}Sr	675.7	729	210.9
^{85}Sr in SGW	^{85}Sr	24.5	20	6.5
RC	^{85}Sr	209.6	222	56.2
	B75	1.32	0.6	0.10

Note: B75, bentonite colloids; ^{85}Sr in DW, $SrCl_2$ solution spiked by ^{85}Sr in deionized water (DW); ^{85}Sr in SGW, $SrCl_2$ solution spiked by ^{85}Sr in synthetic granitic water (SGW); RC, radiocolloids.

magnitude higher than K_d for strongly saline water (Crawford *et al.* 2006).

The obtained BTCs are shown in Figure 4, and clearly illustrate the different behaviour of the tracers used and their mutual influence. BTC B75 and B75-RC describe the bentonite colloid migration in granitic rock under different conditions (see Fig. 4, BTC B75 v. B75-RC). BTC B75 corresponds with the bentonite colloid transport in deionized water, BTC B75-RC describes the transport of bentonite colloids that originated from the radiocolloid suspension. The transport of colloids differed from each other. We can see that the n_{PV} (c_{rel} = 0.5) values of bentonite colloids, which are important in the evaluation of the retardation factor, are very similar (1.17 v. 1.32): differences are the patterns in the shape of the BTC. The lower slope of BTC B75-RC suggests a possible slight retardation (however, not retention) of colloid particles in granite in the presence of radionuclide. The reason for the colloid delay is not fully clear from the results of the dynamic experiment.

The retention of bentonite colloids in granitic fractures has been noted by, for example, Albarran *et al.* (2011). They observed that under lower flow rates (0.017 ml min^{-1}) the granite surface was covered sparsely with bentonite particles with absorbed strontium. The retention of colloids can be explained by colloid filtration in granite fractures, and the colloids are retained as long as the hydrodynamic (or chemical) conditions do not vary.

The strontium transport through crushed granite was significantly faster in the presence of bentonite colloids than without the bentonite colloids being present (see Fig. 4: cf. BTC ^{85}Sr-RC and ^{85}Sr in DW). The explanation for this observation could be that the colloid particles acted as a radionuclide carrier. They carried strontium further into the column, but, finally, the strontium dissociated from the radiocolloid complex and sorbed onto the granite surface. The BTC of ^{85}Sr-RC was earlier than that of ^{85}Sr in DW. Free colloid particles were able to leave the column (see Fig. 4, B75-RC). The dynamic experiments were carried out under a low flow rate for the liquid phase (0.03 ml min^{-1}, which equates to 1.8 ml h^{-1}). If we take into account the pore volume of the column filled with crushed granite (3 ml), the residence time is 100 min. This is enough time for strontium to dissociate from the radiocolloid complex.

The colloid-facilitated transport of radionuclides was also shown by Möri *et al.* (2003). The radionuclide transport in the presence of bentonite colloids was faster than the transport of dissolved species. Hölttä *et al.* (2014) published results of experiments under dynamic conditions carried out with strontium through Olkiluoto tonalite with and without the presence of bentonite colloids in Olso reference water. They observed similar strontium behaviour: strontium was strongly retarded without colloids

Table 9. K_d *values of strontium on granite in various liquid phases*

K_d (ml g^{-1})	Liquid phase	Reference
5	Groundwater	Andersson & Allard (1983)
10	Non-saline groundwater	Carbol & Engkvist (1997)
0.2	Saline groundwater	Carbol & Engkvist (1997)
20	Groundwater	Andersson *et al.* (1983)
5	Groundwater	Crawford (2010)
0.1	Groundwater	Cronstrand (2005)
7.5*	Groundwater, bentonite colloids	Hölttä *et al.* (2014)
56*	Groundwater	Hölttä *et al.* (2014)
210.9	Deionized water	This study
6.5	Groundwater	This study
56.2	Deionized water, bentonite colloids	This study

*Value of the retardation factor, *R*.

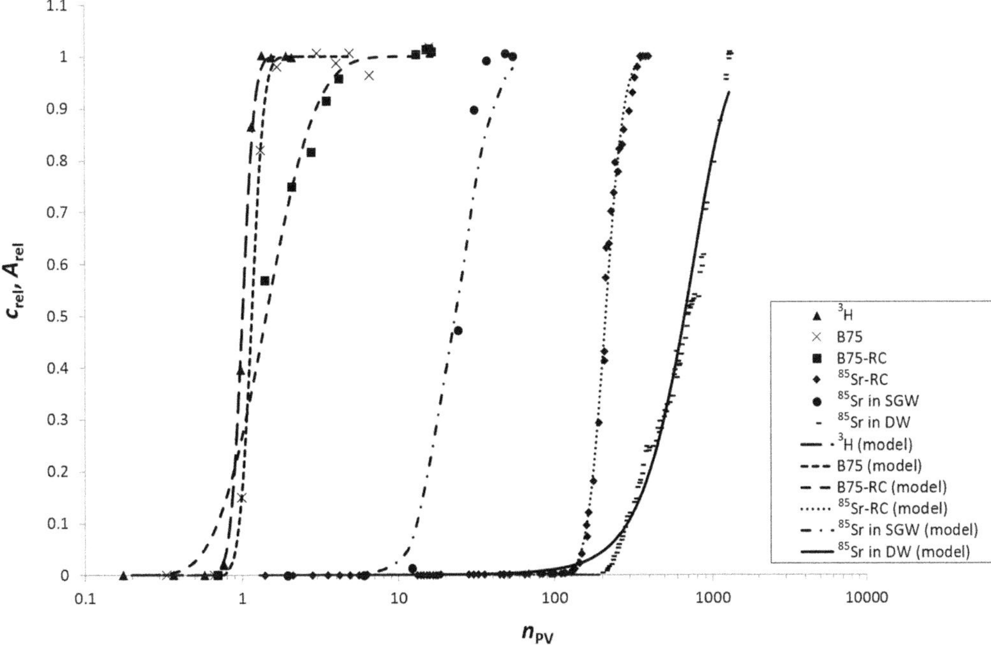

Fig. 4. Experimental and theoretical BTC of individual tracers used in the dynamic experiments.

and the slow elution of strontium was obtained in the presence of bentonite in suspension. The retardation factor without bentonite colloids was 56, and was 7 with bentonite colloids. The generally lower values of R in comparison with our results are caused by the different liquid phases used.

The experimental BTC were fitted using the CXTFIT code, the fitted parameters were flow velocity, v, and retardation factor, R. The theoretical values of the retardation factor, R_{theor}, are summarized in Table 8 and show a good agreement with experimental data.

The sorption capacity of crushed granite was evaluated for each column. The capacity calculation was based on the integration of the BTC; the calculated area under the BTC corresponds to the sorption capacity of the solid material. The results are summarized in Table 10.

Table 10. *The Sr sorption capacity (mol kg^{-1}) of columns filled with crushed granite and the occupancy of sorption sites by Sr (%)*

Column	Sorption capacity (mol kg^{-1})	Occupancy of sorption sites
^{85}Sr in DW	2.11×10^{-4}	100%
RC	8.34×10^{-5}	42%
^{85}Sr in SGW	7.64×10^{-6}	4%

Note: ^{85}Sr in DW, $SrCl_2$ solution spiked by ^{85}Sr in deionized water (DW); RC, radiocolloids; ^{85}Sr in SGW, $SrCl_2$ solution spiked by ^{85}Sr in synthetic granitic water (SGW).

The highest sorption capacity was observed in the case of ^{85}Sr in deionized water (column ^{85}Sr in the DW). Assuming that there were the minimum of competitive ions in the system Sr-deionized water, we can presume that most of the sorption sites in the granite are accessible for strontium sorption. From the BTC of column RC, we can deduce that only 42% of the total sorption capacity was used for sorption of strontium. The granite capacity was not fulfilled, probably due to the colloid-mediated transport of strontium. As mentioned above, strontium transport is influenced by colloid particles, which carried the strontium inside the column, and the upper part of column is not probably utilized for sorption. The significant influence of competitive ions on strontium sorption was observed in the column denoted as ^{85}Sr in the SGW. Only 4% of the sorption capacity was exploited for strontium sorption, most of the sorption sites were occupied by ions present in the SGW (probably calcium and magnesium).

Conclusions

The static and dynamic methods were applied to study the sorption of strontium onto granitic rock.

The experiments were focused on the description of the influences of the mineralogical composition of the granitic rock, the chemical composition of the liquid phase and the effect of the presence of bentonite colloids on the strontium behaviour in crushed granite. The results stress that the composition of the liquid phase is a significant parameter influencing strontium sorption on granite. The dominant impact of the liquid-phase chemistry was noted: the sorption of strontium decreased with the increasing concentration of competition ions in the liquid phase. The sorption experiments showed that the retention of strontium on granite is caused by the sorption of strontium on more than one specific mineral in the crystalline rock; the results suggested that muscovite is not the dominant sorbent of strontium in the studied granitic rock.

The bentonite colloids influenced the migration of radiocnulides in granitic rock. The transport of strontium was significantly faster in the presence of bentonite colloids, and bentonite colloids took the role of radionuclide carrier in the studied system. However, the dissociation of strontium from the radiocolloid complex was fast, and the affinity of strontium to the granitic surface was higher than that to the bentonite colloids, so the transport was only influenced for short distances. The dissociation kinetics and residence time are, hence, important factors when estimating the relevance of colloids in the long-term evaluation of DGR safety.

The research leading to these results received funding from the European Atomic Energy Community's Seventh Framework Programme (FP7/2007–2011) under grant agreement 295487 - project BELBaR and from SÚRAO (CZ). We would like to thank H. Vodičková for excellent support in the laboratory. Two anonymous reviewers helped to improve the original manuscript.

References

ALBARRAN, N., MISSANA, T., GARCÍA-GUTIÉRREZ, M., ALONSO, U. & MINGARRO, M. 2011. Strontium migration in a crystalline medium: effects of the presence of bentonite colloids. *Journal of Contaminant Hydrology*, **122**, 76–85.

ALI KHAN, S., RIAZ-UR-REHMAN & ALI KHAN, M. 1995. Sorption of strontium on bentonite. *Waste Management*, **15**, 641–650.

ALONSO, U., MISSANA, T., PATTELI, A., RIGATO, V. & RIVAS, P. 2003. Study of the contaminant transport into granite microfractures using nuclear ion beam techniques. *Journal of Contaminant Hydrology*, **61**, 95–105.

ANDERSSON, K. & ALLARD, B. 1983. *Sorption of Radionuclides on Geologic Media – A Literature Survey. 1: Fission Products*. SKB TR-83-07. Svensk Kärnbränslehantering (Swedish Nuclear Fuel and Waste Management), Stockholm.

ANDERSSON, K., TORSTENFELT, B. & ALLARD, B. 1983. *Sorption of Radionuclides in Geologic Systems*. SKB TR-83-63. Svensk Kärnbränslehantering (Swedish Nuclear Fuel and Waste Management), Stockholm.

BAIK, M.-H., LEE, S.-Y., LEE, J.-K., KIM, S.-S., PARK, C.-K. & CHOI, J.-W. 2008. Review and compilation of data on radionuclide migration and retardation for the performance assessment of a HLW repository in Korea. *Nuclear Engineering and Technology*, **40**, 593–606.

BAŞÇETIN, E. & ATUN, G. 2006. Adsorption behavior of strontium on binary mineral mixtures of Montmorillonite and Kaolinite. *Applied Radiation and Isotopes*, **64**, 957–964.

CARBOL, P. & ENGKVIST, I. 1997. *Compilation of Radionuclide Sorption Coeffi Cients for Performance Assessment*. SKB R-97-13. Svensk Kärnbränslehantering (Swedish Nuclear Fuel and Waste Management), Stockholm.

CRAWFORD, J. 2010. *Bedrock Kd Data and Uncertainty Assessment for Application in SR-Site Geosphere Transport Calculations*. SKB R-10-48. Svensk Kärnbränslehantering (Swedish Nuclear Fuel and Waste Management), Stockholm.

CRAWFORD, J., NERETNIEKS, I. & MALMSTRÖM, M. 2006. *Data and Uncertainty Assessment for Radionuclide Kd Partitioning Coefficients in Granitic Rock for Use in SR-Can Calculations*. SKB R-06-75. Svensk Kärnbränslehantering (Swedish Nuclear Fuel and Waste Management), Stockholm.

CRONSTRAND, P. 2005. *Assessment of Uncertainty Intervals for Sorption Coefficients*. SFR-1 uppföljning av SAFE. SKB R-05-75. Svensk Kärnbränslehantering (Swedish Nuclear Fuel and Waste Management), Stockholm.

DELOS, A., WALTHER, C., SCHÄFER, T. & BÜCHNER, S. 2008. Size dispersion and colloid mediated radionuclide transport in a synthetic porous media. *Journal of Colloid and Interface Science*, **324**, 212–215.

EPA 1999. *Understanding Variation in Partition Coefficient, K_d, Values – Volume II: Review of Geochemistry and Available K_d Values for Cadmium, Cesium, Chromium, Lead, Plutonium, Radon, Strontium, Thorium, Tritium (^3H), and Uranium*. United States Environmental Protection Agency, Washington, DC.

GASCOYNE, M., STROES-GASCOYNE, S. & SARGENT, F.P. 1995. Geochemical influences on the design, construction and operation of a nuclear waste vault. *Applied Geochemistry*, **10**, 657–671.

HAVLOVÁ, V., HOLEČEK, J., VEJSADA, J., VEČERNÍK, P. & ČERVINKA, R. 2010. *Metodika přípravy syntetické podzemní vody pro laboratorní experimentální práce – technická zpráva v projektu 'Výzkum vlivu mezizrnné propustnosti granitů na bezpečnost hlubinného ukládání do geologických formací a vývoj metodiky a měřící aparatury'*. FR-TI1/367. Arcadis Geotechnika, Prague.

HÖLTTÄ, P., VIDENSKÁ, K. & ČERVINKA, R. 2014. *Macroscale Investigations on Colloid Mobility in Near-Natural Systems*. Deliverable (D-No. 3.5), BELBaR Project. FP7 296487. European Commission, Luxembourg.

HUBER, F., KUNZE, P., GECKEIS, H. & SCHÄFER, T. 2011. Sorption reversibility kinetics in the ternary system radionuclide–bentonite colloids/nanoparticles–granite fracture filling material. *Applied Geochemistry*, **26**, 2226–2237.

KRETZSCHMAR, R. & SCHÄFER, T. 2005. Metal retention and transport on colloidal particles in the environment. *Elements*, **1**, 205–210.

KÜTAHYALI, C., CETINKAYA, B., BAHADIR ACAR, M., OLCAY ISIK, N. & CIRELI, I. 2012. Investigation of strontium sorption onto Kula volcanics using Central Composite Design. *Journal of Hazardous Materials*, **201–202**, 115–124.

LISZEWSKI, M.J., BUNDE, R.L., HEMMING, C., ROSENTRETER, J. & WELHAN, J. 1998. The use of synthesized aqueous solutions for determining strontium sorption isotherms. *Journal of Contaminant Hydrology*, **29**, 93–108.

MEIER, L.P. & KAHR, G. 1999. Determination of the cation exchange capacity (CEC) of clay minerals using the complexes of copper(II) ion with triethylenetetramine and tetraetylenepentamine. *Clays and Clay Minerals*, **47**, 386–388.

MISSANA, T., GARCÍA-GUTIÉRREZ, M. & ALONSO, U. 2004. Kinetics and irreversibility of cesium and uranium sorption onto bentonite colloids in a deep granitic environment. *Applied Clay Science*, **26**, 137–150.

MISSANA, T., ALONSO, U., GARCÍA-GUTIÉRREZ, M. & MINGARRO, M. 2008. Role of bentonite colloids on europium and plutonium migration in a granite fracture. *Applied Geochemistry*, **23**, 1484–1497.

MÖRI, A., ALEXANDER, W.R. *ET AL*. 2003. The colloid and radionuclide retardation experiment at the Grimsel Test Site: influence of bentonite colloids on radionuclide migration in a fractured rock. *Colloids and Surfaces A: Physicochemical and Engineering Aspects*, **217**, 33–47.

TORIDE, N., LEIJ, F.J. & VAN GENUCHTEN, M.TH. 1995. *The CXTFIT Code for Estimating Transport Parameters from Laboratory or Field Tracer Experiments, version 2.0*. Research Report No. 137. United States Department of Agriculture, Riverside, CA.

WALLACE, S.H., SHAW, S., MORRIS, K., SMALL, J.S., FULLER, A.J. & BURKE, I.T. 2012. Effect of groundwater pH and ionic strength on strontium sorption in aquiefer sediments: implications for ^{90}Sr mobility at contaminated nuclear sites. *Applied Geochemistry*, **27**, 1482–1491.

An improved model for through-diffusion experiments: application to strontium and tritiated water (HTO) diffusion in Boom Clay and compacted illite

M. AERTSENS*, L. VAN LAER, N. MAES & J. GOVAERTS

Institute for Environment, Health and Safety, Belgian Nuclear Research Centre, SCK•CEN, Boeretang 200, B-2400 Mol, Belgium

**Correspondence: marc.aertsens@sckcen.be*

Abstract: Through-diffusion experiments are conventional experiments to measure the transport parameters of radionuclides in clays. Typically, a regular replacement of the outlet volume by a tracer-free volume is performed. In the classical approach, this type of through-diffusion experiment is modelled by assuming a zero concentration in the outlet volume. Nonetheless, this assumption is not always correct, usually because the outlet volume is insufficiently large or the time between two consecutive replacements of the outlet volume is too long. Therefore, a model was developed disregarding this assumption and, instead, considers the tracer concentration in the outlet volume to evolve, as in the experiments: the flux into the outlet volume increases the tracer concentration and, at each replacement, the tracer concentration in the outlet volume is set to zero. The model was used to reproduce the diffusion of strontium (Sr) and tritiated water (HTO) in illite and Boom Clay. Model results yielded good matches with the tracer evolution in the inlet and the outlet, and the tracer profile in the core at the end of the experiment.

In Belgium, clay formations (e.g. Boom Clay) are studied as potential host formations for the final disposal of high-level and/or long-lived radioactive waste. In Boom Clay, transport is dominated by diffusion. To evaluate the performance of Boom Clay as a barrier, the values of (1) the capacity factor, ηR (η is the diffusion accessible porosity and R is the retardation factor), and (2) the apparent diffusion coefficient, D_{app}, are needed to determine the diffusive flux in the clay.

Several types of experiments exist to determine these transport parameters in clay (Aertsens *et al.* 2008). For this study, through-diffusion experiments were performed to measure the Sr^{85} and HTO transport parameters in Boom Clay and compacted purified Na-illite, which serves as a model component for Boom Clay. In a through-diffusion experiment, a clay core is confined between two well-mixed water compartments. Initially, tracer is added to the inlet compartment, from where it diffuses through the clay towards the outlet compartment. Between the clay and both water compartments, confining filters are needed to prevent the swelling and disintegration of the clay core. The initial tracer concentration in the clay core, in both filters and in the outlet, is zero.

Traditionally, through-diffusion experiments aim to maintain a constant tracer concentration in the inlet compartment and an approximately zero concentration in the outlet compartment (e.g. by regularly replacing both reservoirs). In this way, a steady state can be reached, allowing the determination of the migration parameters with simple mathematical expressions (Put & Henrion 1988; Takeda *et al.* 2008). Maintaining a constant tracer concentration (CC) is experimentally cumbersome and, therefore, this method, called CC–CC (constant concentration inlet–CC outlet: Takeda *et al.* 2008) has become obsolete. It is also possible to extract the transport parameters using variable concentration (VC) conditions (Zhang *et al.* 2006; Takeda *et al.* 2008; Aertsens *et al.* 2011). In the VC–VC set-up, in both water compartments, the tracer concentration changes according to the diffusive flux into (inlet) or out of (outlet) the clay. Hybrid forms of these set-ups are also possible: (i) a variable inlet concentration with a zero outlet concentration (VC–CC); and (ii) a constant inlet concentration with a variable outlet concentration (CC–VC). The analytical solutions for these four cases are summarized in Takeda *et al.* (2008).

The through-diffusion experiments discussed in the present paper are intermediate between the VC–CC and the VC–VC types. Similar to VC–CC, the concentration in the inlet compartment decreases according to the diffusive flux into the clay, and the outlet compartment is regularly replaced by a zero concentration solution. At the end of a measurement, the concentration in the outlet compartment is sometimes too high (with respect

From: NORRIS, S., BRUNO, J., VAN GEET, M. & VERHOEF, E. (eds) 2017. *Radioactive Waste Confinement: Clays in Natural and Engineered Barriers*. Geological Society, London, Special Publications, **443**, 205–210.
First published online June 24, 2016, https://doi.org/10.1144/SP443.9

to the inlet concentration) to be approximated by a constant zero concentration. Therefore, a model was developed: (i) taking into account, in the outlet boundary condition, the outlet concentration increase during a measurement (as in the VC–VC type); and (ii) after each measurement, the outlet concentration is put to zero again. Another version of the model includes diffusion through the filters confining the clay. Depending on the type and dimensions of clay and filters, as well as on the tracer type (neutral, anion or cation), it is essential to take into account diffusion through the filters in order to predict correctly the values of the transport parameters through the clay (Glaus *et al.* 2007, 2008; Birgersson & Karnland 2009).

Experiments

Through-diffusion experiments with ^{85}Sr and HTO were performed in purified Na-form illite (Illite du Puy, dry density 1.7 g cm^{-3}) and in Boom Clay (see Table 1 for a summary). The experiments with Illite du Puy were performed with a 0.1 M $NaClO_4$ buffered solution with MOPS (3-(*N*-morpholino)propanesulphonic acid) to pH 7. For the experiments with Boom Clay, real porewater extracted from the clay with piezometers was used. This porewater has a background Sr concentration of 1.5×10^{-6} M. In all experiments, the outlet volume (30 ml in both FUNMIG experiments, around 15–20 ml in the other experiments) was regularly replaced by a tracer-free volume. The length of most cores was 15 mm. Exceptions are experiments 'FUNMIG 30 mm' (30 mm) and CATCLAY 1 (31.9 mm). The diameter of the clay cores was 38 mm, apart from idp16 and idp17, where it was 20 mm. For experiment idp17, flushed confining filters (Glaus *et al.* 2015) with a thickness of 1.57 mm were used. For the other experiments, the filters are static and made of stainless steel (thickness 2 mm). More details are given in Table 1, as well as in Glaus *et al.* (2015) (illite experiments) and Aertsens *et al.* (2011) (FUNMIG experiments). The CATCLAY 1 experiment (Boom Clay) was executed in a similar manner to both FUNMIG experiments.

Modelling

Before modelling, all activities were recalculated to zero time.

In the VC–CC model, the assumption of a zero tracer concentration in the outlet volume is an approximation, which is valid when the real concentration $C_{out}(t_i)$ (unit: Bq ml^{-1}) at the end of a measurement, i, is much lower than the concentration at the inlet volume, $C_{in}(t_i)$ (unit: Bq ml^{-1}). The concentration $C_{out}(t_i)$ is given by:

$$C_{out}(t_i) = \frac{\int_{t_{i-1}}^{t_i} J_{out}(t')dt'}{L_{out}} \quad (1)$$

Table 1. *Summary of the fit results with the refresh model*

Experiment	pH	Inlet volume (ml)	Duration (days)	$D_{app,c}$ (m^2 s^{-1})	$(\eta R)_c$ (−)	$D_{app,f}$ (m^2 s^{-1})
Sr in illite						
idp14	5.0	2500	97	$1.2 \times 10^{-11} \pm 4 \times 10^{-13}$	91 ± 5	$1.4 \times 10^{-9} \pm 1 \times 10^{-10}$
idp17	7.0	500	169	$8.7 \times 10^{-12} \pm 3 \times 10^{-13}$	186 ± 7	$2.3 \times 10^{-9} \pm 2 \times 10^{-10}$
idp19	7.0	2500	161	$1.4 \times 10^{-11} \pm 2 \times 10^{-13}$	170 ± 4	$5.3 \times 10^{-10} \pm 8 \times 10^{-12}$
idp6	7.1	150	203	$1.1 \times 10^{-11} \pm 1 \times 10^{-12}$	110 ± 4	$8.8 \times 10^{-10} \pm 1 \times 10^{-10}$
idp5	7.2	150	129	$1.2 \times 10^{-11} \pm 7 \times 10^{-13}$	98 ± 3	$1.7 \times 10^{-9} \pm 2 \times 10^{-10}$
idp7	8.2	150	140	$1.3 \times 10^{-11} \pm 2 \times 10^{-13}$	200 ± 3	$2.4 \times 10^{-10} \pm 3 \times 10^{-12}$
idp8	8.2	150	140	$1.1 \times 10^{-11} \pm 4 \times 10^{-13}$	200 ± 8	$4.5 \times 10^{-10} \pm 2 \times 10^{-11}$
idp4	8.9	150	141	$9.1 \times 10^{-12} \pm 8 \times 10^{-13}$	169 ± 8	$2.8 \times 10^{-10} \pm 1 \times 10^{-11}$
idp18	9.0	2500	162	$1.3 \times 10^{-11} \pm 4 \times 10^{-13}$	315 ± 14	$5.3 \times 10^{-10} \pm 1 \times 10^{-11}$
Average				$1.2 \times 10^{-11} \pm 2 \times 10^{-12}$	171 ± 69	$9.1 \times 10^{-10} \pm 9 \times 10^{-11}$
Sr in Boom Clay						
FUNMIG 15 mm	8.5	200	185	$7.4 \times 10^{-12} \pm 1 \times 10^{-13}$	6830 ± 526	$3.7 \times 10^{-9} \pm 5 \times 10^{-10}$
FUNMIG 30 mm	8.5	200	186	$7.6 \times 10^{-12} \pm 5 \times 10^{-13}$	4640 ± 766	$2.7 \times 10^{-9} \pm 4 \times 10^{-10}$
CATCLAY 1	8.5	500	217	$1.1 \times 10^{-11} \pm 2 \times 10^{-13}$	865 ± 31	$3.0 \times 10^{-10} \pm 5 \times 10^{-12}$
HTO in illite						
idp5	7.2	200	14	$3.6 \times 10^{-10} \pm 9 \times 10^{-12}$	0.57 ± 0.03	$3.6 \times 10^{-10} \pm 9 \times 10^{-12}$

For Sr diffusion in illite, the last line contains the average of the experiments, followed by the SD.

where t_{i-1} (unit: day or s) is the start of a measurement, t_i is the end of this measurement (when the outlet solution is replaced), $J_{out}(t)$ (unit: Bq day^{-1} cm^{-2}) the diffusive flux into the outlet volume, and L_{out} (unit: cm) is the ratio of the outlet reservoir volume and the clay sample cross-section. According to equation (1), during a measurement (t_{i-1}, t_i), the concentration $C_{out}(t)$ monotonously increases, becoming zero again after replacing the outlet reservoir at time t_i. The problem with the VC–CC model is that, if the duration ($t_i - t_{i-1}$) of an experiment is sufficiently large, the calculated outlet concentration $C_{out}(t_i)$ can become larger than the calculated inlet concentration $C_{in}(t_i)$, clearly violating the initial assumption that $0 \approx C_{out} \ll C_{in}$. This was observed, for example, when fitting with the VC–CC model (extended with diffusion through the confining filters) in experiment idp5 (see Fig. 1): for the measurement between 85 and 95 days, this fit (not shown here) leads to an outlet concentration about twice the inlet concentration ($C_{out} \approx 2C_{in}$), even with experimental values $C_{in} \approx 1.6C_{out}$ (basically not allowing the zero concentration approximation either). Figure 1 illustrates this as well (but for transport parameter values other than the optimal values obtained with the VC–CC model with filters, code name Dfit36_var_C0_ met_filters): at the end of the interval between 85 and 95 days, C_{out} calculated with the VC–CC model with filters is approximately 580 Bq ml^{-1}, which is clearly larger than the corresponding computed value of C_{in} of approximately 160 Bq ml^{-1}.

A model was developed using, during a measurement, the VC condition at the outlet instead of the CC condition. The VC condition at the outlet

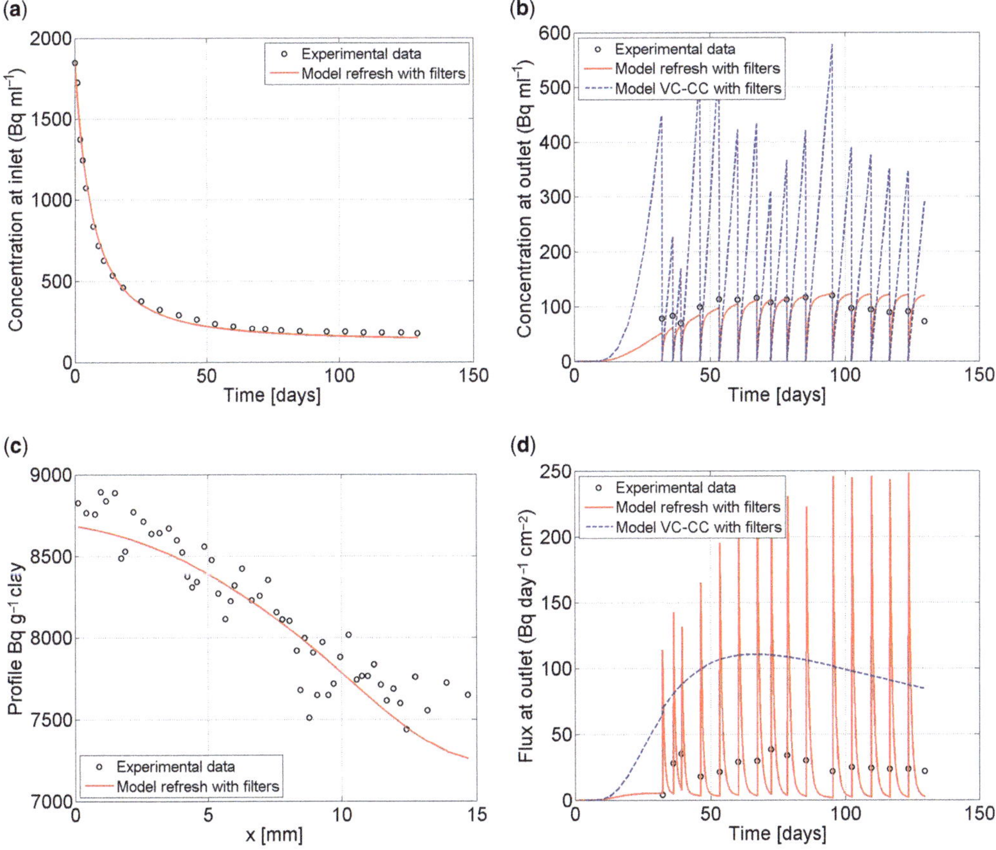

Fig. 1. Experimental data and model prediction of Sr diffusion in illite (0.1 M $NaClO_4$ – pH 7, idp5) obtained by fitting simultaneously: (**a**) the evolution of the inlet concentration, (**b**) the evolution of the outlet concentration and (**c**) the tracer profile in the clay core (note that the *y*-axis starts at 7200 Bq g^{-1} clay). The plot of the outlet concentration also shows the prediction with the VC–CC model, using the same transport parameter values. (**d**) The outlet fluxes according to both models, as well as to the experimental data.

ensures that the flux into the outlet volume decreases to zero when the concentration in the outlet volume approaches the concentration at the outlet side of the outlet filter: preventing that, the calculated concentration in the outlet volume becomes larger than the concentration in the outlet filter or even the inlet. In the new model (code name D2_diff_vol2_met_ filters_refresh, abbreviated here to 'refresh'), at time t_i of replacing the outlet volume, the outlet concentration is put to zero again ('refreshed'). The model is numerically solved using the COMSOL package, coupled to Matlab for fitting. For numerical stability, the tracer profile in the outlet filter was slightly smoothed, leading to a small loss of tracer. A mass balance verification at the end of the fits shows that this loss is negligible (e.g. 0.1% loss for the fit of Fig. 1). In both the VC–CC and the refresh model, flushed filters were considered as traditional static filters (Glaus *et al.* 2015).

The evolution of the tracer concentration at the inlet and at the outlet, as well as the tracer profile in the clay at the end of the experiment, are fitted simultaneously (see Fig. 1). Figure 1 also shows the predicted evolution of the tracer concentration at the inlet, as well as at the outlet, when using the VC–CC model (with filters) with the same transport parameter values. Clearly, the predictions of the outlet concentration and outlet flux by the VC–CC model and the refresh model differ a lot. The basic difference between both models is clearly illustrated by the time evolution of the flux into the outlet during a measurement. After breakthrough of the tracer, the flux according to the VC–CC model remains approximately constant during a measurement because it does not depend on the actual tracer concentration in the outlet (assumed to be constantly zero owing to the CC condition). In the refresh model, the initially (start of a measurement) infinitely high flux monotonously decreases, as the tracer concentration in the outlet approaches the tracer concentration in the outlet filter. Contrary to Sr diffusion, for through-diffusion of HTO through the same core, the predictions of the outlet concentration by both models are very similar (idp5: see Fig. 2). Note the nearly linear increase (as a function of time) of the outlet concentration during a measurement, differing substantially from the concentration evolution during a measurement in Figure 1. As no tracer profile was measured in this experiment (Fig. 2: the core was still needed for other uses afterwards), only the inlet (not shown) and outlet evolution were simultaneously fitted by the refresh model. Assuming a quasi-stationary state, a criterion indicating that a zero concentration outlet boundary is a good approximation or not is derived in Glaus *et al.* (2015). According to this criterion, a zero concentration outlet boundary is a good approximation for the HTO diffusion experiment (Fig. 2) but cannot be used for the Sr diffusion experiment (Fig. 1). Basically, a zero concentration outlet boundary can always be made a good approximation by choosing the measurement times sufficiently short and/or the outlet volumes sufficiently large.

The refresh model has four adjustable parameters: the apparent diffusion coefficients in clay and in the filters ($D_{app,c}$ and $D_{app,f}$) and the corresponding capacity factors ($(\eta R)_c$ and $(\eta R)_f$). For Sr diffusion in illite and Boom Clay, the filter transport

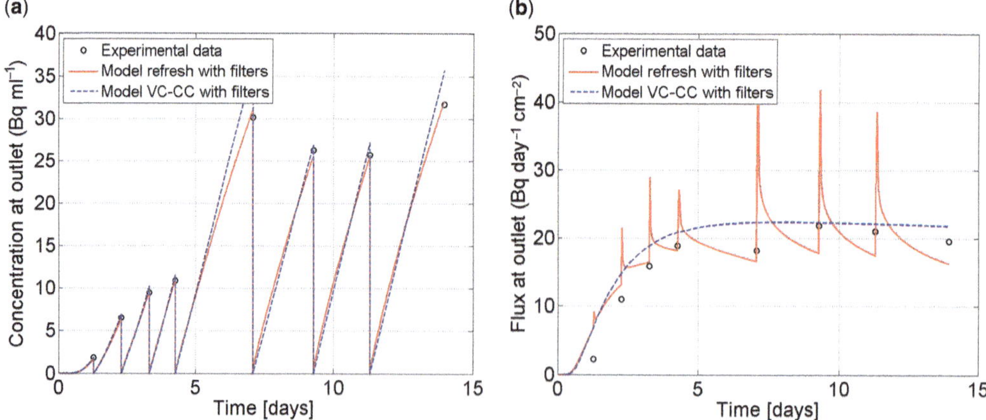

Fig. 2. Experimental data and model predictions for: (**a**) the outlet concentration and (**b**) the outlet flux of HTO diffusion through the same illite core as in Figure 1, obtained by reproducing simultaneously the evolution of the inlet concentration and the outlet concentration. Using the values obtained by fitting the refresh model, the predictions between this model and the VC–CC model agree much better than in Figure 1.

parameters $D_{app,f}$ and $(\eta R)_f$ are fully correlated: only their product $D_{eff,f} = D_{app,f}\,(\eta R)_f$ can be determined. To allow comparison with $D_{app,c}$, Table 1 lists $D_{app,f}$, derived from $D_{eff,f}$ and, because no sorption on (unused) filters was observed ($R_f = 1$), the fixed value $(\eta R)_f = \eta_f = 0.3$ (the filter porosity). The Sr apparent diffusion coefficients in the filter mentioned in Table 1 are all greater than 1×10^{-10} m^2 s^{-1}, which is the value of the Sr diffusion coefficient in the unused filters (Aertsens *et al.* 2011). These higher values are attributed to sorption of Sr on small clay particles infiltrating the pores of the filter, and making $(\eta R)_f = 0.3$ for used filters is only an approximation. Consistent with Glaus *et al.* (2015), the experiment with advectively flushed filters (idp17) has the highest filter diffusion coefficient (although even larger values are observed for two Boom Clay experiments, but here $(\eta R)_c$ is also higher). The scatter in $D_{app,f}$ values is large, and hard to explain. The apparent diffusion coefficient of Sr in illite is roughly the same for all experiments, with an average value of 1.2×10^{-11} m^2 s^{-1} (standard deviation (SD) of 2.7×10^{-13} m^2 s^{-1}). Apart from a much higher value for experiment idp18, the capacity factors are also similar, with an average value of 171 (SD of 69).

For HTO diffusion in illite, a different fit strategy is used because: (i) there are fewer experimental data (no profile in the clay); (ii) the correlation between the filter transport parameters is not as strong as for Sr; and (iii) the transport parameters of HTO in the filters and in Boom Clay do not differ by much (Aertsens *et al.* 2011). Considering the high degree of uncertainty on the measured values of the (unused) filter transport parameters in Aertsens *et al.* (2011), it is assumed that, for HTO, the transport parameters in the filters are the same as in Boom Clay, reducing the number of adjustable parameters to two. The fitted apparent diffusion coefficient of 3.6×10^{-10} m^2 s^{-1} agrees well with the measured value for the filters (around 3.8×10^{-10} m^2 s^{-1}: Aertsens *et al.* 2011), but the capacity factor $(\eta R)_c = 0.57$ is much higher than expected. The reason for the difference with the measured water content of 27% is unclear. In this experiment, the VC–CC model does not lead to the measured water content either, but does produce similar optimal values to the refresh model: a capacity factor of $(\eta R)_c = 0.52$ and an apparent diffusion coefficient of $D_{app} = 3.7 \times 10^{-10}$ m^2 s^{-1}.

For Sr diffusion in Boom Clay, (re)fitting both FUNMIG experiments with the refresh model leads to good fits for the outlet concentration and the tracer profile, but the predicted inlet concentration does not agree well with all the experimental data. For both experiments, the refresh fits are much better (lower χ^2 value) than the fits with the VC–CC models mentioned in Aertsens *et al.* (2011). Also, the initial illite experiments (up to idp14) were reproduced with both models. In each case, the refresh model leads to a better or similar (in the case where both models give similar predictions, as in Fig. 2) χ^2 value than the VC–CC model. The apparent diffusion coefficient of both

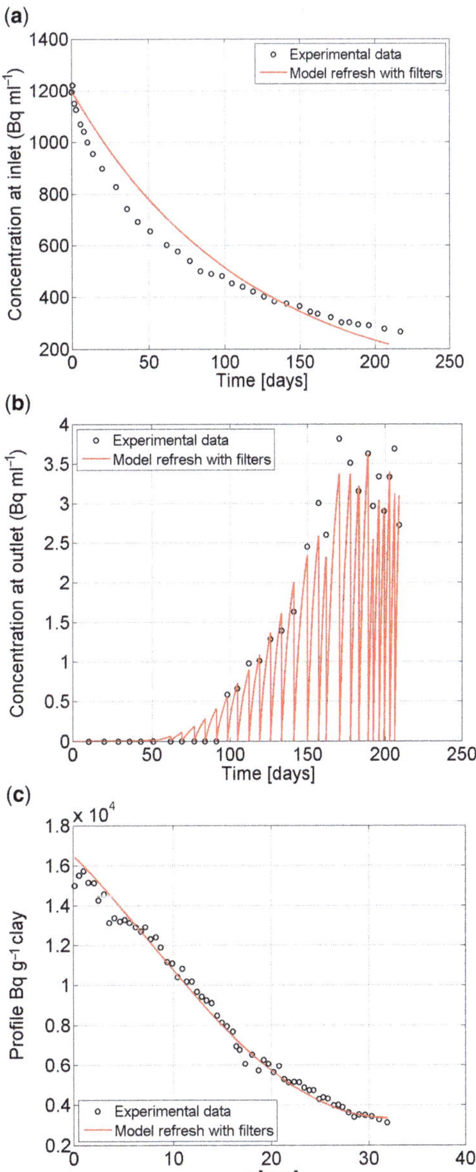

Fig. 3. Experimental data and model prediction of Sr diffusion in Boom Clay: (**a**) evolution of the inlet concentration, (**b**) evolution of the outlet concentration and (**c**) the tracer profile in the clay.

FUNMIG experiments is the same and in line with the values obtained from other types of experiments (Aertsens *et al.* 2009). For the CATCLAY 1 experiment, in particular, the prediction of the inlet concentration is better (see Fig. 3), but the fitted apparent diffusion coefficient of 1.1×10^{-11} $m^2\ s^{-1}$ is (slightly) higher than expected. Still larger are the differences in the capacity factor: 6830 (FUNMIG 15 mm), 4640 FUNMIG (30 mm) and 858 (CATCLAY 1). Apart from a possible heterogeneity between the clay cores, we cannot explain these differences.

Conclusions

For modelling through-diffusion experiments where the outlet volume is regularly replaced by a tracer-free volume, the tracer concentration in the outlet volume is commonly assumed to be zero at all times. In the case of small outlet volumes and/or large measurement times, this assumption is often invalid during almost the entire measurement interval. Therefore, the presented model avoids this simplifying assumption: (i) by taking into account the increase in tracer concentration in the outlet volume during a measurement; and (ii) the tracer concentration is put to zero at the end of a measurement when the outlet volume is replaced by a tracer-free concentration. This model simultaneously reproduces well the evolution of the inlet concentration, the outlet concentration and the tracer concentration profile in the clay core at the end of the experiment. Depending on the type of tracer and clay, as well as on the dimensions of the clay core, the filters, the inlet and outlet volumes, and the duration of an experimental measurement, this model leads to similar or improved fits than the traditional model assuming a zero outlet concentration.

This work is performed in close co-operation with, and with the financial support of, ONDRAF/NIRAS, the Belgian Agency for Radioactive Waste and Fissile Materials, as part of the programme on geological disposal of high-level/long-lived radioactive waste. This work also has received funding from the EURATOM 7th Framework |Programme FP7/2007-2011 under grant agreement No. 249624 (CATCLAY project).

References

AERTSENS, M., DE CANNIÈRE, P., LEMMENS, K., MAES, N. & MOORS, H. 2008. Overview and consistency of migration experiments in clay. *Physics and Chemistry of the Earth, Parts A/B/C*, **33**, 1019–1025.

AERTSENS, M., MAES, N. & VAN GOMPEL, M. 2009. Consistency of the strontium transport parameters in Boom Clay obtained from different types of migration experiments. *Materials Research Society Symposium Proceedings*, **1193**, 421–428.

AERTSENS, M., GOVAERTS, J., MAES, N. & VAN LAER, L. 2011. Consistency of the strontium transport parameters in Boom Clay obtained from different types of migration experiments: accounting for the filter plates. *Materials Research Society Symposium Proceedings*, **1475**, 583–588.

BIRGERSSON, M. & KARNLAND, O. 2009. Ion equilibrium between montmorillonite interlayer space and an external solution – consequences for diffusional transport. *Geochimica et Cosmochimica Acta*, **73**, 1908–1923.

GLAUS, M., BAYENS, B., BRADBURY, M., JAKOB, A., VAN LOON, L. & YAROSHCHUK, A. 2007. Diffusion of ^{22}Na and ^{85}Sr in Montmorillonte: evidence of interlayer diffusion being the dominant pathway at high compaction. *Environmental Science and Technology*, **41**, 478–485.

GLAUS, M., ROSSÉ, R., VAN LOON, L. & YAROSHCHUK, E. 2008. Tracer diffusion in sintered stainless steel filters: measurement of effective diffusion coefficients and implications for diffusion studies with compacted clays. *Clays and Clay Minerals*, **56**, 677–685.

GLAUS, M., AERTSENS, M., MAES, N., VAN LAER, L. & VAN LOON, L. 2015. Treatment of boundary conditions in through diffusion: a case study of $^{85}Sr^{2+}$ diffusion in compacted illite. *Journal of Contaminant Hydrology*, **177–178**, 239–248.

PUT, M. & HENRION, P. 1988. An improved method to evaluate radionuclide migration model parameters from flow-through diffusion tests in reconsolidated clay plugs. *Radiochimica Acta*, **44/45**, 343–347.

TAKEDA, M., NAKAJIMA, H., ZHANG, M. & HIRATSHUKA, T. 2008. Laboratory longitudinal diffusion tests: 1. Dimensionless formulations and validity of simplified solutions. *Journal of Contaminant Hydrology*, **97**, 117–134.

ZHANG, M., TAKEDA, M. & NAKAJIMA, H. 2006. Determining the transport properties of rock specimens using an improved laboratory through-diffusion technique. *Materials Research Society Symposium Proceedings*, **932**, 135–142.

Multispecies random walk simulations in radial symmetry: model concept, benchmark, and application to HTO, ^{22}Na and ^{36}Cl diffusion in clay

SHUO MENG & WILFRIED PFINGSTEN*

Paul Scherrer Institut, 5232 Villigen-PSI, Switzerland

**Correspondence: wilfried.pfingsten@psi.ch*

Abstract: Modelling of radionuclide transport in clay can be simplified and accelerated by exploiting radial symmetry of laboratory experiments or nuclear waste repositories design. Consequently, the multispecies reactive transport code MCOTAC has been extended to radial symmetry, exploiting the advantages of its random walk transport description. Random walk in radial r-symmetry is mimicked by two-dimensional (2D) random walk projected to a radial coordinate for geochemical calculations. This guarantees fast $2D(x, y)$ transport calculations, together with complex geochemical calculations in $1D(r)$ only. The new model concept has been benchmarked for simple geochemical systems and applied to laboratory diffusion experiments with HTO, ^{22}Na and ^{36}Cl in Opalinus Clay using higher spatial resolution for modelling than reported in the literature. This makes it possible to distinguish between the transport properties of the sample and the filters, and yields more careful determination of radionuclide transport parameters: for example, there was a difference of a factor of 2 between the diffusion coefficient and log K_d sorption coefficient in the case of ^{22}Na compared to the former best-fit analysis.

Several European countries consider deep lying argillaceous formations as potential host rocks for the safe disposal of radioactive waste. Owing to the absence of fractures, even on the microscopic scale, and due to the low hydraulic permeability of the clay, the main transport process for radionuclides is molecular diffusion within the clay porewater. It is, therefore, essential to understand, on a fundamental level, diffusion and sorption of radionuclides in such compacted argillaceous formations, and to determine carefully radionuclide-dependent values for diffusion coefficients and for sorption parameters for argillaceous materials. Smaller-scale laboratory in- and through-diffusion experiments and their modelling are used to investigate and determine radionuclide diffusion and sorption parameters in addition to batch sorption experiments. These parameters are then used to predict radionuclide migration for geological nuclear waste repositories for performance assessment purposes.

Generally, several simulation tools are available for modelling such reactive transport processes (e.g. PHREEQC (USGS 2015), TOUGH (LBNL 2015), OpenGeoSYS (OpenGeoSys 2015) and others), which include different complexities with respect to the modelling of transport processes (1D, 2D and 3D), geochemical processes and their coupling. Here we describe the extension and application of the reactive transport code MCOTAC ('Multi-species Coupling Of Transport And Chemistry': Pfingsten 1994, 2006). It accounts for the transport of all ions in solution coupled with geochemical equilibrium and kinetic calculations for chemical reactions in solution, on surfaces and with solid phases at the same time. For MCOTAC, the random walk method is used for solving the transport equations. The advantage of the random walk approach over other numerical schemes is, for example, that negative species concentrations are excluded during transport calculations, which generates numerical problems, if transport and geochemical equilibrium-kinetic modules are directly coupled. Also, it allows quick calculations with only a few particles during a fitting procedure resulting in very scattered concentration distributions or breakthrough curves and slower 'production runs' for final fit curves using much more particles. In addition, the random walk method allows any kind of boundary condition to be included very easily. The original MCOTAC code was designed for Cartesian geometry in 1D and 2D. Nevertheless, several diffusion experiments with Opalinus Clay (OPA) have been performed with disk-like samples in radial symmetry to investigate diffusion processes parallel to the clay layering, which cannot be modelled adequately by the linear MCOTAC in reasonable computing time. In order to model real experimental set-ups more accurately, linear MCOTAC was extended in order to be applicable for radial symmetry, which also includes the geometry

From: NORRIS, S., BRUNO, J., VAN GEET, M. & VERHOEF, E. (eds) 2017. *Radioactive Waste Confinement: Clays in Natural and Engineered Barriers*. Geological Society, London, Special Publications, **443**, 211–224.
First published online August 18, 2016, https://doi.org/10.1144/SP443.15

of foreseen underground tunnels for nuclear waste storage within clay rocks. Recently, MCOTAC has been used to include the sophisticated 2SPNE CE/SC sorption model by Pfingsten *et al.* (2011), describing radionuclide migration in bentonite and clays taking into account, for example, sorption competition reactions between radionuclides and major ions in the porewater, as well as to assess the influence of stable elements on ^{59}Ni migration from a nuclear waste repository (Pfingsten 2014). In order to apply such complex sorption models, the spatial geometry should not be too complex in order to achieve reasonable calculation times for radionuclide migration scenarios. Therefore, the extension to 1D radial geometry is straightforward so that complex geochemical sorption models within simplified geometry can be applied efficiently without too many simplifications. Even non-isotropic diffusion could be described with this newly developed random walk procedure shown below. It should be noted that no direct random walk procedure representing 1D radial symmetrical systems has been found in the literature,

Model concept for random walk in radial geometry

The model concept consists of a multi-grid approach. Random walk transport simulations were performed on a 2D equidistant grid as described in Pfingsten (2002), whereas the calculations of chemical equilibrium were performed in 1D using radial symmetry, reducing the calculation time considerably.

The method used to solve the transport equation for a fixed set of species concentrations in the liquid phase is the random walk method extended to a multi-species transport problem. A spatial, radially symmetrical grid is used to define the location of the immobile solid concentrations and mobile liquid concentrations of all species. (For simplicity, the range of coordinates is $0 \le x \le x_{\max}$ and $0 \le y \le y_{\max}$. The radial coordinate r is given by $r(x, y)$.) Then, at the beginning of the transport calculation, $N_{\max}$ particles are distributed along the model domain of interest (Fig. 1) similar to the method of characteristics (Konikow & Bredehoeft 1978). The particles are used to represent the distribution of concentrations of all species in the liquid phase. The particle masses are related to the particle location $r(x, y)$ in the form of a particle mass vector at a location $r(x_n, y_n)$, which fits into a grid cell with defined species concentrations at time t (equation 1). The particle location, r_n, is continuous in (x, y) and not related to a random walk on a regular lattice, as described by Karapiperis & Blankleider (1992).

The particle mass vector $\vec{n}$ characterizes the particle properties. The vector $\vec{n}$ has $N_j + 2$ components and is defined by:

$$\vec{n} = (m_1, m_2, m_3, \ldots, m_{N_j}, r_n, t). \qquad (1)$$

The vector components m_j ($j = 1, \ldots, N_j$), which are species masses per particle (and location and time), are defined by the location r_n. N_j is the number of species j. The actual particles masses (components m_j) depend on the calculated equilibrium concentrations or total masses $m_j|_{\text{cell } l}$ of species j in the grid cell (volume V) and the actual number of particles $nb(l)$ in the grid cell l, which are both a function of time:

$$m_j|_{\text{cell } l} = \frac{C_j V}{nb(l)}|_{\text{cell } l} \ldots \quad j = 1, \ldots, N_j. \qquad (2)$$

During the transport time step, new particle positions in the x and y directions (equations 3 and 4) are calculated depending on the transport processes diffusion (and/or advection) and dispersion (Fig. 1) (see Kinzelbach 1987):

$$x_n(t + \Delta t) = x_n(t) + v_x \Delta t + Z_x \sqrt{2D_x \Delta t} \qquad (3)$$

and

$$y_n(t + \Delta t) = y_n(t) + v_y \Delta t + Z_y \sqrt{2D_y \Delta t} \qquad (4)$$

where $x_n(t + \Delta t)$ and $y_n(t + \Delta t)$ are the Cartesian coordinates of the radial particle coordinate $r_n(t + \Delta t) = r_n\ (x_n(t + \Delta t), y_n(t + \Delta t))$ of particle n after time step Δt; v_x, v_y and D_x, D_y are flow velocity and diffusion coefficient in the x and y directions, respectively; and Z_x and Z_y are random numbers. The radial particle coordinate is related to the radial cell index l.

After each time step, at $t + \Delta t$, a new particle distribution and the related species concentration distributions in solution $C_j(l)$ are calculated:

$$C_j(l) = \frac{1}{V_l} \sum_{\text{particle in cell } l} m_j \qquad (5)$$

where $C_j(l)$ is the concentration of species j and V_l is the volume of cell l. Then these concentrations in the liquid phase in cell l have to be equilibrated according to the concentrations of the solids in cell l before the new transport step is calculated.

Compared to a 2D Cartesian (e.g. equidistant grid with constant grid cell volume), the 2D radial symmetrical grid has varying grid cell volumes for rings of the same thickness and the cell volumes V_l have to be modified accordingly.

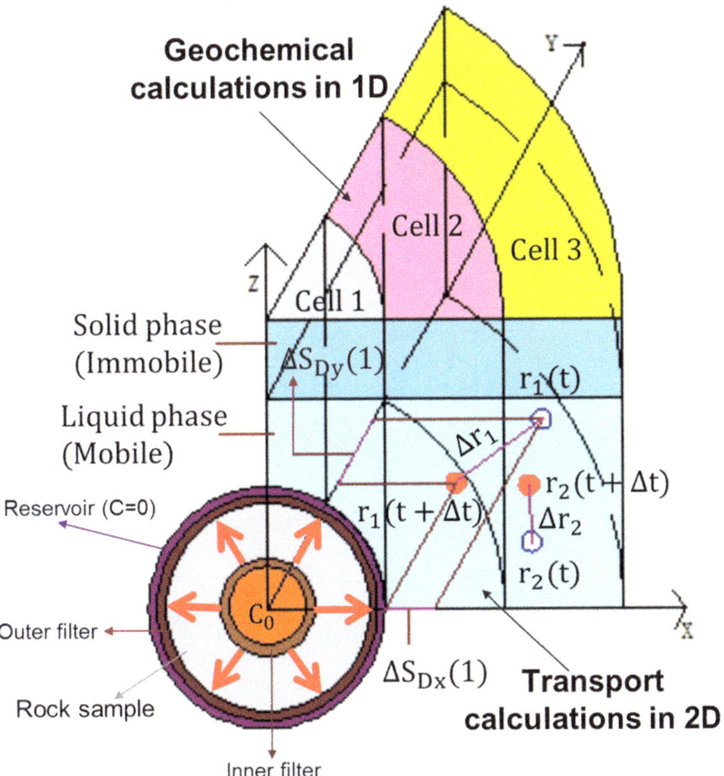

Fig. 1. Model concept for the description of ion diffusion in clay for radial symmetrical systems (laboratory or repository set-up) using a new random walk procedure. Transport during time step Δt is calculated by superposition of random walk (of particle 1) in the x and y directions from location $r_1(t)$ to $r_1(t+\Delta t)$ by $\Delta S_{D_x}(1) = Z_1\sqrt{2D_x\Delta t}$ and $\Delta S_{D_y}(1) = Z_2\sqrt{2D_y\Delta t}$: that is, two independent random movements in x and y, respectively (Z_1 and Z_2 are random numbers, and D_x and D_y are diffusion coefficients in the x and y directions). The resulting particle position $r_1(t+\Delta t)$ is then projected onto the r direction for coupled geochemical equilibrium calculations. This procedure is performed for particles 1, 2, … in the model area, yielding new species concentrations after a transport step Δt.

When discretizing the 2D area into rings of equal thickness with the centre at the origin (see Fig. 1), the outer radius of cell l (or ring l) equals $l\,dx$ and the inner radius is $(l-1)dx$ (for simplicity $dx = dy = dr$ has been chosen for discretization in the x, y and r directions). Then the area S of cell l (ring l) is given by:

$$S[l] = \pi(l\,dx)^2 - \pi[(l-1)\,dx]^2 = \pi\,dx\,dx\,(2l-1). \quad (6)$$

According to equations (2) and (5), the mass vector is related to the volumes of the cells (i.e. the cell area times the cylinder height). Because of the variable (pore) volumes of the cells (see equation 6), which are different compared to a 2D Cartesian grid, the calculation of the particle masses becomes dependent on the individual (pore) volumes of each cell (ring) l. Hence, as (cylinder) height is equal for all rings, an area correction factor has to be introduced compared to the pure Cartesian approach. Each particle has to be related to a variable cell volume (i.e. a variable cell area S(l)), as calculated by equation (6). For this method, it is necessary to count the number of particles in each cell, as for Cartesian coordinates, and it is also necessary to know from which cell (volume) particles enter a specific cell, since the particle mass calculation depends on the cell volume and where it comes from. (For random walk, particles transfer the mass of a species in solution during a time step Δt from one grid cell to the other. This mass is calculated according to cell l with cell volume V_l, where each particle is located before this time step (i.e. $m_j = C_j(l) * V_l$). From this, it is obvious that for adjacent cells in radial symmetry with identical concentrations, but with different cell volumes (equation 6) and with the same number of particles in each cell, the mass of a particle entering a cell from an 'inner' cell has to be higher than the mass of a particle entering the cell from an 'outer' cell, as the concentration in all three contributing cells should remain constant after the particle movements. This means that it is necessary to know from which cell (volume) the particles enter a specific cell. The same is valid for concentration gradients across three adjacent cells. All this is not necessary for an equidistant Cartesian grid, where all cell volumes (V_l) are identical and can be left out of the calculations (equation 7) to save computational time.) Therefore, equation (5) has to be replaced by equation (7) to calculate new concentrations in solution after the transport step (equation 7 is also valid for the Cartesian approach):

$$C_{l,j}^{t+\Delta t} = \frac{\sum_0^{nl} c_{p,m}^t(l)\, V_p(l)^t}{\sum V_p(l)^t} \quad (7)$$

where $C_{l,j}^{t+\Delta t}$ is the new concentration of species j for cell l (in mol l^{-1}) at $t + \Delta t$ calculated from all particles in cell l at $t + \Delta t$, where particles p have particle masses equal to $C_{p,j}^t(l)$ (in mol l^{-1}) multiplied by V_l (in litres) of species j at time t. Therefore, the calculated particle masses depend on the cell index from which they enter cell l. The sum has to be normalized to all contributing particles in cell l. n_l is the particle number in cell l; $V_p(l)^t$ is the cell volume (in litres) of cell l. This procedure guarantees that the ring area/volume is correctly taken into account for the random walk procedure for radial symmetry.

For the particle movements, independent random walk in the x and y directions (equations 3 and 4, which allow also non-isotropic diffusion) is mapped onto the radial coordinate r (see Fig. 1), and chemical equilibrium calculations are performed along r instead of x and y. Therefore, it is not necessary to design and calculate within a 2D 'chemical' grid, which reduces computation time greatly – l calculations instead of l^2 calculations for equidistant discretization in the x, y and r directions with l cells in each direction.

Benchmark

In order to verify this new random walk approach to radial symmetry, we compared MCOTAC radial symmetrical calculations for single-species diffusion in clay with and without assuming linear (K_d) sorption with results from COMSOL Multiphysics (COMSOL 2015) calculations for the same system. The set-up of this diffusion scenario is shown in Figure 1, and assumes a constant concentration (C_0) in an area around the origin up to radius r_i, initially zero concentration in between r_i (inner radius) and r_o (outer radius), and a fixed zero concentration at r_o (no filters are assumed). Diffusion for the time period T is calculated and the related concentration profiles are compared. The agreement for the pure diffusion scenario is very good, even on the logarithmic scale covering a large concentration range (see Fig. 2).

A further benchmark that includes a simple sorption reaction has also been performed. For the COMSOL Multiphysics calculation, only a linear sorption coefficient (K_d) could be used within a single-species transport calculation, whereas in MCOTAC two species were defined: a mobile species A and an immobile species A_s, which are related by the mass action law $A/A_s = R_d$. R_d is also called the distribution coefficient (l kg^{-1}), indicating the ratio of the sorbed amount of a species (in mol kg^{-1}) to the amount of dissolved species (in mol l^{-1}) at equilibrium. Both calculations are shown in Figure 3.

Also for this (simple) reactive transport scenario, the agreement between both calculations, COMSOL Multiphysics and MCOTAC, is very good. It should be noted that both codes used a different model set-up (e.g. spatial discretization, one or two species, respectively), and that the definition of a $C = 0$ mol l^{-1} concentration boundary condition at $r = r_o$ was a little more tricky for COMSOL. Nevertheless, the achieved agreement was very good. Therefore, MCOTAC with the new radial symmetrical transport module could be applied to radial symmetrical systems with more complex

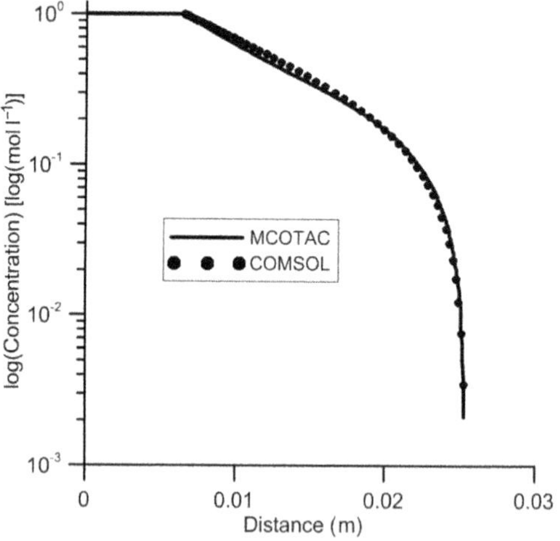

Fig. 2. Comparison of MCOTAC and COMSOL profiles calculated for single-species diffusion into a cylindrical clay sample at 80 347 s. Model parameters: inner radius = 0.00658 m with C_{0in} = constant = 1 M, outer radius = 0.0254 m with C_{0out} = constant = 0.0 M, $D_e = 3.125\ 10^{-10}$ m^2 s^{-1} and porosity = 0.5.

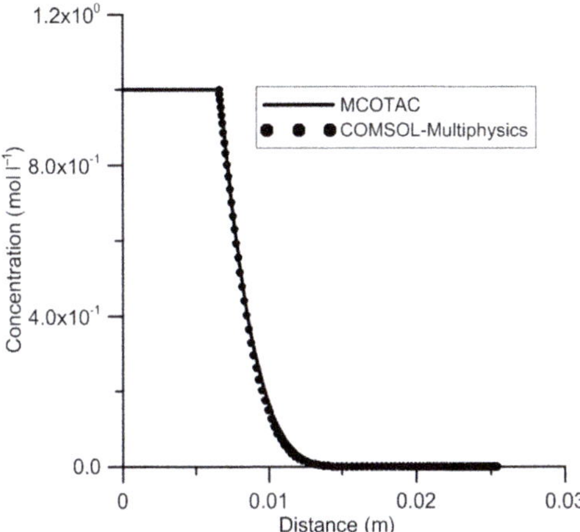

Fig. 3. Comparison between MCOTAC and COMSOL profiles calculated for single-species diffusion with additional K_d sorption at 22570 s. Model parameters: inner radius = 0.00658 m with C_{0in} = constant = 1M, outer radius = 0.0254 m with C_{0out} = constant = 0.0 M, $D_e = 3.125 \times 10^{-10}$ m^2 s^{-1}, porosity = 0.5, $K_d = 0.001$ m^3 kg^{-1}, $\rho_{bulk} = 1100$ kg m^{-3}.

(multispecies–multisite sorption) geochemical systems, which is not possible within COMSOL Multiphysics.

Application to laboratory experiments for the determination of clay-specific transport parameters

The extended MCOTAC is applied to laboratory through-diffusion experiments performed with HTO, ^{22}Na and ^{36}Cl in Opalinus Clay reported in the literature (Van Loon *et al.* 2004), where the related modelling of the experiments included a coarse spatial discretization: that is, there was no differentiation between the transport parameters for filters and the clay sample. Instead, a single parameter value identical for both filters and sample was used. In addition, constant C_0 concentration boundary conditions for the tracers were assumed for the reservoirs.

Here we performed a more detailed modelling of the experiments, taking into account different transport parameters for the filters and sample explicitly (i.e. porosities of 0.418, 0.15 and 0.367 for the inner filter, clay sample and outer filter, respectively), and related the different effective diffusion coefficients $D_e = D_p\varepsilon$ (D_p is the pore diffusion coefficient in m^2 s^{-1} and ε is the porosity) to the filters and the clay sample, as given in Van Loon *et al.* (2004). We also take into account the decreasing tracer concentration within the 'high concentration reservoir' with time as a result of tracer in-diffusion into the clay; however, we stick on the outer C = 0 concentration boundary condition due to frequent water exchange within this outer boundary concentration loop.

A sketch of the radial diffusion experimental set-up is given in Figure 4. A concentration gradient is set up between the centre of the sample (i.e. the high concentration boundary) and the outer border of the sample (i.e. the low concentration boundary). Because the concentration gradient is set up to be parallel to the layering of the clay sample, diffusion will take place along the clay layering. Detailed parameters about the experimental set-up can be found in Table 1 and Figure 4, and in Van Loon *et al.* (2004). (The cylindrical clay sample (diameter 25.4 mm) has a central hole (diameter 6.58 mm) along the cylinder axes and a height of 52 mm. The inner hole and outer cylindrical sample surface are stabilized with stainless steel filters 1.8 mm and 1.6 m in thickness. The inner and outer filters are connected to a high and a low concentration loop, respectively, establishing a concentration gradient from the centre hole of the cylinder towards its outer rim. The sample holder consists of two end pieces at the bottom and at the top of the cylindrical sample, with filters and an outer steel ring all around the outer filter. Bolts held together the two end pieces of the sample holder and allowed the application of a defined load onto the sample, as described in detail in Van Loon *et al.* (2004).)

Because experimental data were recorded as flux (mol m^{-2} s^{-1}), and concentrations (mol l^{-1}) were

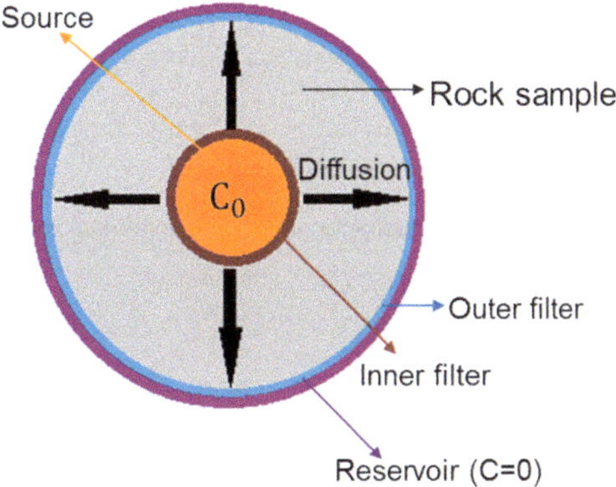

Fig. 4. Sketch of the model set-up for radial through-diffusion experiments. C_0 indicates a high concentration solution loop connected to a large reservoir; inner and outer filters separate the clay rock sample from the two solution loops. The diffusion is from the centre to the reservoir (C = 0) solution loop. Parameters for the experimental set-up are given in the text.

Table 1. *Summary of the comparison between MCOTAC best-fit transport parameters for HTO, ^{22}Na and ^{36}Cl diffusion in Opalinus Clay and parameter values from the literature**

		Sample effective diffusion coefficient ($m^2 s^{-1}$)	Filter effective diffusion coefficient ($m^2 s^{-1}$)	Other parameter
HTO-1	MCOTAC	5.4×10^{-11}	10.9×10^{-11}	
	Literature model	5.4×10^{-11}	5.4×10^{-11}	
HTO-2	MCOTAC	5.4×10^{-11}	10.0×10^{-11}	
	Literature model	5.4×10^{-11}	5.4×10^{-11}	
^{36}Cl	MCOTAC	$1.3E \times 10^{-11}$	10.0×10^{-11}	
	Literature model	1.6×10^{-11}	1.6×10^{-11}	
^{22}Na sorption model	MCOTAC	4.4×10^{-11}	12.5×10^{-11}	$\text{Log}(K_c) = -1.26$
	Literature model	7.2×10^{-11}	7.2×10^{-11}	$\text{Log}(K_c) = -2.10$
^{22}Na K_d model	MCOTAC	4.4×10^{-11}	12.5×10^{-11}	$K_d = 1.34$
	Literature model	7.2×10^{-11}	7.2×10^{-11}	$K_d = 3.41$

*Taken from Van Loon *et al.* (2004) and Appelo *et al.* (2010).

provided by MCOTAC as a function of time or location, the concentrations calculated by MCOTAC had to be recalculated to flux data for comparison with the experimental data. The relationship between flux and concentration is written as:

$$\text{Flux}\,(\text{mol}\,\text{m}^{-2}\,\text{s}^{-1}) = -D_{\text{e\_outer}} \frac{\partial C}{\partial r}\left[\text{m}^2\,\text{s}^{-1}\,\frac{(\text{mol}\,\text{m}^{-3})}{\text{m}}\right] \quad (8)$$

$D_{\text{e\_outer}}$ is the local effective diffusion coefficient: that is, the outer filter effective diffusion coefficient (the inner and outer filters have different porosities). $\partial C/\partial r$ is the concentration gradient between the filter and the right-hand boundary. The filter properties are given in Table 2.

Four different through-diffusion experiments were analysed, and best-fit (by eye) parameter values for the effective diffusion coefficient, sample porosity and/or sorption parameters were deduced from the experiments. Two HTO experiments (HTO-1 and HTO-2) were performed with different reservoir volumes that had an influence on the evolution of the concentration at the high concentration boundary. For HTO, pure diffusion processes are assumed, as well as for ^{36}Cl through-diffusion experiments. However, as ^{36}Cl is negatively charged in solution, lower accessible porosity should be considered due to anion-exclusion processes. Finally, a ^{22}Na through-diffusion experiment was analysed using two different sorption model approaches. One was a K_d approach using single-species transport. The second was a sorption model approach that includes multispecies transport and sorption on several exchange sites, as proposed in Jakob *et al.* (2009). The related cation-exchange reactions (equations R1–R5) are given in Table 3. Here, reactions between ^{22}Na and exchange sites, stable Na and other ions (K, Ca, Mg) in solution were individually taken into account in order to guarantee that isotope exchange is also taken into account. Therefore, equations (R1)–(R5) (Table 3) had to be modified and extended to equations (R6)–(R11), as exemplified for planar sites in Table 4. It is also worthwhile to note that no sorption was assumed for the filters in both approaches calculated by MCOTAC, since no sorption was

Table 2. *Overview of the tracers and their related parameters given in Van Loon* et al. *(2004) and Glaus* et al. *(2008)*

Tracer	D_e sample ($m^2 s^{-1}$)	α sample*	D_e filter ($m^2 s^{-1}$)	V_{high}/V_{low} † ($m^3 cm^{-3}$)
HTO-1	$(5.4 \pm 0.4) \times 10^{-11}$	0.15 ± 0.03	$(10.9 \pm 2.2) \times 10^{-11}$	200/20
HTO-2	$(5.4 \pm 0.4) \times 10^{-11}$	0.17 ± 0.02	$(10.9 \pm 2.2) \times 10^{-11}$	1000/200
$^{36}Cl^-$	$(1.6 \pm 0.1) \times 10^{-11}$	0.08 ± 0.01	Not measured	500/50
$^{22}Na^+$	$(7.2 \pm 0.5) \times 10^{-11}$	0.62 ± 0.05	$(9.1 \pm 1.8) \times 10^{-11}$	500/50

*$\alpha = \varepsilon + \rho_b R_d$ (rock capacity factor), where $\tilde{n}_b$ is the dry bulk density (kg m^{-3}) and R_d is the sorption distribution ratio ($m^3 kg^{-1}$).
†V_{high}, V_{low}, reservoir volume of the high and low concentration loop.

Table 3. *Cation-exchange reactions and values for the selectivity coefficient K_c for three sorption site types $\equiv S_{plan}$, $\equiv S_{Type-II}$ and $\equiv S_{FES}$ for major ions according to the model by Bradbury & Baeyens (1998, 2000)*

Cation exchange reaction		Log K_c
$\equiv S_{plan}Na^+ + K^+ \leftrightarrow \equiv S_{plan}K^+ + Na^+$	(R1)	0.7
$\equiv S_{plan}2Na^+ + Ca^{2+} <=> \equiv S_{plan}Ca^{2+} + 2Na^+$	(R2)	0.67
$\equiv S_{plan}2Na^+ + Mg^{2+} <=> \equiv S_{plan}Mg^{2+} + 2Na^+$	(R3)	0.59
$\equiv S_{Type-II}K^+ + Na^+ <=> \equiv S_{Type-II}Na^+ + K^+$	(R4)	−2.1
$\equiv S_{FES}K^+ + Na^+ <=> \equiv S_{FES}Na^+ + K^+$	(R5)	−2.4

observed for ^{22}Na on the filters. This was different to that found in Van Loon *et al.* (2004).

For the MCOTAC modelling, time-dependent source concentration (C_0) was taken into account as measured in the experiments: that is, a decreasing tracer concentration in the reservoir due to diffusion into the filter and sample. The discretization was set up so that filters included up to 10 grid cells in order to have a reasonable spatial resolution. For the clay sample, many more grid cells were used. The fitting procedure was performed first using literature data from Van Loon *et al.* (2004) for clay and filters as a first trial, then measured filter properties were introduced from Glaus *et al.* (2008) and, finally, the effective diffusion coefficient for the clay sample or the sorption parameter was modified to get a final best fit. The comparison of the calculated and experimental data is then shown on the linear and logarithmic scale to allow a reasonable judgement about the quality of the individual best fits.

HTO-1 modelling

Figure 5 shows the best fit for the HTO-1 through-diffusion experiment with a small volume on the high concentration side. We deduced, apparently by chance, the same best-fit transport parameters as reported in the literature, where mean parameters identical for sample and filters, and a constant HTO concentration on the high concentration side, were assumed. In order to test the influence of the concentration boundary condition, we fixed it to the maximal (initial) and minimal (final) boundary concentration obtained from the experiment for two further calculations. The result indicated about a $\pm 15\%$ difference calculated for the related

Table 4. *Modified cation-exchange reactions using sorbed potassium instead of sorbed sodium for planar sites*

Cation exchange reaction		Log K_c
$\equiv S_{plan}K^+ + Na^+ <=> \equiv S_{plan}Na^+ + K^+$ for stable Na	(R6)	−0.7
$\equiv S_{plan}2K^+ + Ca^{2+} <=> \equiv S_{plan}Ca^{2+} + 2K^+$ for stable Na	(R7)	0.71
$\equiv S_{plan}2K^+ + Mg^{2+} <=> \equiv S_{plan}Mg^{2+} + 2K^+$ for stable Na	(R8)	0.81
$\equiv S_{plan}K^+ + Na^+ <=> \equiv S_{plan}Na^+ + K^+$ for ^{22}Na	(R9)	−0.7
$\equiv S_{plan}2K^+ + Ca^{2+} <=> \equiv S_{plan}Ca^{2+} + 2K^+$ for ^{22}Na	(R10)	0.71
$\equiv S_{plan}2K^+ + Mg^{2+} <=> \equiv S_{plan}Mg^{2+} + 2K^+$ for ^{22}Na	(R11)	0.81

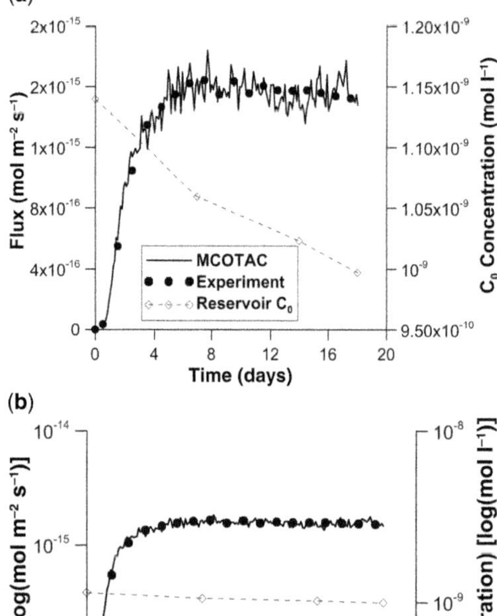

Fig. 5. Comparison between the simulation results and experimental results for HTO-1 on (**a**) the linear and (**b**) the logarithmic scale with best-fit parameters obtained in Table 1.

breakthrough fluxes. In addition, when we used a constant high (C_0) concentration for the MCOTAC modeling, as done by Van Loon *et al.* (2004), with our best-fit parameters, we could not reproduce sufficiently the experimental data owing to the fact that a constant concentration boundary does not fit with the experimental conditions. Therefore, experimental conditions should be integrated into the modelling in detail in order not to achieve an arbitrary fitting as a result of the model simplifications.

In the calculated flux curves of HTO-1, some fluctuations can be observed. This phenomenon was also observed for the other simulations (HTO-2, ^{22}Na and ^{36}Cl). It was partially produced by the interpolated flux calculation method and partly due to the random walk procedure. These fluctuations could be reduced if more than 10 000 particles, as used for all calculations here, were involved in the simulations. But this resulted in additional computational time, as shown in Appendix A.

HTO-2 modelling

For the HTO-2 experiment, with the large volume reservoir, it was not necessary to take into account a decreasing HTO concentration in the reservoir – the same best-fit parameters were deduced for the clay sample as reported by Van Loon *et al.* (2004). From Figure 6, it is obvious that there exist small differences between the simulation results and the experimental results (i.e. the HTO plateau on the linear scale). In order to improve this agreement, some modifications were tested and, after several trials, a best-fit result was obtained (see Fig. 7). Without any modifications of the other parameters, only the filter effective diffusion coefficient was slightly reduced.

Here, the influence of the constant concentration boundary condition was also tested, as for the HTO-1 experiment. However, no significant difference could be observed, as the concentration

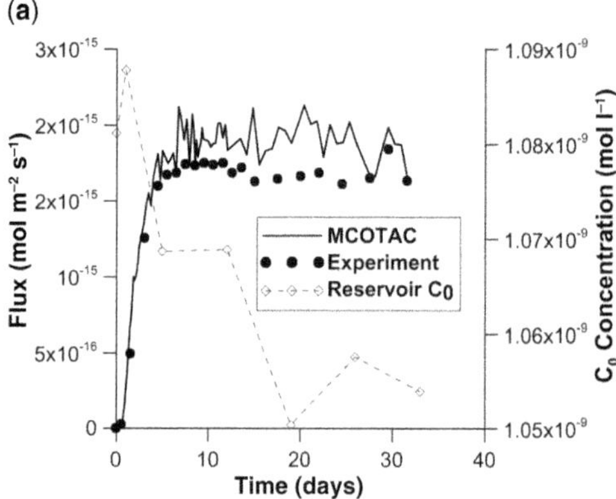

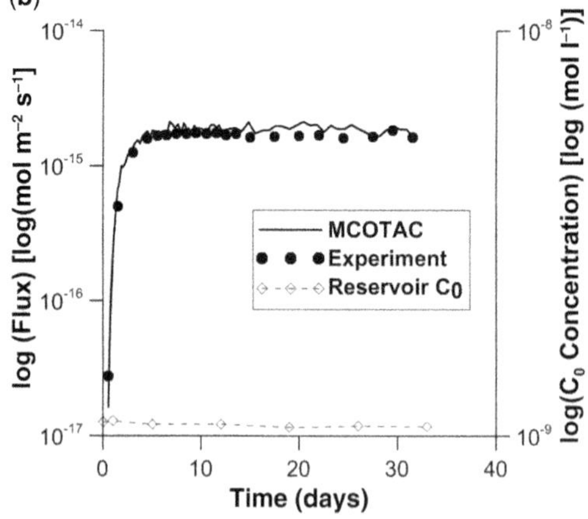

Fig. 6. Comparison between the simulation results and experimental results for HTO-2 on (**a**) the linear and (**b**) the logarithmic scale with parameters obtained in Table 2.

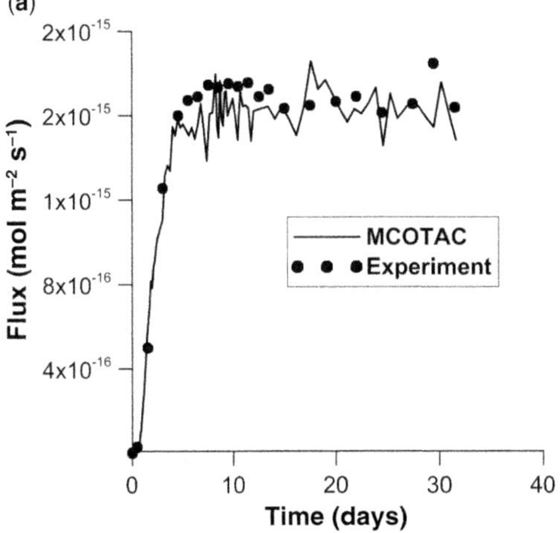

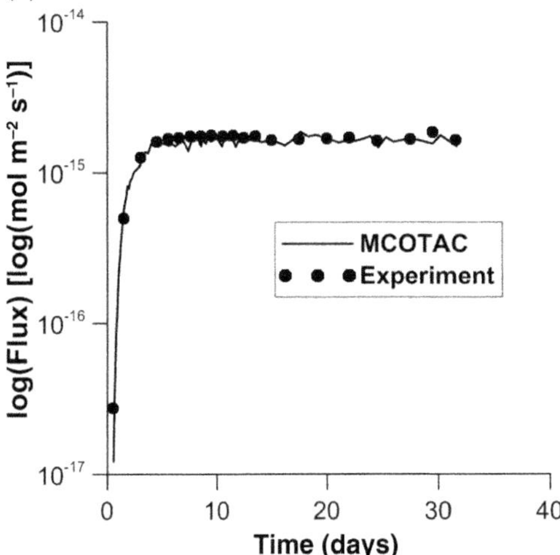

Fig. 7. Comparison between MCOTAC and experimental results for HTO-2 with best-fit parameters used in MCOTAC: (**a**) linear scale; (**b**) logarithmic scale.

decrease at the high concentration boundary was much smaller than for the experiment HTO-1, which could be explained by the differences in the reservoirs used for HTO-1 (200 ml) and HTO-2 (1000 ml) (see Figs 5 & 6).

^{22}Na modelling

Two sorption model approaches (non-linear sorption) and K_d approach were applied to reproduce the ^{22}Na experimental data. First, the non-linear sorption model was tested where all species in solution diffuse into the porewater and react with surfaces. ^{22}Na and stable Na in the clay porewater were explicitly taken into account to include isotopic exchange effects in the modelling. When using the reference data from the literature (Van Loon *et al.* 2004), a breakthrough curve resulted that did not correlate very well with the experimental data (see Fig. 8). Large differences can be observed clearly at both the early stage (or transitional) and the steady-state flux. The steady-state flux was calculated to be only half as high as the experimental value, thereby reaching steady state earlier.

In order to get a best-fit for ^{22}Na using a complex sorption model approach, more parameters might need to be changed to better fit the experimental data than for a simple K_d approach. Filter and sample diffusion coefficients, their porosities and several log K_c values (Tables 3 & 4) could be specific to the sample mineralogy (see Van Loon *et al.* 2005) and can be varied. The log K_c value affects the transitional state, while the steady-state concentration level is determined by both the

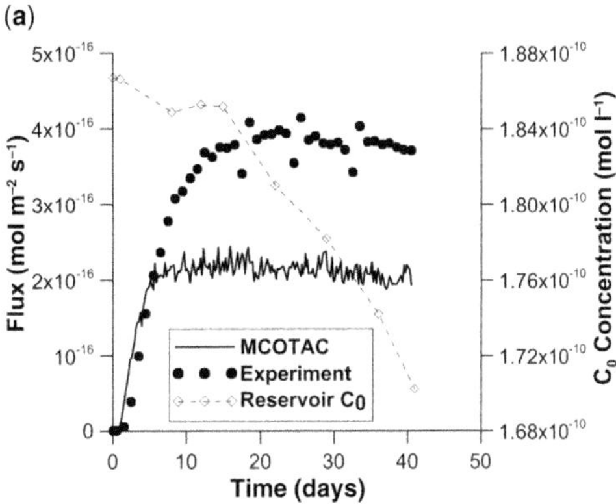

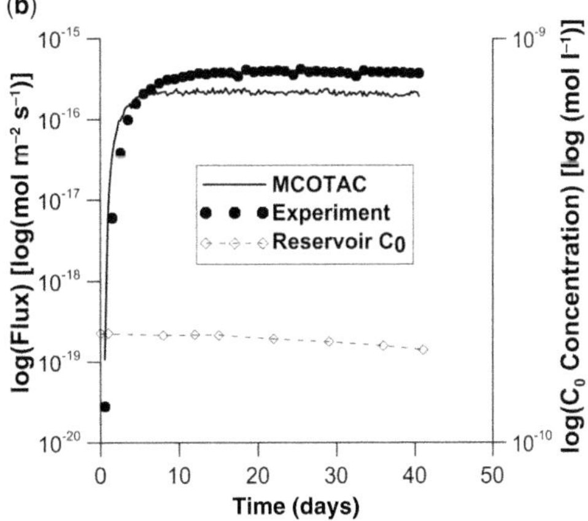

Fig. 8. Comparison between the simulation results and experimental results for ^{22}Na on (**a**) the linear and (**b**) the logarithmic scale with parameters obtained in Table 2.

diffusion coefficient and porosity. After some tests, the Type II reaction between potassium and sodium (Reaction R4 in Table 3) showed the largest influence on the calculated breakthrough curve. At the same time, the effective diffusion coefficients of sample and filters needed to be adjusted to reproduce the experimental steady-state flux. The related best-fit curves are shown in Figure 9.

For the K_d approach, only a single species has to be modelled. In this simulation, the same basic parameters as for the sorption model approach were used. The K_d value was recalculated from the rock capacity factor α (Table 2). The results are shown in Figure 10.

Again, the agreement between measured and calculated breakthrough curve is not very good. This indicates that the effective diffusion coefficients of the sample and filters, the porosity of the sample, and the K_d value have to be adjusted compared to earlier modelling. According to the real experimental set-up, sorption reactions did not take place in filters: in other words, $K_d = 0$ for

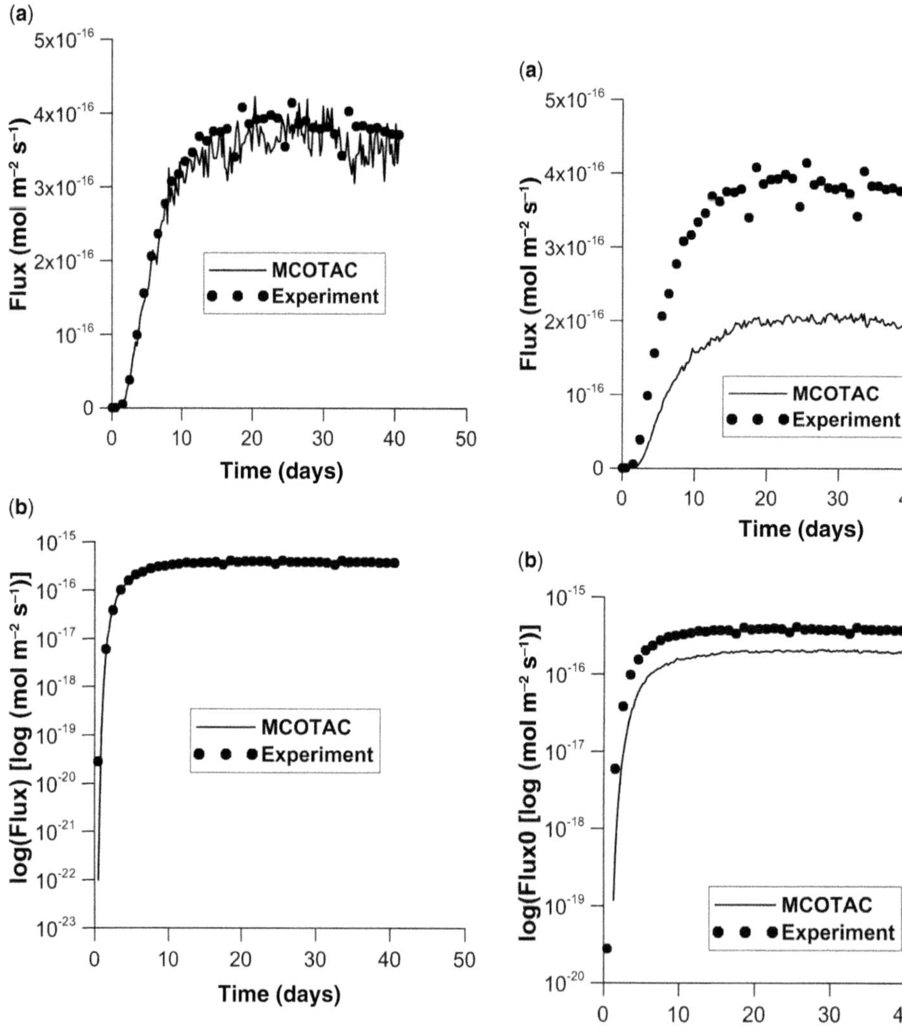

Fig. 9. Comparison between simulation results and experimental results for ^{22}Na with modified parameters used in MCOTAC when the log (K_c) value of reaction (R4) in Table 3 equals -1.26: (**a**) linear scale; (**b**) logarithmic scale. Here the sample and filter effective diffusion coefficients are equal to 4.6×10^{-11} ($m^2 s^{-1}$) and 12.5×10^{-11} ($m^2 s^{-1}$), respectively.

Fig. 10. Comparison between simulation results and experimental results for ^{22}Na with parameters obtained from Table 2, which were used in MCOTAC when log (K_d) equals 0.5328 [log((mol l^{-1})/(mol l^{-1}))]. (**a**) Linear scale and (**b**) logarithmic scale.

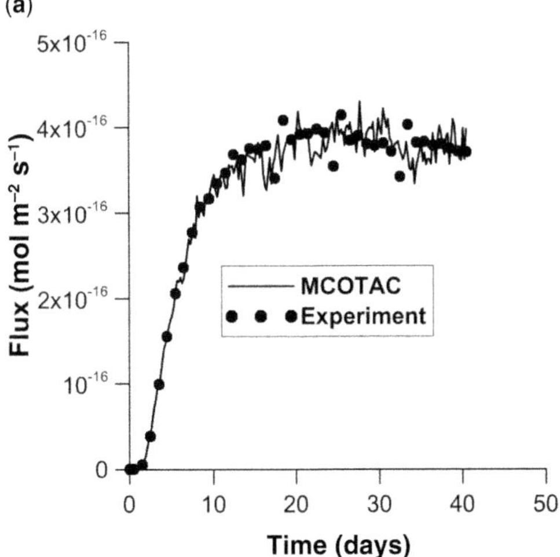

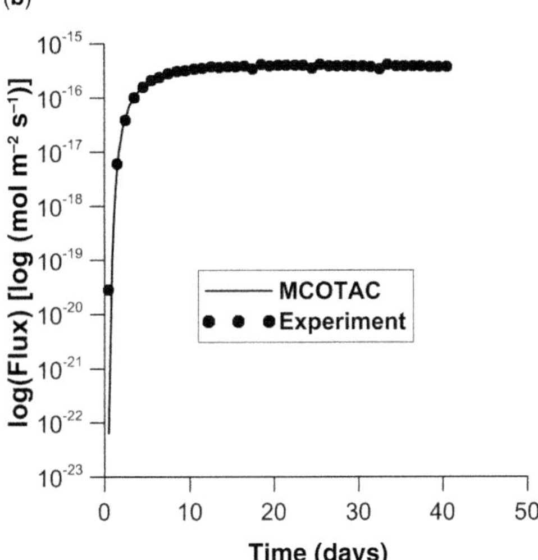

Fig. 11. Comparison between MCOTAC and experimental results for ^{22}Na on (**a**) the linear scale and (**b**) the logarithmic scale when log (K_d) equals 0.13 [log((mol l^{-1})/(mol l^{-1}))]. Here sample and filter effective diffusion coefficients are equal to 4.4×10^{-11} ($m^2 s^{-1}$) and 12.5×10^{-11} ($m^2 s^{-1}$), respectively.

the filters, which was ignored in Van Loon *et al.* (2004). Since their model could not take into account spatially varying K_d values, they used an average K_d for the clay and filters. For the best fit, we achieved $K_d = 1.34$ ((mol l^{-1})/(mol l^{-1})) for clay. The results are shown in Figure 11.

^{36}Cl modelling

For ^{36}Cl modelling, no measured filter parameters were available: therefore, we used initially the filters values for HTO. However, the reproduction of the flux curves using literature parameters for clay from Van Loon *et al.* (2004) was not successful (Fig. 12). The calculated breakthrough is a bit higher than the experimental curve. In order to better fit the experimental curve, the porosity of the sample and filter, as well as the filter effective diffusion coefficient and the sample effective diffusion coefficient, have been varied. Considering that information about ^{36}Cl was not that consistent compared to HTO or Na, parameters here have been varied over a wider range. In addition, the measured boundary concentration was not as smooth as that for HTO-1, HTO-2 and ^{22}Na (see Fig. 12), which might have induced some 'strange' constraint into the modelling. Finally, a best-fit curve (see Fig. 13) was obtained with the parameters given in Table 2.

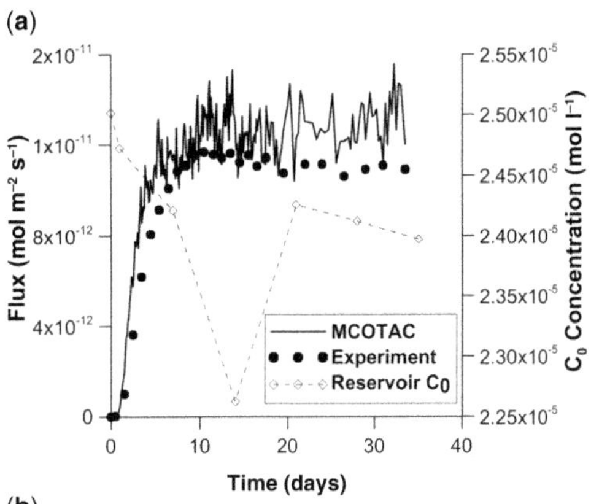

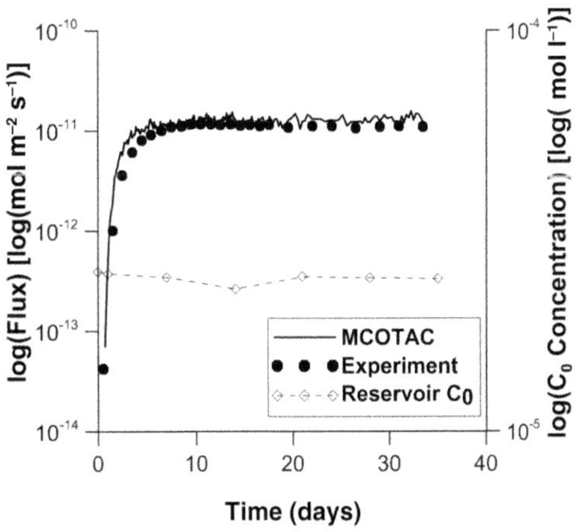

Fig. 12. Comparison between simulation results and experimental results for ^{36}Cl on (**a**) the linear scale and (**b**) the logarithmic scale with parameters obtained in Table 2.

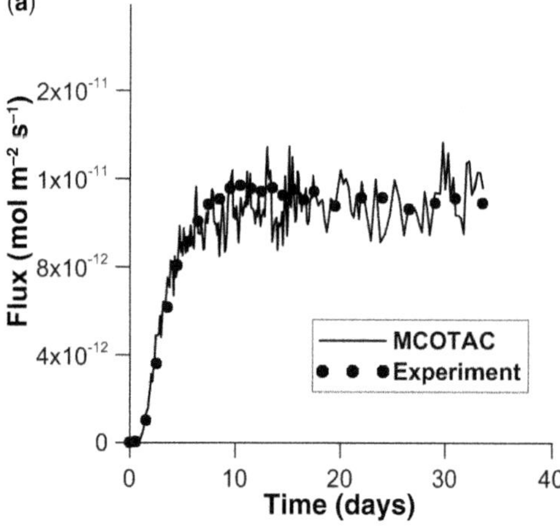

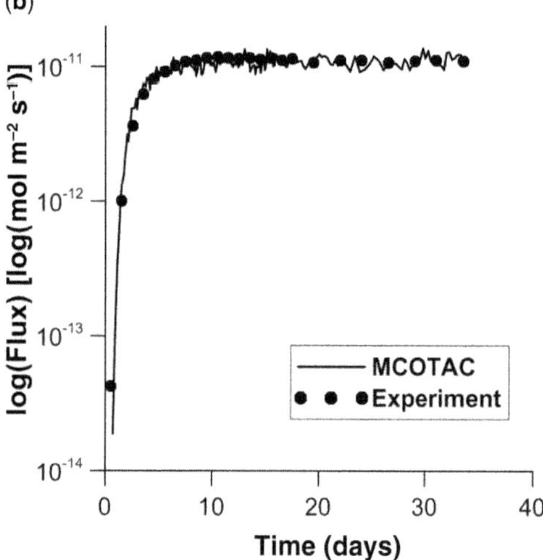

Fig. 13. Comparison between MCOTAC and experimental results for ^{36}Cl with best-fit parameters used in MCOTAC. (**a**) Linear scale and (**b**) logarithmic scale.

Summary

The reactive transport code has been successfully extended by a new random walk procedure for radial symmetrical geometry. Random walk transport in 2D(x, y) is projected onto 1D(r) radial symmetry and coupled to geochemical reactions, which allows computationally efficient modelling of radial symmetrical systems. For the spatial grid used and the geochemical systems modelled here, up to a factor of 10 faster computational times could be achieved using the radial symmetrical set-up 1D(r) compared to a 2D(x, y) set-up, as time-consuming geochemical calculations were minimized in 1D(r).

The procedure has been benchmarked successfully with the commercial software 'COMSOL Multiphysics' for simple chemical systems: that is, pure tracer diffusion and single-species diffusion with K_d sorption.

Finally, MCOTAC has been applied to radial symmetrical experimental set-ups for diffusion of HTO, ^{22}Na and ^{36}Cl in clay. In order to improve the modelling of these experiments, material-specific transport parameters for filters and clay samples were used to distinguish between their individual influences. This was possible due to finer discretization of the spatial model domain in 1D(r) geometry. The results were compared to previous modelling, where mean identical transport parameters were used both for samples and filters. Major differences were observed when analysing sorbing tracer breakthrough curves. It was shown that with a higher spatial resolution used for the modelling here, the diffusion coefficient and linear sorption coefficient (log K_d) were about a factor of about 2 different compared to the coarse modelling of the ^{22}Na experiment reported in the literature. Therefore, the application of MCOTAC to radionuclide transport experiments in radial symmetry with appropriate spatial resolution for filters and samples yields more carefully determined transport parameters. Compared to earlier but coarser modelling of these experiments, it should be emphasized that model simplifications do not always yield the same transport parameters, especially for sorbing radionuclides.

Therefore, it seemed necessary to analyse laboratory radionuclide migration experiments with a sufficient spatial resolution to keep track of important material heterogeneities, as well as to use reasonable chemical complexity to incorporate all relevant chemical reactions/processes and also to take into account experimental boundary conditions correctly, if they are available.

The detailed comments made by two anonymous reviewers led to an improved version of the manuscript and are gratefully acknowledged. Partial financial support by the Swiss National Cooperative for the Disposal of Radioactive Waste (Nagra) is gratefully acknowledged.

Appendix A

In order to test the random walk transport property of achieving smoother concentration distributions or breakthrough curves just by increasing the number of particles used within the model calculation, we used the 'HTO-1 modelling' case as an example. At the same time, we tested the computational efficiency when using different numbers of particles in the modelling. For the original 'HTO-1 modelling' (Fig. 5), we used 10 000 particles within the model area and the calculation took less than

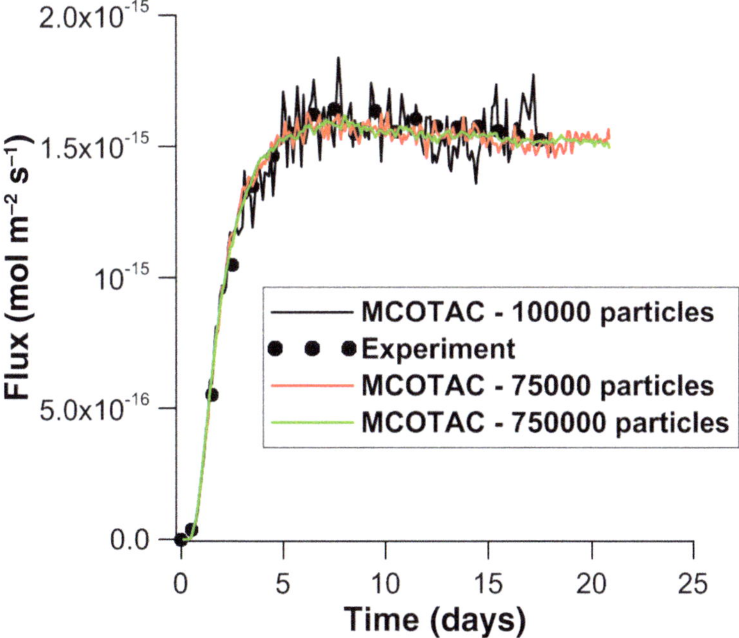

Fig. A1. The influence of the number of particles used in the MCOTAC modelling on the smoothness of the calculated HTO flux curves for the 'HTO-1 modelling' case (see also Fig. 5).

1 min on a 'reference PC'. When increasing the number of particles to 75 000 and 750 000, we obtained smoother and smoother flux curves for HTO-1 (see Fig. A1), but the calculation time on the same computer increased to about 4 and 35 min, respectively: the time increasing approximately linearly with the increase in the number of particles. Owing to this property of the random walk procedure, it is already possible to obtain 'rough' results within quite short computation times.

References

Appelo, C.A.J., Van Loon, & Wersin, L.R.P. 2010. Multicomponent diffusion of a suite of tracers (HTO, Cl, Br, I, Na, Sr, Cs) in a single sample of Opalinus Clay. *Geochimica et Cosmochimica, Acta*, **74**, 1201–1219.

Bradbury, M.H. & Baeyens, B. 1998. A physicochemical characterisation and geochemical modelling approach for determining porewater chemistries in argillaceous rocks. *Geochimica et Cosmochimica Acta*, **62**, 783–795.

Bradbury, M.H. & Baeyens, B. 2000. A generalised sorption model for the concentration dependent uptake of caesium by argillaceous rocks. *Journal of Contaminant Hydrology*, **42**, 141–163.

COMSOL 2015. *COMSOL Multiphysics®. The Platform for Physics-Based Modeling and Simulation*. COMSOL, Inc., Burlington, MA, http://www.comsol.com/comsol-multiphysics

Glaus, M.A., Rossé, R., Van Loon, L.R. & Yaroshchuk, A.E. 2008. Tracer diffusion in sintered stainless steel filters: measurement of effective diffusion coefficients and implications for diffusion studies with compacted clays. *Clays and Clay Minerals*, **56**, 677–685.

Jakob, A., Pfingsten, W. & Van Loon, L. 2009. Effects of sorption competition on caesium diffusion through compacted argillaceous rock. *Geochimica et Cosmochimica Acta*, **73**, 2441–2456.

Karapiperis, T. & Blankleider, B. 1992. *Comparison of Cellular Automata and Differential Equation Approaches to Reaction-Transport Process*. Internal Report TM-41-91-41. Paul Scherrer Institute, Villigen, Switzerland.

Kinzelbach, W. 1987. *Numerische Methoden zur Modellierung des Transports von Schadstoffen im Grundwasser*. Schriftenreihe gwf Wasser-Abwasser, **21**.

Konikow, L.F. & Bredehoeft, J.D. 1978. Chapter C2. Computer model of two-dimensional solute transport and dispersion in groundwater. *In*: Konikow, L.F. & Bredehoeft, J.D. (eds) *Techniques of Water-Resources Investigations of the United States Geological Survey*. United States Geological Survey, Washington, DC.

LBNL. 2015. *TOUGH: Suite of Simulators for Nonisothermal Multiphase Flow and Transport in Fractured Porous Media*. Lawrence Berkeley National Laboratory, Berkeley, CA, http://esd.lbl.gov/research/projects/tough/

OpenGeoSys. 2015. OpenGeoSys Platform, http://www.opengeosys.org/

PFINGSTEN, W. 1994. *Modular Coupling of Transport and Chemistry: Theory and Model Applications.* PSI Bericht No. 94-15. Paul Scherrer Institut, Switzerland.

PFINGSTEN, W. 2002. Experimental and modelling indications for self-sealing of a cementitious L&ILW repository by calcite precipitation. *Nuclear Technology*, **140**, 63–82.

PFINGSTEN, W. 2006. *MCOTAC Manual 1.0*, PSI Internal Report AN-44-06-09, Paul Scherrer Institut, Switzerland.

PFINGSTEN, W. 2014. The influence of stable element inventory on the migration of radionuclides in the vicinity of a high level nuclear waste repository exemplified for ^{59}Ni. *Applied Geochemistry*, **49** , 103–115.

PFINGSTEN, W., BRADBURY, M.H. & BAEYENS, B. 2011. The influence of Fe(II) competition on the sorption and migration of Ni(ii) in MX-80 bentonite. *Applied Geochemistry*, **26**, 1414–1422.

USGS. 2015. *PHREEQC (Version 3) – A Computer Program for Speciation, Batch-Reaction, One-Dimensional Transport, and Inverse Geochemical Calculations.* United States Geological Survey, Washington, DC, http://wwwbrr.cr.usgs.gov/projects/GWC_coupled/phreeqc/

VAN LOON, L.R., SOLER, J.M., MÜLLER, W. & BRADBURY, M.H. 2004. Anisotropic diffusion in layered argillaceous rocks: a case study with Opalinus Clay. *Environmental Science & Technology*, **38**, 5721–5728.

VAN LOON, L.R., BAEYENS, B. & BRADBURY, M.H. 2005. Diffusion and retention of sodium and strontium in Opalinus clay: comparison of sorption data from diffusion and batch sorption measurements, and geochemical calculations. *Applied Geochemistry*, **20**, 2351–2363.

Combining high-resolution two-phase with simplified single-phase simulations in order to optimize the performance of PA/SA simulations for a deep geological repository for radioactive waste

PHILIPP SCHAEDLE[1,2], THOMAS KAEMPFER[1]*, GUILLAUME PÉPIN[3], JACQUES WENDLING[3] & JUERGEN BROMMUNDT[1]

[1]*AF-Consult Switzerland Ltd, Groundwater Protection and Waste Disposal, Täfernstrasse 26, CH-5405 Baden, Switzerland*

[2]*Present address: ETH Zürich, Rämistrasse 101, 8092 Zürich, Switzerland*

[3]*ANDRA, 1–7 rue Jean Monnet, Parc de la Croix-Blanche, F-92298 Châtenay-Malabry Cedex, France*

**Correspondence: thomas.kaempfer@afconsult.com*

Abstract: The transport of a radioactive solute during the transient thermo-hydraulic regime with gas generation in and around a disposal cell depends on complex multi-phase processes. Numerical simulations can improve the understanding of the system by providing detailed information on the temporal and spatial distribution of the radionuclides. In particular, their fluxes can be computed under the given transient conditions considering radionuclide, heat and gas release from the waste. However, such detailed multi-phase simulations are very demanding with respect to computational resources and time. Based on the knowledge gained from such complex simulations, we have developed a robust simplified single-phase approach for performance and safety assessment, the improved efficiency of which enables extensive parameter studies. The simplified approach comprises, on the one hand, homogenization of features of high detail and, on the other hand, the employment of two-phase simulation results that are used to deduce equivalent single-phase parameterizations. The results have been validated with various benchmark criteria at well-defined interfaces in the modelled disposal cells based on the simulated radionuclides fluxes.

For the performance analysis and safety assessment (PA/SA) of a planned deep geological repository for high-level long-lived radioactive waste (HLW-LL) and intermediate-level long-lived radioactive waste (ILW-LL), it is international practice to develop models that are capable of simulating two-phase flow and radionuclide transport through the repository and its host-rock formation. Many of these models require a high spatial resolution, including all significant structures and features. Furthermore, all relevant physical flow and transport processes have to be parameterized and modelled as accurately as possible. These circumstances lead to non-linearities in the governing equations and, hence, the resulting discretized mathematical equations have to be solved iteratively. Additional challenges are posed both by the spatial and temporal scales. The planned repository analysed in this work consists of several thousand disposal cells with relevant details on the decimetre scale, as well as access tunnels and shafts with kilometre-scale lengths. The potential radionuclide release in and from the repository, as well as the transport to the biosphere, must be simulated for up to 1 myr. For this, large-scale numerical models have been designed over the last couple years (Brommundt *et al.* 2014; Enssle *et al.* 2014). Furthermore, many modern PA/SA include probabilistic studies to assess uncertainties. These may involve several hundred or, even, thousands of simulation runs. All of these factors combined suggest that fundamental optimizations of the model are essential in order to achieve computation times that allow the required number of simulations to be conducted within a reasonable study duration, while still ensuring high simulation quality in terms of results.

We tested an approach to increase the efficiency of the simulations by decoupling the flow and transport computations. The idea is to perform the time-consuming two-phase flow simulations only once, followed by many less-demanding single-phase unsaturated transport simulations that reuse the two-phase flow results. The main limitation for the validity of this approach is that the considered radionuclides need to be highly soluble in water and their influence on flow has to be negligible.

From: NORRIS, S., BRUNO, J., VAN GEET, M. & VERHOEF, E. (eds) 2017. *Radioactive Waste Confinement: Clays in Natural and Engineered Barriers*. Geological Society, London, Special Publications, **443**, 225–234.
First published online June 22, 2016, https://doi.org/10.1144/SP443.4

However, these requirements are fulfilled by many radionuclides in a repository. In addition, various optimizations of the spatial discretization have been tested to further reduce the computational effort.

A similar decoupling approach was described in Schaedle *et al.* (2014). In this previous work, a different transport code (TRACES) was used to that used for the present study (TOUGH2) and a different model set-up has been considered.

Methodology

This study had two main objectives: first, the impact on accuracy resulting from decoupling the hydraulic and transport simulation was investigated; and, second, the sensitivity to spatial discretization was analysed.

Decoupling of flow and transport

The first part consists of decoupling the flow part and the transport part of the computation (sequential PA/SA approach). The computationally intensive calculation of the two-phase flow field has to be performed only once and can then be used as an input for various single-phase transport calculations. For the subsequent decoupled transport calculation, the computational effort is much less. This sequential approach can lead to a significant reduction in the required computational resources when performing probabilistic studies with regard to uncertainties in transport parameters. Furthermore, the specific needs of each model in terms of temporal discretization, restricted by the maximum temporal resolution in the two-phase flow and transport simulation, can be considered.

In practice, the decoupling is implemented according to the methodology shown in Figure 1. First, the reference results are calculated in a detailed two-phase flow and transport model, and specific hydraulic results are stored at every time step. Next, a single-phase transport model mimicking transport properties from the reference case is set up. For this, the hydraulic results from the reference case are used. The accuracy of the results is verified by several benchmarking criteria.

Theoretical background of the decoupling

In this subsection, the sequential approach is described in detail for a case where the release of a radionuclide from a disposal cell is modelled. The coupled two-phase flow and transport simulation is performed with the TOUGH2-MP EOS75R module. EOS75R is a derivate of EOS7R (Pruess *et al.* 1999; Zhang *et al.* 2008) where air has been replaced by hydrogen (Kämpfer *et al.* 2014). For the single-phase transport simulation, the module EOS9nT (Moridis *et al.* 1999) of TOUGH2 was used. The compatibility between two-phase hydraulics and single-phase unsaturated transport has been studied carefully and is illustrated in the following subsections.

Compatibilities between two-phase hydraulic and unsaturated transport

The mass conservation equation used in TOUGH2 is written as (Pruess *et al.* 1999):

$$\frac{\partial}{\partial t}\int_{V_n} M^{\kappa}\,\mathrm{d}V_n = \int_{\Gamma_n} \boldsymbol{F}^{\kappa}\boldsymbol{n}\,\mathrm{d}\Gamma_n + \int_{V_n} Q^{\kappa}\,\mathrm{d}V_n \qquad (1)$$

where the mass accumulation term M^{κ} for component κ is:

$$M^{\kappa} = \phi\sum_{\alpha} S_{\alpha}\rho_{\alpha}X_{\alpha}^{\kappa} \qquad (2)$$

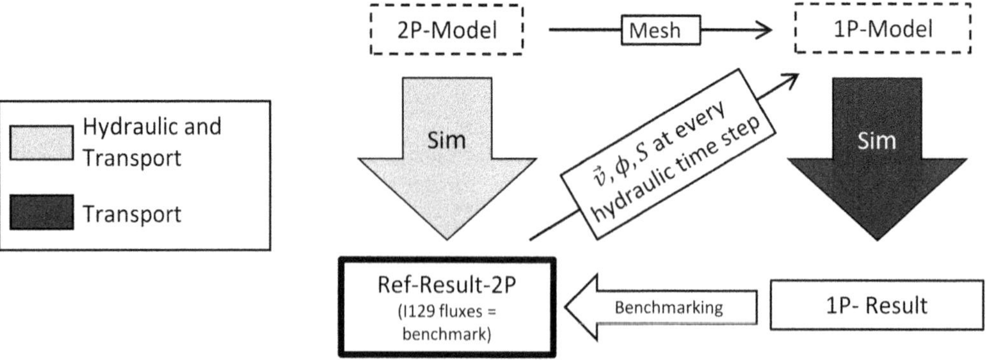

Fig. 1. Schematic representation of a decoupled sequential approach for hydraulic and transport simulation.

and the advective transport term, $\boldsymbol{F}^{\kappa}$, is defined as:

$$\boldsymbol{F}^{\kappa}|_{\mathrm{adv}} = \sum_{\alpha} X_{\alpha}^{\kappa} \rho_{\alpha} \boldsymbol{v}_{\alpha} \quad (3)$$

where V_n is an arbitrary subdomain, Γ_n is a closed surface, $\boldsymbol{n}$ is an inwards-pointed normal vector on surface element $\mathrm{d}\Gamma_n$, Q^{κ} are sinks and sources of component κ, ϕ is the porosity, S_{α} is the saturation of phase α, ρ_{α} is the density of phase α, X_{α}^{κ} is mass fraction of component κ in phase α and $\boldsymbol{v}_{\alpha}$ is the Darcy velocity of the phase α.

Darcy's law, as used in TOUGH2 for the computation of the multiphase flow, is described by (Pruess *et al.* 1999):

$$\boldsymbol{v}_{\alpha} = -k \frac{k_{\mathrm{r}\alpha}}{\mu_{\alpha}} (\nabla p_{\alpha} - \rho_{\alpha} \mathbf{g}) \quad (4)$$

where k is the permeability, $k_{\mathrm{r}\alpha}$ is the relative permeability, μ_{α} is the viscosity, $\mathbf{g}$ is the standard gravity vector and p_{α} is the fluid pressure in phase α.

The flux of a component in the liquid phase may have contributions from advection, diffusion/dispersion and surface diffusion (of an adsorbed component), and is given by (Moridis *et al.* 1999):

$$\boldsymbol{F}^{\kappa} = \boldsymbol{F}^{\mathrm{w}} X_{\mathrm{w}}^{\kappa} - \rho_{\mathrm{w}} \boldsymbol{D}^{\kappa} \nabla X_{\mathrm{w}}^{\kappa} - \boldsymbol{F}_{\mathbf{s}}^{\boldsymbol{\kappa}} \quad (5)$$

where $\boldsymbol{F}^{\mathrm{w}}$ is the advective flux of liquid water, $\boldsymbol{D}^{\kappa}$ is the dispersion tensor, which also accounts for diffusion, and $\boldsymbol{F}_{\mathbf{s}}^{\boldsymbol{\kappa}}$ is surface diffusion.

To solve the transport problem, the Darcy velocity of the liquid phase, $\boldsymbol{v}_{\mathrm{w}}$, the liquid saturation field, S_{w}, and the porosity, ϕ, are needed. The water flux, $\boldsymbol{F}^{\mathrm{w}}$, and the dispersion tensor, $\boldsymbol{D}^{\kappa}$, are functions of the Darcy velocity, $\boldsymbol{v}_{\mathrm{w}}$. The dispersion tensor, $\boldsymbol{D}^{\kappa}$, itself and the mass fraction, X_{w}^{κ}, are functions of the liquid saturation, S_{w}, and the porosity, ϕ. The surface diffusion, $\boldsymbol{F}_{\mathbf{s}}^{\boldsymbol{\kappa}}$, accounts for the migration of adsorbed species and is a function of the saturation, S_{w} and the porosity, ϕ.

Physically, there are no inconsistencies between formulating the transport problem either based on two-phase flow and transport equations or on a single-phase transport equation using hydraulic data from the two-phase simulation. The only restriction is that radionuclide transport in the gas phase remains negligible and the considered radionuclide has no impact on the fluxes.

Implementation of the sequential approach

During the two-phase gas–water flow and transport simulation, the following data were saved at each time step:

- The Darcy velocity field of the liquid phase, $\boldsymbol{v}_{\mathrm{w}}$, which determines the advective transport of the solute radionuclides.
- The liquid saturation field (S_{l}). The storage term of the transport equation is computed from the liquid saturation field (S_{l}), the pore volume and the solute mass.
- The porosity field (ϕ). The porosity may vary in response to pressure and temperature changes, and therefore the porosity field has to be recorded. If both effects are neglected, the porosity does not have to be recorded.
- The temperature field (T) has to be saved if conditions are non-isothermal and the temperature-dependency of solute diffusion is taken into account.

The single-phase transport simulation was set up as described in the following. As mentioned above, the single-phase transport simulation uses the results from the coupled two-phase flow and transport simulation. The single-phase transport code was modified in order to read the two-phase results. In particular, the time-stepping of the transport simulation was constrained by the available two-phase results. If the transport simulation required finer time-stepping, then the two-phase data were interpolated in-between two given two-phase results. In order to be able to compare the sequential transport simulations to the two-phase coupled simulation, the diffusion models in both the TOUGH2 modules (EOS9nT and EOS75R) were extended and matched. Furthermore, it is crucial to use equal sorption models.

Optimization of the spatial discretization

In the second part of the study, the sensitivity of the results to the spatial discretization was evaluated. We used the result from the coupled two-phase flow and transport simulation to define locations that allow for a coarsening without significantly reducing the accuracy. The aim of the coarsening was to reproduce relevant key features (as discussed below) of the results of the reference case as exactly as possible. For this study, multiple coupled two-phase flow and transport simulations were performed.

The coarsening was performed stepwise: meaning that the number of elements is reduced further in each step. Every coarsening step is based on the previous one and derived from previous simulation results.

In some coarsening steps, a merging of different materials is necessary. If different materials are merged, then their properties need to be lumped. The lumping was carried out individually for each combination of materials, and for each of their properties depending on the dominant physical processes (which may depend on the radionuclide to be considered) at the specific location in the mesh. Two different averaging laws were applied in order to

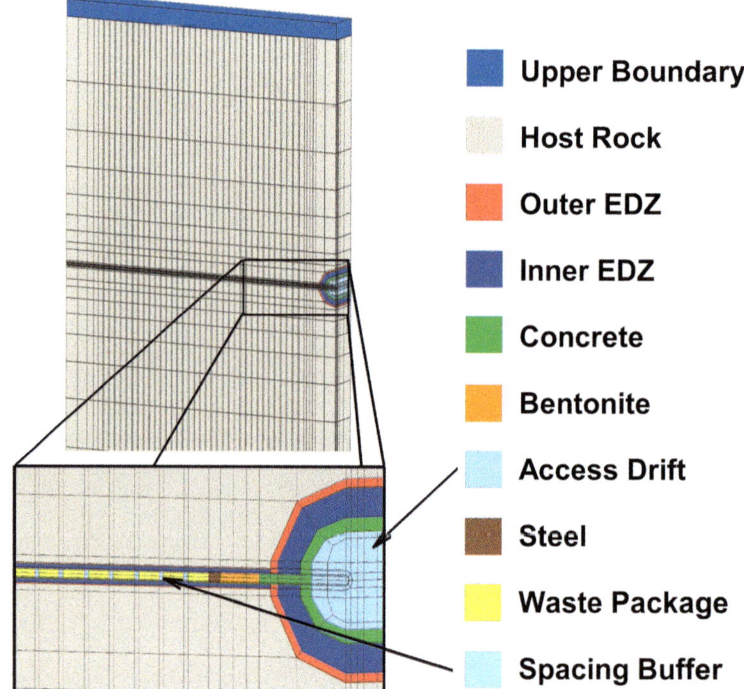

Fig. 2. Mesh with considered materials in the HLW-LL model.

lump the material properties. For parameters characterizing the storage of mass or heat, a weighted mean was used. The weighting factor for each parameter was derived such that the effective storage property of the lumped material parameter equals the sum of the storage properties of the individual material parameters to be lumped. However, for parameters characterizing the material's resistance with respect to the transfer of mass or heat, the weighting law depends on the spatial arrangement of the materials to be lumped with respect to the dominant transfer direction. Here, either the weighted

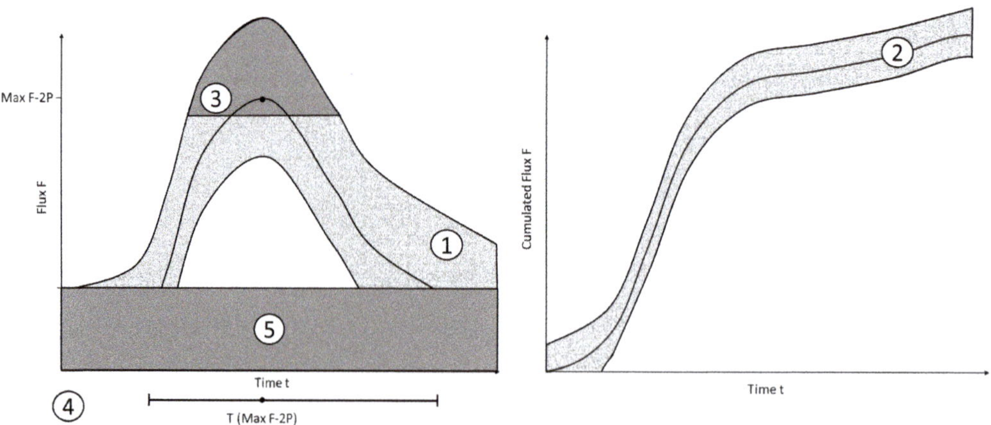

Fig. 3. Benchmark criteria for radionuclide flux over time (left) and cumulated flux over time (right). The following criteria are shown here: shape of the flux curve (1) and the cumulative flux curve (2); the maximum flux (3) and its time (4) of occurrence; negligibly small fluxes are not considered (5).

Table 1. *Definition of benchmark surfaces*

Surface	Description
BS01	Flux out of the excavation damaged zone of the disposal cell into the host rock
BS02	Flux out of the disposal cell into the excavation damaged zone at the disposal cell
BS03	Flux into the bentonite plug out of the steel plug
BS04	Flux out of the bentonite plug into the concrete plug
BS05	Flux out of the concrete plug into the access drift
BS06	Flux into the access drift out of the concrete lining

mean or the weighted harmonic mean is applied. For each property and configuration, the lumping law has to be derived individually based on the knowledge gained from the reference simulation. A more general law was not derived here.

Test case for the given study

The methodology was tested with the radionuclide iodine-129 (^{129}I), which is highly soluble. ^{129}I is released from both HLW-LL and ILW-LL in a deep geological repository situated in the Callovo-Oxfordian Clay Formation in France. The numerical simulations were executed using TOUGH2.

A reference model was built using a detailed representation of the geometry and by accounting for all relevant physical processes governing the radionuclide transport from the time of closure of the repository up to 1 myr. Both optimizations, the decoupling and the spatial optimization, were tested for an emplacement drift with surrounding host rock in two different areas in the repository: the HLW-LL and the ILW-LL. Furthermore, various sensitivity tests were performed for both areas. The sensitivities with respect to variations of the flow and transport parameters were investigated. In one additional study, the source term of the radionuclide was modified. The sensitivity study is not described in detail: however, a brief overview of the results is given in the course of this paper.

For brevity, we address only the HLW-LL cell, which is shown in Figure 2. The HLW-LL cell is surrounded by the host rock and connected to the repository system by an access drift. The disposal cell is a dead-end tunnel, which connects to the access drift and has a pipe-like shape. The lateral boundaries of the model domain are located in the host rock besides the access drift and the dead end, respectively. The diameter of the disposal cell is about 0.7 m and has a usable length of around 80 m. It extends for another 10 m, enclosing the different sealing components at the end of the cell closest to the access drift. The disposal cell is drilled from the access drift and lined with a metal sleeve. After the installation, the disposal cell is filled intermittently with several barrel-shaped disposal packages carrying waste and thermal spacing buffers. Two different materials, namely concrete and bentonite, seal off the disposal head.

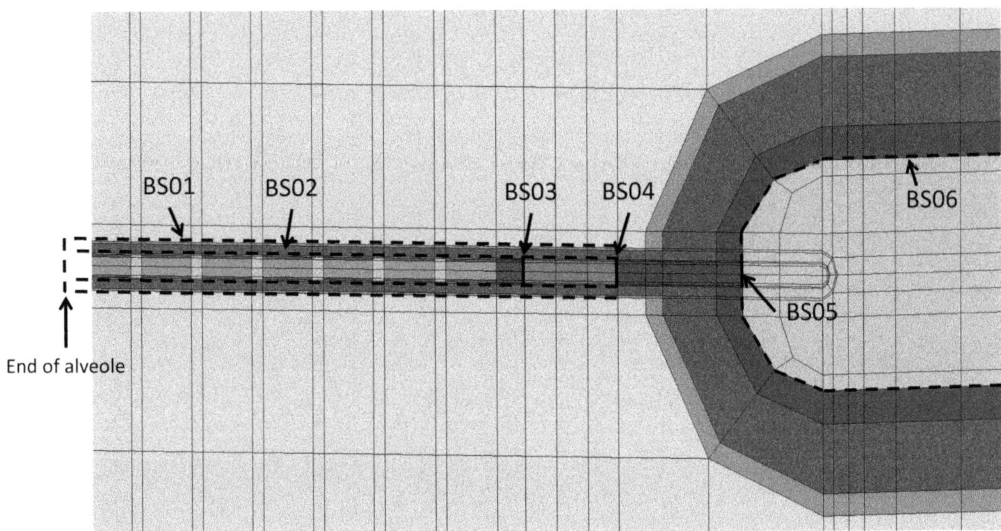

Fig. 4. Benchmark surfaces defined for the HLW-LL disposal drift.

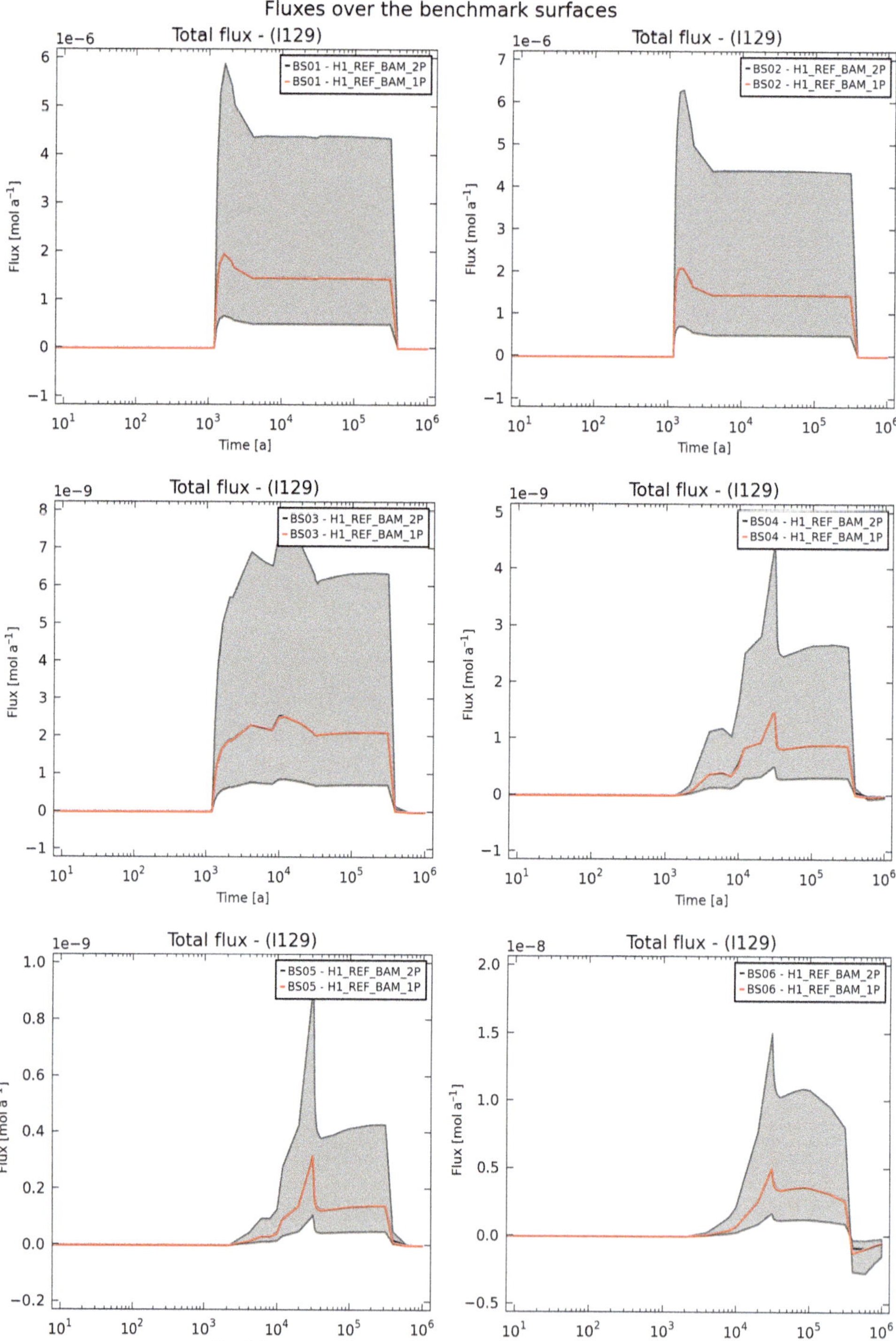

Fig. 5. Flux at the benchmark surfaces for the fully coupled (H1_REF_BAM_2P) and the sequential simulations (H1_REF_BAM_1P). The grey shaded area represents the benchmark criteria 1. The red and black lines are overlapping (and, thus, the black line is not visible).

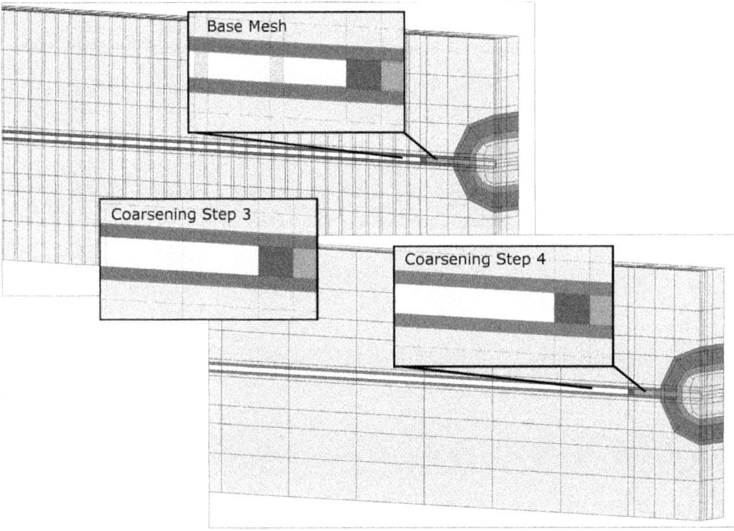

Fig. 6. Illustration of the coarsening steps (base mesh as an overview and zoom; coarsening step 3 only as zoom; coarsening step 4 as an overview and zoom).

Owing to identical geometries, identical waste inventories, and more-or-less synchronous drilling, loading and sealing of the disposal cells, the thermo-hydraulic condition and the radionuclide transfer are considered to be symmetrical. Therefore, it is reasonable to position the vertical model boundaries: (i) along the longitudinal disposal cell axis; (ii) at half the distance between two neighbouring HLW-LL disposal cells; (iii) along the longitudinal access drift axis; and (iv) at half the distance between two opposing disposal cell ends. The four vertical boundaries are considered to be no-flow boundaries. The upper and lower boundaries of the model domain are given by the upper and lower limit of the Callovo-Oxfordian host rock. They are defined as Dirichlet boundary conditions with constant hydraulic head and temperature, and fully saturated conditions with a radionuclide concentration equal to zero.

Table 2. *Coarsening steps with number of elements and computation time*

Mesh	Short	Elements	Computation time (up to 16 cores)	
Base	BAM	12.246	37.845 s	10.5 h
Coarse 1	CS01	11.821	15.525 s	4.3 h
Coarse 2	CS02	6.816	6.844 s	1.9 h
Coarse 3	CS03	6.616	5.911 s	1.6 h
Coarse 4	CS04	6.416	1.938 s	0.5 h
Coarse 5	CS05	2.636	892 s	15 min
Coarse 6	CS06	2.585	811 s	13.5 min

Ventilation in the access drift is considered only during the operational phase in the first 10 years of the simulation. The ventilation leads to a desaturation in the engineered compartments and the surrounding host rock. Heat production is considered as soon as the disposal packages have been placed in the disposal drift. Thus, the model is non-isothermal. Owing to corrosion of the metallic compartments in the disposal cell, hydrogen is produced and this leads to further desaturation of the system. The considered radionuclide ^{129}I can be adsorbed by the concrete compartments.

Benchmarking

To judge the quality of the results, benchmarking criteria need to be defined, which check that the results respect the accuracy requirements for the PA/SA study in question. Since common PA/SA indicators are defined based on fluxes, cumulative fluxes or maximum fluxes, the benchmarking criteria were chosen to evaluate the quality of these entities. The exact definition of the benchmarking criteria depends on the exact goals of the PA/SA study and on how conservative results are allowed to be. The acceptable range for the present study for the single-phase transport results is illustrated in Figure 3 by the grey shaded areas. The benchmark criteria are fulfilled if the results of the single-phase transport simulation lie within this area.

The defined benchmarking criteria consider fluxes, cumulative fluxes over time and maximal fluxes. These fluxes are observed at characteristic

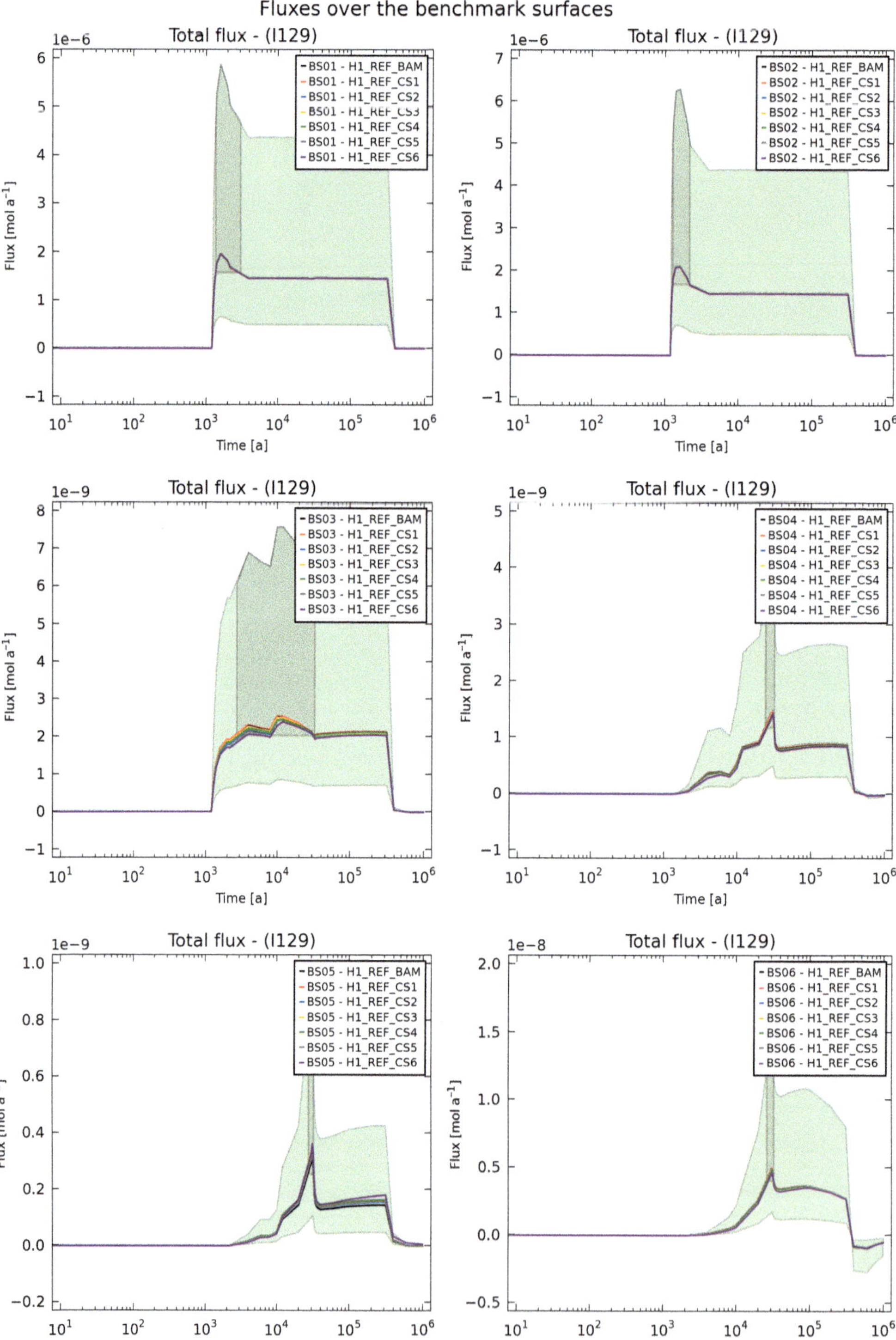

Fig. 7. Flux at all six benchmark surfaces for coarsening steps 1–6, with the benchmark criteria 3 and 4 in dark green, and benchmark criterion 1 in light green.

interfaces in the model domain. The single-phase flux has to lie within a range of factor 3 of the two-phase flux (1). Further, the cumulative flux has to lie within a range of 20% of the two-phase cumulative flux (2). The maximum single-phase flux (3) and its time of occurrence (4) are further important safety indicators in PA/SA studies. Therefore, the peak of the single-phase transport results is required to be a minimum 80% of the two-phase peak (Max F-2P) and to occur within a range of factor 2. Negligibly small fluxes are generally not considered (5).

The characteristic interfaces where we evaluated the benchmark criteria for the HLW-LL cell are listed in Table 1 and are shown in Figure 4.

Results of the sequential approach

The sequential approach was verified using the described benchmark criteria.

Figure 5 shows that the radionuclide fluxes across all the benchmarking surfaces are in almost exact agreement with the reference case. The minor discrepancies in the resulting curves are due to small differences in the output times, resulting from numerical error.

Results of the spatial optimization

For the test case 'HLW-LL cell', six coarsening steps were defined as illustrated selectively in Figure 6.

The number of processors used was chosen to optimize performance. For a small number of mesh elements and a large number of processors, the cost of communication between the processors can exceed the cost of the actual simulation.

Table 2 illustrates that a slight decrease in the number of elements can lead to a significant reduction in computation time. This is not only due to a reduction in the number of elements but also to an optimization of the representation with respect to the numerical algorithms.

If, for example, we take a look at coarsening steps 3 and 4, we can see that a reduction of only 200 elements leads to a reduction in the computation time of two-thirds. This significant acceleration results from an optimized representation near the interface between the waste canister and steel liner, and reduces numerical convergence issues with TOUGH2 when a phase change resulting from a transition between the two-phase and single-phase conditions in high-porosity materials occurs.

It is important to verify that such optimizations do not impact on the quality of the results. Thus, the spatial optimization has also been verified using the defined benchmark criteria.

Figure 7 shows that the fluxes at all six benchmarking surfaces are in good agreement for each coarsening step. Therefore, for this particular model, we have shown that the computation time can be reduced significantly by local grid coarsening without losing accuracy.

Conclusions

Our results show that the applied methodology for going from a two-phase flow simulation to a single-phase transport calculation is very accurate. This implies that this methodology can be applied for PA/SA studies, given that the radionuclides are highly soluble. Furthermore, our results suggest that local grid coarsening can significantly reduce computational costs. Both aspects of the study, the decoupling and the optimized spatial discretization, have been verified for different areas in the repository, although this paper has presented results for only one area.

With the sensitivity analyses, we demonstrated that the sequential approach provides accurate results for a wide range of simulation parameters, such as the diffusion coefficient. However, after local grid coarsening, the results are more sensitive to variations in the simulation parameters.

The presented sequential approach leads to a significant reduction in the computational effort, whilst maintaining very good accuracy. This developed approach is used and pursued in ANDRA's PA/SA programme.

References

Brommundt, J., Kaempfer, Th.U., Enssle, C.P., Mayer, G. & Wendling, J. 2014. Full-scale 3D modelling of a nuclear waste repository in the Callovo-Oxfordian clay. Part 1: thermo-hydraulic two-phase transport of water and hydrogen. *In*: Norris, S., Bruno, J. et al. (eds) *Clays in Natural and Engineered Barriers for Radioactive Waste Confinement.* Geological Society, London, Special Publications, **400**, 443–467, https://doi.org/10.1144/SP400.34

Enssle, C.P., Brommundt, J., Kaempfer, Th.U., Mayer, G. & Wendling, J. 2014. Full-scale 3D modelling of a nuclear waste repository in the Callovo-Oxfordian clay. Part 2: thermo-hydraulic two-phase transport of water, hydrogen, ^{14}C and ^{129}I. *In*: Norris, S., Bruno, J. et al. (eds) *Clays in Natural and Engineered Barriers for Radioactive Waste Confinement.* Geological Society, London, Special Publications, **400**, 469–481, https://doi.org/10.1144/SP400.35

Kämpfer, Th.U., Mishin, Y., Brommundt, J., Roger, J., Treille, E. & Hubschwerlen, N. 2014. Extension and tuning of TOUGH2-MP EOS7R for the assessments of deep geological repositories for nuclear waste: hydrogen, arbitrarily long decay chains, and solubility limits. *Nuclear Technology*, **187**, 131–146, https://doi.org/10.13182/NT13-80

Moridis, G., Wu, Y.-S. & Pruess, K. 1999. *EOS9nT: A TOUGH2 Module for the Simulation of Water Flow and Solute/Colloid Transport in the Subsurface*. LBNL-42351. Lawrence Berkeley National Laboratory, Berkeley, CA.

Pruess, K., Oldenburg, C. & Moridis, G. 1999. *TOUGH2 User's Guide, Version 2.0*. LBNL-43134. Lawrence Berkeley National Laboratory, Berkeley, CA.

Schaedle, P., Hubschwerlen, N. & Class, H. 2014. Optimizing the modeling performance for safety assessments of nuclear waste repositories by approximating two-phase flow and transport by single-phase transport simulations. *Nuclear Technology*, **187**, 188–197, https://doi.org/10.13182/NT13-82

Zhang, K., Wu, Y.-S. & Pruess, K. 2008. *User's Guide for TOUGH2-MP – A Massively Parallel Version of the TOUGH2 Code*. Earth Sciences Division, Lawrence Berkeley National Laboratory, Berkeley, CA.

Coupled THM modelling of engineered barriers for the final disposal of spent nuclear fuel isolation

ERDEM TOPRAK[1]*, SEBASTIA OLIVELLA[2] & XAVIER PINTADO[3]

[1]*International Centre for Numerical Methods in Engineering (CIMNE), Universidad Politécnica de Cataluña, Campus Norte UPC, 08034 Barcelona, Spain*

[2]*Department of Geotechnical Engineering and Geosciences, Universidad Politécnica de Cataluña, Campus Norte UPC, 08034 Barcelona, Spain*

[3]*B + TECH Oy, Laulukuja 4, 00420 Helsinki, Finland*

**Correspondence: erdem.toprak@upc.edu*

Abstract: This paper describes the thermohydromechanical (THM) simulation of engineered barrier systems (EBS) for the final disposal of nuclear spent fuel in Finland. The bentonite barriers were simulated with the Barcelona Basic Model and the model was calibrated from laboratory tests. The evolution of gap closure and the presence of a fracture intersecting the disposal were analysed. The simulations were performed in 2D axisymmetrical geometries. Full 3D simulations were carried out in order to check the effect of the third dimension. The time required for the barriers to reach full saturation, the maximum temperature, deformations and displacements at the buffer–backfill interface and the homogenization of components both locally and globally are the main interests. The effect of rock fracture and the hydraulic conductivity of the rock are subjected to 2D sensitivity analyses.

The final disposal of nuclear spent fuel in crystalline bedrock is developing in Finland. Olkiluoto bedrock was proposed as the site for the repository (Posiva 2013). The disposal is based on the use of multiple release barriers, which ensures that the nuclear spent fuel cannot be released into the biosphere or become accessible to humans. The release barriers include the physical state of the fuel, the disposal canister, the bentonite buffer, the backfilling of the tunnels and the surrounding rock.

The KBS-3V concept (Posiva 2013) consists of deposition holes spaced between 7.5 and 11 m of each other in backfilling tunnels excavated at a depth of approximately 400–450 m and located at approximately 25 m from each other (Fig. 1). The spent nuclear fuel will be encapsulated in final disposal canisters made of cast iron and enclosed in a copper shell. These canisters will be placed in the deposition holes and surrounded with bentonite clay, which protects the canister from any potential jolt in the bedrock and slows down the movement of water in the proximity of the canister.

Mechanical constitutive model for clay-based materials

In this section, constitutive model calibration is described. The calibration was performed by modelling the results from oedometer tests and infiltration tests in buffer and backfill materials: MX-80 bentonite (buffer blocks: Kiviranta & Kumpulainen 2011; Juvankoski *et al.* 2012; Pintado & Rautioaho 2013), Friedland clay (backfill blocks: Keto *et al.* 2013; Kiviranta *et al.* 2016) and pellets (Juvankoski *et al.* 2012).

In this paper, buffer (MX-80 bentonite), backfill (Friedland clay) and pellets are modelled using the Barcelona Basic Model (BBM) originally described in Alonso *et al.* (1990). This model is briefly described here. Elastic, isotropic, non-isothermal volumetric strains are defined by:

$$d\varepsilon_v^e = \frac{\kappa_i}{1+e}\frac{dp'}{p'} + \frac{\kappa_s}{1+e}\frac{ds}{s+0.1} + \alpha\, dT, \quad (1)$$

where e is the void ratio, κ_i, κ_s and α are parameters, p' is net mean stress, s is suction and T is temperature. For deviatoric elastic strains, a constant Poisson ratio is used.

Plasticity is accounted for using a Modified Cam-Clay yield surface (and plastic potential) with the following equation:

$$F = q^2 - M^2(p' + p_s)(p_0 - p') = 0, \quad (2)$$

where $p_s = ks$ and p_0 correspond to the intersection of the ellipse with the p'-axis, q is the equivalent

From: NORRIS, S., BRUNO, J., VAN GEET, M. & VERHOEF, E. (eds) 2017. *Radioactive Waste Confinement: Clays in Natural and Engineered Barriers*. Geological Society, London, Special Publications, **443**, 235–251.
First published online September 26, 2016, https://doi.org/10.1144/SP443.19

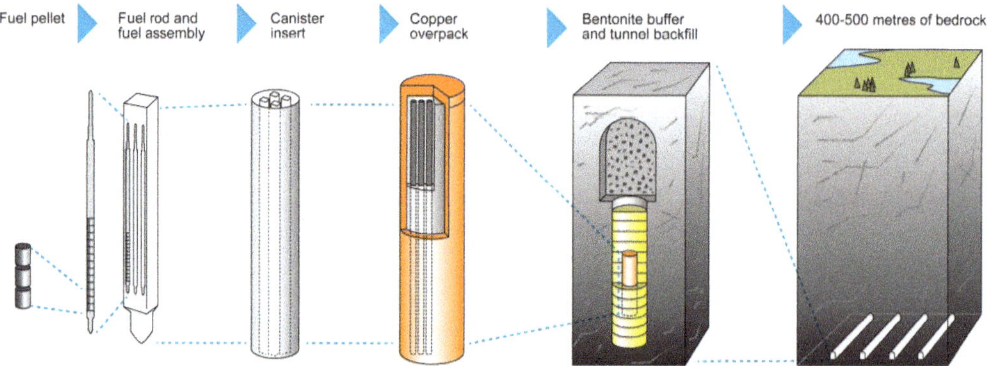

Fig. 1. A possible design of the final disposal facility (KBS-3) planned to be constructed at the Olkiluoto site.

shear stress, and M and k are parameters. The preconsolidation mean net stress (p_o) is a function of suction in the BBM with the following form:

$$p_o(s) = p^c \left(\frac{p_0^*}{p^c} \right)^{\frac{\lambda(0)-\kappa}{\lambda(s)-\kappa}}, \quad (3)$$

where p^c is the reference stress of the loading-collapse curve, p_0^* is the initial yield mean net stress, $\lambda(0)$ is the slope of the virgin elastoplastic compressibility for saturated conditions of the soil and κ is the slope of the unload–reload line (elastic response). The function $\lambda(s)$ is the slope of the virgin elastoplastic compressibility for a given suction (volumetric compressibility index), written as:

$$\lambda(s) = \lambda(0)[(1-r)\exp(-\beta s) + r], \quad (4)$$

where r and β are parameters.

Finally, the hardening law for the BBM is given as:

$$\mathrm{d}p_0^* = \frac{1+e}{\lambda(0)-\kappa_i} p_0^* \, \mathrm{d}\varepsilon_v^p, \quad (5)$$

where $\mathrm{d}\varepsilon_v^p$ is the plastic volumetric strain increment.

Hydraulic and thermal constitutive models

The fluid flow is governed by Darcy's law, given as:

$$\mathbf{q}_l = -\frac{\mathbf{k}k_{rl}}{\mu_l}(\nabla P_l - \rho_l \mathbf{g}), \quad (6)$$

where $\mathbf{q}_l$ is the volumetric flux of liquid, $\mathbf{k}$ is the intrinsic permeability tensor, k_{rl} is the phase relative permeability, μ_l is the viscosity of the fluid, P_l is the pressure of the fluid and ρ_l is the density of the fluid. Relative permeability is calculated as a power of the degree of saturation:

$$k_{rl} = S_l^m, \quad (7)$$

where S_l is the degree of saturation and m is a parameter. The degree of saturation for the liquid phase is calculated using the retention curve with the relationship of Van Genuchten (1980); this is written as:

$$S_e = \frac{S_l - S_{rl}}{S_{ls} - S_{rl}} = \left[1 + \left(\frac{P_g - P_l}{P} \right)^{1/(1-\lambda)} \right]^{-\lambda}, \quad (8)$$

where S_e is the effective degree of saturation of porous media, S_l is the degree of saturation of liquid, S_{rl} is the residual degree of saturation, S_{ls} is the maximum degree of saturation, P_g is the gas pressure, P_l is the liquid pressure, λ is the shape function coefficient for the retention curve and P is a parameter that can be interpreted as the pressure of air entrance.

The molecular diffusion of vapour is governed by Fick's law:

$$\mathbf{i}_g^w = -\left(\tau\phi\rho_g S_g D_g^w \mathbf{I} \right) \nabla\omega_g^w, \quad (9)$$

where $\mathbf{i}_g^w$ is the non-advective mass flux vector, ϕ is the porosity of porous media, ρ_g is the density of the gas phase, S_g is the degree of saturation of the gas phase, D_g^w is the molecular diffusion coefficient for vapour in the gas phase, ω_g^w is the mass fraction of vapour in the gas phase, $\mathbf{I}$ is the identity tensor and τ is a tortuosity coefficient.

The diffusion coefficient of vapour is given by:

$$D_g^w = D^v \frac{(273.15+T)^n}{P_g}, \quad (10)$$

where D^v is the coefficient of diffusion (5.9×10^{-6} $\mathrm{m^2\,s^{-1}\,K^{-n}\,Pa}$) and n is a parameter ($n = 2.3$).

For the air gap element, it is convenient that retention properties evolve as the gap closes. In order to represent this response using the available functions in CODE_BRIGHT, the dependence of the retention curve on porosity has been taken into account when considering the following equation:

$$P(\phi) = P\exp\left(a(\phi_0 - \phi)\right), \quad (11)$$

where a is a parameter. This function allows a low retention capacity for the gap to be used when it is open (very small saturation), which evolves into a higher retention capacity for the gap as it closes (closer or similar to what occurs in the surrounding clay).

The heat transfer process is governed by Fourier's law, given as the heat flux vector:

$$\mathbf{i}_c = -\lambda \nabla T, \quad (12)$$

where thermal conductivity depends on the degree of saturation in the following way:

$$\lambda = \lambda_{sat} S_l + \lambda_{dry}(1 - S_l), \quad (13)$$

where $\mathbf{i}_c$ is the conductive flux vector of heat, T is the temperature, λ is the thermal conductivity, λ_{sat} is the thermal conductivity of the water-saturated porous medium, λ_{dry} is the thermal conductivity of the dry porous medium and S_l is the degree of saturation.

For the thermal analysis, the required parameters for the materials are the thermal conductivity (λ), the specific heat (c_s) and the solid density (ρ_s).

Material properties

Friedland clay was chosen as the reference material for manufacturing the backfill blocks (Keto *et al.* 2013). Although the backfill consists of different components (i.e. a foundation bed, blocks and pellets) (Fig. 2), it is treated as a single material.

It was planned that the pellets be installed between the buffer blocks and rock (Juvankoski *et al.* 2012), and that the pellets would occupy a volume of 10–20% of the excavation volume associated with the deposition holes. It is also anticipated that the pre-compacted buffer blocks would swell and compress the pellet component. As a result of this, it was anticipated that the buffer block–pellet material in the tunnel should meet the performance requirements of the buffer described in Juvankoski *et al.* (2012).

In many aspects, it can be said that the pellet component is, in fact, a simple filler of the space between the buffer and the rock. There are several reasons for using fillers in gap. Fillers can secure better thermal conductivity and prevent spalling of the surrounding rock into the deposition hole. Also fillers might be needed in order to retain an adequate average density (Marjavaara & Kivikoski 2011).

In order to determine the hydromechanical parameters of the buffer (MX-80) and backfill (Friedland clay), an experimental programme was followed. Table 1 summarizes the initial properties of the materials for each test considered. For each material, an oedometer and an infiltration test was performed.

Figure 3 shows the calibration results from the oedometer test data for Friedland clay (backfill) (Fig. 3a) and MX-80 bentonite (buffer) (Fig. 3b). As the samples were initially unsaturated, there was a stabilization (flooding) process at the beginning of the oedometer tests. It can be seen that both materials showed considerable expansive behaviour before loading. Once the saturation took place, samples were exposed to loading and unloading phases. In this study, the Barcelona Basic Model (Alonso *et al.* 1990) parameters used in Toprak *et al.* (2013) for MX-80 bentonite clay have been updated. The BBM parameters for MX-80 bentonite and Friedland clay were obtained by calibration of the oedometer tests. For the pellets, a preliminary set of parameters that take into account non-linear elasticity was established. The pellets represent a small volume fraction of the total volume of the clay barrier. The calibrated BBM parameters that were used in the 2D and 3D modelling of MX-80 bentonite, Friedland clay and the pellets are given in Table 2.

Figure 4 shows the final distribution of the water content and the infiltration tests set up for Friedland clay (Fig. 4a) and MX-80 bentonite (Fig. 4b) at the end of the experiment. The infiltration cell was made of stainless steel. The inner diameter of the infiltration cell was 50 mm and the height of the sample was approximately 63 mm. There were two sintered porous frits made of stainless-steel balls, with a diameter of 10 μm, at the top and bottom of the sample. The materials (MX-80 bentonite and Friedland clay) were placed directly inside the cells. During the tests, a target water pressure was imposed at the bottom of the sample using a pressure–volume controller manufactured by GDS (http://www.gdsinstruments.com). On the other side, the water pressure was zero (the atmospheric pressure was fixed as the origin). The tests were performed at laboratory temperature (approximately 24°C). The suction of the materials was measured before the tests using a chilled mirror psychrometer WP-4 (http://www.decagon.com). The sample volume was kept constant during the test. The water inlet flow was measured continuously. At the end of the test, the sample was cut into slices with a saw. After cutting the sample, the water content was measured.

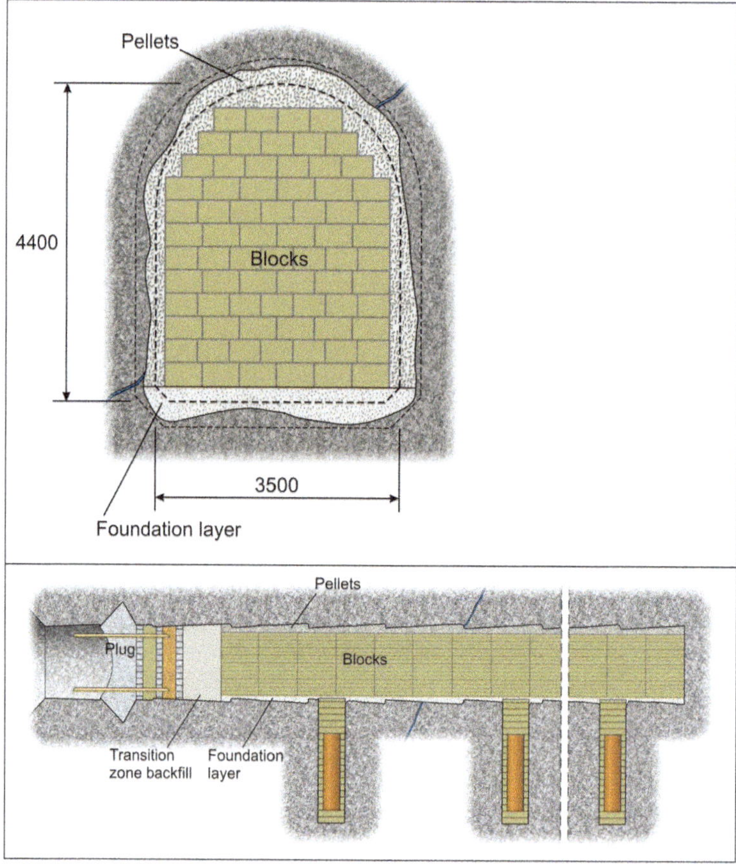

Fig. 2. Schematic figures showing the main backfill components (foundation layer, backfill blocks and pellets).

The swelling near the injection zone is possible because other parts of the sample undergo compression. Some level of suction is still measured at the end of the test. This could be due to the fact that the sample was not fully saturated, and dismantling of the sample with an associated unloading induces some suction. The hydraulic parameters of MX-80 bentonite and Friedland clay were calibrated from these infiltration tests and are listed in Table 3.

The host rock (Hagros *et al.* 2003) and the canister (Raiko 2012) are considered to be linear elastic with parameters E, ν and α, respectively, for Young's (elastic) modulus, Poisson's ratio and the coefficient of thermal expansion. Hydraulic and mechanical properties of the host rock and canister are given in Tables 3 and 4, together with the properties of the gap.

Mechanical parameters for the air gap were proposed in Toprak *et al.* (2013). The 10 mm air gap

Table 1. *Initial conditions for materials in the infiltration and oedometer tests*

Material	Test	Initial dry density (kg m^{-3})	Initial water content (%)	Initial degree of saturation (%)	Test duration (days)
MX-80 bentonite	Oedometer	1600	6.04	25	191
Friedland clay	Oedometer	1950	8.9	59	243
MX-80 bentonite	Infiltration	1700	5.33	23.34	68
Friedland clay	Infiltration	1750	7.7	37	13

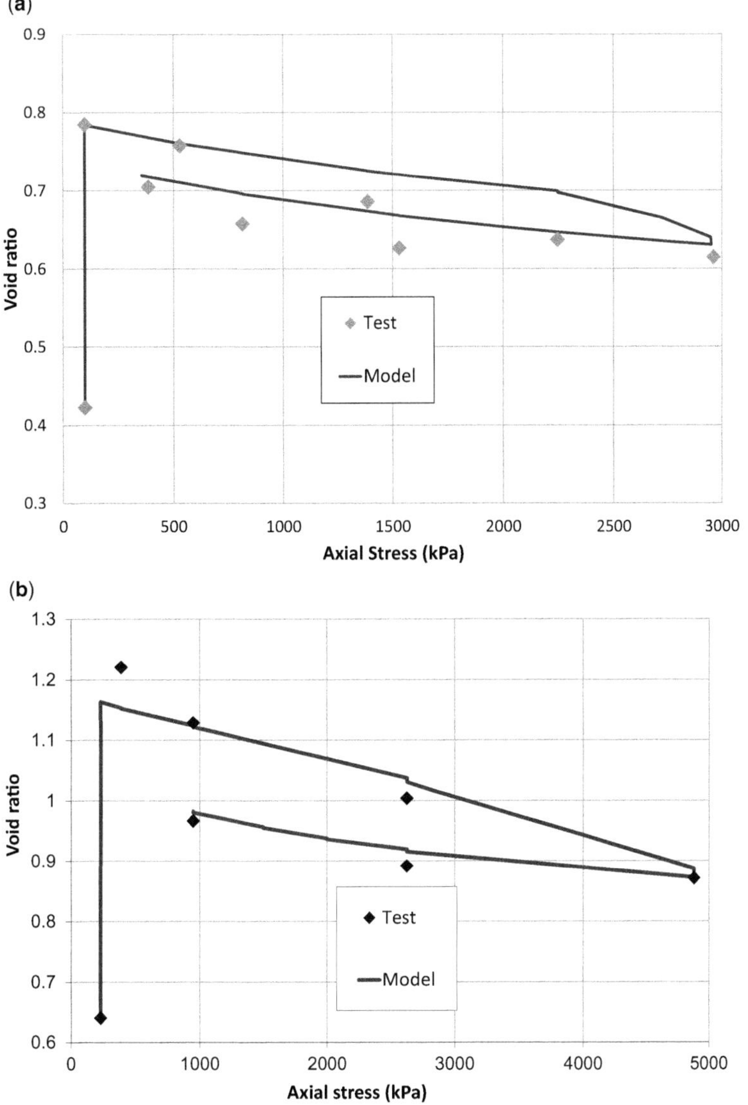

Fig. 3. Void ratio v. effective axial stress in (**a**) Friedland clay and (**b**) MX-80 bentonite. During an oedometer test, inundation of the sample produces a large initial swelling.

(between the canister and buffer) was modelled for simplicity using a bi-linear elasticity model that uses two values of Young's modulus: one for the opened gap and the other for the closed gap. A large value of Young's modulus (E_c) was used for the closed gap (representing the contact between gap surfaces) and a relatively low value (E_o) was used for the open gap. Compression strain was used to check whether the gap was open or closed. As, during the calculations, the gap undergoes a path from an initially open state towards a closed state as the buffer swells, the approximation considered was deemed to be adequate.

Finally, typical thermal properties for all the materials are given in Table 5.

Governing equations

The equation of equilibrium of forces (or stresses), the equation of mass balance of water and the equations of mass balance of energy were solved

Table 2. *BBM parameters for MX-80 bentonite, Friedland clay and pellets*

Parameter	Symbol	MX-80 bentonite	Friedland clay	Pellets
Poisson ratio (−)	ν	0.3	0.3	0.3
Minimum bulk module (MPa)	K_{min}	10	10	10
Parameters for elastic volumetric compressibility against mean net stress change (−)	κ_{i0}	0.09	0.05	0.03
Parameters for elastic volumetric compressibility against suction change (−)	κ_{s0}	0.09	0.05	0.03
Slope of void ratio – mean net stress curve at zero suction (−)	$\lambda(0)$	0.25	0.18	–
Parameters for the slope void ratio – mean net stress at variable suction (−, MPa^{-1})	r	0.8	0.8	–
	β	0.02	0.02	–
Reference pressure for the P_0 function (MPa)	p^c	0.1	0.1	–
Pre-consolidation mean stress for saturated soil (MPa)	P_0^*	2	2	–
Critical state line (−)	M	1.07	1.07	–

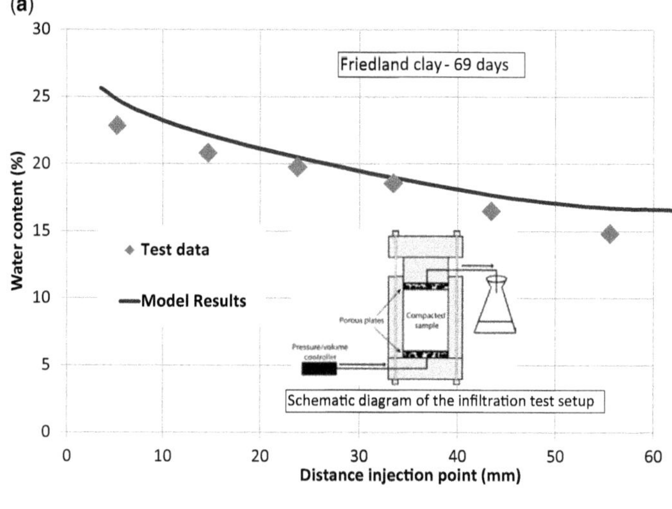

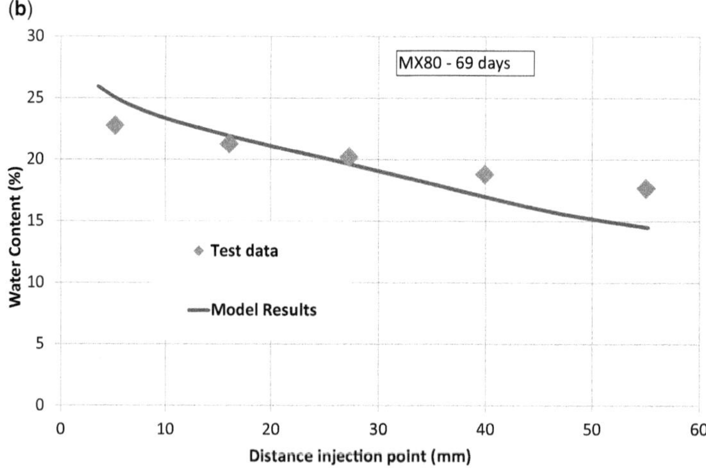

Fig. 4. Water content v. distance to the injection point for infiltration tests on (**a**) Friedland clay and (**b**) MX-80 bentonite.

Table 3. *Hydraulic parameters for materials*

Equation	Parameter	Rock	Friedland clay	MX-80 bentonite	Pellets	Gap element
Van Genuchten retention curve	P (MPa)	1.5	25	31.25	31.25	0.001
	λ (−)	0.3	0.4	0.5	0.5	0.5
	a (−)	–	–	–	–	10
Advective flux (Darcy)	k (m^2)	1.5×10^{-19} or 3.5×10^{-20}	1.6×10^{-20}	5.6×10^{-21}	5.6×10^{-20}	10^{-16}
	m (−)	3	3	3	3	3
Diffusive flux (Fick)	τ (−)	0.4	0.4	0.4	0.4	0.4

simultaneously to obtain displacements, pressure and temperature at each point and time of the model. The formulation can be found elsewhere: for instance, in Olivella *et al.* (1994, 1996).

Modelling of vertical disposal schemes

The model geometry and materials are shown in Figure 5a. This figure shows boundary conditions that are both hydraulic (Fig. 5b) and mechanical (Fig. 5c). The analysis assumed under axisymmetrical conditions.

The initial water pressure for all materials in the deposition hole was −41 MPa. The initial temperature was 10.5°C throughout the domain modelled. The rock had a hydrostatic water pressure (with $P_l = 4.36$ MPa at $z = 436$ m and $P_l = 3.98$ MPa at $z = 398$ m, where z is the depth below the surface), which was maintained along with corresponding boundary conditions. The temperature at the boundaries is seen to evolve with time (Fig. 6a). This temperature can be calculated from an analytical solution that takes into account the presence of all canisters (Ikonen 2003). An initial confining stress of 10.6 MPa was considered for the host rock. This confining pressure was also used as a boundary condition (applied on the top boundary).The lateral and bottom boundaries had prescribed normal displacements. During the excavation process (assumed to be 1 year), a prescribed liquid pressure (−5 MPa) was applied to the excavation surface to represent the process of ventilation. This boundary condition was removed when the buffer and backfill materials were emplaced. During the simulated time period, the hydrostatic liquid pressure was imposed on the top and bottom boundaries.

Regarding the power evolution of the canister, there are two main parameters playing a fundamental role, these are the residual power at the time of deposition and the decay rate. The work by Hökmark *et al.* (2009) was taken as a reference for the calculation of the power and the decay heat rate. The power as a function of time for an individual canister can be expressed as:

$$P(t) = P(0) \sum_{i=1}^{7} a_i \exp(-t/t_i). \qquad (14)$$

In this expression, $P(0)$ is the canister power at the time of deposition, and a_i, and t_i are parameters. Two parameter sets from SKB of power data relative to the time of cooling prior to disposal are given in Table 6. A burnup of 38 MWd/kg U (megawatt days per kg of uranium) is used as a reference.

The coefficients given in Table 6 are valid for an initial power level of 1837.3 W (in the case of 30 year-old fuel) and an initial power level of 1545.3 W (in the case of 40 year-old fuel). The work presently being performed is targeting a 1700 W initial power level at the time of deposition. The power function and the prescribed temperatures used in the models presented here are shown in Figure 6.

Table 4. *Mechanical parameters for air-gap element, rock and canister*

Parameter	Rock	Canister	Parameters	Gap element
E (MPa)	65 000	21 000	E_c (MPa)	1000
ν	0.25	0.3	ν	0.3
α (°C^{-1}), linear	10^{-5}	10^{-5}	α (°C^{-1}), linear	–
			E_o (MPa)	1
			Strain limit	0.95

Table 5. *Thermal parameters for materials*

Parameter	Rock	Canister	Backfill	Bentonite	Pellets	Gap element
ρ_s (kg m^{-3})	2749	7800	2780	2780	2780	–
ϕ_o	0.02	0.01	0.4604	0.438	0.669	0.8
c_s (J kg^{-1} K^{-1})	784	450	800	800	800	–
λ_{dry} (W m^{-1} K^{-1})	2.61	390	0.3	0.3	0.3	0.045
λ_{sat} (W m^{-1} K^{-1})	2.61	390	1.3	1.3	1.3	0.6

Values of 1700 W of canister power, 25 m of tunnel spacing and 11 m of canister spacing, and variable temperatures at the boundaries (Fig. 6a, b) were used to perform the thermohydromechanical (THM) calculations (see Toprak *et al.* 2013 for detailed information on thermal calculations under axisymmetrical conditions).

With the formulation and parameters decided, a series of THM calculations were carried out using a 2D axisymmetrical configuration, as well as using a 3D configuration. For the 2D axisymmetrical configuration, three cases described in Keto *et al.* (2013) were been considered. Each one has been given a name depending on the hydraulic properties chosen for the host rock, and one case includes a higher permeability zone representing a fracture. The cases are:

- Normal deposition hole (Normal_Case). The rock hydraulic conductivity is 1.52×10^{-12} m s^{-1} (equivalent to 1.52×10^{-19} m^2 of intrinsic permeability). Assuming water density of $\rho = 1000$ kg m^{-3}, water viscosity $\mu = 0.001$ Pa s and gravity $g = 10$ N kg^{-1}, gives the following conversion: $K = (\rho g/\mu)k = 1000 \times 10/0.001\ k$: that is, K (m s^{-1}) $= 10^7 k$ (m^2), where K is hydraulic conductivity and k is intrinsic permeability. The rock is considered

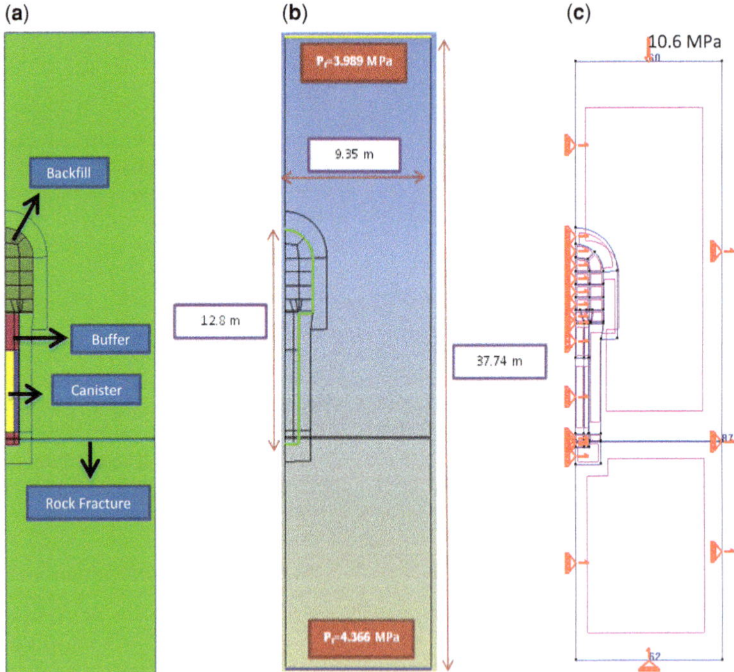

Fig. 5. (**a**) Geometry and materials, and (**b**) and (**c**) boundary conditions for the 2D models. Fracture is considered only in the wet case. The buffer consists of buffer blocks, buffer ring and pellets. Water pressure is prescribed only on the top and bottom of the model in (b), while vertical surfaces are assumed impermeable due to symmetry. Normal displacements are prescribed on the bottom and lateral surfaces, while a constant stress in applied on the top in (c).

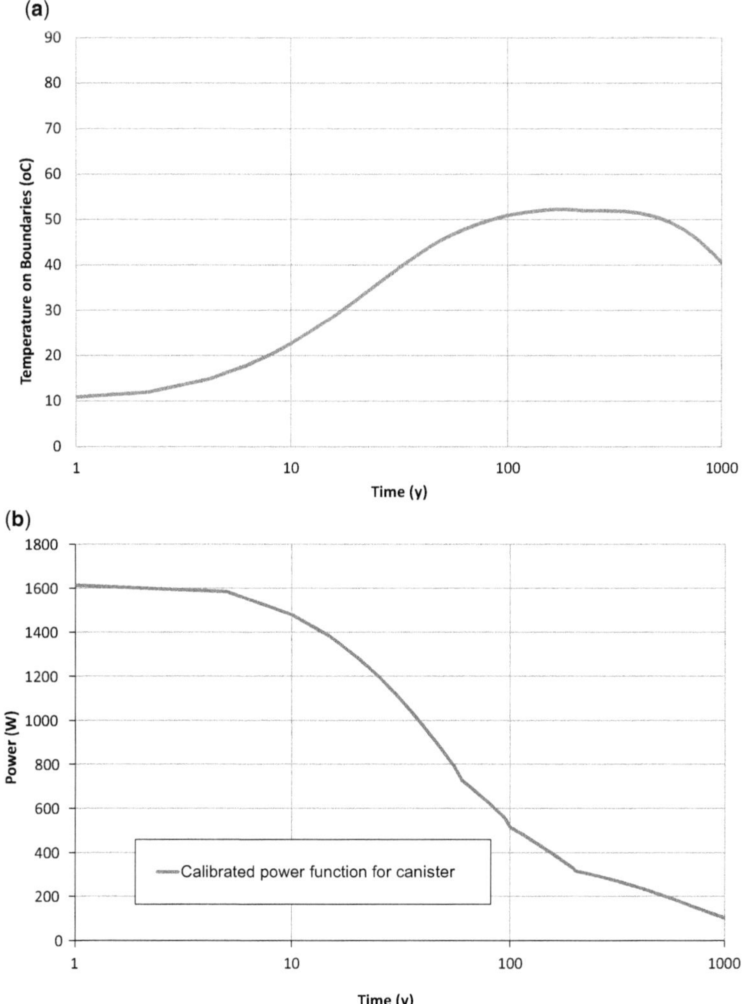

Fig. 6. (**a**) Temperature calculated from an analytical model (Ikonen 2003) prescribed on the top and bottom boundaries, and (**b**) power of the canister considered.

Table 6. *Parameters for the exponential equation for canister power calculations*

I	t_i (years)	a_i (30 years)	a_i (40 years)
1	20	0.070	0.049
2	50	0.713	0.696
3	200	−0.051	−0.059
4	500	0.231	0.271
5	2000	0.024	0.027
6	5000	−0.009	−0.010
7	20 000	0.022	0.026

For 30 year-old fuel, $P(0) = 1837.3$ W. For 40 year-old fuel, $P(0) = 1545.3$ W.

homogeneous: that is, no fracture is considered. This value of intrinsic permeability is considered to be a standard value for the rock at the repository conditions.

- Wet deposition hole (Wet_Case). In this case, the rock hydraulic conductivity is 1.52×10^{-12} m s^{-1} (equivalent to 1.52×10^{-19} m^2 of intrinsic permeability) and a predefined fracture was considered with a transmissivity of 1.1×10^{-9} m^2 s^{-1}. The fracture (high conductive zone) is horizontal because for the axisymmetrical geometry, it is not possible to represent inclined surfaces. The fracture thickness considered is 8 cm in order to avoid elements that are too small. The fracture is modelled with continuum elements. The equivalent hydraulic conductivity of the fracture is 1.37×10^{-8} m s^{-1} (1.37×10^{-15} m^2 of intrinsic permeability).

- Dry deposition hole (Dry_Case). The rock hydraulic conductivity is 3.53×10^{-13} m s^{-1} (equivalent to 3.53×10^{-20} m^2 of intrinsic permeability). In order to get drier conditions, the intrinsic permeability of the rock is reduced by approximately one order of magnitude, compared to the so-called normal case.

Figure 7 shows the hydraulic boundary conditions and intrinsic permeability for the three cases. In the case of the wet deposition tunnel, there is a zone with higher conductivity representing a fracture, and an appropriate boundary condition is applied at the end of the fracture. The temperature of water inflow through the fracture is the temperature in the lateral boundary (no heat-flow boundary condition). As indicated in Figure 7, the rock is less permeable in the dry deposition case compared to the wet and normal cases.

There are three critical zones that are considered relevant in this THM model, these are: (1) 'the buffer backfill interface' (Åkesson *et al.* 2010; Leoni 2013), where the vertical displacements show the largest values; (2) 'canister wall in contact with the buffer', the maximum temperature of the engineered barrier is reached on this surface (Ikonen 2003), and desaturation of the buffer takes place because of intense heating from the canister; and (3) 'buffer blocks under the canister', the maximum stresses are achieved here owing to the combination of the canister weight and swelling of the clay materials.

Results for 2D axisymmetrical modelling

Figure 8 shows the temperature evolution in the canister. The temperature increases due to heat

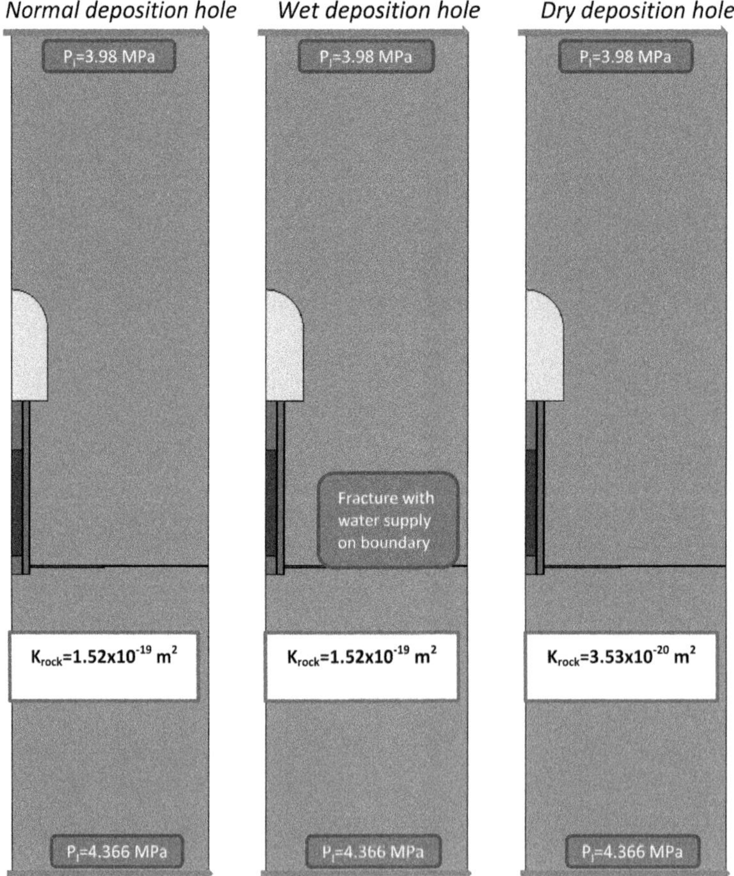

Fig. 7. Properties and boundary conditions for the fracture in three cases.

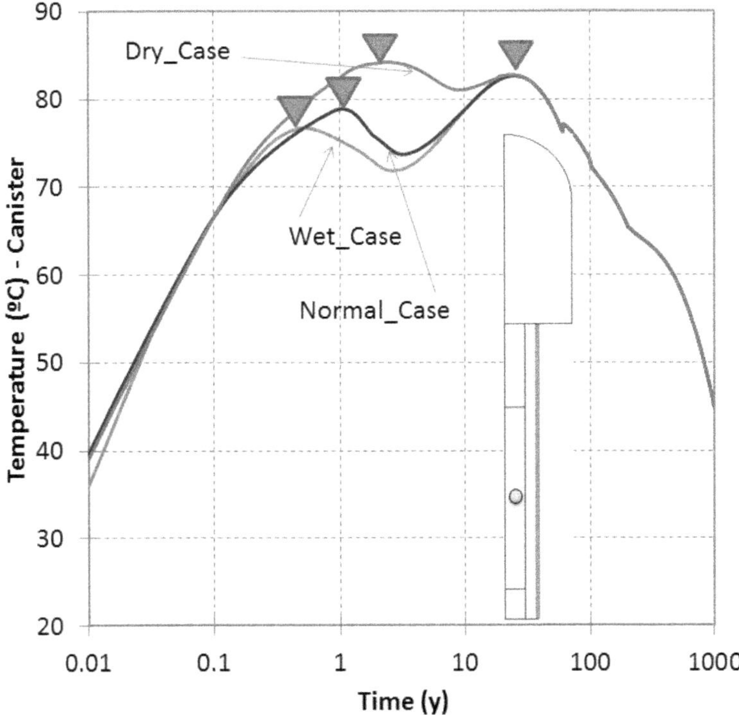

Fig. 8 Evolution of temperature in three cases (location: canister). Time from canister emplacement. The triangles indicate the local and global maxima for each curve.

generation in the canister, which, as shown in Figure 6, is characterized by decay to half-value in about 70–80 years. For the geometry considered, a peak value above 80°C takes place and the temperature decreases afterwards. However, the shape of the temperature evolution in the canister is influenced by the water saturation of the different components of the barrier and the presence of the gap. Both heat capacity and thermal conductivity depend on the degree of saturation.

The maximum temperature is not significantly influenced by the hydration conditions in general, but the evolution at earlier times is different due to the effect of the gap and saturation conditions of the barrier components. The gap under dry conditions produces thermal isolation as it remains open (air has lower heat transport capacity than water or soil). When the rock intrinsic permeability is very low (Dry_Case), the gap closure is delayed and the temperature in the canister increases due to the isolation effect. For the Dry_Case, the maximum temperature in the canister is higher, by a few degrees compared to the other cases and occurs in a different peak that occurs earlier. The peak that occurs around 1 year is caused by the presence of the gap, so the temperature decrease after that peak is motivated by the increase in conductivity induced by the combined effect of saturation and closure. The peak caused by the gap closure occurs earlier than the peak caused by the decaying power function.

Figure 9 shows the evolution of the liquid pressure for the three cases in the bentonite ring adjacent to the canister. In the case of the wet deposition hole, as the global intrinsic permeability (including the fracture) is higher compared with the other cases, the saturation of materials takes place faster. Desa turation of the bentonite ring is much greater in the Dry_Case because the rock intrinsic permeability is lower than in the other cases, and this implies a lower thermal conductivity of the dried-up buffer (Börgesson *et al.* 1994; Tang *et al.* 2007; Pintado *et al.* 2013).

With regard to stress development, the most critical zone is below the canister where the stresses become greater. Figure 10 shows the swelling pressure of buffer blocks under the canister for the three cases. Stress increases as liquid pressure evolves from extremely negative values caused by heating towards zero and positive values (i.e. hydration produced by a water supply from the host rock). The effect of the fracture is clearly observed in Figure 10,

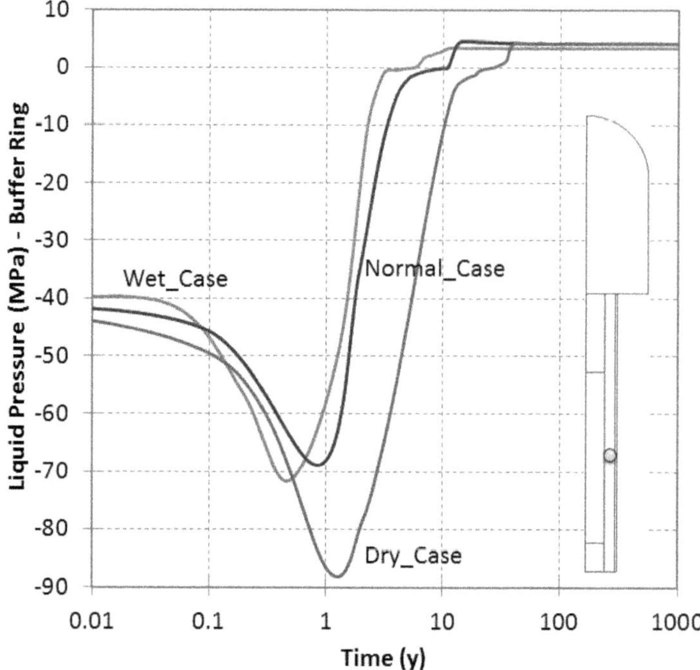

Fig. 9 Evolution of liquid pressure in three cases (location: buffer ring). Time from canister emplacement.

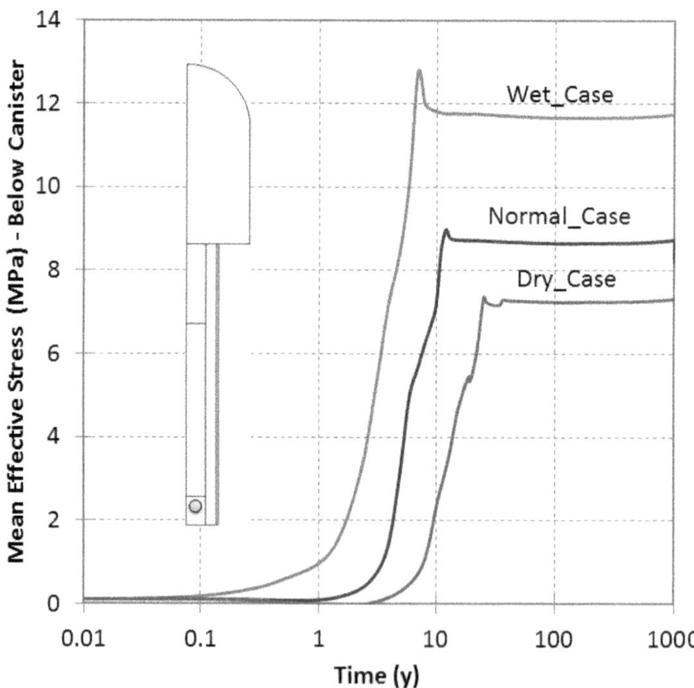

Fig. 10 Evolution of the mean effective stress in three cases (location: under the canister). Time from canister emplacement.

as the section where stress is plotted is close to the fracture. In the case of high intrinsic permeability of rock, water flows from the rock to the buffer in a much faster way. The time to reach a stabilized stress and the values achieved are different between the three cases. A lower intrinsic permeability of rock delays the process of swelling and, hence, the effective stress development. A larger effective stress is developed below the canister as saturation takes place faster. This can be explained by the fact that backfill is more rigid during buffer swelling in the Wet_Case compared to the Normal_Case. The same behaviour is observed if the Normal_Case is compared with the Dry_Case.

Figure 11 shows the swelling pressure of the backfill (central zone) for the three cases. As this zone is far away from the fracture, the time to achieve the maximum swelling pressure of the central backfill mainly depends on the intrinsic permeability of the rock and not on the water entry from the fracture. In the Wet_Case, the swelling pressure of the central backfill reaches a value of 7.5 MPa and stabilizes at 6 MPa. Except for the Dry_Case, the other two cases reach a peak value of mean effective stress at the same time. The reason for this is that the intrinsic permeability of the rock is of greater importance with respect to the time required to reach full saturation of the backfill. However, the swelling of the buffer blocks and the ring is greater in the Wet_Case. As a result, the achieved stress in the backfill is greater in the Wet_Case, despite it taking the same time to reach full saturation as the Normal_Case.

The peak that appears in the evolution of effective stress is caused by the effect of unloading after saturation. Once saturation is reached, the maximum effective pressure is achieved. Then, pore pressure becomes higher than atmospheric pressure and, as the material is saturated, effective stresses come into play. The increase in pore pressure implies a decrease in effective stresses. This is represented in Figure 12, which shows the suction–mean effective stress path. Although suction is a positive definite variable, the negative values have been represented in order to highlight the fact that effective stress decreases after full saturation of the medium.

Figure 13 shows the evolution of vertical displacements at the buffer–backfill interface for the three cases. The vertical displacements are generated by the swelling of the buffer blocks. Displacements are mainly concentrated at the

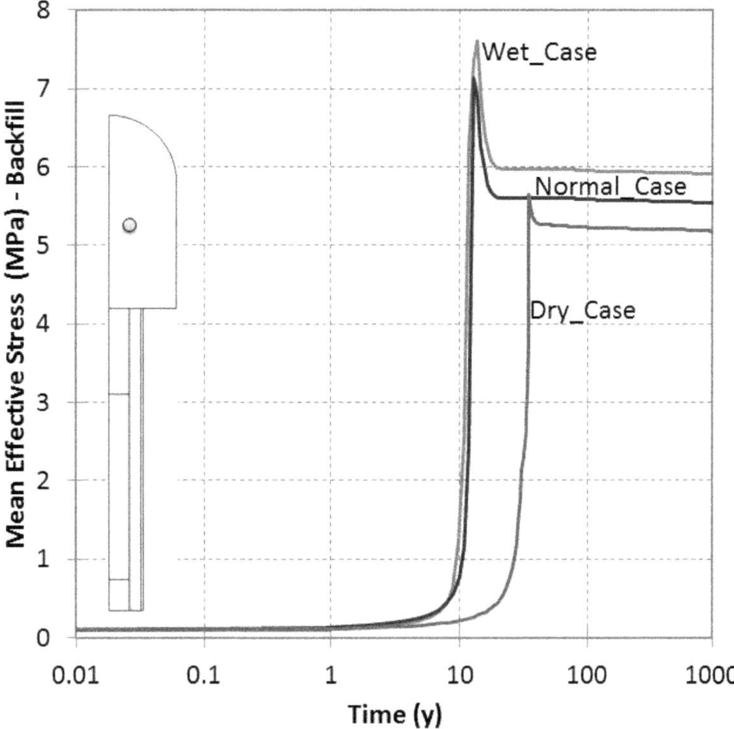

Fig. 11 Evolution of the mean effective stress in three cases (location: central backfill). Time from canister emplacement.

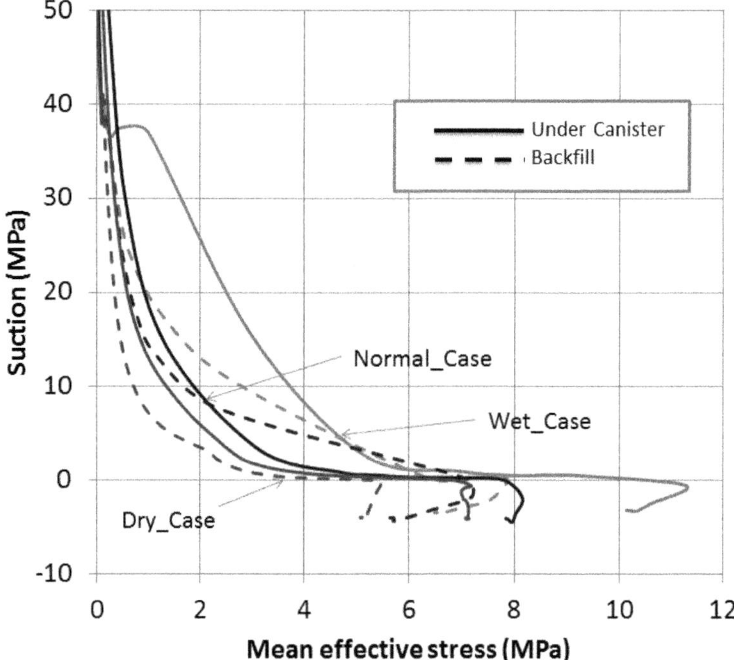

Fig. 12 Suction v. mean effective stress path for a point below the canister and a point in the backfill. Negative suctions are included to show that after saturation the mean effective stress decreases due to the development positive pore pressure.

buffer–backfill interface. Owing to the swelling, the buffer blocks push the backfill up. In the case of the dry deposition hole (Dry_Case), as the swelling of buffer blocks takes place slowly, the achieved displacements at the interface show a lower value. In other words, the buffer and backfill swell in a more simultaneous way for the Dry_Case compared to the Normal_Case and Wet_Case for which the buffer hydrates more rapidly than the backfill.

As a conclusion, it can be said that owing to the delay in water supply in the Dry_Case, displacements take place later compared to the other two cases. The maximum buffer–backfill interface displacement peak is about 5–6 cm in the Dry_Case. In the Wet_Case, vertical displacements reach to 9–10 cm, and for the Normal_Case they are the range of 7–8 cm.

Table 7 summarizes the results in terms of the maximum temperatures achieved, times to reach saturation, saturated density of the buffer and backfill, stress development, and maximum displacements. The Dry_Case curve for temperature shows an early peak (absolute maximum) that is caused by the lower thermal conductivity of the open gap, which remains open for a longer time compared to the other cases. Earlier saturation is motivated by the presence of the highly conductive zone that simulates a fracture. Final densities show little variation. As the global volumetric deformation of the engineered barrier is nearly zero (except for the deformation of the rock, which is very small), mean effective stress development is of the same order of magnitude as the swelling pressure. There are differences, however, as the saturation of the engineered barrier takes place in different ways depending on the conditions of hydration (intrinsic permeability and fractures, and the shape of the buffer–backfill system). A greater development of stresses is calculated with the Wet_Case. Finally, displacement at the interface is influenced by the earlier or later expansion of the buffer.

Results for 3D modelling

In this section, some results for 3D modelling are presented in order to check the influence of the simplification performed in the axisymmetrical models. For simplicity, the 3D geometry does not take into account the presence of either the pellets or the air-gap element. Besides this, backfill tunnel has a greater volume compared to the 2D calculations. The BBM parameters used for the buffer and backfill are the same as in the previous 2D

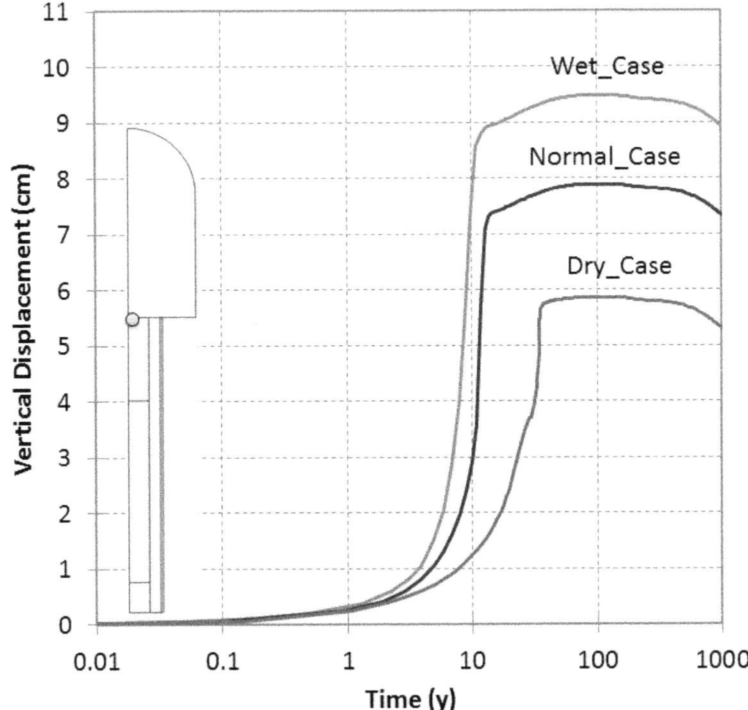

Fig. 13. Evolution of vertical displacement in three cases (location: the buffer–backfill interface). Time from canister emplacement.

axisymmetrical calculations. The value of the intrinsic permeability for the rock corresponds roughly to the Dry_Case (1.52×10^{-20} m^2), the intrinsic permeability of the backfill is the same as in the previous cases, and the intrinsic permeability of the buffer is higher but is in the range of typical values for bentonite. So, the values chosen fall in the same order of magnitude of intrinsic permeability and this is favourable for the numerical solution, especially for the 3D model with a relatively simple mesh. Further work on mesh optimization is required in order to obtain 3D solutions of the THM problem, with barriers having greater contrasts in permeability.

Figure 14 shows a comparison between the 2D and the 3D case in terms of the total mean stress. As the backfill tunnel has a greater volume for the 3D geometry, the whole system needs more time to reach full saturation. Therefore, the generated stresses reach the steady-state conditions in faster times in the case of 2D calculations. The values achieved for the total mean stresses are in the same range (i.e. from 8 to 12 MPa).

Table 7. *Comparison of three cases (2D axisymmetrical cases)*

Analysed parameters	Units	Wet_Case	Normal_ Case	Dry_Case
Temperature maximum at the bentonite ring	°C	82.5	82.5	82.5
Temperature maximum at the canister	°C	82.5	82.5	84.0
Time to reach full saturation of the bentonite discs – under the backfill	Years	10.0	12.7	32.5
Time to reach full saturation of the backfill – central zone	Years	10.0	12.7	34.0
Saturated density of the backfill (average)	kg m^{-3}	2156	2156	2156
Saturated density of the buffer (average)	kg m^{-3}	1999	1991	1994
Swelling pressure of the bentonite discs under the canister	MPa	11.7	8.7	7.2
Swelling pressure of the backfill – central	MPa	6.2	5.5	5.0
Displacements at the buffer–backfill interface	cm	9.0	7.4	5.3

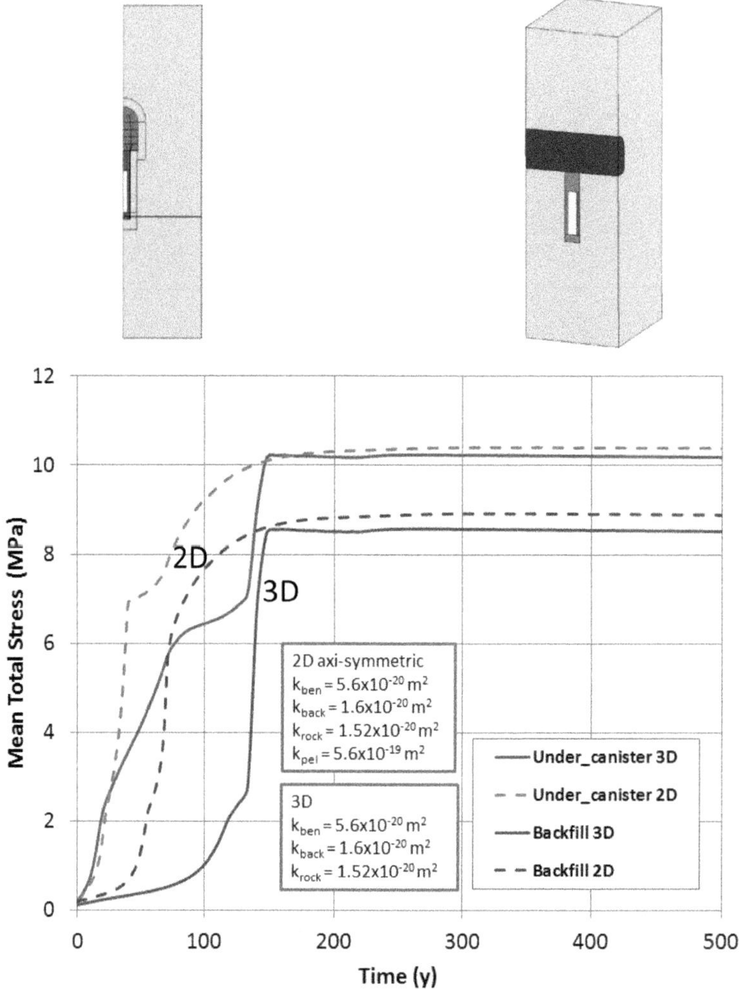

Fig. 14 Evolution of the mean total stresses in 2D and 3D modelling. Time from canister emplacement.

Conclusions

The laboratory tests (oedometer and infiltration) performed by B + Tech Oy (Finland) were modelled using the finite-element code Code_Bright. The Barcelona Basic Model (BBM) was used to model the mechanical constitutive behaviour of the MX-80 bentonite and Friedland clay, together with Darcy's law for the flow. The parameters obtained from the modelling of laboratory tests were used to perform 2D and 3D sensitivity analyses for the final disposal of nuclear spent fuel.

The values in the Table 7 correspond to representative points for the axisymmetrical cases. In the Dry_Case, the gap between the canister and the buffer closes later compared to the other cases. The maximum temperature in the canister reaches a value of 84°C for the Dry_Case (compared to 82.5°C), which is a clear effect of the air-gap element. As discussed throughout this paper, the fractures and hydraulic conductivity of rock have an impact on the results obtained in terms of maximum temperatures, swelling stress and displacements.

The buffer–backfill interface has been analysed and the maximum vertical displacement has the same order of magnitude as the one calculated previously by Leoni (2013).

With respect to the 3D calculations, the system reaches steady-state conditions later than in the 2D axisymmetrical models, mainly because the backfill volume is larger. However, the values obtained for stresses in the 3D models are in the same range as in the 2D calculations.

The present research work was financed in part by B + Tech Oy (Finland) under a POSIVA Oy project. The authors especially thank Jorma Autio from B + Tech and Kari Koskinen from Posiva for their support with the administrative tasks necessary to carry out this project, and for their comments and suggestions during this work.

References

Åkesson, M., Kristensson, O. & Börgesson, L. 2010. *SR-Site Data Report. THM Modelling of Buffer, Backfill and Other System Components. Critical Processes and Scenarios*. SKB TR-10-11. Svensk Kärnbränslehantering AB (SKB), Stockholm, Sweden, http://www.skb.com/publication/2095121/TR-10-11.pdf

Alonso, E.E., Gens, A. & Josa, A. 1990. A constitutive model for partially saturated soils. *Géotechnique*, **40**, 405–430.

Börgesson, L., Fredrikson, A. & Johannesson, L.-E. 1994. *Heat Conductivity of Buffer Materials*. SKB TR 94-29. Svensk Kärnbränslehantering AB (SKB), Stockholm, Sweden, http://www.skb.com/publication/11249/TR94-29.pdf

Hagros, A., Aikas, K., McEvven, T. & Anttila, P. 2003. *Host Rock Classification*. Working Report 2003-04. Posiva, Eurajoki, Finland, http://www.posiva.fi/files/2183/POSIVA-2003-04_Working-report_web.pdf

Hökmark, H., Lönnqvist, M., Kristensson, O., Sundberg, J. & Hellström, G. 2009. *Strategy for Thermal Dimensioning of the Final Repository for Spent Nuclear Fuel*. SKB Report R-09-04. Svensk Kärnbränslehantering AB (SKB), Stockholm, Sweden.

Ikonen, K. 2003. *Thermal Analyses of Spent Nuclear Fuel Repository*. Posiva Report 2003-04. Posiva, Eurajoki, Finland.

Juvankoski, M., Ikonen, K. & Jalonen, T. 2012. *Buffer Production Line 2012. Design, Production and Initial State of the Buffer*. Posiva Report 2012-17. Posiva, Eurajoki, Finland.

Keto, P., Hassan, M. *et al.* 2013. Backfill production line 2012. *In*: Keto, P. (ed.) *Design, Production and Initial State of the Deposition Tunnel Backfill and Plug*. Posiva Report 2012-18. Posiva, Eurajoki, Finland.

Kiviranta, L. & Kumpulainen, S. 2011. *Quality Control and Characterization of Bentonite Materials*. Posiva Working Report 2011-84. Posiva, Eurajoki, Finland, http://www.posiva.fi/files/1994/WR_2011-84_web.pdf

Kiviranta, L., Kumpulainen, S., Pintado, X., Karttunen, P. & Schatz, T. 2016. *B + Tech Oy, Characterization of Bentonite and Clay Materials 2012–2015*. Posiva Working Report WR 2016-05. Posiva, Eurajoki, Finland.

Leoni, M. 2013. *2D and 3D Finite Element Analysis of Buffer–Backfill Interaction*. Posiva Report 2012-25. Posiva, Eurajoki, Finland.

Marjavaara, P. & Kivikoski, H. 2011. *Filling the Gap Between Buffer and Rock in the Deposition Hole*. Posiva Working Report 2011-33. Posiva, Eurajoki, Finland, http://www.posiva.fi/files/1923/WR_2011-33web.pdf

Olivella, S., Carrera, J., Gens, A. & Alonso, E.E. 1994. Nonisothermal multiphase flow of brine and gas through saline media. *Transport in Porous Media*, **15**, 271–293.

Olivella, S., Gens, A., Carrera, J. & Alonso, E.E. 1996. Numerical formulation for a simulator (CODE BRIGHT) for the coupled analysis of saline media. *Engineering Computations*, **13**, 87–112.

Pintado, X. & Rautioaho, E. 2013. *Thermo-Hydraulic Modelling of Buffer and Backfill*. Posiva Report 2012-48. Posiva, Eurajoki, Finland.

Pintado, X., Mamunul, H.Md. & Martikainen, J. 2013. *Thermo-Hydro-Mechanical Tests of Buffer Material*. Posiva Report 2012-49. Posiva, Eurajoki, Finland.

Posiva 2013. *Safety Case for the Disposal of Nuclear Spent Nuclear Fuel at Olkiluoto*. Posiva Report 2013-01. Posiva, Eurajoki, Finland.

Raiko, H. 2012. *Canister Design*. Posiva Report 2012–13. Posiva, Eurajoki, Finland, http://www.posiva.fi/files/3093/POSIVA_2012-13.pdf

Tang, A.-M., Cui, Y.-J. & Le, T.-T. 2007. A study on the thermal conductivity of compacted bentonites. *Applied Clay Science*, **41**, 181–189.

Toprak, E., Mokni, N., Olivella, S. & Pintado, X. 2013. *Thermo-Hydro-Mechanical Modelling of Buffer, Synthesis Report*. Posiva Report 2012-47. Posiva, Eurajoki, Finland.

van Genuchten, M.Th. 1980. A closed equation for predicting the hydraulic conductivity of unsaturated soils. *Soil Science Society of America Journal*, **44**, 892–898.

Gaining insight into corrosion processes from numerical simulations of an integrated iron-claystone experiment

OLIVIER BILDSTEIN[1]*, JEAN-ÉRIC LARTIGUE[1], MICHEL L. SCHLEGEL[2], CHRISTIAN BATAILLON[2], BENOÎT COCHEPIN[3], ISABELLE MUNIER[3] & NICOLAS MICHAU[3]

[1]*CEA, DEN, DTN, 13108 Saint Paul-lez-Durance, France*

[2]*CEA, DEN, DPC, 91191 Gif-sur-Yvette, France*

[3]*French National Radioactive Waste Management Agency (ANDRA), DRD, Châtenay-Malabry, France*

**Correspondence: olivier.bildstein@cea.fr*

Abstract: Numerical simulations concerning iron–clay interactions in the conditions of an integrated experiment were conducted within a Callovo-Oxfordian claystone block at 90°C for 2 years. The calculations aim at determining the configurations and the parameters for which the simulation reproduce the mineral paragenesis observed at the end of the experiment. This paragenesis suggests that a thin magnetite layer precipitates at the surface of the corroding iron and dissolves on the claystone side to promote precipitation of Fe-silicate and Fe-carbonate minerals. The claystone is also altered close to the interface with iron via a significant precipitation of a Fe-carbonate mineral.

The results obtained using the coupled reactive transport code Crunchflow show that adjusting both the kinetics of magnetite dissolution/precipitation and the properties of the corroded layer (considered as a diffusive barrier) was required in order: (i) to model the destabilization of the magnetite layer formed at the original iron–claystone interface early in the corrosion process and the precipitation of other iron-bearing minerals; and (ii) to isolate the chemical conditions at the iron surface from the conditions in the clay environment in order to favour magnetite precipitation at the iron surface. With these assumptions, the model closely reproduced the mineral paragenesis observed in the experiment.

In order to predict iron–clay interactions (i.e. iron corrosion in clay environment) occurring in deep geological disposal of high-level long-lived radioactive waste, a global scientific approach is attempted by combining results from batch experiments dedicated to the transformation kinetics of clay minerals in the presence of iron or steel and integrated/*in situ* experiments with massive claystone and iron/steel components. The interpretation of the results often involves numerical modelling with reactive transport codes so as to validate the models and determine the set of parameters required for the different phenomenological processes (e.g. De Combarieu *et al.* 2007; Ngo *et al.* 2015). The simulation results are compared with observations of metallic archaeological analogues that evolved over hundreds to thousands of years buried in soils in order to extend the validity of the parameters determined in the experiments over such periods of time (Neff *et al.* 2006; Saheb *et al.* 2010), although *in situ* data have to be carefully interpreted. These parameters are then integrated into numerical simulations in order to predict the evolution of such systems on the timescale of the repository (e.g. Bildstein *et al.* 2006; Samper *et al.* 2008; Marty *et al.* 2010; King *et al.* 2014; Wersin & Birgersson 2014; Bildstein & Claret 2015).

In this framework, a series of calculations were performed to simulate iron corrosion in contact with claystone within the 'Arcorr2008' integrated experiment conducted under conditions close to those prevailing in deep geological disposal (Schlegel *et al.* 2014). The corrosion rate was monitored during the experience using *in situ* electrochemical impedance spectroscopy (EIS), and a thorough post-mortem mineral characterization of the corrosion zone was conducted. The purpose of the calculations was to adjust model parameters for mineral dissolution/precipitation kinetics and for diffusional transport properties, so that the paragenesis predicted by the model matches the nature and sequence of minerals observed in the experiment.

The integrated corrosion experiment in claystone: 'Arcorr2008'

Description of the experiment

The experimental set-up is fully detailed in Schlegel *et al.* (2014). Briefly, a massive block of

From: NORRIS, S., BRUNO, J., VAN GEET, M. & VERHOEF, E. (eds) 2017. *Radioactive Waste Confinement: Clays in Natural and Engineered Barriers*. Geological Society, London, Special Publications, **443**, 253–267.
First published online June 22, 2016, https://doi.org/10.1144/SP443.2

Fig. 1. View of the COx claystone brick with the three different probes, from top to bottom: the gold probe, the glass–iron 'micro-container' and the massive Armco iron rod (Schlegel *et al.* 2014).

Callovo-Oxfordian claystone (COx) drilled at −445 m depth at the Bure site (France) was machined, and three probes (12 mm in diameter and 38 mm long) were inserted: a gold probe used as a pseudo-reference for the EIS measurements, a glass–iron 'micro-container', and a massive Armco iron rod (Fig. 1). The 58 mm cubic block was placed in a triaxial cell, fully saturated with synthetic COx porewater at 90°C and maintained at a confining pressure of 40 bar; the corrosion experiment was conducted for 2 years in anoxic conditions. The probe of interest in this modelling study, for iron corrosion and identification of secondary minerals, is the iron rod. At the end of the experiment, the average thickness of the oxidized iron was about 13 μm (with a maximal corrosion rate of 90 $\mu m\ a^{-1}$ in the first month, decreasing to less than 1 $\mu m\ a^{-1}$ at the end of the experiment).

Post-mortem mineral characterization

A discontinuous layer composed of magnetite was identified directly at the contact with iron, followed by a layer of a Fe-silicate solid in the dense corrosion product layer (Fig. 2), and sometimes an intermediate layer of chukanovite ($Fe_2(CO_3)(OH)_2$). A layer of Ca–Fe-carbonate mineral (Ca-enriched siderite) was also observed in the transformed clay matrix close to the original iron surface. The total thickness of the alteration zone was less than 0.1 mm.

In the absence of continuous monitoring of mineral precipitation at the interface, it is hypothesized that the layer of magnetite formed early in the corrosion process and that this layer, or at least a part of it, acted as a passivating layer controlling the corrosion rate by limiting the diffusion of oxygen towards iron. This layer is thought to be formed at the iron–metal-oxide interface and to be progressively destabilized on the claystone side so as to constitute a moving barrier adhering to the iron surface and retreating during the corrosion process (Schlegel *et al.* 2014). The Fe-silicate layer may also have played a role in isolating the iron–magnetite surface from water.

A fine characterization of clay minerals in the claystone near-field is currently not available for this experiment. The evolution of claystone minerals may be inferred from other experiments in

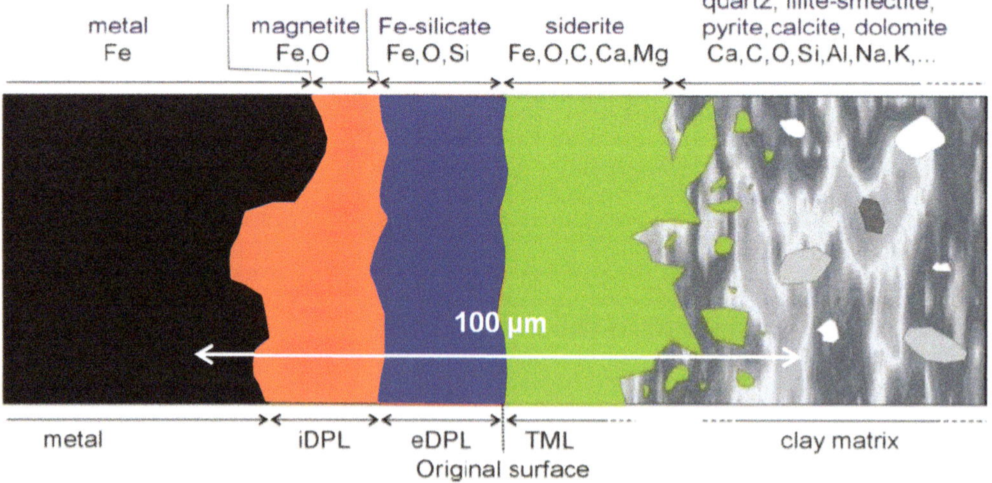

Fig. 2. Sequence of corrosion products forming the internal and external dense product layer (DPL) and the transformed clay matrix layer (TML) after 2 years (Schlegel *et al.* 2014): magnetite is in red, Fe-phyllosilicate is in blue and Ca-enriched doped siderite is in green.

which clay and iron powders are used (sometimes with iron plates) in the presence of larger water volumes than in the present compact system (e.g. De Combarieu *et al.* 2007; Bourdelle *et al.* 2014; Rivard *et al.* 2015): under conditions close to the Arcorr 2008 experiment, the authors found that quartz is mainly dissolved along with some the interstratified illite–smectite phase, and Fe-serpentines precipitate. Concerning carbonate minerals, dolomite is unstable and calcite is the most stable primary mineral.

Model description

Physical and geochemical conditions

The numerical simulations were performed with the coupled reactive transport code Crunchflow (Steefel *et al.* 2014) under the same conditions as in the experiment (90°C, 40 bars). The initial water composition in the claystone was calculated by progressively increasing the temperature from 25°C to 90°C and maintaining thermodynamic equilibrium with the claystone primary minerals (and also goethite, which precipitates during the process) (Table 1). Special care was given to the consistency of this water composition with that measured *in situ* at 25°C in the COx claystone. The claystone mineral composition was therefore constructed using minerals available in the Thermochimie database version 8 (Giffaut *et al.* 2014) and adjusting some equilibrium constants (as well as adding a new mineral for the Ca-enriched siderite, chukanovite, with the data from Lee & Wilkin 2010) in order to converge with the measured porewater composition (Tables 2 & 3). Note that ion exchange and surface complexation are not considered in these calculations, nor is sulphate reduction (which is considered not to occur in massive clayey environments at this temperature).

Table 1. *Initial water composition at 25°C and 90°C*

Chemical parameter	Composition at 25°C	Composition at 90°C
pH	7.12	6.97
Ionic strength	0.10	0.11
Eh (V)	−0.172	−0.295
pE	−2.91	−4.09
Si	1.79×10^{-4}	8.40×10^{-4}
Ca	8.26×10^{-3}	3.55×10^{-4}
Carbonates	3.57×10^{-3}	2.13×10^{-2}
Al	6.02×10^{-9}	1.93×10^{-7}
K	9.99×10^{-4}	8.07×10^{-3}
Mg	5.55×10^{-3}	8.50×10^{-5}
Fe	1.44×10^{-5}	1.38×10^{-6}
Sr	1.77×10^{-4}	2.43×10^{-4}
Cl	4.16×10^{-2}	4.16×10^{-2}
Na	5.34×10^{-2}	8.73×10^{-2}
pCO_2 (bar)	0.01	0.23
S total	1.88×10^{-2}	1.75×10^{-2}
Sulphates	1.88×10^{-2}	1.75×10^{-2}
Sulphides	8.90×10^{-10}	3.60×10^{-7}

Concentration in mol l^{-1}.

Model geometry and material properties

Since the thickness of the perturbed zone at the interface between iron and claystone is very small (*c.* 100 μm) compared to the size of the claystone block, a one-dimensional (1D) radial domain was used to model the processes occurring at the interface (Fig. 3). The claystone zone is discretized into 160 grid cells of increasing thickness from 10 μm (at the interface with iron) to 500 μm (at the outer boundary at 29 mm). The clay matrix has a porosity of 18% and the diffusion coefficient is set to 8×10^{-11} $m^2\ s^{-1}$ (at 90°C) for all aqueous species (ANDRA 2005). The iron zone is discretized only at the interface with the claystone (300 μm): a technological gap of 40 μm is also considered, which all together is represented by 34 grid cells (each 10 μm in length). Note that reactive transport codes require non-zero porosity in order to calculate the aqueous chemistry in each grid cell. This constraint is satisfied by considering these grid cells to be composed of iron with a porosity of 11.3%, averaging the technological void volume in each grid cell. The same porosity structure as in the claystone is arbitrarily assumed in the iron zone: using Archie's law with a cementation coefficient of 2.5, the resulting diffusion coefficient is 2.6×10^{-11} $m^2\ s^{-1}$. These assumptions will be challenged in the sensitivity calculations.

In order to prevent numerical difficulties associated with porosity clogging (i.e. the calculation stops when porosity decreases to zero), the porosity is set constant throughout the calculation. Thus, porosity is not updated as a function of mineral dissolution/precipitation volume balance. This rather unrealistic assumption is an actual limitation of the modelling approach and is discussed below.

The iron corrosion process is represented in the model with the following reaction involving the dissolution of a 'solid iron' mineral and the production of aqueous hydrogen: $Fe(s) + 2H_2O \rightarrow Fe^{2+} + 2\ OH^- + H_{2(aq)}$. The corrosion rate used in the simulations is deduced from electrochemical experimental data (Fig. 4a), and the time-dependent values are directly fed to the model. Considering this rate, the corroded iron thickness is 30 μm at the end of the experiment, which corresponds to the complete corrosion of iron in the first three grid cells in contact with the claystone. This behaviour is modelled using a sequential, node-by-node corrosion mode

Table 2. *Primary minerals selected for the calculations with equilibrium constants, kinetics constants and reactive surface areas – dissolution rates*

Primary minerals dissolution rates	Equilibrium constant (log K at 90°C)*	Kinetics constant k_diss (mol m^{-2} s^{-1})	Reactive surface area S (m^2 g^{-1})	Reactive surface area S (m^2 m^{-3})	k_diss × S (mol m^{-3} s^{-1})
Montmorillonite-MgCa	−1.84	1.00×10^{-14}	9.30×10^{-2}	1.12×10^{4}	1.12×10^{-10}
Montmorillonite-MgCa	−1.84	†5.75×10^{-14}	9.30×10^{-2}	1.12×10^{4}	6.47×10^{-10}
Montmorillonite-MgNa	−1.67	1.00×10^{-14}	9.30×10^{-2}	3.85×10^{3}	3.85×10^{-11}
Montmorillonite-MgNa	−1.67	†5.75×10^{-14}	9.30×10^{-2}	3.85×10^{3}	2.22×10^{-10}
Illite-Mg	4.43‡	1.00×10^{-14}	9.30×10^{-3}	5.59×10^{3}	5.59×10^{-11}
Kaolinite	1.92‡	6.61×10^{-14}	9.70×10^{-3}	2.16×10^{4}	1.43×10^{-9}
Kaolinite	1.92‡	†6.61×10^{-9}	9.70×10^{-3}	2.16×10^{4}	1.43×10^{-4}
Microcline	−1.72	3.89×10^{-13}	2.30×10^{-4}	3.28×10^{1}	1.27×10^{-11}
Quartz	−3.08	1.00×10^{-14}	2.30×10^{-4}	1.13×10^{2}	1.13×10^{-12}
Quartz	−3.08	†1.62×10^{-12}	2.30×10^{-4}	1.13×10^{2}	1.84×10^{-10}
Calcite	−8.81	1.58×10^{-6}	2.20×10^{-4}	1.57×10^{2}	2.48×10^{-4}
Dolomite	−18.25	2.95×10^{-8}	2.10×10^{-4}	7.17	2.12×10^{-7}
Ca-Siderite	−11.25‡	2.95×10^{-8}	2.20×10^{-2}	4.88×10^{2}	1.44×10^{-5}
Pyrite	−48.60	1.66×10^{-13}	4.00×10^{-3}	1.34×10^{2}	2.22×10^{-11}
Celestite	−6.70	2.19×10^{-6}	2.20×10^{-4}	4.89×10^{-1}	1.07×10^{-6}

Equilibrium constants from the Thermochimie database version 8 (downloadable at https://www.thermochimie-tdb.com) and kinetics constants based on a compilation from Palandri & Kharaka (2004).
*With Thermochimie_v8 basis components.
†Alkaline domain (value calculated for pH = 9).
‡Modified log K values to converge with the measured concentration in solution.

Table 3. *Primary minerals selected for the calculations with equilibrium constants, kinetics constants and reactive surface areas – precipitation rates*

Primary minerals precipitation rates	Kinetics constant k_prec (mol m^{-2} s^{-1})	Reactive surface area S (m^2 g^{-1})	Reactive surface area S (m^2 m^{-3})	k_diss × S (mol m^{-3} s^{-1})
Montmorillonite-MgCa	1.00×10^{-14}	9.30×10^{-2}	1.12×10^{4}	1.12×10^{-10}
Montmorillonite-MgNa	1.00×10^{-14}	9.30×10^{-2}	3.85×10^{3}	3.85×10^{-11}
Illite-Mg	1.00×10^{-14}	9.30×10^{-3}	5.59×10^{3}	5.59×10^{-11}
Kaolinite	6.61×10^{-16}	9.70×10^{-3}	2.16×10^{4}	1.43×10^{-11}
Quartz (prec.)	1.00×10^{-14}	2.30×10^{-4}	1.13×10^{2}	1.13×10^{-12}
Calcite	1.58×10^{-6}	2.20×10^{-4}	1.57×10^{2}	2.48×10^{-4}
Dolomite	2.95×10^{-8}	2.10×10^{-4}	7.17	2.12×10^{-7}
Siderite_085	2.95×10^{-8}	2.20×10^{-2}	4.88×10^{2}	1.44×10^{-5}
Pyrite	1.66×10^{-13}	4.00×10^{-3}	1.34×10^{2}	2.22×10^{-11}
Celestite	2.19×10^{-6}	2.20×10^{-4}	4.89×10^{-1}	1.07×10^{-6}

Equilibrium constants from the Thermochimie database version 8 (downloadable at https://www.thermochimie-tdb.com) and kinetics constants based on a compilation from Palandri & Kharaka (2004).

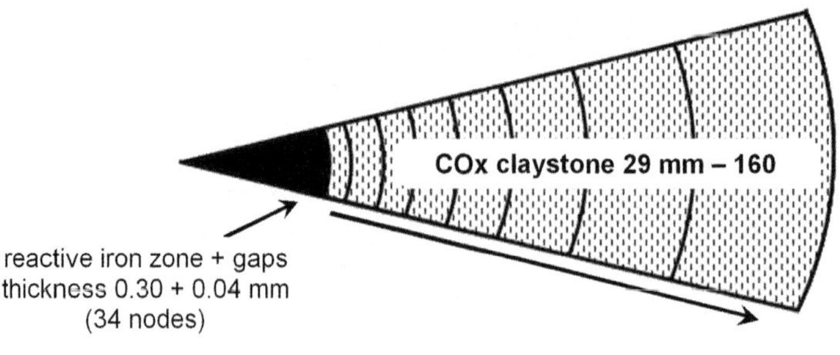

Fig. 3. One-dimensional radial domain and discretization used in the numerical simulations.

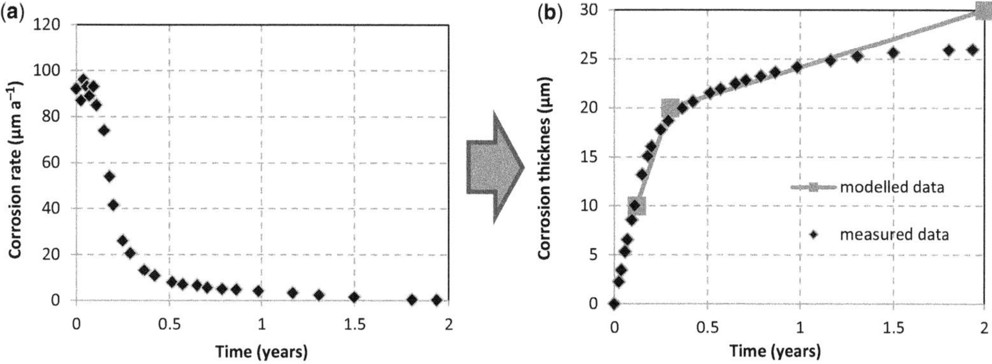

Fig. 4. Time evolution according to (**a**) the corrosion rate and (**b**) the thickness of corroded iron as calculated from the corrosion rate (curve with diamonds; shown as a blue curve in the online version) and the corresponding values used in the simulations (curve with squares; shown as a grey curve in the online version).

(the 'modelled data' curve in Fig. 4b; shown as the grey curve in the online version): corrosion starts in a cell only when the neighbouring cell (to the right) is completely corroded, in order to simulate a corrosion front. The decrease in the corrosion rate due to a passivating layer is thus taken into account in an implicit way.

Definition of a base case and sensitivity calculations

The selection of secondary minerals in the database is the key to the evolution of the geochemical system. This selection is based, in particular, on the list of minerals identified in experiments performed in conditions close to the Arcorr2008 system, and also on exclusion criteria such as the temperature and pressure of formation (e.g. micas, evaporites). At the end of this step, the selected list included Fe-bearing serpentines, Fe-oxides and Fe-carbonates (including chukanovite), and different types of clay minerals (saponites, vermiculites and nontronites) (Table 4).

Kinetic constants and reactive surface areas for the dissolution and precipitation of primary and secondary minerals also represent crucial parameters for this study. A base case was built with the kinetics constants compiled by Palandri & Kharaka (2004) and Marty *et al.* (2014) and Brunauer–Emmett–Teller (BET) values for the reactive surface areas. When no kinetics data are available in the literature for a particular mineral, a value is chosen using an analogy with another mineral with similar structural units (e.g. montmorillonite values for saponites, nontronites and vermiculites). In the base case, precipitation rate constants for secondary minerals were initially decreased by two orders of magnitude compared to dissolution rate constant and will be further revised in sensitivity calculations. The surface area for secondary minerals are set to the default value (100 m^2 m^{-3}). The sensitivity to reactive surface area was not specifically investigated here but was imbedded into the sensitivity to kinetics (as the product with the kinetics constant in the rate law: Lasaga 1998).

Sensitivity cases were built from the base case, considering the large uncertainties in the model parameters, by changing: (i) some parameters or a group of parameters in the initial set of kinetic constants; and (ii) the diffusional transport properties in the corroded zone. The sensitivity calculations also aim at progressively improving the match with experimental observations. A great number of calculations were performed in order to investigate the effects of the kinetic parameters (dissolution of certain primary phases and precipitation of secondary phases) but only the results obtained with the most influential parameters are presented below.

As far as transport properties are concerned, no data are available to link the evolution of porosity–diffusivity and the evolution of the mineral composition in the corrosion zone. Thus, the protective character of the corrosion products was simulated using a pre-existing magnetite layer to which different porosity and diffusion coefficients were applied so as to act as a diffusive barrier.

Results and discussion

Base case

In this configuration, iron corrosion results in an initial rapid increase in pH, up to a maximum value of 9.5 in the first iron cell at the interface with claystone. The pH values then progressively decrease with time (Fig. 5), mainly due to the decrease in the corrosion rate. The alkaline perturbation is

Table 4. *Secondary minerals selected for calculations with equilibrium constants, kinetics constants for the base case and reactive surface areas*

Secondary minerals rate constants	Equilibrium constant (log K at 90°C)	Kinetics constant k (mol m^{-2} s^{-1})	k_diss × S (mol m^{-3} s^{-1})
Greenalite	16.21	1.15×10^{-10}	1.15×10^{-8}
Saponite-FeCa	18.51	1.00×10^{-14}	1.00×10^{-12}
Saponite-FeNa	18.30	1.00×10^{-14}	1.00×10^{-12}
Saponite-Ca	19.82	1.00×10^{-14}	1.00×10^{-12}
Saponite-Na	19.60	1.00×10^{-14}	1.00×10^{-12}
Saponite-K	18.80	1.00×10^{-14}	1.00×10^{-12}
Saponite-Mg	18.90	1.00×10^{-14}	1.00×10^{-12}
Vermiculite-Ca	28.20	1.00×10^{-14}	1.00×10^{-12}
Vermiculite-Na	26.98	1.00×10^{-14}	1.00×10^{-12}
Vermiculite-K	25.18	1.00×10^{-14}	1.00×10^{-12}
Vermiculite-Mg	26.83	1.00×10^{-14}	1.00×10^{-12}
Nontronite-Ca	−17.97	1.00×10^{-14}	1.00×10^{-12}
Nontronite-Na	−18.18	1.00×10^{-14}	1.00×10^{-12}
Nontronite-K	−19.00	1.00×10^{-14}	1.00×10^{-12}
Nontronite-Mg	−17.80	1.00×10^{-14}	1.00×10^{-12}
Magnetite	−7.13	1.66×10^{-13}	1.66×10^{-11}
Goethite	−7.23	1.15×10^{-10}	1.15×10^{-8}
Pyrrhotite	−9.96	1.66×10^{-13}	1.66×10^{-11}
Strontianite	−9.28	2.95×10^{-10}	2.95×10^{-8}
Calcedoine	−2.84	3.80×10^{-12}	3.80×10^{-10}
SiO_2(am.)	−2.29	3.80×10^{-10}	3.80×10^{-8}
Lizardite	25.14	2.00×10^{-14}	2.00×10^{-12}
Glauconite	−11.67	7.94×10^{-10}	7.94×10^{-8}
Cronstedtite	11.53	1.00×10^{-14}	1.00×10^{-12}
Berthierine	17.31	1.00×10^{-14}	1.00×10^{-12}
Siderite	−11.17	2.95×10^{-10}	2.95×10^{-8}
Chukanovite	−28.47*	2.95×10^{-10}	2.95×10^{-8}

Equilibrium constants based on Thermochimie version 8 and kinetics constants based on a compilation from Palandri & Kharaka (2004).
*From Lee & Wilkin (2010).

quickly neutralized by claystone and penetrates only very little into the rock (*c.* 5 mm). The partial pressure of H_2 increases in the zone of corrosion and slightly exceeds the prescribed maximum pressure (40 bars) at the beginning of the simulation due to kinetics effects, it then stabilizes at 40 bars from 5 months up until the end of the simulation (Fig. 6). This result is consistent with the fact that H_2 gas was also identified in the experiment.

The distribution of the corrosion products is dominated by magnetite in the first two cells of the iron zone in contact with the claystone (Fig. 7). Greenalite precipitates preferentially at the surface of iron, but in smaller amounts than in the experiment, whereas Ca-siderite strongly precipitates in claystone, where it becomes the dominating Fe product. These results compare well with the observed experimental paragenesis, but they differ in terms of the nature and order of solid layers. Specifically, the model predicts magnetite formation not directly at the metal surface but at the interface with the claystone.

Experimentally, greenalite is present at this claystone interface (Fig. 2), and is thought to form via magnetite destabilization (Schlegel *et al.* 2014). The discrepancy can be interpreted as an insufficient supply of silica simulated by the code, resulting from the alteration of primary silicate minerals in the claystone, such as quartz (Fig. 8). This behaviour contradicts observations in other experiments where quartz is significantly dissolved in iron-clay systems (e.g. De Combarieu *et al.* 2007; Bourdelle *et al.* 2014; Rivard *et al.* 2015). Regarding the precipitation of other secondary minerals in claystone, only small amounts of vermiculite and saponite are predicted close to interface with iron (<1% in a few millimetres, not shown). Finally, it can be noted that the amount of Ca-siderite precipitating in the in the first 0.1 mm of the transformed clay matrix zone exceeds the pore volume available (up to 160%), due to the constant porosity assumption, and should therefore be considered to precipitate as a thicker layer. Chukanovite is never predicted to precipitate in these simulations.

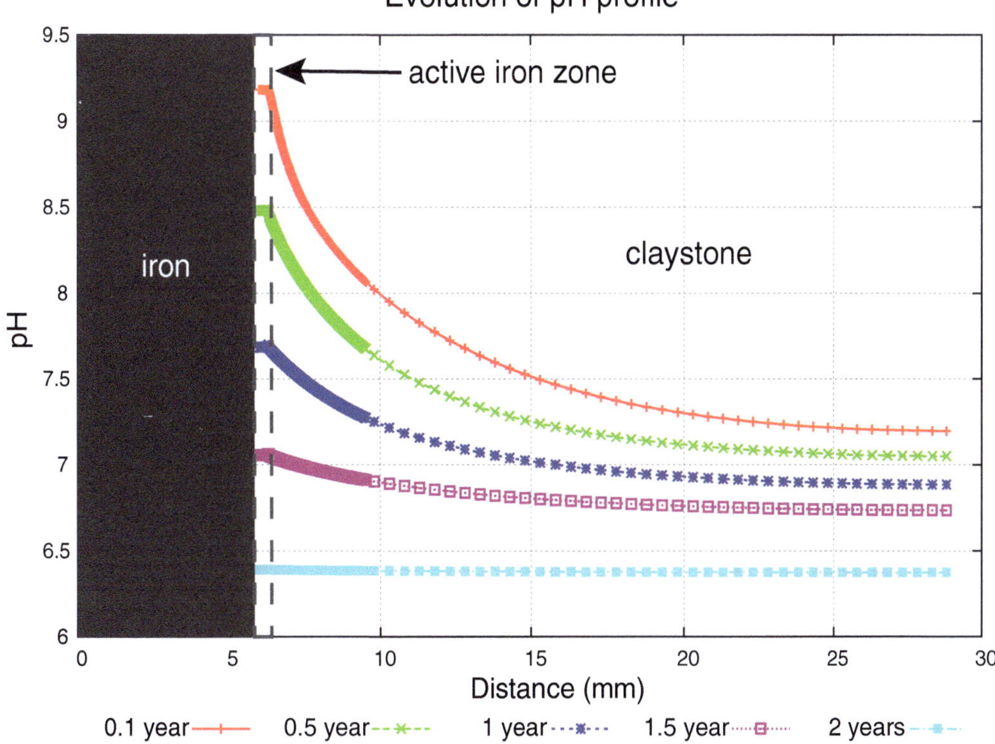

Fig. 5. pH profiles as a function of time (base case simulation).

Sensitivity to quartz dissolution kinetics

In order to test the effect of increased quartz reactivity on the secondary mineral sequence at the iron–claystone interface, the quartz dissolution rate was increased by one and two orders of magnitude. This increase results in the precipitation of larger amounts of greenalite (in the iron zone) and Ca-siderite (in the claystone) at the expense of magnetite (Fig. 9). In this case, quartz dissolves significantly at the interface, which concomitantly limits the dissolution of other primary minerals containing silica (illite, montmorillonite and kaolinite). The concentration of aqueous silica actually decreases dramatically at the contact with iron during the first 2 months of the simulations as a result of the high corrosion rate and the large amount of greenalite precipitation. The simulations also show that the evolution of the pH profile is similar to that found in the base case, but the maximal pH values tends to decrease with increasing quartz dissolution rate (from 9.5 in the base case to 9, and to 8.3, when the rate is increased by one and two orders of magnitude, respectively).

The simulated corrosion product paragenesis approaches the experimental observations, but the mineral sequence still does not conform to observations. Magnetite precipitates at the beginning of the simulation when the corrosion rate is high, and persists throughout the simulation: however, precipitation should be followed by dissolution, with the magnetite layer following the 'retreating' iron surface. Magnetite is, indeed, slightly undersaturated in the iron cell at the interface with claystone, but the dissolution rate calculated by the code is too slow.

Sensitivity to overall kinetics parameters

Starting from the conclusion of the previous calculations and taking into account recent results from the literature obtained using high reaction rates in the same geochemical system (e.g. precipitation at equilibrium in Ngo *et al.* 2015), a series of calculations were carried out by increasing the reaction kinetics of the most influential minerals. For this purpose, factors of 10 and 100 were applied to the dissolution and precipitation kinetics for quartz and for the dominating corrosion products (magnetite, greenalite and Ca-siderite).

As a result of the increase in the reaction kinetics, greenalite and Ca-siderite precipitation is favoured at the expense of magnetite (Fig. 10). The evolution of the pH is similar in the various cases,

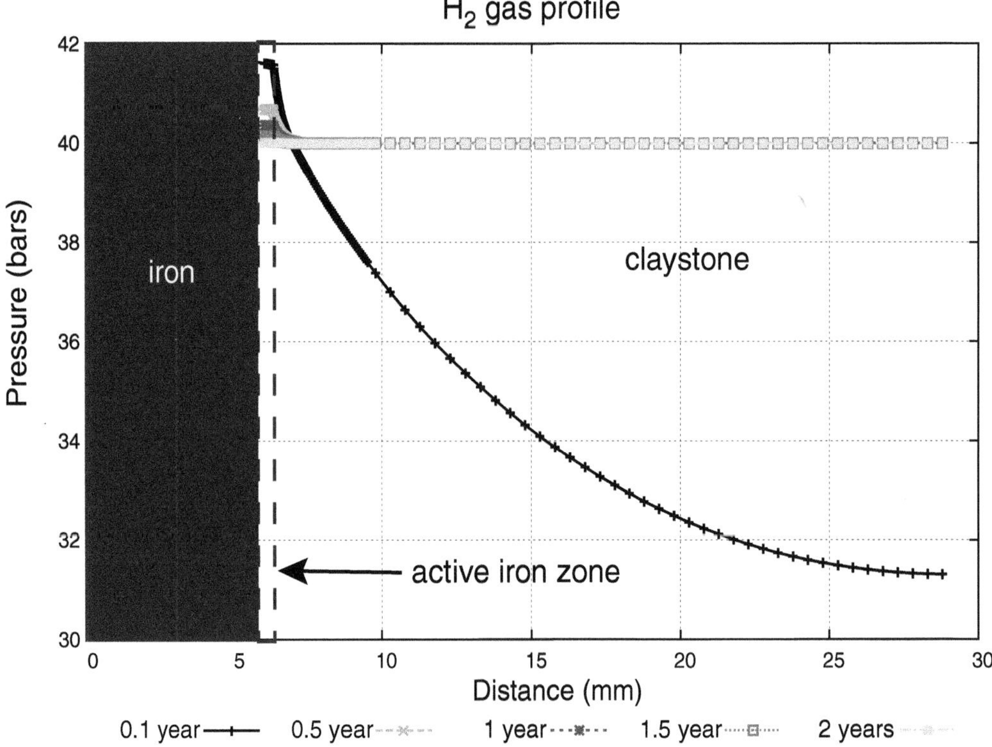

Fig. 6. H_2 pressure profiles as a function of time (base case simulation).

with a maximum value of pH decreasing from 9.5 (in the reference case) to 8.8 and 8.1 (for kinetics $\times 10$ and $\times 100$), respectively, when reaction kinetics increase. This pH increase parallels the progressive precipitation of greenalite instead of magnetite.

In an additional case, with kinetics parameters increased by three orders of magnitude (not shown here), magnetite does not form at all and pH remains below 8. These results are consistent with recent observations in batch experiments, in which steel and claystone are used as powders, resulting in high reactive surface areas (Bourdelle *et al.* 2014). Nevertheless, in all the simulated cases, when magnetite forms at the beginning of the simulation, it persists until the end. In other words, the respective quantities of corrosion products can be adjusted by modifying the kinetics parameters used in the simulation (even preventing the formation of magnetite), but, so far, the mineral sequence cannot be entirely reconciled with the ArCorr2008 experimental observations.

Sensitivity to corrosion rate

In the previous calculations, the corrosion rate was set to the values measured in the experiment. All simulations show that greenalite systematically precipitates at the surface of the corroding iron after 3–6 months of simulation. This behaviour cannot be attributable to an additional supply of silica at this time of simulation because aqueous silica is transported by diffusion processes from the claystone. The local geochemical conditions imposed by the corrosion rate could be responsible for the stabilization of a mineral phase at the expense of another. Sensitivity calculations were thus performed by decreasing the corrosion rate by both a factor of 2 and 10 with respect to the experimental values in order to verify this hypothesis.

The results show a threshold effect of the corrosion rate on the nature of precipitated mineral phases: when the corrosion rate is divided by a factor of 2, more Ca-siderite precipitates at the expense of magnetite and greenalite. With a corrosion rate 10 times lower than the experimental value, magnetite does not form at all (Fig. 11). One outcome of these evolutions is that magnetite tends to form at the beginning (when the corrosion rate is high) and the greenalite at the end of simulations (when the rate is lower). In addition, the ratio between the amount of dissolved Fe migrating into the claystone and the amount precipitating in the iron zone is

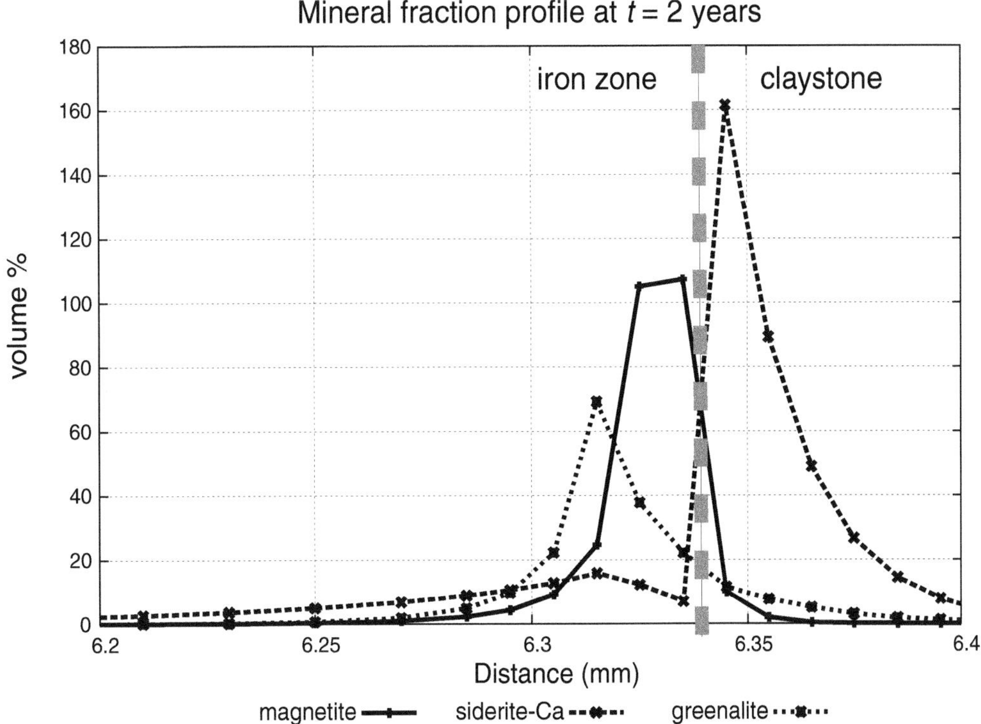

Fig. 7. Corrosion products profile at the end of the simulation (base case).

much higher when the corrosion rate is decreased compared to the base case.

Sensitivity to transport parameters

The persistence of magnetite in the iron zone in the interface with claystone is not observed at the end of the experiments, but it was probably formed at this location earlier in the experiment (when the corrosion was high), as testified by other shorter experiments performed with the Arcorr set-up (Schlegel *et al.* 2008, 2010). As the Arcorr2008 experiment suggests that this layer is formed continuously at the contact with iron and dissolves on the other side as corrosion proceeds, a series of simulations were conducted to investigate the assumption of a magnetite layer that isolates the corroding iron surface from the internal and external dense product layer (iDPL and eDPL in Fig. 2). This process is confirmed by electrochemical corrosion models at the nanometric scale in the internal dense product layer (Bataillon *et al.* 2010), which also attribute the decrease in the corrosion rate to limited diffusion of oxygen in an interfacial dense layer of magnetite-like oxide.

A series of calculations were thus carried out in a configuration in which one or two grid cells contain corrosion products at the start of the simulation (i.e. 1 or 3 months after the beginning of the corrosion process). Low diffusivity properties were assigned to the 'pre-corroded' iron layers in order to decrease the diffusion of aqueous species in solution. Only the final case with two grid cells containing magnetite is presented hereafter.

In this case, the two pre-corroded cells are assigned: (i) a value of 0.01 for porosity (99% magnetite) and a coefficient of diffusion decreased by two orders of magnitude compared to the preceding cases; and (ii) a value of 0.001 for porosity (99.9% of magnetite) and a diffusion coefficient decreased by four orders of magnitude to simulate the dense layer of corrosion products (Fig. 12). Magnetite reactivity was also enhanced (three orders of magnitude) in order to favour dissolution in the second stage of the corrosion process. This configuration was implemented in order to check the efficiency of such a barrier in limiting the iron diffusion in cell 2, thus allowing the destabilization of magnetite initially present in cell 1 and the precipitation of iron-silicate at this location. Note that iron corrosion only takes place in cell 3 during the simulation, starting with a rate of approximately 20 $\mu m\ a^{-1}$ and decreasing, with time, to less than 1 $\mu m\ a^{-1}$ (see the corrosion rate values from 3 months to 2 years in Fig. 4).

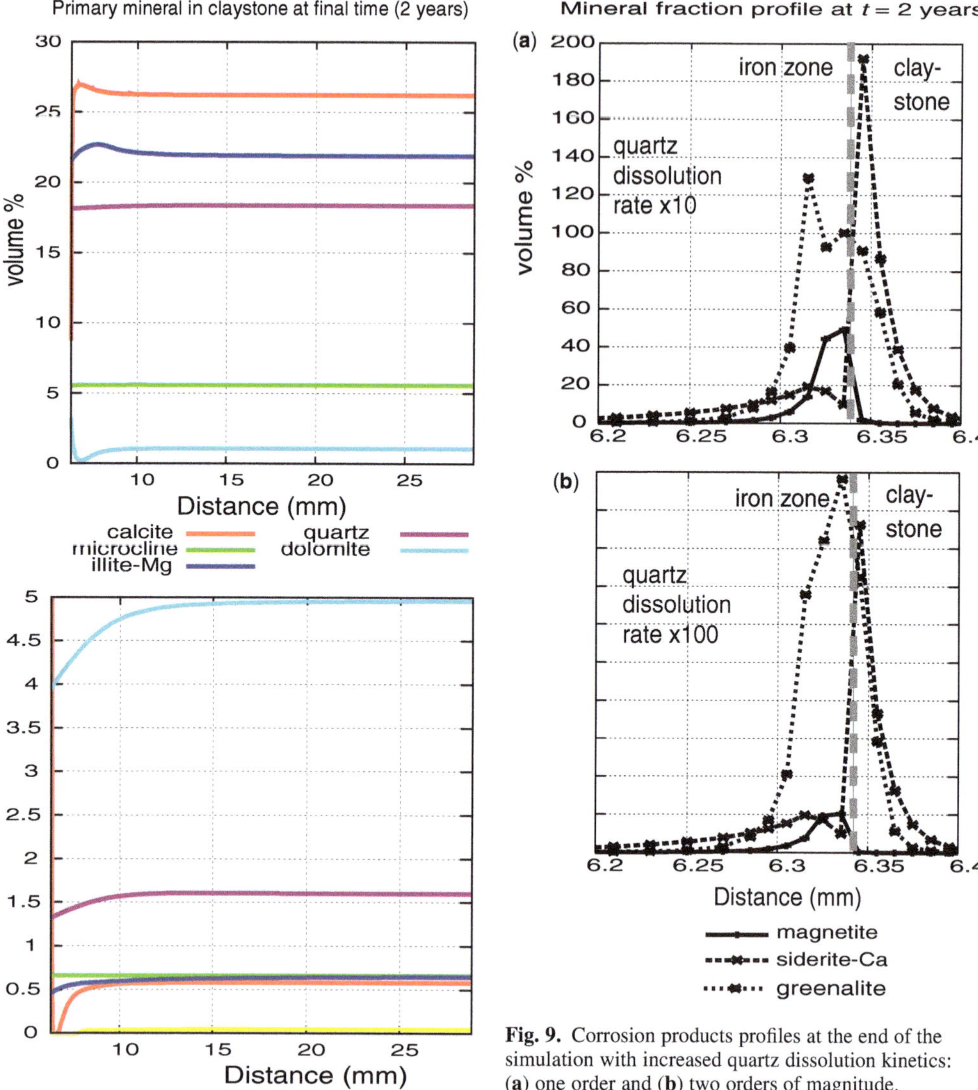

Fig. 8. Profiles of primary minerals as a function of time in the claystone zone (base case).

Fig. 9. Corrosion products profiles at the end of the simulation with increased quartz dissolution kinetics: (**a**) one order and (**b**) two orders of magnitude.

In these conditions, the sequence of corrosion products obtained at the end of the simulation better conforms to that observed in the experiment (Fig. 12). Magnetite precipitates at the contact with iron, greenalite becomes the dominant corrosion product at the contact with claystone and Ca-siderite massively precipitates in the transformed claystone layer.

The evolution of corrosion products profiles with time in the three iron cells shows that the pre-existing magnetite is strongly destabilized and disappears at 1.2 years, concomitantly with greenalite precipitation in the first cell, right at the beginning of the simulation (Fig. 13, right-hand plot). Small amounts of Ca-siderite also precipitate in this cell. In cell 2, magnetite is still forming until about 1.3 years, when the corrosion rate reaches approximately 2 $\mu m\ a^{-1}$, and then starts to dissolve in favour of greenalite and Ca-siderite precipitation (Fig. 13, centre plot). In cell 3, at the contact with the corroding iron, magnetite precipitates until 1.3 years and remains the main corrosion product until the end of the simulation (Fig. 13, left-hand plot). In the last 8 months of the simulation, magnetite

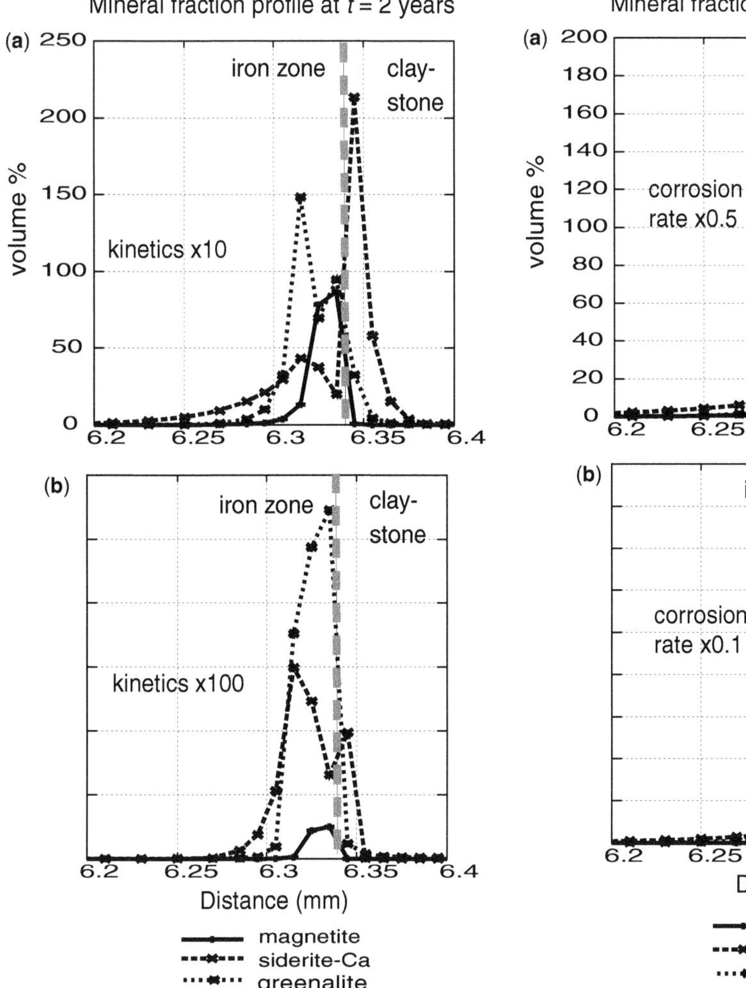

Fig. 10. Corrosion products profiles at the end of the simulation with increased quartz and corrosion product kinetics: (**a**) one order and (**b**) two orders of magnitude.

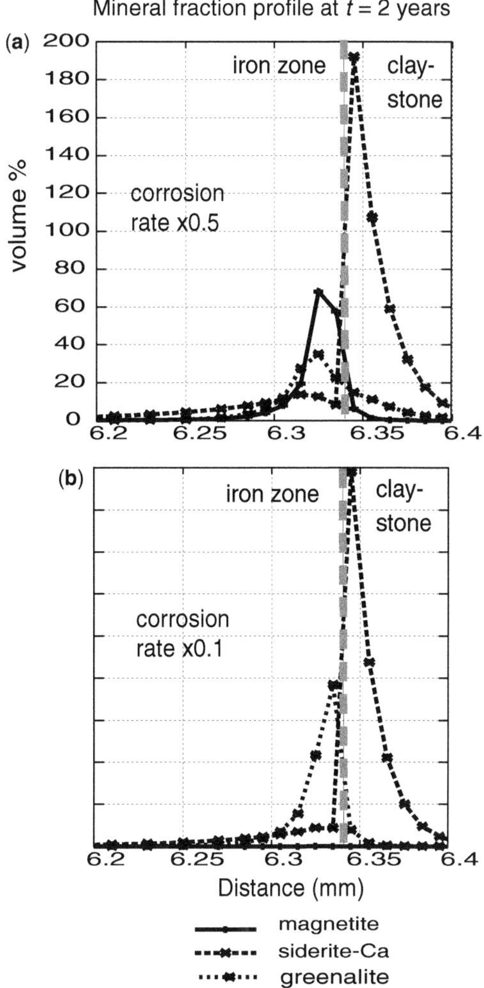

Fig. 11. Corrosion products profiles at the end of the simulation, with corrosion rates decreased by a factor of (**a**) 0.5 and (**b**) 0.1.

is slightly destabilized, and precipitation of Ca-siderite and greenalite takes place.

The behaviour of magnetite can be directly linked to the corrosion rate and the evolution of pH in the iron zone: the pH value remains systematically <8 in cell 1 during the simulation, whereas it is maintained at values >8 when the corrosion rate is >2 $\mu m\ a^{-1}$ in cells 2 and 3 up until 1.3 years (Fig. 14).

These results show that it is possible to model magnetite precipitation at the surface of the corroding iron by assuming the presence of a transient magnetite layer that acts as a diffusive barrier despite the decrease in the iron corrosion rate. This layer creates favourable local geochemical conditions (pH, Fe concentration), and limits the diffusion of aqueous silica and carbonates.

Summary and conclusions

A reactive transport model of iron corrosion and clay transformation was developed based on kinetics parameters from the literature and tested against the results of the Arcorr2008 experiment. Calculations aimed at determining the configurations and the parameters that make it possible to approach the mineral paragenesis observed in the experiment. The first results show that a global increase in the kinetics parameters found in the literature, in

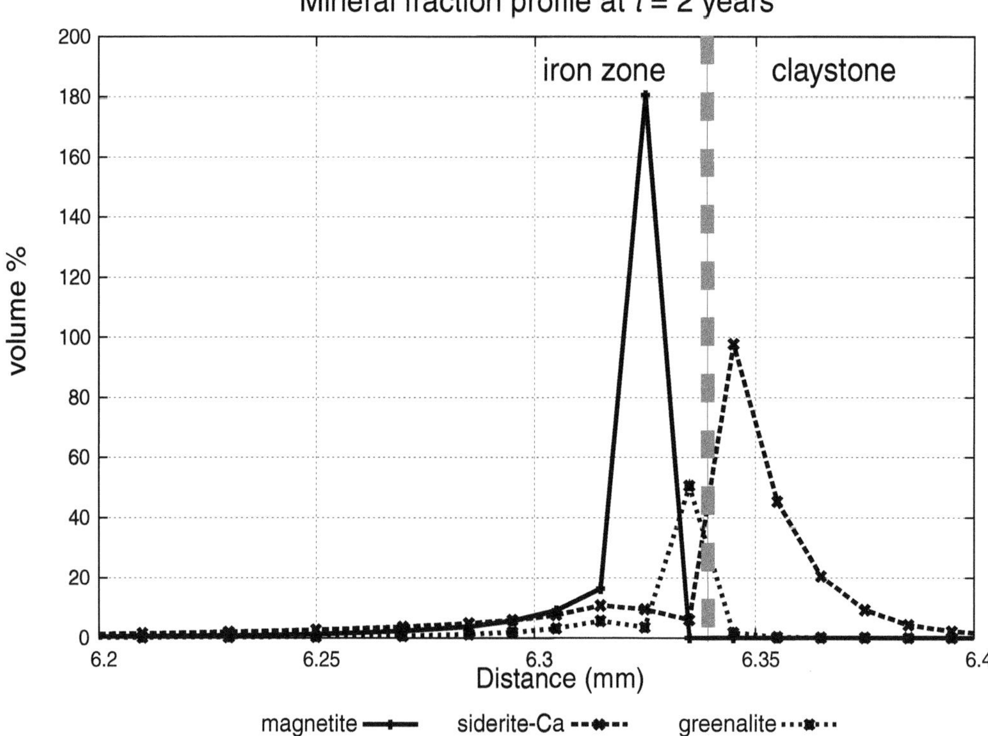

Fig. 12. Corrosion product profiles at the end of the simulation, with two 'pre-corroded' cells at the contact with the claystone (cells 1 and 2).

particular those affecting the supply of dissolved silica (such as quartz dissolution), is necessary to precipitate Fe-phyllosilicate and carbonate minerals in the iron zone and at the interface with the claystone. The decrease in the corrosion rate with time tends to favour the precipitation of greenalite and Ca-siderite at the expense of magnetite after 2–3 months in the simulations. However, magnetite, when formed early in the simulations and especially at the iron–claystone interface, tends to persist later on. This prediction is at variance with the experiment, which shows that magnetite is present only at the iron surface after 2 years of corrosion and should gradually dissolve at the claystone side. Sensitivity calculations on kinetics parameters show that changing the precipitation/dissolution kinetics of minerals primarily affects the amount of precipitated minerals, whereas the mineral sequence remains identical (unless very high rates are used for magnetite).

A diffusive barrier between the iron surface and the dense corrosion product layer is therefore invoked to explain the stability of magnetite at the corroding iron surface. This hypothesis has been tested with simulations implementing an initial layer of magnetite (simulating a 'pre-corroded' iron layer) with low diffusion properties. In these test cases, the migration of iron and other aqueous species in solution is dramatically decreased. Calculations implementing such a diffusive barrier were carried out with one and two layers of pre-existing magnetite. The results obtained with this assumption show a mineral paragenesis closely matching the one observed in the Arcorr2008 experiment. Magnetite dissolves at the interface with the claystone and precipitates close to iron. Ca-siderite precipitates in the disturbed claystone zone, and greenalite is formed between the two zones. The diffusive barrier in the iron zone thus effectively supports magnetite precipitation in contact with iron, but a high reaction rate for magnetite is also necessary to redissolve the initial magnetite layer at the interface with the claystone. This latter assumption is corroborated by other modelling studies: for example, Ngo *et al.* (2015) used an equilibrium assumption to manage the precipitation of secondary minerals and showed a good match with iron–claystone interaction experimental results from Bourdelle *et al.* (2014), which were obtained in a batch system with iron and claystone in the form of powder.

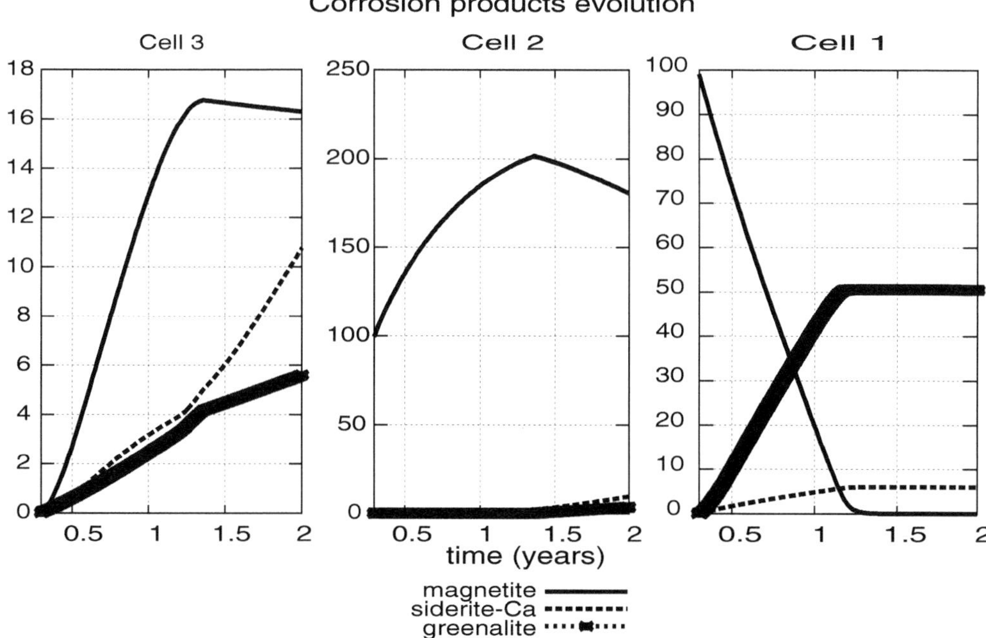

Fig. 13. Profiles of corrosion products as a function of time in the iron zone: simulation with 'pre-corroded' cells 1 and 2 and corroding cell 3 starting at 0.3 years.

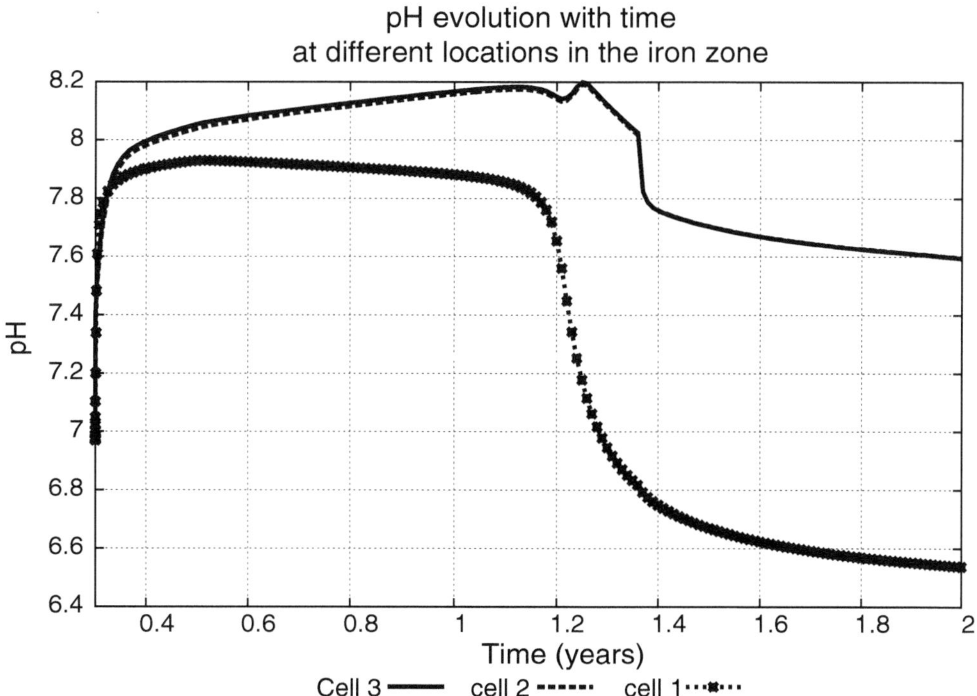

Fig. 14. pH profiles as a function of time in the different grid cells in the iron zone (simulation with two 'pre-corroded' cells).

A persistent limitation in reactive transport codes is the difficulty of taking into account porosity variations in systems where massive precipitation of mineral phases takes place in a grid cell. The allowance for solid precipitation in the calculations, sometimes at volume fractions higher than the available porosity in the cell, also results in an overestimated Fe precipitation in the iron zone and in the very first millimetres of claystone. Oxidized Fe is therefore less 'mobile' in the simulations than in the experiment, where 50% of oxidized Fe is estimated to migrate into the claystone (Schlegel *et al.* 2014).

An additional perspective of this study with regard to model development is the implementation of a control on the cementation factor by the nature and amount of the precipitated mineral phases. New simulations are therefore being carried out that focus on diffusivity and reactivity in the magnetite layer, with a dynamic updating of the transport properties in the corrosion zone.

We acknowledge fruitful and constructive discussions with Didier Crusset, Jean-Marie Gras and François Foct on aspects of corrosion. Financial support by ANDRA is also acknowledged.

References

ANDRA 2005. *Dossier 2005 Argile Synthesis. Evaluation of the Feasibility of a Geological Repository in an Argillaceous Formation Meuse/Haute-Marne Site*. ANDRA (National Agency for Radioactive Waste Management), Paris.

Bataillon, C., Bouchon, F. *et al.* 2010. Corrosion modelling of iron based alloy in nuclear waste repository. *Electrochimica Acta*, **55**, 4451–4467.

Bildstein, O. & Claret, F. 2015. Chapter 5 – Stability of clay barriers under chemical perturbations. *In*: Tournassat, C., Steefal, C.I., Bourg, I.C. & Bergaya, F. (eds) *Natural and Engineered Clay Barriers*. Developments in Clay Science, **6**. Elsevier, Amsterdam, 155–188.

Bildstein, O., Trotignon, L., Perronnet, M. & Jullien, M. 2006. Modelling iron–clay interactions in deep geological disposal conditions. *Physics and Chemistry of the Earth*, **31**, 618–625.

Bourdelle, F., Truche, L., Pignatelli, I., Mosser-Ruck, R., Lorgeoux, C., Roszypal, C. & Michau, N. 2014. Iron–clay interactions under hydrothermal conditions: impact of iron specific surface area on reaction pathway revealed by mineralogy and a continuous monitoring of pH, solution and gas compositions. *Chemical Geology*, **381**, 194–205.

De Combarieu, G., Barboux, P. & Minet, Y. 2007. Iron corrosion in Callovo–Oxfordian argilite: from experiments to thermodynamic/kinetic modelling. *Physics and Chemistry of the Earth*, **32**, 346–358.

Giffaut, E., Grivé, M. *et al.* 2014. ANDRA thermodynamic database for performance assessment: ThermoChimie. *Applied Geochemistry*, **49**, 225–236.

King, F., Kolar, M. & Keech, P.G. 2014. Simulations of long-term anaerobic corrosion of carbon steel containers in Canadian deep geological repository. *Corrosion Engineering, Science and Technology*, **49**, 455–459.

Lasaga, A.C. 1998. *Kinetic Theory in the Earth Sciences*. Princeton University Press, Princeton, NJ.

Lee, T. & Wilkin, R. 2010. Iron hydroxy carbonate formation in zerovalent iron permeable reactive barriers: characterization and evaluation of phase stability. *Journal of Contaminant Hydrology*, **116**, 47–57.

Marty, N.C.M., Fritz, B., Clément, A. & Michau, N. 2010. Modelling the long term alteration of the engineered bentonite barrier in an underground radioactive waste repository. *Applied Clay Science*, **47**, 82–90.

Marty, N.C.M., Claret, F. *et al.* 2014. A database of dissolution and precipitation rates for clay-rocks minerals. *Applied Geochemistry*, **55**, 108–118, https://doi.org/10.1016/j.apgeochem.2014.10.012

Neff, D., Dillmann, P., Descostes, M. & Beranger, G. 2006. Corrosion of iron archaeological artefacts in soil: estimation of the average corrosion rates involving analytical techniques and thermodynamic calculations. *Corrosion Science*, **48**, 2947–2970.

Ngo, V.V., Clément, A., Michau, N. & Fritz, B. 2015. Kinetic modeling of interactions between iron, clay and water: comparison with data from batch experiments. *Applied Geochemistry*, **53**, 13–26, https://doi.org/10.1016/j.apgeochem.2014.12.003

Palandri, J.L. & Kharaka, Y.K. 2004. *A Compilation of Rate Parameters of Water–Mineral Interaction Kinetics for Application to Geochemical Modeling*. United States Geological Survey, Open-File Report, **2004-1068**.

Rivard, C., Pelletier, M., Michau, N., Razafitianamaharavo, A., Abdelmoula, M., Ghanbaja, J. & Villiéras, F. 2015. Reactivity of Callovo-Oxfordian claystone and its clay fraction with metallic iron: role of non-clay minerals in the interaction mechanism. *Clays and Clay Minerals*, **63**, 290–310.

Saheb, M., Descostes, M., Neff, D., Matthiesen, H., Michelin, A. & Dillmann, P. 2010. Iron corrosion in an anoxic soil: comparison between thermodynamic modelling and ferrous archaeological artefacts characterised along with the local in situ geochemical conditions. *Applied Geochemistry*, **25**, 1937–1948.

Samper, J., Lu, C. & Montenegro, L. 2008. Reactive transport model of interactions of corrosion products and bentonite. *Physics and Chemistry of the Earth*, **33**, 306–316.

Schlegel, M.L., Bataillon, C., Benhamida, K., Blanc, C., Menut, D. & Lacour, J.-L. 2008. Metal corrosion and argillite transformation at the water-saturated, high temperature iron–clay interface: a microscopic scale study. *Applied Geochemistry*, **23**, 2619–2633.

Schlegel, M.L., Bataillon, C., Blanc, C., Prêt, D. & Foy, E. 2010. Anodic activation of iron corrosion in clay media under water-saturated conditions at 90°C: characterization of the corrosion interface. *Environmental Science & Technology*, **44**, 1503–1508.

Schlegel, M.L., Bataillon, C., Brucker, F., Blanc, C., Prêt, D., Foy, E. & Chorro, M. 2014. Corrosion of metal iron in contact with anoxic clay at 90°C: characterization of the corrosion products after two years of interaction. *Applied Geochemistry*, **51**, 1–14, https://doi.org/10.1016/j.apgeochem.2014.09.002

Steefel, C.I., Appelo, C.A.J. *et al.* 2014. Reactive transport codes for subsurface environmental simulation. *Computational Geosciences*, **19**, 445–478.

Wersin, P. & Birgersson, M. 2014. Reactive transport modelling of iron–bentonite interaction within the KBS-3H disposal concept – the Olkiluoto site as case study. *In*: Norris, S., Bruno, J. *et al.* (eds) *Clays in Natural and Engineered Barriers for Radioactive Waste Confinement*. Geological Society, London, Special Publications, **400**, 237–250, https://doi.org/10.1144/SP400.24

Bentonite re-saturation: different conceptual models – similar mathematical descriptions

KLAUS-PETER KRÖHN

Gesellschaft für Anlagen- und Reaktorsicherheit (GRS) mbH, PO Box 38122, Braunschweig, Germany

Klaus-Peter.Kroehn@grs.de

Abstract: Thermohydromechanically (THM) coupled models based on two-phase flow formulations for the hydraulics are presently the most established tools for simulating bentonite re-saturation. Alternatively, the extended vapour diffusion (EVD) model, involving basically diffusion of water vapour and of hydrated water, has been developed specifically for this purpose. These two model concepts are obviously based on entirely different hydraulic processes.

However, applying some acceptable simplifications to the underlying mathematical models leads to an identical formulation for the two approaches that looks formally like Fick's second law with a saturation-dependent 'diffusion coefficient'. Concerning the hydraulics, the THM codes hence state a mathematical problem that is similar to the one resulting from the EVD model.

A subsequent quantitative comparison of the two hydraulic approaches requires only the respective diffusion coefficients to be compared. This has been done on the basis of benchmark calculations for two different water-uptake tests with compacted MX-80 at largely the same conditions. Even for variations in temperature and dry density, the resulting diffusion coefficients for the two approaches showed remarkable agreement.

Different physical models have thus led to a similar mathematical problem. The mathematics indicates a diffusion-like process, although the exact nature of this process still remains to be determined.

The history of modelling bentonite re-saturation in the framework of repositories for radioactive waste began in the early 1980s. The first attempts to describe the process mathematically led to an empirical model formally based on Fick's second law (Pusch 1980; Börgesson 1984) with a constant 'diffusion coefficient' and the local water content as the unknown. This approach worked fine for isothermal problems at atmospheric pressure but failed for conditions that were more complex.

In the early 1990s, the first thermohydromechanically coupled (THM) models, as we know them today, were developed to tackle this problem. An informative compilation of work from that period can be found in Stephansson *et al.* (2001). It shows that, for historical reasons, a two-phase-flow (TPF) formulation was integrated as the hydraulic part of these models. In some of these cases, the gas phase was neglected conceptually. TPF was then replaced by a formulation for unsaturated flow.

However, as with the empirical 'diffusion' model, again this approach did not represent the physical processes that are actually taking place on the micro-scale. The classic TPF approach that is usually applied in THM models is based on a single porosity (e.g. Helmig 1997), while in clays there is, in addition, the interlamellar space containing hydrated water (e.g. Jasmund & Lagaly 1993; Pusch & Yong 2006). Furthermore, the retention curve replacing the capillary pressure–saturation relationship of the classical TPF approach represents suction in THM models and not capillary forces. Finally, the concept of relative permeability developed for rigid porous media (e.g. Brooks & Corey 1964) does not apply to re-saturating bentonite, where pore size clearly change with saturation.

At the turn of the century, Gesellschaft für Anlagen- und Reaktorsicherheit (GRS) had performed an uptake experiment using water vapour as the source for re-saturation. It showed a water uptake that compared quantitatively rather well with experiments using liquid water. This observation triggered the development of a vapour diffusion model that explained bentonite re-saturation solely by binary vapour diffusion in the pore space and by hydration (Kröhn 2004).

However, in the course of qualifying the referring code VIPER (Kröhn 2011), it turned out that *in situ* water-uptake tests with an initially high saturation could not satisfactorily be explained by this very simple model approach. The problem was solved by adding diffusion of hydrated water in the interlamellar space onto the model concept, which led to the extended vapour diffusion (EVD) model (for details, see Kröhn 2011). In order to concentrate on the hydraulic aspect of re-saturation, the EVD was developed under the premise of a fully bentonite-filled confined space, which is a typical

From: NORRIS, S., BRUNO, J., VAN GEET, M. & VERHOEF, E. (eds) 2017. *Radioactive Waste Confinement: Clays in Natural and Engineered Barriers*. Geological Society, London, Special Publications, **443**, 269–279.
First published online July 6, 2016, https://doi.org/10.1144/SP443.12

condition for the buffer in a repository. This allowed mechanical calculations to be dropped and thereby accelerated model calculations considerably. At present, simulation of heat transport is not yet incorporated into the model. Instead, at the present development stage, predetermined or measured temperature data are used as input. In the following, only hydraulic processes will thus be addressed.

Despite their highly differing physical bases, the TPF and EVD approaches have been shown to yield good results for simulating the re-saturation of bentonite under various conditions (e.g. Kröhn 2011; Gens 2016). The objective of the work presented here is, therefore, to compare these approaches not on the basis of the assumed physical concept, but on the basis of the resulting mathematical models.

Comparison of approaches

Despite the conceptual differences between VIPER and the conventional thermohydromechanically (THM) coupled codes, an interesting relation should be noted. Two-phase water flow in the THM codes is obviously based on advection. However, two-phase-flow equations can be transformed into Richards' equation if the influence of gas is neglected. The Richards' equation can, in turn, be formally interpreted as a diffusion equation whose only coefficient depends eventually on the degree of saturation. Evaluating this coefficient based on constitutive relationships actually used for simulating bentonite re-saturation – meaning the relative permeability–saturation relationship and the retention curve – shows that the saturation-dependent coefficient in the Richards' equation varies little in the range between 20 and 80% saturation. This statement also holds true more or less independently of the chosen constitutive relationships. If the gas phase is of negligible influence on the water migration, the THM codes therefore solve a mathematical problem closely resembling a diffusion equation for the hydraulic part.

As the EVD model also basically describes diffusion processes, it appears that both mathematical formulations can be brought into the same form where only the resulting 'diffusion coefficient' has different meanings. In this case, a quantitative comparison of the two hydraulic models can be reduced to a comparison of this diffusion coefficient. Based on two comparable water-uptake tests, characteristic parameters for the respective models were compiled for this purpose. Where the empirical model is applicable, even the related diffusion coefficients are included in the comparison. The required simplifications, as well as the subsequent comparison of the resulting diffusion coefficients, are the subjects of this paper.

Simplifications of mathematical models

A general starting point for deriving a balance equation is the definition of a time-dependent extensive state variable $Z(t)$ in a moving domain $B(t)$ where $z(\boldsymbol{x}, t)$ is the related density (please refer to Table 1 for a list the symbols used in the equations and their definitions):

$$Z(t) = \int_{B(t)} z(\boldsymbol{x}, t)\,\mathrm{d}V. \qquad (1)$$

Allowing for a flux J of quantity Z across the moving surface of B, and including a source r of quantity Z within B, Reynolds' transport theorem yields a general balance equation for a fixed domain G (e.g. Gärtner 1987):

$$\int_G \left[\frac{\partial z}{\partial t} + \nabla(\boldsymbol{v}z + \boldsymbol{J}) - r\right] \mathrm{d}V = 0. \qquad (2)$$

Each mathematical formulation describing water migration in the re-saturating bentonite that follows from one of the three conceptual models considered here is either directly derived from equation (2) or can be interpreted as ensuing from equation (2).

The two-phase-flow (TPF) model

In the case of two-phase flow, two extensive quantities $Z_i(t)$, $i = 1, 2$ are considered: the masses of gas and of liquid, respectively (Helmig 1997). The related mass densities $z_i(\boldsymbol{x}, t)$, as well as the generalized Darcy's law, are inserted in equation (2). The diffusive flux $\boldsymbol{J}$ is neglected, leading to:

$$\frac{\partial(\rho_i \Phi S_i)}{\partial t} + \nabla\left(\rho_i \frac{k_{ri}}{\eta_i}\boldsymbol{k}(\nabla p_i - \rho_i g)\right) = \rho_i q_i. \qquad (3)$$

The balance equation for the gas (index g) can be reduced to the storage term:

$$\frac{\partial(\rho_g S_g \Phi)}{\partial t} = 0 \qquad (4)$$

applying the following assumptions:

- very fast gas flow in comparison to all other processes;
- negligible influence of gravity on gas;
- no sinks or sources for gas.

No temporal change in the storage term means that there is a local conservation of mass where the bulk density of gas ($\rho_g S_g \Phi$) remains constant. Therefore, equation (4) is of no further interest. Note that Gens (2016, p. 124) concluded, with a view to calculating a non-isothermal uptake test with THM models, that the results 'with and without air balance equations' were quite similar.

Table 1. *List of nomenclature used in this paper*

	Description	Dimension
Symbols		
$\boldsymbol{g}$	Gravitational acceleration	m s^{-2}
k_r	Relative permeability	–
$\boldsymbol{k}$	Absolute permeability	m^2
p	Pressure	Pa
p_c	Capillary pressure/suction	Pa
q	Production rate of a sink/source	kg m^{-3} s^{-1}
r	Sink/source of Z in G	<dim*> m^{-3} s^{-1}
r_h	Relative humidity	–
$\tilde{r}$	Source of vapour	kg m^{-3} s^{-1}
$\bar{r}$	Source of interlamellar water	kg m^{-3} s^{-1}
w	Gravimetric water content of the bentonite	$kg_{water}\ kg_{solids}^{-1}$
t	Time	s
$\boldsymbol{v}$	Advective flow velocity	m s^{-1}
$\boldsymbol{x}$, x	Position vector, x-coordinate	m
z	Density of Z	<dim*> m^{-3}
B	time-dependent 3D domain	m^3
D_m	Coefficient of binary vapour diffusion in air	m^2 s^{-1}
D'	Coefficient of diffusion of the interlamellar water	m^2 s^{-1}
$\tilde{D}$	Diffusion coefficient for the empirical approach	m^2 s^{-1}
$\hat{D}$	Diffusion coefficient derived from the TPF approach	m^2 s^{-1}
$\bar{D}$	Diffusion coefficient derived from the EVD approach	m^2 s^{-1}
G	3D domain fixed in space	m^3
$\boldsymbol{J}$	Non-advective flux of Z across the surface of G	<dim*> m^2 s^{-1}
S	Degree of saturation	–
T	Temperature	K
Z	extensive state variable in B with the dimension <dim>	<dim*>
β_p	Compressibility of water defined as $\beta_p = (1/\rho_l)(\partial\rho_l/\partial p_l)$	Pa^{-1}
β_T	Volumetric coefficient of thermal expansion of water defined as $\beta_T = (1/\rho_l)(\partial\rho_l/\partial T)$	K^{-1}
η	Viscosity	Pa s
ρ	Density	kg m^{-3}
σ	Surface tension	N m^{-1}
τ	Tortuosity of the pore space	–
τ_{hyd}	Tortuosity of the interlamellar space	–
Δ	Laplacian operator	m^{-2}
Φ	Porosity	–
Indices		
a	Macroscopic, apparent	
d	Dry state of the bentonite	
g	Gas	
l	Liquid	
ini	Initial state of the bentonite (i.e. partially saturated state)	
max	Fully saturated state of the bentonite	
s	Solid	
w	Water	
v	Vapour	
12	At completion of one hydrate layer	
23	At completion of two hydrate layers	
hyd i	At forming the i-th hydrate layer	
sat	At vapour saturation	

*dim denotes the dimension of the not yet specified quantity Z.

A further set of assumptions is based on a direct relationship between physical assumptions for the liquid phase and the referring mathematical formulations:

(a) pressure- and temperature-dependent density;
(b) instantaneous local thermal equilibrium;
(c) confined conditions for the bentonite;
(d) negligible changes of porosity;

(e) negligible influence of gravity on liquid phase;
(f) capillary pressure can be equated with suction;
(g) negligible temporal change of the gas pressure;
(h) negligible spatial variations of the liquid density.

With the help of assumptions (a)–(h), equation (3) can be transformed for the liquid phase (index l) into:

$$\left(1 + S_l\beta_p\frac{\partial p_c}{\partial S_l}\right)\frac{\partial S_l}{\partial t} + S_l\left(\beta_T + \beta_p\frac{\partial p_c}{\partial T}\right)\frac{\partial T}{\partial t} + \nabla\left(\frac{k_{rl}}{\eta_l}\frac{\partial p_c}{\partial S_l}\frac{\boldsymbol{k}}{\Phi}\nabla S_l\right) = \frac{q_l}{\Phi}. \quad (5)$$

Further simplifications of equation (5) require a deeper insight into the properties of water and bentonite. The referring data have been compiled and analysed (Kröhn 2016), which led to additional assumptions:

(i) the term $S_l\beta_p\frac{\partial p_c}{\partial S_l}$ is negligible (for $S_l > 0.3$);
(j) slow temperature changes;
(k) no sinks or sources for the liquid phase;
(l) negligible spatial variations of $\hat{D}$;
(m) homogeneous permeability;
(n) negligible compressibility of water;

The additional assumptions (i)–(n) allow simplification of equation (5). In the case of isotropic permeability, it can be written as:

$$\frac{\partial S_l}{\partial t} - \hat{D}\Delta S_l = 0 \quad \text{with } \hat{D} = -\frac{k_{rl}}{\eta_l}\frac{\partial p_c}{\partial S_l}\frac{k}{\Phi}. \quad (6)$$

Note that all these assumptions are largely compatible with the process of re-saturation of a confined bentonite buffer. Note further that equation (6) is formally one way to formulate the Richards' equation for unsaturated flow.

The extended vapour diffusion (EVD) model

Based on equation (2), a balance equation for the mass of vapour in the pore space when considering binary gas diffusion in the pore space and mass loss to the interlamellar space by hydration can be derived (Kröhn 2011):

$$\frac{\partial(\Phi\rho_v)}{\partial t} - \nabla(\Phi\tau D_m\nabla\rho_v) = \tilde{r}. \quad (7)$$

A similar balance equation for the interlayer water when considering diffusion of interlamellar water, as well as mass gain from the pore space by hydration, reads:

$$\rho_d\frac{\partial w}{\partial t} - \rho_d\nabla(\tau_{hyd}D'\,\nabla w) = \bar{r}. \quad (8)$$

Coupling equations (7) and (8) on the condition of instantaneous hydration, meaning that $\tilde{r} = -\bar{r}$, allows the water balance equation to be written as:

$$\frac{\partial(\Phi\rho_v)}{\partial t} - \nabla(\Phi\tau D_m\nabla\rho_v) + \rho_d\frac{\partial w}{\partial t} - \rho_d\nabla(\tau_{hyd}D'\nabla w) = 0 \quad (9)$$

Several substitutions and simplifications described in detail in Kröhn (2011) lead to a partial differential equation with the vapour partial density and temperature as the only variables. This equation is rather complex as it consists of seven summands containing time derivatives, gradients and Laplacians of both variables. Again, a deeper insight into the physical properties of water and bentonite was required to identify negligible terms. The referring considerations leading to the following assumptions can be found in detail in Kröhn (2016):

(i) assumptions leading to the non-isothermal balance equation: see Kröhn (2011);
(ii) slow temperature changes;
(iii) weak temperature-dependency of both diffusion coefficients (vapour and interlamellar water);
(iv) rough approximation of the measured isotherm by a linear function;
(v) small curvature in the temperature distribution.

These assumptions are also compatible with the re-saturation process in a bentonite, and lead to the following form of the balance equation:

$$\frac{\partial S}{\partial t} - \bar{D}\Delta S = 0 \quad \text{with} \quad \bar{D} = \frac{\rho_{v\,sat}}{\rho_d w_{max}}\Phi_{ini}\tau_{ini}D_m + \tau_{hyd}D' \quad (10)$$

where the degree of saturation, S, is defined as:

$$S = \frac{w}{w_{max}}. \quad (11)$$

Note that the interlamellar water, which constitutes the water in the bentonite by a majority, is assumed to have a constant density independent of the state of hydration.

The empirical re-saturation model

A working hypothesis underlying the empirical re-saturation models was that the gradient of the water content is the driving force for water uptake. While uptake experiments evaluated by this equation were usually one-dimensional, there exists also an uptake test showing that this relationship holds even in three dimensions (Gattermann 1998). The mathematical formulation thus reads:

$$\frac{\partial w}{\partial t} - \tilde{D}\Delta w = 0. \quad (12)$$

The same formulation emerges from the general balance equation, equation (2). The mass of water is again balanced, as described in the earlier subsection on 'The two-phase-flow (TPF) model', but, instead of allowing for advective flow in addition to sinks and sources, only a diffusive water flux is taken into account:

$$\frac{\partial(\rho_i \Phi S_i)}{\partial t} - \nabla(\Phi\tilde{D}[\nabla(\rho_i S_i)]) = 0. \quad (13)$$

Water density and porosity are again considered to be constant. The same applies to the diffusion coefficient. Equation (13) can thus be written as:

$$\frac{\partial S}{\partial t} - \tilde{D}\Delta S = 0. \quad (14)$$

Inserting relationship (11) into equation (14) shows that equations (12) and (14) are equivalent.

Summary

For all concepts considered here, the evolving mathematical models are based on a mass-balance equation for the water contained in the bentonite. All formulations originate from, or can be traced back to, the general balance equation (2). After certain simplifications, the resulting equations (6), (10) and (14), respectively, are formally identical and all are based on the degree of saturation as the primary variable. Note that the density of the water in the bentonite is assumed to be constant for all practical purposes. It does not matter, therefore, whether the degree of saturation is based on a ratio of volumes or on a mass ratio.

While initial and boundary conditions of the models are thus equivalent, the models differ in the understanding of the nature of the water in the bentonite and in the underlying processes of water migration. We assume advection of liquid water in the case of two-phase flow, binary gas diffusion and diffusion of interlayer water in the case of the EVD, and an unspecified diffusion process of liquid water in the case of the empirical model.

Comparison of resulting diffusivities

Methodological approach

Because of the formal identity of equations (6), (10) and (14), a comparison of the approaches can be reduced to a comparison of the related diffusion coefficients: $\hat{D}$ for the two-phase-flow (TPF) approach, $\bar{D}$ for the extended vapour diffusion (EVD) approach; and $\tilde{D}$ for the empirical approach. While $\tilde{D}$ represents just a constant value, $\hat{D}$ and $\bar{D}$ depend on material parameters and constitutive relationships that may, in turn, be dependent on specific conditions of the bentonite.

In case of the TPF, the required parameters and constitutive relationships (i.e. the relative permeability–saturation relationship and the retention curve) are not yet clearly established. In most cases, however, a cubic relationship between relative permeability and saturation is used. The comparison of diffusion coefficients is therefore based here on two independent non-isothermal water-uptake tests with MX-80 bentonite at comparable dry densities: one performed in the laboratory at CEA (Gatabin & Billaud 2005); and one performed *in situ* at the Hard Rock Laboratory at Äspö (Kristensson & Börgesson 2007). Characteristic parameters for these tests are compiled in Table 2. These two experiments have been used by the International Task Force on Engineered Barrier Systems of SKB and UPC as a basis for a modelling exercise (e.g. Gens 2016). This exercise includes contributions from AECL (Canada), BGR (Germany), CIMNE (Spain), CRIEPI (Japan), Clay Technology (Sweden) and Cardiff University (the UK). It provides a selection of parameters and constitutive relationships that were individually chosen by the participants to match the modelling results with the measurements. The resulting data form the basis of a quantitative formulation of the diffusion coefficient and, at the same time, also indicate the uncertainty of the input in the framework of the TPF.

The three diffusion coefficients, including their respective ranges of uncertainties, are then quantified. As a reference state for the comparison, a bentonite at a dry density of 1750 kg m^{-3}, a

Table 2. *Characteristic parameters for the tests providing the basis for comparison*

Quantity	Laboratory test, sample 1	Laboratory test, sample 2	*In situ* test
Dry density (kg m^{-3})	1791	1735	1782
Maximum temperature (°C)	150	150	95
Initial water content (%)	13.7	17.9	17
Initial saturation (%)	75.5	89.7	85

temperature of 20°C and an initial water content of 16% was chosen. In order to ensure the validity of the results for varying conditions, two additional comparisons were also performed. As the bentonite dry density had been rather high in the two experiments, the comparison was repeated with a significantly lower, but still relevant, value and with accordingly recalculated diffusion coefficients $\hat{D}$ and $\bar{D}$ (cf. equations 6 and 10). Finally, a variant in the comparison with increased temperature was also provided. All relevant parameter data are listed in Table 3. For the comparison at elevated temperatures, however, the coefficient $\tilde{D}$ had to be excluded as the empirical approach had failed for increased temperatures in the bentonite (L. Börgesson pers. comm. 2001).

Quantification of the diffusion coefficients

For most simple cases of water uptake at room temperature, even the diffusion coefficient $\tilde{D}$ found by various authors using the first empirical models can be taken into account. They are compiled in Table 4. These coefficients are obviously independent of the degree of saturation.

In all the THM models considered here, a cubic relationship between relative permeability and saturation has been adopted. There is no established retention curve, however. The choice depends on the code, the choice of other parameters, material models and, eventually, even the judgement of the modeller. The remaining uncertainty can be appreciated from the variety of retention curves that have been used in the benchmark exercise (Gens 2016), as depicted in Figure 1. The resulting curves for the coefficient $\hat{D}$ based on the retention curves in Figure 1 are compiled in Figure 2. These curves span a band with a width of only one order of magnitude. The slope is rather low in the region, between about 20 and 95% saturation, but quite high outside this range.

In many cases, a singularity appears at full saturation. This singularity is caused by using the van Genuchten approach for the retention curve, which is customarily done in the framework of the TPF approach. Note that modified retention curves have been used in several models to improve the modelling results. While the modifications are hardly noticeable in Figure 1, they become obvious as jumps at saturations above 0.6 in the ensuing curves in Figure 2. With one exception, all of them resulted in a related diffusion coefficient $\hat{D}$ that has a finite value at full saturation. In fact, the singularity could easily have been avoided by switching to an equivalent curve based on the Brooks–Corey approach (Brooks & Corey 1964). It might also prove to be useful to turn to more recent formulations of retention curves that are specifically derived to cope with clayey materials (e.g. Revil & Lu 2013).

The saturation range below 20% is not of interest in the case of isothermal re-saturation, as air-dry bentonite typically already shows a degree of saturation of about 40%. While water uptake under non-isothermal conditions can theoretically lead to saturation values below 20%, this was not the case in either of the experiments considered here. This range is therefore excluded from the comparison.

In contrast, the coefficient $\bar{D}$ for the EVD approach is clearly defined by the initial saturation, the degree of bentonite compaction and temperature. Only the tortuosity of the pore space and the interlamellar space introduce an uncertainty. The values found typical for the two experiments in question ($\tau = 0.6$ and $\tau_{hyd} = 0.28$) were used to quantify $\bar{D}$. To get an impression of the possible variability, both tortuosity values are also tentatively set to 0.2 and 0.8.

Comparison for uptake tests at room temperature

The diffusion coefficients for the TPF, the EVD and the empirical approaches show a very good fit in the reference case, as depicted in Figure 3. The curves for the EVD approach and the values for the empirical approach are within the bandwidth of the different curves for the TPF approach. Note that the diffusion coefficients for the empirical approach have been determined for bentonites with very different degrees of compaction (dry densities of between 1430 and 1900 kg m^{-3}: cf. Table 4).

In addition, a trend for increasing values with increasing saturation can be observed in $\hat{D}$, as well as in $\bar{D}$. While the diffusion coefficient increases continuously in the case of the TPF approach, this increase occurs in steps in the case of the EVD approach. This seemingly curious behaviour can be explained by the fact that the diffusion of the interlayer water depends not only on the gradient of the water content, but also on the number of completed hydrate layers in the interlamellar space (Skipper *et al.* 2006). At the related water content, the coefficient for interlayer water diffusion changes spontaneously in the model. In reality, however, a somewhat smoother transition is expected.

A final observation is that the diffusion coefficient $\bar{D}$ for the EVD approach has a finite value at full saturation. This property can also be found in most examples of diffusion coefficient $\hat{D}$ for the TPF approach that were based on modified retention curves. The significant number of modified retention curves used in the framework of the TPF approach therefore seems to indicate certain shortcomings of the approach after van Genuchten in the range close to full saturation.

Table 3. *Bentonite conditions for the comparison of diffusion coefficients*

Quantity	Dimension	Value	Source
Material parameters			
ρ_s	kg m^{-3}	2780	For example, Kristensson & Börgesson (2007)
ρ_w	kg m^{-3}	1000	(Commonly used)
ρ_d	kg m^{-3}	1750/1500*	(Defined)
$\Phi_d(\rho_d)$	–	0.371/0.460	For example, after Kröhn (2004)
$w_{max}(\rho_d)$	–	0.212/0.307	For example, after Kröhn (2004)
w_{ini}	–	0.160	(Reference value; defined)
TPF parameters			
$k(\rho_d)$	m^2	3.33×10^{-21}/ 1.41×10^{-20}*	For example, after Kröhn (2004)
η ($T = 20°C$)	Pa s	0.001002	After IAPWS (2003)
η ($T = 80°C$)		0.0003544	
σ ($T = 20°C$)	N m^{-1}	0.072736	After IAPWS (1994)
σ ($T = 80°C$)		0.0626714	
EVD parameters: pore space			
$\Phi_{ini}(\rho_d, w_{ini})$	–	0.091/0.220*	For example, after Kröhn (2004)
τ	–	0.2, 0.8, 0.6†	(Estimated)
D_m ($T = 20°C$)	m^2 s^{-1}	2.45×10^{-05}	After Vargaftik *et al.* (1996)
D_m ($T = 80°C$)		3.43×10^{-05}	
D_a (reference)	10^{-11} m^2 s^{-1}	1.25, 5.01, 3.76†	After Kröhn (2011)
D_a ($\rho_d = 1500$ kg m^{-3})		3.56, 14.2, 10.7†	
D_a ($T = 80°C$)		29.5, 118, 88.5†	
p_{sat} ($T = 20°C$)	Pa	2339	After IAPWS (1997)
p_{sat} ($T = 80°C$)		47 415	
$\rho_{v\ sat}$ ($T = 20°C$)	kg m^{-3}	0.0173	After IAPWS (1997)
$\rho_{v\ sat}$ ($T = 80°C$)		0.2912	
EVD parameters: interlamellar space			
τ_{hyd}	–	0.2, 0.8, 0.28†	(Estimated)
w_{12}	–	0.075	Kröhn (2011)
w_{23}	–	0.170	Kröhn (2011)
$S_{12}(w_{12}, w_{max})$	–	0.35/0.24*	After Kröhn (2011)
$S_{23}(w_{23}, w_{max})$	–	0.80/0.55*	After Kröhn (2011)
D'_{hyd1}	m^2 s^{-1}	2.00×10^{-10}	See Kröhn (2011)
D'_{hyd2}	m^2 s^{-1}	1.00×10^{-9}	See Kröhn (2011)
D'_{hyd3}	m^2 s^{-1}	2.00×10^{-9}	See Kröhn (2011)
D'_a ($T = 20°C$, D'_{hyd1})	10^{-11} m^2 s^{-1}	4.00, 16.0, 5.60*	After Kröhn (2011)
D'_a ($T = 80°C$, D'_{hyd1})		5.09, 20.4, 7.13*	
EVD parameters: hydration			
$\partial w/\partial r_h$	–	w_{max} (linear isotherm)	Kröhn (2011)

*Reference value/variation.
†Minimum, maximum and reference value.

Table 4. *Empirical diffusion coefficient $\tilde{D}$ for MX-80 at room temperature*

$\tilde{D}$ (m^2 s^{-1})	Dry density (kg m^{-3})	Source
1.0×10^{-9}	*c.* 1900	Pusch (1980)
2.2×10^{-10}; 3.0×10^{-10}	1430;1750	Börgesson (1984)
3.4×10^{-10}	?	Kahr *et al.* (1986)
3.5×10^{-10}	'Highly compacted'	Bucher & Müller-Vonmoos (1987)
3.5×10^{-10}	1500	Kröhn (2005)

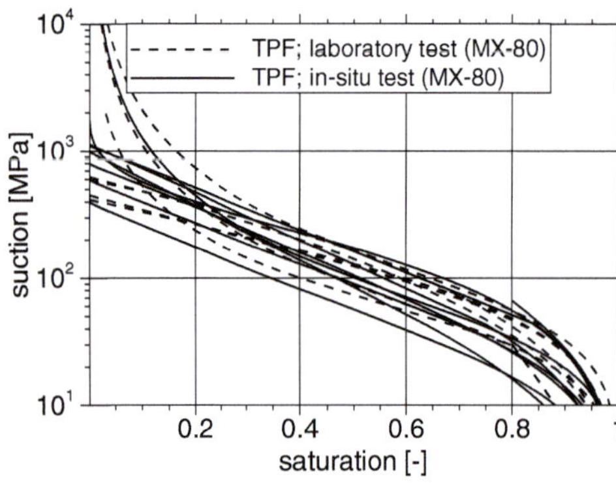

Fig. 1. Retention curves for two tests with MX-80 at a dry density of about 1750 kg m^{-3}.

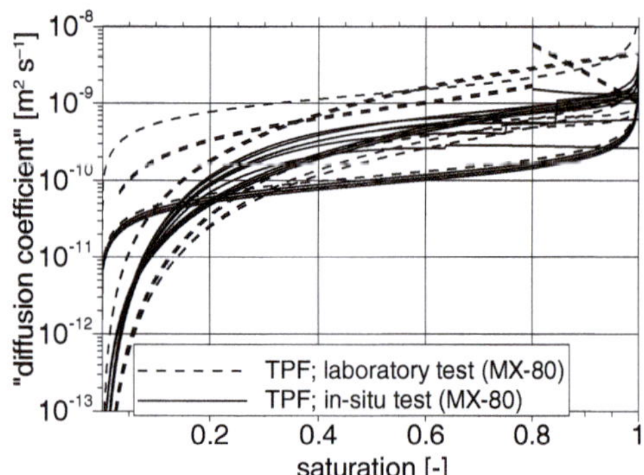

Fig. 2. Diffusion coefficient based on the retention curves in Figure 1.

Impact of dry density

To validate the results of conditions different to the reference case, the influence of dry density on the coefficients $\hat{D}$ and $\bar{D}$ was investigated, too. According to equations (6) and (10), a decrease in dry density leads either directly or indirectly to an increase in $\hat{D}$ and $\bar{D}$ via the related porosity.

A possible influence of the dry density on the retention curve is neglected here, however. Suction data for FEBEX bentonite at dry densities of between 1550 and 1750 kg m^{-3} have been measured under constant volume conditions and indicate a certain dependency of the dry density for saturations above approximately 80% (Sánchez & Gens 2006). This result, however, is not confirmed by suction data for dry densities between 1600 and 1700 kg m^{-3}, which are presented by Lloret *et al.* (2004). The resulting curves apparently show qualitatively the same shape if scaled to the maximum water content. In this sense, Dueck & Börgesson (2007) developed a method to describe the effect of swelling under constant volume conditions on the retention curve. This method is based on suction data from free swelling samples where a modification is applied above a saturation of approximately 80%. Since the retention curve appears to be only moderately affected by the dry density at most, this influence is not investigated here.

The repository relevant range for dry density lies roughly between 1500 and 1800 kg m^{-3}, so a value of 1750 kg m^{-3} in the reference case appears to lie at the upper end of the spectrum. As a variant, a value of 1500 kg m^{-3} was therefore chosen. The resulting

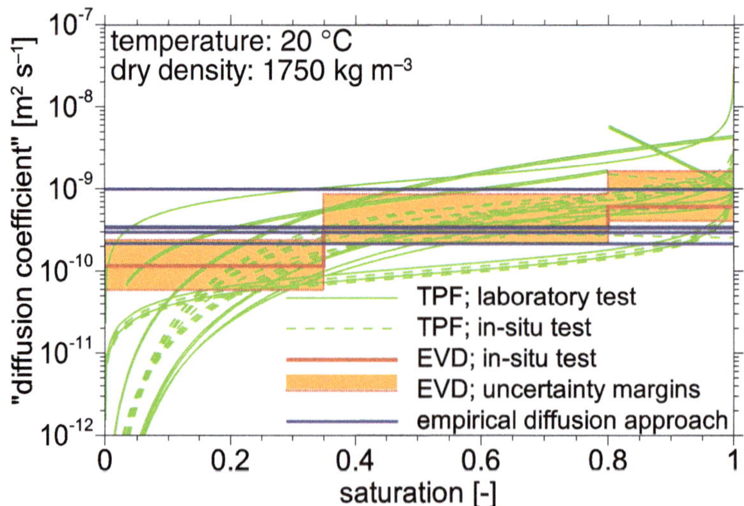

Fig. 3. Diffusion coefficients from different model approaches at $T = 20°C$ and $\rho_d = 1750$ kg m^{-3}.

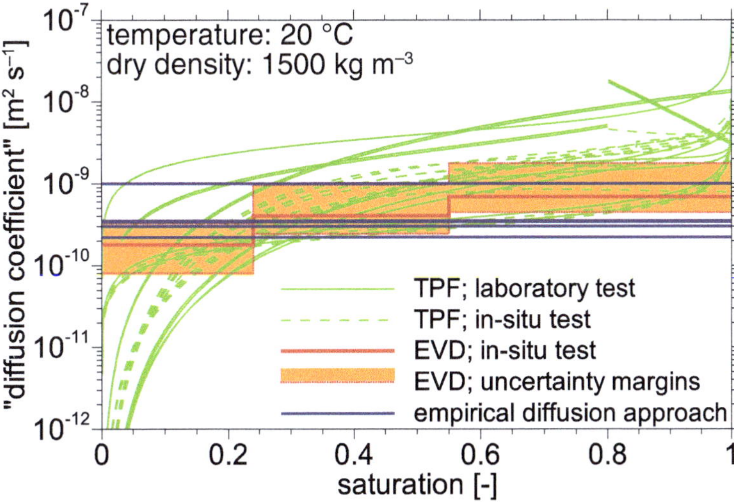

Fig. 4. Diffusion coefficients at $T = 20°C$ and $\rho_d = 1500$ kg m^{-3}.

shifted curves are shown in Figure 4. The match between the empirical coefficient $\tilde{D}$ and the two-phase-flow-related coefficient $\hat{D}$ is not quite as good as in the reference case, but is still rather convincing.

Note that the state of completion of a hydrate layer relates to a fixed water content w_{crit}. By switching from water content to the degree of saturation, however, this characteristic value gets lost because the degree of saturation is scaled to the maximum water content, which, in turn, depends on the dry density. The degree of saturation that represents the specific water content w_{crit} thus varies with the dry density.

Impact of temperature

The influence of temperature on the coefficients $\hat{D}$ and $\bar{D}$ is shown exemplarily by repeating the coefficient comparison for a temperature of 80°C. The TPF-related coefficient $\hat{D}$ is affected by the temperature-dependent water viscosity, as well as by the surface tension that controls the capillary pressure. The coefficient $\bar{D}$ from the EVD approach is changed by the temperature-dependent saturation partial density of the vapour and the diffusion coefficient for binary gas diffusion in the pore space, as well as the diffusion coefficient of the interlayer water. Despite the different temperature-dependent

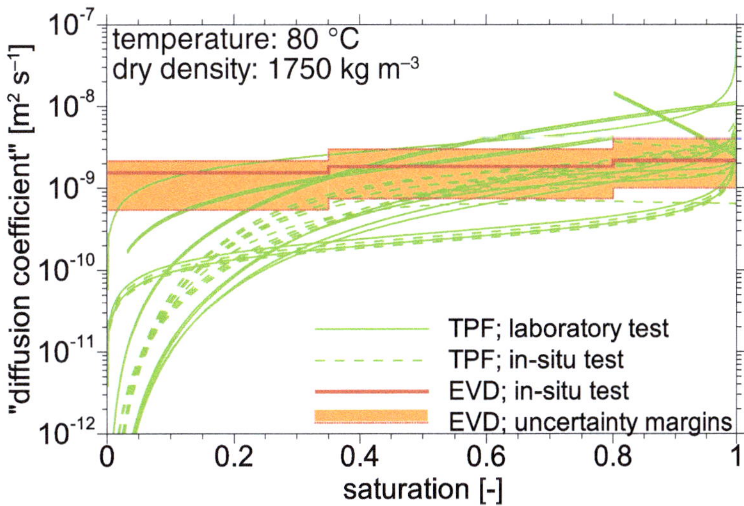

Fig. 5. Diffusion coefficients at $T = 80°C$ and $\rho_d = 1750$ kg m^{-3}.

physical properties changing the diffusion coefficients $\hat{D}$ and $\bar{D}$ with temperature, Figure 5 never theless, again, shows a good match between the resulting curves.

Summary and conclusions

From a conceptual and physical point of view, several entirely different approaches for the hydraulics of the re-saturation of bentonite have been derived over the last few decades. Among them are

- the empirical Fickian 'diffusion' approach;
- the two-phase-flow (TPF) approach utilized in thermohydromechanically coupled models
- the extended vapour diffusion (EVD) approach for a confined buffer.

However, all three formulations originate from, or can be traced back to, a general balance equation for the water in the bentonite. Applying a number of simplifications to the mathematical formulations for the latter two approaches reveals a basic similarity of the mathematical models for all three concepts. Formally, the simplified formulations look like Fick's second law or a special form of Richards' equation.

But the resemblance of the mathematical models goes even beyond the already remarkable qualitative similarity. Formulations for the only parameter – the resulting 'diffusion coefficient' – were derived to provide the means of a quantitative comparison. Exemplarily, these coefficients have been calculated based on two independent non-isothermal water-uptake tests with MX-80 bentonite with a dry density of about 1750 kg m^{-3}: one performed in the laboratory at CEA; and one *in situ* at the Hard Rock Laboratory at Äspö. In the framework of an international benchmark for modelling re-saturation, each of the participants used different constitutive relationships to explain the measurements in these tests. The resulting variety of constitutive relationships formed an excellent base for this comparison because, in the end, it also provided the means to indicate the uncertainty in the resulting distributions of the diffusion coefficients.

During the benchmark exercise, considerable effort had gone into modifying the retention curve for the TPF approach to achieve a reasonable fit with the measurements. It is interesting to note that all modifications, except one, resulted in a finite value for the diffusion coefficient at full saturation rather than the infinite value that is provided by the original approach for the retention curve after van Genuchten. This, on the one hand, indicates certain shortcomings of the van Genuchten approach and, on the other hand, confirms the EVD approach where the diffusion coefficient is *a priori* finite at full saturation.

As it turned out, the resulting diffusion coefficients plotted as functions of the degree of saturation agreed surprisingly well. Even the coefficients from the empirical approach fell into the covered range.

To broaden the validity of this conclusion, the influence of temperature and of the degree of compaction (expressed as the dry density) were also investigated. All parameters in the formulations for the diffusion coefficients that depend on temperature and dry density were changed accordingly for that purpose. The match of the distributions for the TPF approach and for the EVD formulation was, again, fairly good.

Despite all similarities of the diffusion coefficients, it has to be conceded, however, that the uncertainty margins involved are not negligible. Ideally, the two water-uptake tests should therefore be simulated on the basis of the two simplified approaches optimizing and, eventually, comparing the referring diffusion coefficients. However, these coefficients are highly dependent on saturation and temperature, with the consequence that simple analytical solutions for Fick's second law are not applicable. Putting up a specific framework for this kind of numerical investigation is therefore advisable, but remains to be done in the future.

The mathematical models for the three approaches are based on a quite different conceptual understanding of the processes controlling re-saturation and are therefore certainly not identical. Nevertheless, they show a strong underlying similarity in terms of the resulting mathematical formulations, indicating that re-saturation could be basically a diffusion-like process after all. However, the following question can apparently not be answered yet: Which processes really occur on the micro-scale? This obviously cannot be determined by models alone, but should, without doubt, be the subject of further investigations including theoretical, as well as decisive experimental work.

The author gratefully acknowledges the funding of the various experimental and theoretical investigations within the projects WiGru-7 (FKZ 02 E 11102) and BIGBEN (02 E 11284) by the German Federal Ministry for Economic Affairs and Energy (BMWi).

References

BÖRGESSON, L. 1984. Water flow and swelling pressure in non-saturated bentonite-based clay barriers. Paper presented at the Clay Barriers for Isolation of Toxic Chemical Wastes International Symposium, 28–30 May, Stockholm.

BROOKS, R.H. & COREY, A.T. 1964. *Hydraulic Properties of Porous Media.* Hydrology Paper No. 3 Colorado State University, Fort Collins, CO.

BUCHER, F. & MÜLLER-VONMOOS, M. 1987. Bentonite als technische Barriere bei der Endlagerung hochradioaktiver Abfälle. *Mitteilungen des IGB der ETH Zürich*, **133**, S51–S64.

Dueck, A. & Börgesson, L. 2007. Model suggested for an important part of the hydro-mechanical behavior of a water unsaturated bentonite. *Engineering Geology*, **92**, 3–4.

Gärtner, S. 1987. *Zur diskreten Approximation kontinuumsmechanischer Bilanzgleichungen*. Report No. 24/1987. Institut für Strömungsmechanik und Elektronisches Rechnen im Bauwesen, University of Hanover, Hanover.

Gatabin, C. & Billaud, P. 2005. *Bentonite THM Mock-up Experiments – Sensors Data Report*. CEA Report NT DPC/SCCME 05-300-A. Commissariat à l'énergie atomique et aux énergies alternatives (CEA), Saclay, France.

Gattermann, J.H. 1998. *Theorie und Modellversuch für ein Abdichtungsbauwerk aus hochverdichteten Bentonitformsteinen*. Thesis, TU Aachen.

Gens, A. 2016. *Task Force on Engineered Barrier System (EBS) – Task 1: Laboratory Tests*. Svensk Kärnbränslehantering (SKB) International Report. Swedish Nuclear Fuel and Waste Management (SKB), Stockholm (under review).

Helmig, R. 1997. *Multiphase Flow and Transport Processes in the Subsurface*. Springer, Berlin.

IAPWS 1994. *Release on Surface Tension of Ordinary Water Substance*. International Association for Properties of Water and Steam, http://www.iapws.org

IAPWS 1997. *Release on the IAPWS Industrial Formulation 1997 for the Thermodynamic Properties of Water and Steam*. International Association for Properties of Water and Steam, http://www.iapws.org

IAPWS 2003. *Revised Release on the IAPS Formulation 1985 for the Viscosity of Ordinary Water Substance*. International Association for Properties of Water and Steam, http://www.iapws.org

Jasmund, K. & Lagaly, G. 1993. *Tonminerale und Tone*. Steinkopff, Darmstadt.

Kahr, G., Kraehenbuehl, F., Müller-Vonmoos, M. & Stoeckli, H.F. 1986. *Wasseraufnahme und Wasserbewegung in hochverdichtetem Bentonit*. NAGRA Technischer Bericht 86-14. NAGRA (National Cooperative for the Disposal of Radioactive Waste), Wettingen, Switzerland.

Kristensson, O. & Börgesson, L. 2007. *CRT – Canister Retrieval Test; EBS Task Force Assignments*. Clay Technology, Lund.

Kröhn, K.-P. 2004. *Modelling the Re-Saturation of Bentonite in Final Repositories in Crystalline Rock*. Final Report FKZ 02 E 9430 (BMWA). GRS Report GRS-199. Gesellschaft für Anlagen- und Reaktorsicherheit (GRS), Cologne.

Kröhn, K.-P. 2005. New evidence for the dominance of vapour diffusion during the re-saturation of compacted bentonite. *Engineering Geology*, **82**, 127–132.

Kröhn, K.-P. 2011. *Code VIPER – Theory and Current Status*. Status Report FKZ 02 E 10548 (BMWi). GRS Report GRS-269. Gesellschaft für Anlagen- und Reaktorsicherheit (GRS), Cologne.

Kröhn, K.-P. 2016. *Comparison of Mathematical Models for Bentonite Re-saturation*. FKZ 02 E 11284 (BMWi). Gesellschaft für Anlagen- und Reaktorsicherheit (GRS), Cologne (under review).

Lloret, A., Romero, E. & Villar, M.V. 2004. *FEBEX II Project – Final Report on Thermos-Hydro-Mechanical Laboratory Tests*. Publicación Técnica ENRESA 10/04. Empresa Nacional de Residuos Radiactivos (ENRESA), Madrid.

Pusch, R. 1980. *Water Uptake, Migration and Swelling Characteristics of Unsaturated and Saturated, Highly Compacted Bentonite*. KBS Report 80-11. Swedish Nuclear Fuel and Waste Management (SKB), Stockholm.

Pusch, R. & Yong, R. 2006. *Microstructure of Smectite Clays and Engineering Performance*. Taylor & Francis, London.

Revil, A. & Lu, N. 2013. Unified water isotherms for clayey porous materials. *Water Resources Research*, **49**, 5685–5699, https://doi.org/10.1002/wrcr.20426

Sánchez, M. & Gens, A. 2006. *FEBEX Project – Final Report on Thermo-Hydro-Mechanical Modelling*. Publicación Técnica ENRESA 05-2/2006. Empresa Nacional de Residuos Radiactivos (ENRESA), Madrid.

Skipper, N.T., Lock, P.A., Titiloye, J.O., Swenson, J., Mirza, Z.A., Howells, W.S. & Fernandez-Alonso, F. 2006. The structure and dynamics of 2-dimensional fluids in swelling clays. *Chemical Geology*, **230**, 182–196.

Stephansson, O., Tsang, C.-F. & Kautsky, F. (eds). 2001. DECOVALEX II. *International Journal of Rock Mechanics and Mining Sciences*, **38**, (1) (Special Issue).

Vargaftik, N.B., Vinogradov, Y.K. & Yargin, V.S. 1996. *Handbook of Physical Properties of Liquids and Gases, Pure Substances and Mixtures*. Begell House, New York.

Alteration of MX-80 bentonite backfill material by high-pH cementitious fluids under lithostatic conditions – an experimental approach using core infiltration techniques

F. DOLDER[1]*, U. MÄDER[1], A. JENNI[1] & B. MÜNCH[2]

[1]*RWI, Institute of Geological Sciences, University of Bern, 3012 Bern, Switzerland*

[2]*EMPA, Swiss Federal Institute for Materials Testing and Research, 8600 Dübendorf, Switzerland*

**Correspondence: fdolder@gmail.com*

Abstract: We characterize and quantify processes at a cement–bentonite interface spatially and temporally during a long-term core infiltration experiment.

A young ordinary Portland cement pore fluid (K^+–Na^+–OH^-: pH 13.4) was infiltrated into a MX-80 bentonite core with an initial saturated density of 1920 kg m^{-3} that shifted to 1890–1930 kg m^{-3} after 761 days. A hydrostatic external pressure of 4.1 MPa and an infiltration pressure of 2.1 MPa were applied in a triaxial-type apparatus. A decrease in hydraulic conductivity from approximately 2.2×10^{-13} to approximately 4.2×10^{-15} m s^{-1} was observed passing from advection-dominated flow to a diffusion-dominated regime. Sulphate replaced chloride in the outflow during the high-pH infiltration period controlled by the dissolution of gypsum, the uptake of K^+ by ion exchange, and complex mineral reactions occurred near the inlet. X-ray computed tomography (CT) scans performed repeatedly during the experiment tracked a progressing hemispherical reaction plume in the first millimetres of the bentonite, revealing a zone of bulk density increase. This zone consisted of two distinct, but overlapping, zones of Mg- and Ca-enrichment related to precipitation of saponite and calcite. The experiment attested an effective chemical buffering capacity for bentonite, a progressing coupled hydraulic–chemical sealing process and also the preservation of the physical integrity of the interface region in this set-up with a total pressure boundary condition on the core sample.

Long-term storage of high-level nuclear waste (HLW) and spent fuel (SF) is one of the great challenges of our society. The Swiss concept for a geological repository foresees an engineered barrier system (EBS) to protect the biosphere from SF/HLW (NAGRA 2002). One role of the EBS is to attenuate diffusion of radionuclides. Cement-based materials as tunnel reinforcement and tunnel plugs/seals may affect the EBS (NAGRA 2002). Porewater emanating from concrete or mortar induces a high-pH reaction plume that may influence the chemical and physical properties of the EBS by the dissolution/precipitation of minerals, accompanied by changes in permeability, swelling pressure and radionuclide retention.

Bentonite is considered a possible material for the EBS and consists predominantly of the clay mineral montmorillonite (smectite) (Karnland 2010). Processes like swelling in contact with water and adsorption of cations are related to a negative surface charge of the clay layers, giving rise to a cation-exchange capacity (CEC) and swelling capacity.

Pore fluids of fresh OPC (Ordinary Portland Cement) are strongly enriched in K^+ and Na^+, with an initial pH of 13–14, and controlled by portlandite, calcium silicate hydrates (C–S–H), ettringite and monosulphate saturation (Lothenbach & Winnefeld 2006).

In the case of cement–rock interaction, the period of highest pH at the cement–clay interface is estimated to be relatively short due to buffering reactions. Various experimental and modelling studies predicted relatively rapid and significant cement and rock alterations near the interfaces (Savage *et al.* 2002; De Windt *et al.* 2004; Gaucher & Blanc 2006; Watson *et al.* 2009; Kosakowski *et al.* 2014).

Two main experimental approaches were used in laboratories to perform cement–clay interaction experiments: batch reactors with a high water/solid (w/s) ratio; and flow-through devices using compacted samples (low w/s ratio). A higher w/s ratio led to faster and more prolific reactions, and more available space for mineral precipitation (Eberl *et al.* 1993; De La Villa *et al.* 2001; Ramirez *et al.* 2002*a*; Fernández *et al.* 2006, 2010; Sánchez *et al.* 2006; Karnland *et al.* 2007). A lower w/s ratio meant slower and less extensive reactions. Of direct interest for this study are the results of

From: NORRIS, S., BRUNO, J., VAN GEET, M. & VERHOEF, E. (eds) 2017. *Radioactive Waste Confinement: Clays in Natural and Engineered Barriers*. Geological Society, London, Special Publications, **443**, 281–305.
First published online June 24, 2016, https://doi.org/10.1144/SP443.10

experiments using MX-80 bentonite (Wyoming, USA) or FEBEX bentonite (Almería, Spain) at high-pH conditions. Two main types of high-pH infiltration fluids were used for these experiments: alkaline (K–Na–OH)-based fluids, simulating early cement porewaters (pH 13–14); and (Ca–OH)-based fluids, simulating more evolved pore fluids (pH $\leq$ 12.5) buffered initially by portlandite and C–S–H phases. It was shown that the Ca-dominated pore fluids induced less extensive reactions, even at elevated temperatures (Karnland 1997; Ramirez *et al.* 2002*b*; Karnland *et al.* 2007; Fernández *et al.* 2009, 2010). Hyperalkaline pore fluids were shown to strongly react with bentonite, associated with changing physical and chemical properties. Experiments did show that the temperature was of great importance for the durability of the bentonite barrier near a cement interface. At temperatures above 60°C, the reactivity is strongly increased mainly by pH-dependent faster kinetics. Therefore, results of high-temperature experiments can be used for extrapolation to low temperatures in case the kinetics should be too slow to observe any reaction at feasible timescales. At temperatures >60°C, strong mineral alterations with mineral precipitation/dissolution and a reduction in swelling pressures were observed (Dauzeres *et al.* 2010). Precipitation of zeolites was reported at temperatures >75°C, with analcime occurring in Na-dominated and phillipsite in K-dominated solutions. Calcium (aluminate) silicate hydrates (C–(A)–S–H) and hydrotalcite were also observed. Clays, mainly smectite and illite, were reported to be transformed and interlayered with new mineral phases, as well as being precipitated/dissolved. The dissolution of silicate minerals, like quartz, cristobalite, smectite and feldspar, as well as calcite, gypsum and brucite, was observed. Finally, changes in swelling pressures were reported. At ambient temperatures (20–35°C), most of the studies reported minor changes: the formation of smectite (illite–smectite mixed layers (I–S)), brucite, C–S–H gels and zeolites (phillipsite); and a reduction in swelling pressure (Eberl *et al.* 1993; Karnland 1997; Bauer & Berger 1998; Bauer & Velde 1999; De La Villa *et al.* 2001; Ramirez *et al.* 2002*a*; Fernández *et al.* 2006, 2010; Sánchez *et al.* 2006; Karnland *et al.* 2007).

This work summarizes a core infiltration experiment using a compacted MX-80 bentonite sample and an alkaline cement fluid (pH 13.4) that mimics an OPC pore fluid after full hydration. Computed tomography (CT) measurements allowed tracking of a reaction plume and related density–volume changes over time, while X-ray diffraction (XRD), scanning electron microscopy (SEM), Raman spectroscopy and optical microscopy were used for post-mortem characterization.

Methods and material

Core infiltration experiment

Our experimental approach used a core infiltration method that is characterized by a low w/s ratio. Constant confining pressure is applied to a cylindrical rock sample that may represent lithostatic/hydrostatic pressure conditions. This external pressure is a total-pressure constraint for swelling materials rather than a constant-volume condition used in standard percolation or diffusion equipment. The method was first described in Adler (2001) and Adler *et al.* (2001), and focused on chemical rock–water interaction combined with tracer transport and retardation in Opalinus Clay. The application was continuously extended to porewater extraction by advective displacement in clay rocks (Mäder *et al.* 2004; Mazurek *et al.* 2013) and reactive transport in bentonite (Fernández *et al.* 2011*b*; Mäder *et al.* 2012). Most recently, the analytical capability was extended with an X-ray transparent set-up to monitor intermittently a running experiment (Dolder *et al.* 2014).

A confining fluid pressure was applied to a sealed cylindrical clay rock sample and an infiltration fluid injected into one side, inducing an advective flow by a pressure gradient between inflow and outflow. The two filter pairs on each side of the core guarantee a uniform distribution of the infiltration fluid and collection of the outflowing fluid. A detailed description of the method is given in Dolder *et al.* (2014). The current study was performed in an X-ray transparent core infiltration device consisting of a base station and a detachable pressure vessel, containing the rock sample, both equipped with pressure tanks for autonomous operation (Dolder *et al.* 2014). During the first 136 days (equilibration phase), the bentonite core had time to equilibrate to the new pressure conditions using artificial Opalinus Clay porewater (APW_{OPA}). During the following 625 days, artificial Ordinary Portland Cement porewater (APW_{OPC}) was infiltrated. The experiment was stopped by closing the inlet and reducing the confining fluid pressure stepwise to atmospheric conditions.

After dismantling the experiment, the rock sample was unpacked, measured (length, diameter and mass) and sectioned. Two longitudinal samples of 20 × 50 × 10 mm (Fig. 1: XRD and SEM samples) were cut along its centre. Both samples were exposed to liquid nitrogen (−196°C) in a sealed plastic bag and dried for 2 days in a freeze-dryer at <1 Pa and ambient temperature. Samples for SEM and Raman spectroscopy analysis were vacuum impregnated with resin and polished with ethanol. The second longitudinal sample was dedicated to XRD analysis and cut into 15 subsamples

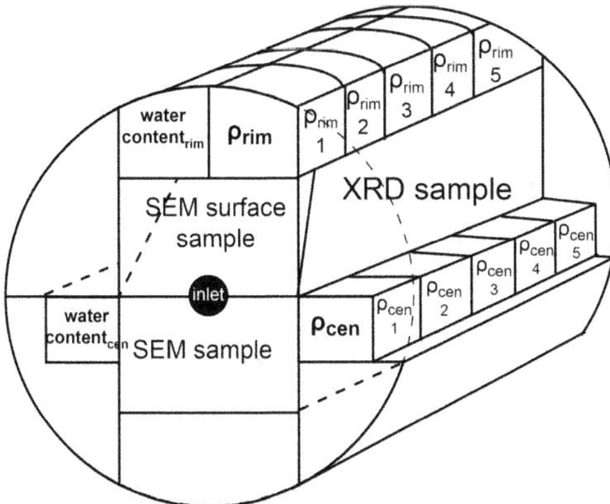

Fig. 1. Post-mortem bentonite core-cutting pattern (diameter and length 50 mm).

of 2–7 mm in thickness perpendicular to the sample axis. The rest of the core was cut into two longitudinal profiles for water content and density (ρ) measurements, five samples of each. Two profiles were located along the outer surface of the cylinder (subscript: rim) and two profiles at 10 mm off-axis (subscript: cen) (see Fig. 1). The filters were immersed in isopropyl alcohol and dried in a nitrogen-gas-filled desiccator to prevent carbonation.

Analytical methods

Ion chromatography (IC, Metrohm 850 Professional) was used to measure cation and anion concentrations in the collected outflow and in the different types of artificial porewaters (APWs) (separation columns C4-150 and A-Supp7-250). The analytical error of these measurements varied between ± 5 and $\pm 20\%$. pH was measured with an Orion PerpHecT Ross combination microelectrode (± 0.02 pH), using three-point calibrations of pH 4, 7 and 9 (outflow and neutral-pH APWs), and pH 9, 11 and 12 (high-pH APW). The alkalinity was measured on a 785 DMP titrino from Metrohm. Inductively coupled plasma atomic emission spectroscopy (ICP-OES) measurements were performed on a Varian 720-ES (RSD $<1\%$). This method was used to measure minor and some trace elements, such as aluminium, barium, silicon and strontium.

The electrical conductivity measurements were performed online with a small-volume flow-through cell (from Metrohm) that was connected to a Jumo industrial meter (ecoTRANS Lf 03) and recorded with a data acquisition system. This device was not accurately calibrated owing to the very small cell volume, but it was stable and served for recording relative changes.

The density and water content of the bentonite core were determined post mortem. The bulk density was measured on saturated samples directly after opening the experiment using a Mettler Toledo density accessory kit (immersion in paraffin oil) mounted on a Mettler Toledo balance (AT261 Delta Range). The grain density of the original MX-80 was measured using a He-gas pycnometer (Micromeritics AccuPyc II 1340). Water content was determined by weighing a sample wet and after 2 weeks of drying at 105°C. The porosity was calculated using the measured wet density, the water content and the grain density of bentonite.

The mineralogy was determined by XRD on an X'Pert PRO PANalytical diffractometer with a Cu–Kα radiation and an X'Celerator detector using a current intensity and voltage of 40 mA and 40 kV, respectively. The XRD runs were performed from 4°– 60° or to 70° 2θ with step times of 1.1 s and step size of 0.017°. The scans were either performed on disorientated powder samples (grain size *c.* 10 μm) or, in case of insufficient material, deposited on a silica wafer (samples directly from the interface). All samples were preconditioned at 33% RH (relative humidity) using a saturated $MgCl_2$ solution. All XRD traces were normalized on the location of the 3.34 Å quartz reflection. Additional samples were measured after saturation with ethylene-glycol (EG). The filters were scanned by emplacing them on a silica wafer.

SEM measurements were performed on uncoated sample surfaces with a Zeiss EVO-50 XVP microscope equipped with an EDAX Sapphire light element detector. The machine was used in low-vacuum mode with a beam acceleration of 20 kV and a working distance of approximately 9 mm. Energy-dispersive spectroscopy (EDX) was used for elemental analyses of points and areas, as well as for element maps. EDX point measurements were acquired for 1 min and semi-quantitatively analysed without standardization using EDAX Genesis software (ZAF matrix correction). EDX element maps with a resolution of 512×400 pixels were acquired using a dwell time of 200 μs and frames were averaged for 1–2 h. Backscattered electron (BSE) images were taken with a resolution of 1024×800 pixels. All detected elements, except for C and O, were normalized to 100 wt%: C and O could not be used as they are also the main constituents of resin. Systematic errors were below 4 wt% absolute in low-vacuum mode, but the random error indicated by the error bars in the results section may be below 1 wt% absolute, depending on the element.

The Raman spectroscopy measurements were performed on polished sample surfaces, where points of interest defined by SEM could be revisited. The Raman microscope is a Jobin Yvon LabRAM-HR800 (800 mm focal-length spectrograph) combined with an Olympus BX41 microscope. A He–Ne laser with an excitation wavelength of 632.817 nm (red) was focused on the sample surface in an approximately 2 µm spot using a ×100 objective and the Raman signal was collected in backscattered mode. The spectra were recorded with Labspec V4.14 software.

The CT measurements were performed on medical CT scanners (Siemens Somatom Emotion 6 and Siemens Somatom Definition AS), both operated by the Institute of Forensic Medicine at the University of Bern, Switzerland. In total, 17 CT scans were performed during the experiment: 16 on the first machine and one on the second after replacement of the original instrument. The parameters used were chosen to optimize the contrast of the images. The generated X-ray energy was 130 keV (polychromatic X-ray beam) with an X-ray tube current of 120 mA for the first machine, and 140 keV and 140 mA for the second. The voxel dimension of the recorded images was 0.109 × 0.109 × 0.5 mm for the first scanner and 0.107 × 0.107 × 0.5 mm for the second. The change of the CT machines made a direct comparison of the different datasets difficult. For this reason, the last dataset measured on the new machine was corrected by a best fit of the bulk absorbance not affected by chemical changes in order to compare the last dataset with previous ones. The CT scanners were calibrated on air and its images are shown in Hounsfield units (HU), a generic unit used in medicine. It is defined by air (−1000 HU) and distilled water at 0 HU. The data were processed and analysed using the open-source software Mango (Multi-Image Analysis GUI: Lancaster & Martinez 2007) and Fiji, an open-source image processing software package based on ImageJ (Schindelin *et al.* 2012). 3D image analysis was performed using Fiji and Matlab. The 3D evolution of the progress of mineral alteration (density/porosity and core volume) was monitored by CT. Image processing included segmentation, quantitative assessment and rendering. Segmentation was achieved by thresholding, after image normalization as a pre-processing step. The reaction plume is defined by an increase in X-ray attenuation, related to a higher bulk density. Normalization was required in order to hide the transition of the filter to the bentonite zone and thereby enabling simple thresholding of the reaction plume without inducing a discontinuous junction. The resulting binary mask of the reaction plume was quantified with respect to its geometrical expanse, as well as being triangulated and Gouraud-shaded for generating 3D views.

Aqueous geochemical modelling

Aqueous speciation calculations and saturation states at 25°C were performed using PhreeqC (V2.18) (Parkhurst & Appelo 1999, 2013). The PhreeqC.dat database was used for neutral-pH fluids, while modelling of the high-pH cementitious fluid required a database containing cement minerals. For this purpose, the CEMDATA07 (version 07.02) for PhreeqC was used (Lothenbach *et al.* 2008; Jacques 2009 and the references therein). Equilibrium modelling in bentonite was performed using PhreeqC (V2.18) in combination with the PhreeqC database, extended by montmorillonite, cristobalite (Thermoddem database: Blanc *et al.* 2007), tridymite (llnl.dat database: Wolery 1992), and cement phases, such as ettringite, portlandite and C–S–H (CEMDATA07: Jacques 2009; Lothenbach *et al.* 2008 and the references therein).

Starting material

Untreated MX-80 bentonite powder was used, manufactured by Volclay Ltd, Merseyside, UK. It consists of Na-montmorillonite (81–85 wt%), feldspars (*c.* 5 wt%), quartz (*c.* 3 wt%), muscovite (*c.* 3 wt%), gypsum (*c.* 1 wt%), cristobalite (*c.* 1 wt%) and calcite (0.2 wt%) (Karnland 2010). The exchanger of montmorillonite is dominated by Na^+ with 75 eq.%, followed by Ca^{2+}, Mg^{2+} and K^+ with 17, 6 and 2 eq.%, respectively (Karnland 2010). The MX-80 bentonite sample was compacted and saturated in a pre-treatment device using 157.66 g of dry rock powder with a residual water content of 11.98 wt%, corresponding to 45.9 g of porewater after saturation (w/s ratio 0.33) (Dolder *et al.* 2014). This amount corresponds to one pore volume (PV) used to convert transport time to number of PV passing through the core. The saturated MX-80 bentonite core was 49.5 mm in length, 50 mm in diameter and had a mass of 186.69 g, corresponding to a saturated density (ρ_{sat}) of 1920 kg m^{-3} (ρ_{dry} 1448 kg m^{-3}) and a water-content porosity of 44.73%.

Saturation and infiltration fluids

$APW_{Äspö}$ and APW_{OPA} are fluids that were used for saturation and during the equilibration phase of the experiment, respectively (Table 1). $APW_{Äspö}$ is based on moderately saline fracture water from the Äspö underground rock laboratory (Karnland *et al.* 2009; Dolder *et al.* 2014). The aqueous modelling showed undersaturation with respect to gypsum, near saturation with calcite and a partial CO_2 pressure (pCO_2) similar to atmospheric conditions. The APW_{OPA} was based on a recipe from Mäder (2009) for porewater from the Opalinus Clay

Table 1. *Measured $APW_{Äspö}$ and APW_{OPA} properties (errors are c. 5%)*

	$APW_{Äspö}$*	APW_{OPA}*
pH	7.2	7.6
Ionic strength (molal)	0.26	0.23
EC (mS cm^{-1})	18	17
Na^+ (mM)	88	166
K^+ (mM)	0.3	2.7
Ca^{2+} (mM)	54	11.9
Si^{4+} (mM)	<0.007	<0.007
Al^{3+} (mM)	4.1×10^{-4}	5.7×10^{-4}
Mg^{2+} (mM)	1.7	9.1
Cl^- (mM)	203	157.6
SO_4^{2-} (mM)	1.8	22.1
HCO_3^- (mM)	0.2	0.7

*Dolder *et al.* (2014).
EC, electrical conductivity (at 25°C).

Formation and is also described in Dolder *et al.* (2014). The fluid was saturated with calcite and gypsum, and the pCO_2 was atmospheric.

The APW_{OPC} represented a young cement pore fluid with a pH of 13.36 and a water/cement ratio of 0.8 (Table 2). The composition was based on Lothenbach & Winnefeld (2006) and Lothenbach (2010), and it represented a thermodynamically modelled OPC pore fluid in agreement with measurements of reference samples after hydration lasting for 625 days, and corresponded to more-or-less complete hydration under closed-system conditions. The corresponding modelling was carried out using the Gibbs free energy minimization program GEMS (Lothenbach & Winnefeld 2006) and the GEMS-PSI thermodynamic database (Hummel *et al.* 2002; Thoenen & Kulik 2003). A detailed description of the APW_{OPC} recipe is given in Dolder *et al.* (2014). The main differences between the modelled solution of Lothenbach & Winnefeld (2006) and the APW_{OPC} are the lower K^+ concentrations, and the higher OH^- and HCO_3^- concentrations in our recipe. In order to simplify the system, only inorganic carbon was used, and a surplus of $Ca(OH)_2$ and $CaCO_3$ were added to buffer any unwanted ingress of atmospheric CO_2 by precipitation of calcite (Mäder *et al.* 2006; Dolder *et al.* 2014). A comparison of the initial APW_{OPC} composition to the fluid from the injection tank after 625 days of infiltration revealed only minor changes (Table 2). All infiltration fluids were prepared under atmospheric oxygen conditions, but remained in a protected He atmosphere during the experiment.

Results

Hydraulic evolution of the experiment

The experiment duration was 761 days. In total, 0.44 PV were flushed through the core: 0.18 PV of the APW_{OPA} and 0.26 PV of the APW_{OPC}. The experiment was carried out in a laboratory with minor variations in temperature (18–24°C: Fig. 2a). The confining fluid pressure was increased stepwise to 4.1 MPa (Fig. 2a). In the first 12 days, 0.26 l

Table 2. *Composition of the APW_{OPC} and comparison to the literature*

	Lothenbach & Winnefeld (2006) and Lothenbach (2010)		Mixed fluid (APW_{OPC})	
	Experimental data (measured)	Modelled data	Start (measured)	End (after 625 days; measured)
Hydration time (days)	360	625		
pH	13.3	13.36	13.36	13.34
Ionic strength (molal)	–	–	0.28	0.27
EC (mS cm^{-1})	–	–	56	49
Percentage charge error	–	–	−0.02	0.19
Na^+ (mM)	90	118.7	115.4	114
K^+ (mM)	208	222	180.7	173.3
Ca^{2+} (mM)	1.9	0.9	1.3	0.5
Si^{4+} (mM)	0.05	0.06	0.17	–
Al^{3+} (mM)	0.01	0.03	0.03	–
Mg^{2+} (mM)	–	–	<0.41	<0.1
Cl^- (mM)	–	–	0.08	<1.6
SO_4^{2-} (mM)	3.4	2.9	2.9	2.9
C (mM)	13 (DOC)	0.22	1.5 (DIC)	–

Note: EC, electrical conductivity (at 25°C); DOC, dissolved organic carbon; DIC, dissolved inorganic carbon; –, not measured or data not available.

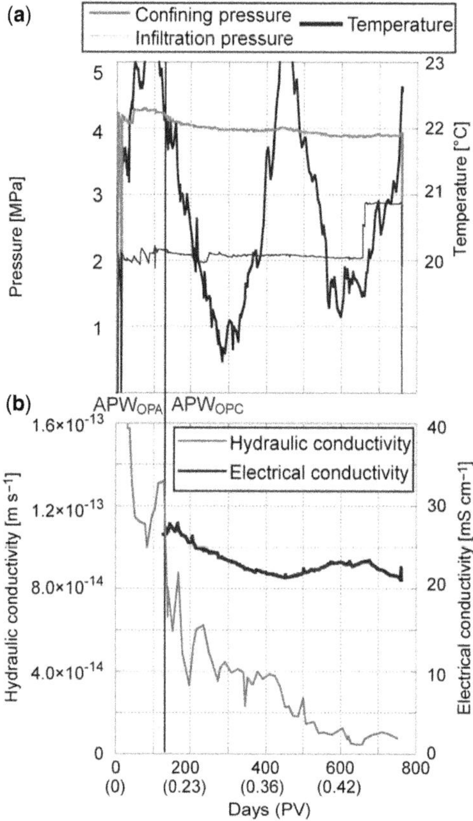

Fig. 2. Evolution of the physical parameters: (**a**) pressure and temperature; and (**b**) hydraulic and electrical conductivity.

of gas were collected in syringes at the outflow, indicating a sealing problem in the exfiltration system. The exceptional gas outflow was related to trapped atmospheric gas in the confining fluid system, leaking continuously into the fluid outflow of the experiment. The experiment had to be stopped and restarted after 13 days, including repacking of the core. Switching to the APW_{OPC} induced a decrease to 3.9 MPa, caused by core volume shrinkage in combination with pressure loss during decoupling for CT scans. The infiltration fluid pressure was set to 2.1 MPa (Fig. 2a). After restarting the experiment, the pressure was, again, readjusted and stayed constant during the experiment with some minor deviations. After 642 days, the infiltration pressure was increased to 2.8 MPa in order to increase the fluid flow.

The hydraulic conductivity shown in Figure 2b was based on averages of four single water-level measurements on syringes at the fluid outflow, with time spans of 2–4 weeks. The hydraulic conductivity was calculated based on Darcy's law (Dolder *et al.* 2014). The hydraulic conductivity decreased during the equilibration period from 2.2×10^{-13} to 1.1×10^{-13} m s^{-1} due to consolidation of the core to the new pressure, initially releasing water. APW_{OPC} infiltration induced a drop to 4.7×10^{-14} m s^{-1} and a more gradual decrease to 2.0×10^{-14} m s^{-1} after 460 days (Fig. 2b). After 643 days, the conductivity reached 4.2×10^{-15} m s^{-1}. The sharp decrease after switching to APW_{OPC} occurred over a period of about 30 days: frequent removal of the experiment for CT scanning perturbated the outflow (Dolder *et al.* 2014), giving rise to some of the fluctuations.

The in-line electrical conductivity measurement of the outflow fluid started after 127 days at 26.5 mS cm^{-1}, decreased to a minimum of 21.4 mS cm^{-1} after 450 days and increased, finally, to 23 mS cm^{-1} (Fig. 2b).

Chemical evolution of the outflow

Twenty outflow aliquots were collected in syringes: five syringes were sampled during the equilibration phase and 15 during the high-pH infiltration phase. The average solution loss in the syringes before and after cold storage was between 1.0×10^{-3} and 2.0×10^{-3} g, corresponding to 0.01–0.05 vol%. Samples were analysed for the inorganic aqueous constituents and pH (Fig. 3). The alkalinity was measured on combined samples of two aliquots each, due to the small fluid volumes. The pH of the first four syringes was approximately 5.2 and increased after 134 days (0.18 PV) to 7.4. During APW_{OPC} infiltration, a gradual increase to a pH of approximately 8.2 after 756 days (0.43 PV) was observed. The values below pH 6 were most probably artefacts, due to the ingress of compressed air mentioned above.

The outflow gradually approached the APW_{OPA} composition during the equilibration phase; a reduction in all ion concentrations was observed, with the exception of sulphate (Fig. 3). High-pH infiltration reduced ion concentrations further, with the exception of potassium, sulphate, silicon, aluminium and bicarbonate. All ion concentrations were approaching that of the APW_{OPC} composition, with the exception of silicon, sulphur and bicarbonate (Fig. 3). After switching to the APW_{OPC}, chloride was replaced as the main anion charge carrier by sulphate after 0.4 PV, which corresponds to the end of a transient-flow phase. Measured ammonium, barium, fluoride and nitrate were mostly below or near the detection limit.

Physical evolution: CT

The calculations of core volume were based on the HU histograms of each dataset. A region of interest

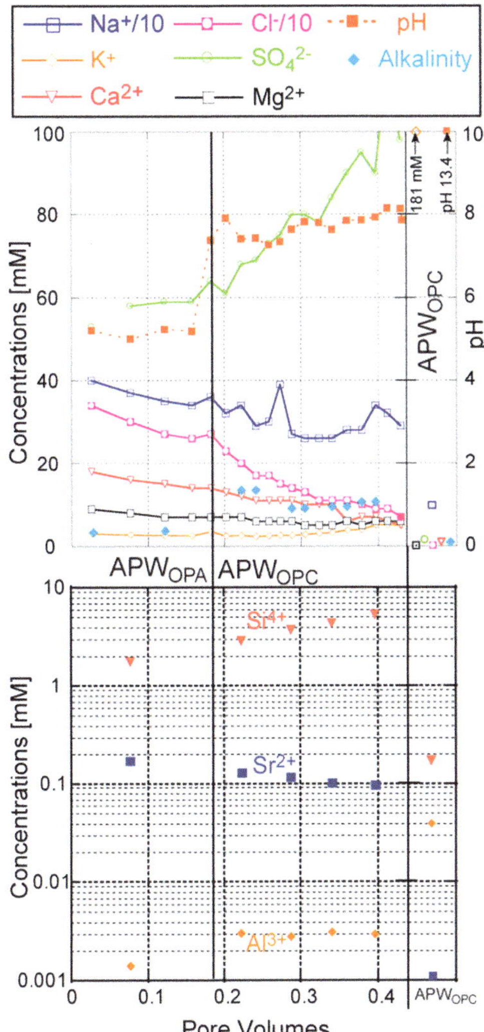

Fig. 3. Ion concentrations, alkalinity and pH of the outflow plotted against pore volume (PV).

(ROI) was experimentally defined; the voxels were counted and multiplied by the voxel dimension. For the first CT scanner, the ROI was 1000–2700 HU, and for the second scanner 960–2800 HU. The bulk density of the core was linearly dependent on the HU. The pre-experiment (pre-exp.) bentonite volume was 97 200 mm^3, and decreased after insertion into the infiltration device and applying pressure by 1 vol% (Fig. 4a). A further volume decrease started on day 207 of high-pH infiltration and ended after 588 days, corresponding to −1.5 vol% relative to the end of the equilibration phase after 136 days of experiment (called time 0). It corresponds to a shrinkage rate of approximately 1.7 mm^3/day and a porewater volume of 2.1×10^4 mm^3. The core diameter calculations were performed in the same way as the volume calculations. By doing this for all CT slices of a dataset, the evolution of the core sectional area and the core diameter over time could be evaluated along the core axis (Fig. 4b). The core diameter was reduced after 107 days of APW_{OPA} infiltration relative to the pre-experimental value by approximately 0.7 mm (inlet), approximately 1 mm (12 mm from the inlet) and approximately 0.3 mm (outlet), respectively (Fig. 4b). This is in agreement with the postulated density increase along the core axis in Dolder *et al.* (2014). The core diameter still exhibited four distinct domains, each approximately 12 mm in length, that were remnants of sample compaction and the pre-saturation process (assembled from four saturated discs).

The differences in diameter, as well as the histogram evolution, of the core during high-pH infiltration relative to a measurement after 136 days of APW_{OPA} infiltration is shown in Figure 5b. The main volume/diameter reduction occurred between 72 and 207 days of APW_{OPC} infiltration and onwards; the maximum core diameter shrinkage occurred between depths of 5 and 16 mm. The maximum stage was reached after 588 days, with a core diameter reduction of 0.7 vol%. The diameter shrinkage reached the end of the core sample between 315 and 364 days of APW_{OPC} infiltration. The last measured diameters of the core were approximately 49.2 mm for the inlet, 48.8 mm for the middle part and 49.7 mm for the outlet, which is in agreement with post-mortem measurements. The comparison of the bentonite transection image and the core diameter evolution in Figure 5a, b, respectively, showed that the zone of maximum diameter decrease matches approximately the zone of whitish discoloration of the sample material. HU histograms of single CT slices at 0.5, 1.0, 8.5 and 12.5 mm from the inlet are plotted against time in Figure 5c. The CT slice at 0.5 mm from the inlet showed a drastic reduction in intensity and HU over time, which can be related to a decrease in bulk density. In the subsequent histograms further away from the inlet, the opposite trend was observed. This means that the rest of the core showed an increasing HU intensity and bulk density over time, which is in agreement with the diameter reduction mentioned before.

Tomographical cross-sections of the bentonite core during the equilibration phase showed a homogeneous material of greyish colour with whitish dots of denser minerals like pyrite (Fig. 6). The shading code of tomographical images is the following: bright shades are low X-ray attenuations or HU, and dark shades represent higher densities. Interestingly, a concentric pattern of circular rings grew out

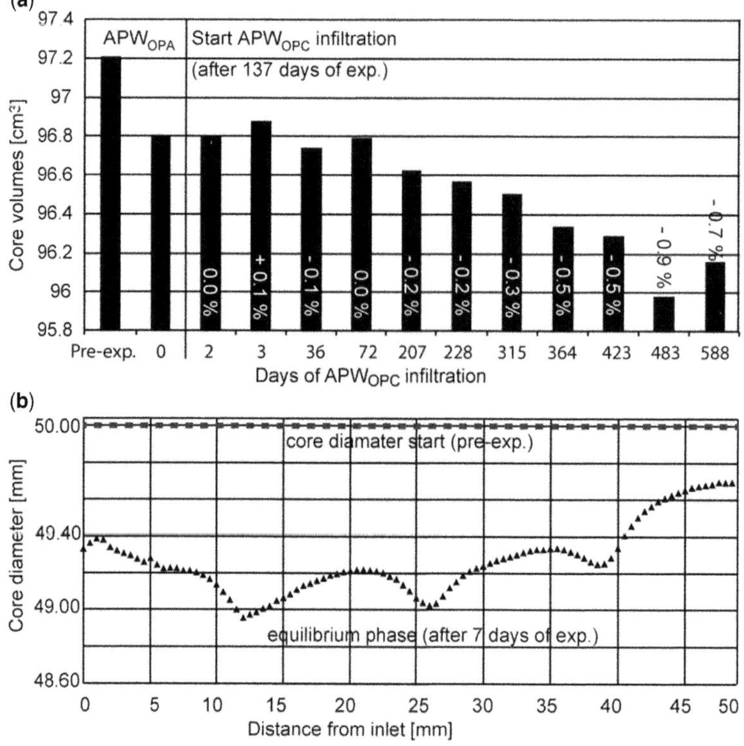

Fig. 4. Core evolution based on CT measurements: (**a**) core volumes including percentage change in core volumes; and (**b**) core diameters. Time 0 corresponds to the end of the equilibration phase (APW_{OPA} infiltration). Exp, experiment.

of the filter into the first millimetre of bentonite. The first ring structure appeared after 137 days of APW_{OPC} infiltration and reached a depth of 1.75 mm. The circular precipitations in the filter remained constant after they reached maximum extension and intensity after 137 days of APW_{OPC} infiltration. The reaction front in the filter had a diameter of approximately 12.5 mm and comprised three distinct circular-shaped zones of increased X-ray attenuation/density. The circular-shaped reaction plume grew cylindrically into the bentonite sample. The maximum extension of the diffuse plume was reached after 588 days of APW_{OPC} infiltration, with a maximum penetration depth of 5–8 mm.

Post-mortem analysis: physical parameters

The bentonite core in a saturated condition at the end weighed 187.74 g (including a small piece of the ripped inlet filter), representing a gain of 1.05 g. The sample length was 51.26 mm and had increased by 1.8 mm. The sample diameter was 49.4 mm near the inlet, 49.4 mm in the middle part and 50.0 mm near the outlet. The water content in both profiles showed a decreasing trend from inlet to outlet, with slightly higher values along the centre (Fig. 7). The saturated density in both profiles showed reduced densities of 1891–1895 kg m^{-3} at the inlet and values of around 1920–1930 kg m^{-3} in the rest of the core, with higher values along the centre. The porosity at the inlet was 49% and decreased towards the outlet to around 46%. The degree of saturation varied between 0.98 and 1.00 in both profiles, indicating full saturation within measurement uncertainties.

XRD analysis

All diffractograms of MX-80 bentonite showed the following major minerals: smectite (montmorillonite), quartz (3.3, 4.2 Å), cristobalite (4.0 Å), alkali feldspar (3.21, 3.22 Å), plagioclase feldspar (3.18 Å), muscovite and/or illite (*c.* 10 Å), gypsum (7.6 Å) and calcite (3.03 Å) (Fig. 8). Variations in the (001) smectite reflection result from different hydration states and cation occupancy. The differentiation of di- and trioctahedral subgroups of

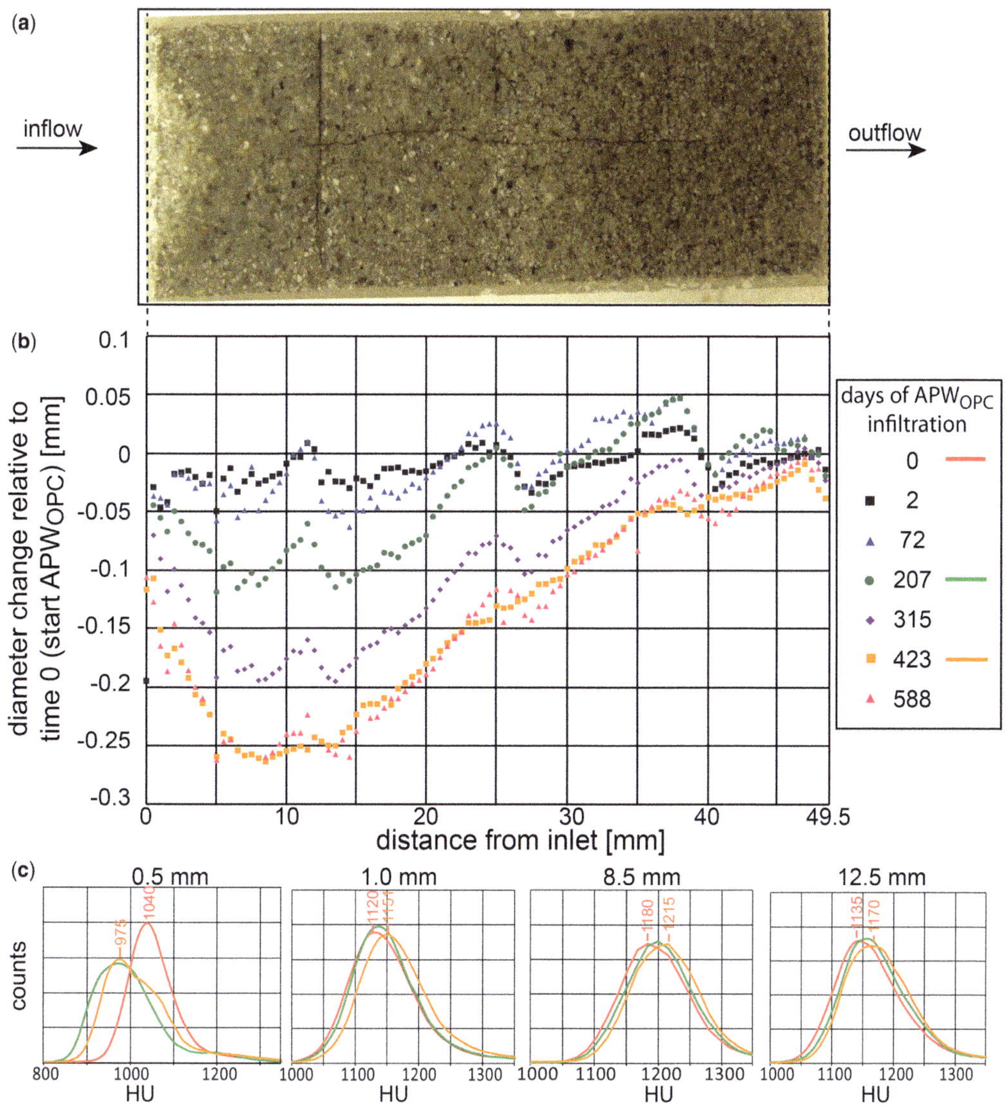

Fig. 5. Bentonite core size evolution: (**a**) transection photograph of the bentonite after the experiment (optical image); (**b**) diameter of the core relative to time 0 (end of equilibration phase, day 136); and (**c**) variations in the HU histograms over time.

smectites was carried out on the (060) reflection (Brindley & Brown 1980). Reflections at approximately 1.49 Å are related to dioctahedral and reflections at approximately 1.52 Å to trioctahedral smectites. The °Δ2Θ((003)–(002)) may be used to identify I–S interlayers (Moore & Reynolds 1989). The (002) reflection of smectite at 4.5 Å was used for intensity comparison along the bentonite core.

The (001)-spacing of the raw MX-80 bentonite was 12.3 Å, which is a typical value for a Na-saturated MX-80 bentonite (Ferrage *et al.* 2005). The MX-80 bentonite sample had a (060) reflection at 1.49 Å, which is characteristic for a dioctahedral smectite. The EG-saturated sample had a (001) reflection at 16.9 Å, a (002) reflection at 8.5 Å and a (003) reflection at 5.7 Å, corresponding to a °Δ2Θ((003)–(002)) of 5.29°, which is, after Moore & Reynolds (1989), a pure montmorillonite with no I–S.

All XRD spectra of the filters showed a broad hump of amorphous material at 22.25 Å from the PVC-filter material (Fig. 9). The inlet filter showed

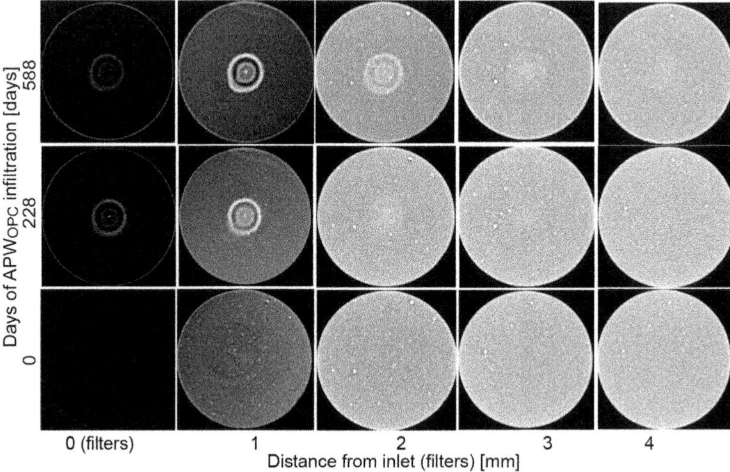

Fig. 6. Evolution of CT slices in the inflow region of the experiment over time.

a strong enrichment in calcite on both sides. On the clay side of the inlet filter, the main bentonite minerals were identified (smectite, muscovite, quartz and feldspar). Talc, gypsum, and brucite could also be identified. The outlet filters showed precipitation of halite with a reflection at 2.8 Å that precipitated during sample drying after the experiment.

XRD measurements of the core sample showed the (001) reflection of smectite in the centre, as well as 5 mm off-centre directly at the interface at a value of 12.1 Å. The rest of the bentonite had this reflection at 11.8 Å with a tendency of reflection broadening. Repetitive sample preparation and scanning showed that the occurrence of reflection broadening and the change in intensities of the (001) smectite reflection varied, and was most likely an artefact of sample preparation (e.g. an incomplete random orientation or rehydration during measurement). The sample directly at the inlet was depleted by 2% in smectite (height of the (002) reflection), whereas the samples up to 5.5 mm were enriched by up to 3% relative to the unreacted bentonite parts. The (060) reflection of smectite was at 1.5 Å in all measurements and, hence, it could be identified as a dioctahedral clay, such as montmorillonite. Samples at the interface and up to 4 mm in depth showed reflections at 1.519 and 1.516 Å that are related to trioctahedral smectite (e.g. saponite). The d-spacing of the smectite was smaller in all samples compared to the starting material and clay porewater-saturated MX-80 bentonite, indicating an exchange of Ca^{2+} and some Na^{+} by K^{+}. This is in agreement with Ferrage *et al.* (2005), who measured and calculated a c-cell distance of 11.36 Å for a K^{+}-saturated MX-80 bentonite and 12.44 Å for a Na^{+}-saturated MX-80 bentonite at approximately 35% RH. All diffractograms of EG-saturated samples showed (001) reflections at 16.9 Å, (002) reflections at 8.5 Å and (003) reflections at 5.6 Å, which correspond to a °Δ2Θ((003)−(002)) value of 5.29 Å, and thus there is no significant illite interstratification.

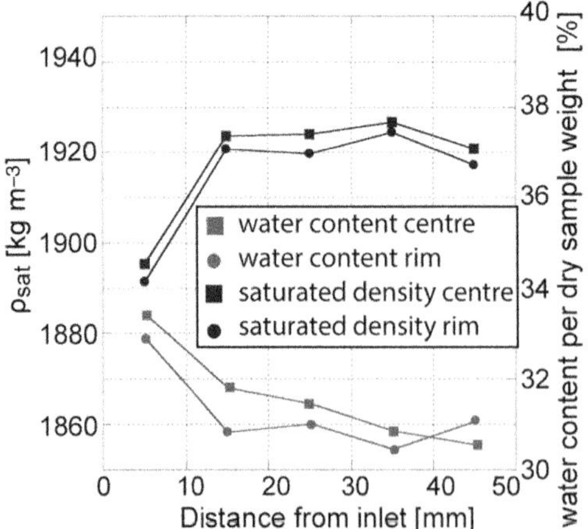

Fig. 7. Saturated densities and water content of the bentonite sample at the end of the experiment.

All samples showed a clear reflection of mica-like muscovite or illite, quartz and feldspar. Samples at depths of between 11.5 and 44.5 mm contained small amounts of gypsum. Cristobalite main reflection intensities were reduced towards the inlet. Directly at the interface, a maximal intensity reduction of 32% was observed, extending up to a depth of 8.5 mm. The most prominent variation occurred

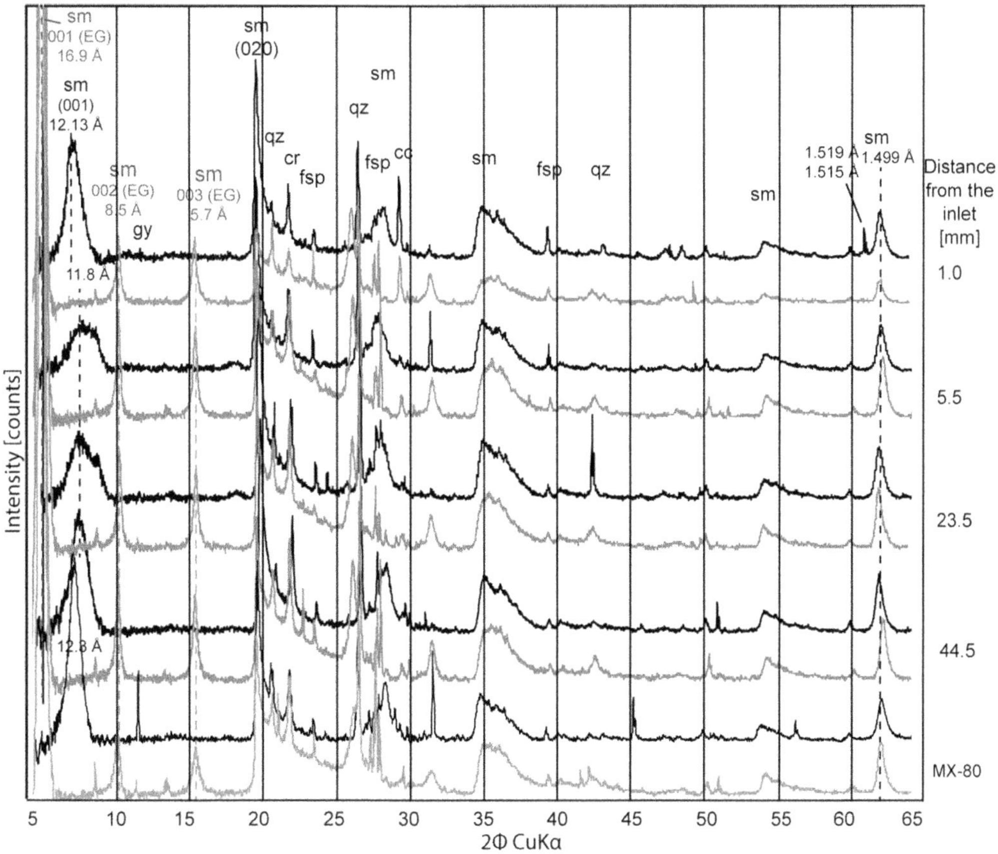

Fig. 8. XRD spectra of the raw MX-80 bentonite and of the bentonite core indicated as distances from the inlet. Sm, smectite; qz, quartz; cc, calcite; cr, cristobalite; fsp, feldspar; gy, gypsum. All curves are normalized on the 3.3 Å qz reflection (normalized on qz). Black curves are measured at 33% Rh; grey curves are EG-saturated samples.

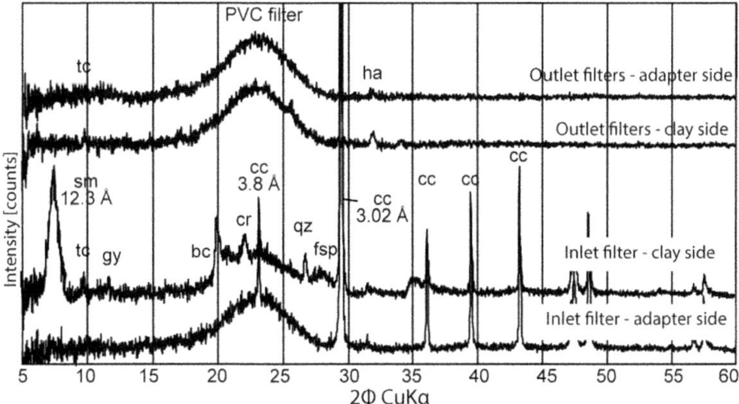

Fig. 9. XRD spectra of the inlet and outlet filters: sm, smectite; mu, muscovite; qz, quartz; cr, cristobalite; fsp, feldspar; cc, calcite; tc, talc; ha, halite; gy, gypsum; bc, brucite (normalized on qz and, in the case where no qz is present, on the amorphous PVC hump).

in the calcite reflection (3.03 Å), which indicated a strong enrichment in the first 3 mm. The calcite reflection height directly at the interface was 6.6 times higher compared to the average sample, and at a distance of 3 mm an increase of 1.7 times was observed.

Microscopy and SEM-EDX analysis

Filters. The inlet filter showed discoloured concentric rings 13.8, 6.5 and 1.4 mm in diameter on the equipment side. The presence of minerals was seen under the microscope in the fibre structure of the two inlet filters. EDX measurements showed Ca-enrichment, which is most likely to be related to newly formed calcite, which was detected by XRD.

Bentonite core. From the inflow surface, a whitish reaction plume penetrated 8 mm into the bentonite in the form of a hemisphere (Figs 5a & 10d). The greyish rims are artefacts of sample preparation – the sample is coated in resin on both sides. BSE images of the inflow region revealed a high density of mostly cracks (resin-filled) that are freeze-drying artefacts from the sample preparation procedure (Fig. 10a). The first 0.5 mm of the bentonite had a darker shade due to an uneven, dipping sample surface, exhibiting a lower signal intensity. The bentonite displayed a clay-dominated mineralogy with randomly embedded accessory mineral grains of higher mean atomic number (Fig. 10a). The minerals were identified by means of SEM-EDX spot analysis. Pre-existing calcite and muscovite grains had diameters of up to 500 μm. Alkali and plagioclase feldspar grains were sharp edged and <50 μm. SiO_2 group minerals, such as quartz, cristobalite and tridymite, could be observed but not distinguished. Pyrite occurred in grains <5 μm.

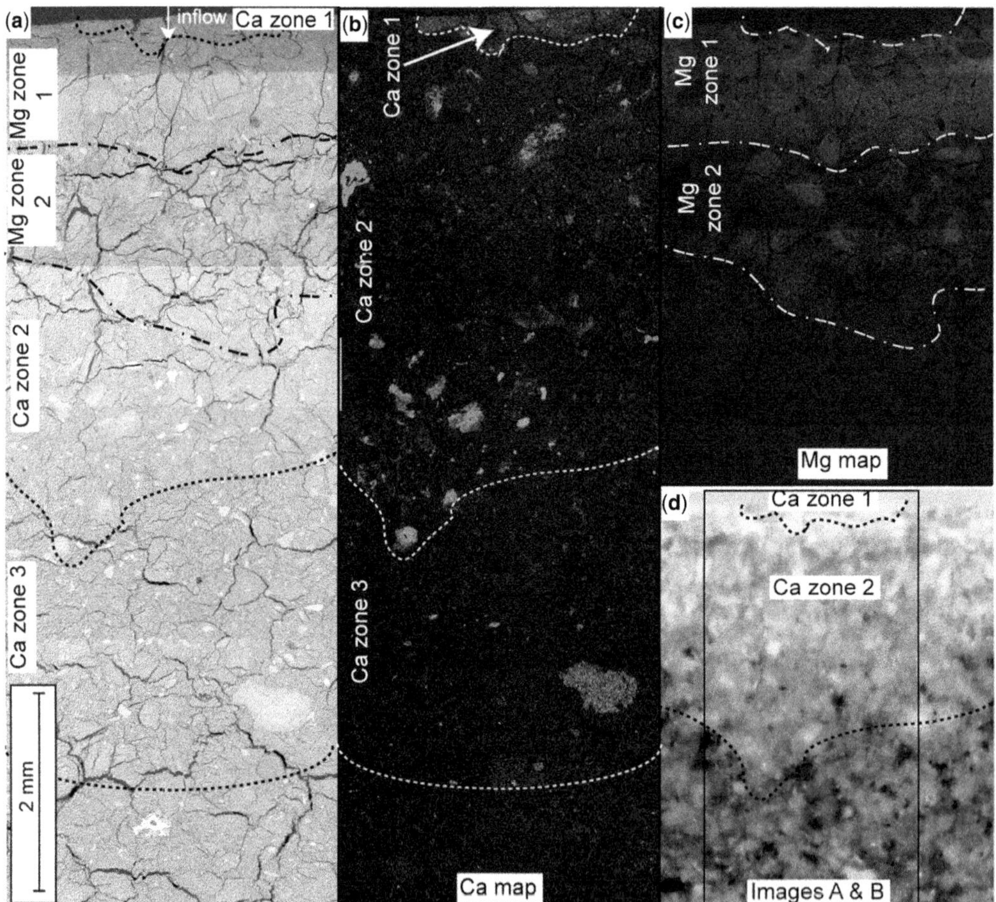

Fig. 10. Detailed section of the first 8 mm of the bentonite sample: (**a**) BSE image; (**b**) Ca element map; (**c**) Mg element map; and (**d**) reflected light micrograph of the part close to the inlet. Borders of distinct zones are indicated.

The average oxide composition of the pristine clay matrix and of the bulk sample is shown in Table 3. The Al_2O_3/SiO_2 ratio is in agreement with measured montmorillonite from MX-80 given in Karnland (1997). This means that the tetrahedral sheet is almost entirely filled with Si^{4+} and contains just traces of Al^{3+}. The octahedral sheet is dominated by Al^{3+} with minor amounts of Mg^{2+} and Fe^{3+}, and the adsorbed cations are dominated by Na^+ with minor amounts of Ca^{2+} and just traces of K^+ and Mg^{2+}. In the inflow region, zones enriched in CaO and MgO were detected (Fig. 10b, c). The CaO-enrichment could be divided into three distinct zones with various amounts of newly precipitated calcite (previously shown by XRD): Ca zones 1–3. Ca zone 1 was a small domain 2.2 mm in diameter and 0.4 mm in depth at the centre of the interface. No distinct mineral phases were visible and bulk SEM-EDX measurement (Fig. 11) revealed only a small CaO-enrichment (Table 3). Ca zone 2 was shaped like a hemispherical shell around the fluid inlet with a width of approximately 4.4 mm (from inner to outer border). It consisted of a weak CaO-enrichment in the clay matrix (see Table 3), containing small (<5 µm), Ca-rich mineral grains of variable size (Fig. 12a, c, d). In some locations, they seemed to grow along preferred flow paths around pre-existing clay stacks, overgrowing the matrix (Fig. 12a). The larger size fraction was between 25 and 400 µm, and consisted of Ca minerals replacing pre-existing mineral grains (Fig. 12b, c). The bulk SEM-EDX signal was clearly clay-dominated, including an increased amount of MgO in the MgO-enriched zone. The CaO-enrichment mainly occurred in small, newly grown minerals that were not clearly detectable in the bulk EDX transect measurements (Fig. 11). In some cases, replacement of gypsum by calcite could be observed based on SEM-EDX spot analyses. Ca zone 3 was marked as a zone of continuous decrease in CaO and ended at a depth of 8 mm (see Table 3). The zone volumes calculated were: 0.5 mm^3 (zone 1), 277 mm^3 (zone 2) and 795 mm^3 (zone 3).

The MgO-enrichment occurred only in the first millimetres of the bentonite, but not directly at the inlet, and could be divided into two zones: Mg zones 1 and 2 (Figs 10c & 11). Mg zone 1 extended from 0.4 to 1.9 mm from the inflow interface and coincided with Ca zone 1. It was a zone of massive and homogeneous MgO-enrichment in the clay matrix (Table 3). The measured MgO concentration was enriched seven-fold, the Al_2O_3/SiO_2 ratio was identical to that of the unreacted clay. The comparison of spot analysis in the MgO-enriched clay matrix with the unreacted bulk bentonite measurements showed clearly larger amounts of MgO in the single-spot analysis. Magnesium was either fixed in smectite crystals or in newly grown very-fine-grained Mg phases (Fig. 12e, f). Mg zone 2 extended from 1.9 to 3.1 mm from the inflow interface. It was characterized by heterogeneously distributed MgO-enriched patches of clay that faded out gradually away from the inlet, indicating a heterogeneous shape of the progressing reaction front (Table 3). The zone volume was calculated to be 62 mm^3.

Dissolution of certain SiO_2-rich grains, which were <20 µm in size, was observed near the inlet (Fig. 12b, g, h). In some cases, SiO_2 phases were replaced by Ca- or sulphate- minerals, most probably calcite or gypsum (Fig. 12b). XRD measurements indicated that the dissolved SiO_2 phase was mainly cristobalite.

Raman spectroscopy

Raman spectroscopy was used to identify mineral phases along the bentonite section (Fig. 5a). Clays tend to be highly fluorescent in the laser light due to associated iron(III) hydroxide minerals and/or organic matter, saturating the detector over the entire wavelength region (Alia *et al.* 1999; Blacksberg *et al.* 2010; Košařová *et al.* 2013). Measurements in the clay matrix near the inlet (Mg-enriched part), 8 mm from the interface, and at the outlet all showed identical patterns, with bands at 254, 371, 707, 819 and 1111 cm^{-1} (Fig. 13a). No trend is observed between Mg-enriched parts and unaltered bentonite, indicating no significant changes in structure. The Raman shift patterns of the newly precipitated Ca phase at the interface showed bands at 281 and 1086 cm^{-1}, which were identified as calcite bands (Fig. 13b).

Discussion

The evolution of the core volume and hydraulic conductivity, as well as the ion concentrations and pH in the outflow, is complex and reflects coupling between physical and chemical processes. The main controlling processes are: ionic strength effects and coupled transient compaction/decompaction (shrinkage or expansion of the core due to changes in the montmorillonite interlayer distance/water content); the internal porewater composition control by minor soluble accessory phases (sulphate and/or carbonate minerals); the cation-exchange processes; carbonate precipitation; and also the silicate mineral dissolution/precipitation of the reactive 'cement phases' after injection of the APW_{OPC}.

Equilibration phase

Evolution of physical properties. The bentonite core volume decreased after applying a confining fluid

Table 3. *SEM-EDX measurements of the bentonite and the CaO- and MgO-enriched zones*

Type of measurement	Bentonite (pristine) Bulk	Clay matrix (pristine) Point	Ca zone 1 Bulk	Ca zone 2 Bulk	Ca zone 3 Bulk	Mg zone 1 Point	Mg zone 2 Point
SiO_2 (wt%)	69.6 ± 0.7	71.1 ± 0.9	62.6 ± 0.7	63.9 ± 3.4	69.2 ± 0.6	61.9 ± 0.7	67.8 ± 1.4
Al_2O_3 (wt%)	20.0 ± 0.4	19.9 ± 0.8	17.6 ± 1.7	18.5 ± 0.9	19.8 ± 0.5	17.7 ± 0.6	17.7 ± 2.5
Fe_2O_3 (wt%)	4.2 ± 0.3	3.8 ± 0.3	2.5 ± 1.1	4.3 ± 0.3	4.4 ± 0.2	3.3 ± 0.2	3.4 ± 0.5
MgO (wt%)	2.2 ± 0.1	2.3 ± 0.2	12.4 ± 2.0	9.0 ± 5.2	2.3 ± 0.2	14.5 ± 1.6	8.0 ± 2.5
CaO (wt%)	1.6 ± 0.4	1.1 ± 0.2	2.8 ± 0.6	2.1 ± 0.5	1.8 ± 0.6	1.1 ± 0.1	1.5 ± 0.8
Na_2O (wt%)	1.7 ± 0.1	1.4 ± 0.1	1.6 ± 0.1	1.5 ± 0.3	1.7 ± 0.1	0.9 ± 0.2	1.1 ± 0.2
K_2O (wt%)	0.6 ± 0.1	0.5 ± 0.2	0.4 ± 0.5	0.7 ± 0.1	0.9 ± 0.1	0.5 ± 0.1	0.6 ± 0.1
Al_2O_3/SiO_2 (wt%)	0.29 ± 0.01	0.28 ± 0.02	0.28 ± 0.03	0.29 ± 0.01	0.29 ± 0.01	0.29 ± 0.01	0.26 ± 0.04
MgO/SiO_2 (wt%)	0.032 ± 0.001	0.033 ± 0.003	0.20 ± 0.03	0.15 ± 0.09	0.033 ± 0.003	0.24 ± 0.03	0.12 ± 0.04

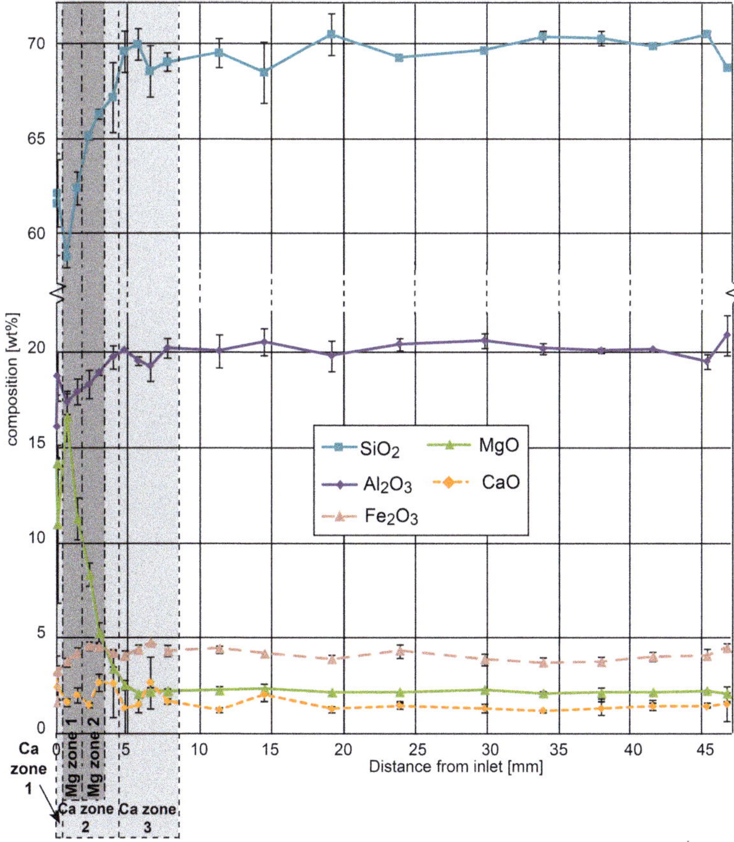

Fig. 11. SEM-EDX oxide compositions of scanned areas of 1 × 0.8 mm along a central transection. The error bars are the standard deviations of three measurements at different locations within the same zone.

pressure at the beginning of the experiment. With this compaction of the core, porewater was released from the porosity, leading to an apparent initially increased hydraulic conductivity, which was calculated based only on the net outflow volume. The hydraulic conductivity decreased continuously and approached an apparent steady-state condition of approximately 1.1×10^{-13} m s^{-1} after about 62 days, which is in agreement with Karnland *et al.* (2006), who measured values of between 3.0×10^{-13} and 6.0×10^{-14} m s^{-1} for a similar density and ionic strength. CT data indicated no mineral reactions during the equilibration phase. The Péclet number (average linear anion velocity multiplied by core length divided by the effective diffusion coefficient of chloride) was approximately 13 at the end of the equilibration phase, which corresponds to an advective-dominated flow. A bentonite core length of 49.5 mm was used as a typical length, and an anion accessible porosity of 13%, based on chloride through-diffusion and chloride breakthrough data, and an effective diffusion coefficient for chloride of 1.6×10^{-11} m^2 s^{-1} were also used (Van Loon *et al.* 2007; Fernández *et al.* 2011*a*, *b*).

Chemical evolution of outflow. At the beginning, a high ionic strength fluid was expelled (0.48 molal). This can be explained either by the saturation and compaction procedure of the bentonite during sample preparation or by an initial compaction of the core. The first option assumes that the saturation of the bentonite led to ion enrichment in the porosity due to dissolution of salts and sulphate minerals. The second option assumes that the initial compaction of the bentonite core in the applied triaxial set-up was due to a lower compaction used for sample preparation and, therefore, initially more saline porewater was expelled from the intergranular pore space where chloride is enriched due to anion-exclusion effects. As reported in Dolder *et al.* (2014), the first pH measurements near pH 5 were distinctly low compared to all later values.

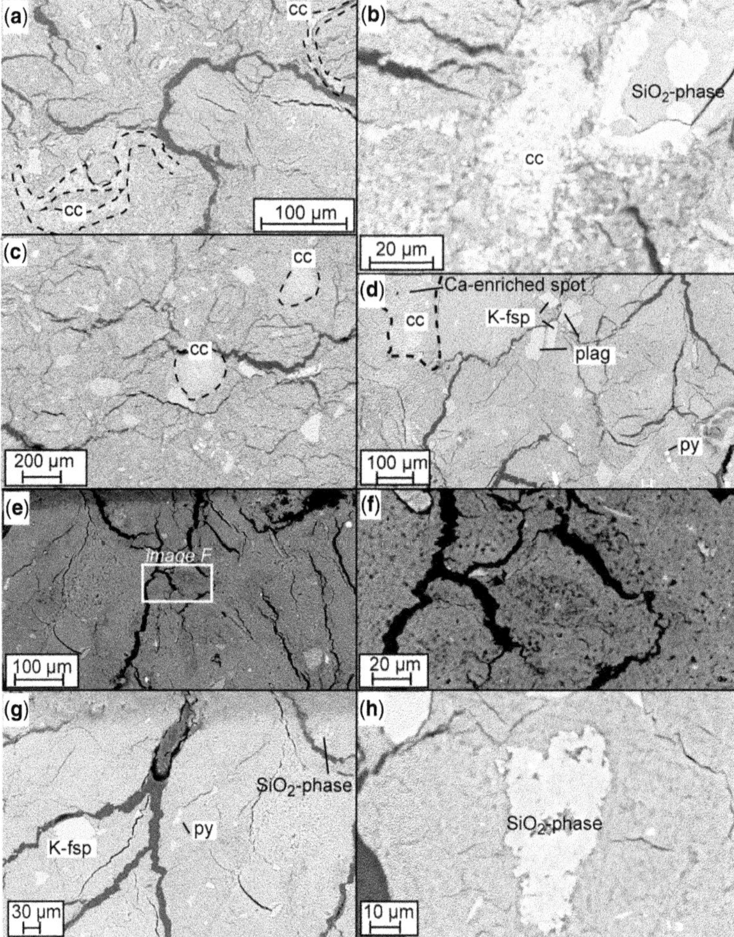

Fig. 12. BSE images of the first 5 mm of the bentonite core: (**a**) calcite (cc) precipitation in the Ca zone 2 at a depth of 7 mm; (**b**) calcite growth along with the dissolution of the SiO_2 phase at a depth of 3.5 mm in Ca zone 2; (**c**) calcite growth in Ca zone 2 at a depth of 6 mm; (**d**)–(**f**) Mg zone 1 between depths of 0.4 and 0.8 mm; (**g**) & (**h**) Mg zone 1 with SiO_2-phase dissolution. Plag, plagioclase; K-fsp, alkali feldspar; py, pyrite. The dashed lines indicate zones of Ca-enrichment.

Such a low pH is not in agreement with proposed values in compacted MX-80 bentonite based on thermodynamic considerations and mineral equilibria (Bradbury & Baeyens 2009). The effect was most likely to be an artefact triggered by an incident at the beginning of the experiment, whereby compressed air leaked into the outflow system, enriching the pore fluid in dissolved CO_2 gas.

The ion concentrations during the equilibration phase with APW_{OPA} infiltration are in agreement with squeezing experiments performed by Muurinen & Lehikoinen (1999). At similar densities and lower w/s ratio, their data showed the same ion enrichment relative to the inflow/external water, controlled mainly by dissolution of carbonates and sulphates and ion-exchange processes. The calculated ionic strengths range from 0.4 to 0.48 molal (Table 4), and are slightly higher compared to the porewater measurements performed on MX-80 by Muurinen *et al.* (2004) at a similar dry density and variable chloride concentrations. This may indicate that our porewater is not yet in equilibrium with the bentonite. Calculations of SIs (Table 4) show saturation of the outflow with respect to gypsum and amorphous silica. The erroneous low pH leads to an apparent undersaturation with respect to calcite in the first aliquots, followed by saturation at a neutral pH.

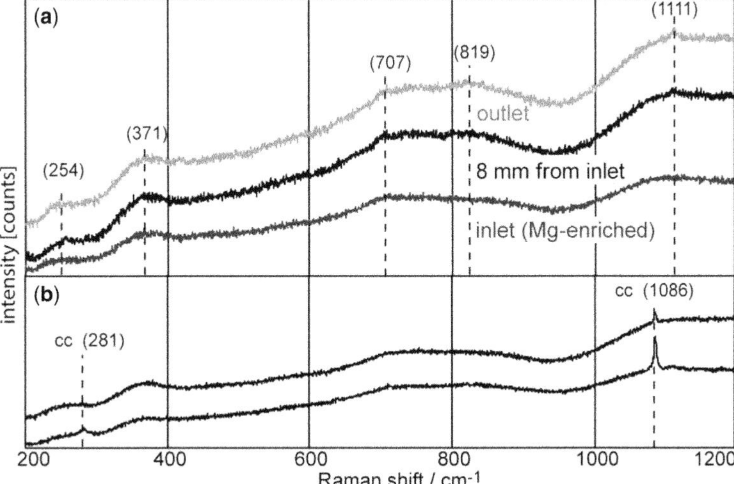

Fig. 13. Raman shift patterns: (**a**) the bentonite matrix; and (**b**) the newly grown calcite near the bentonite interface.

High-pH infiltration phase

Evolution of fluid composition. APW_{OPC} infiltration led to a further reduction in ionic strength in the outflow, which is in agreement with the electrical conductivity measurements (Table 4; Fig. 2b). Chloride as a major anionic charge carrier dropped continuously and was replaced after 0.2 PV, mainly by sulphate (Fig. 3). If chloride is considered as a non-reactive tracer mostly present in intergranular porosity, the chloride decrease represents a breakout curve, whereby the volume required to flush out chloride to half of its initial concentration would represent approximately one volume of flow-active porosity. This volume is equivalent to 0.196 PV relative to the initial full water content and corresponds to approximately 310 days. The mass balance of chloride showed a difference of 2.9 mmol between the cumulative inflow and outflow, indicating that it was still present in the bentonite core at the end of the experiment (Table 5).

Hydroxide, the main anion of the cement fluid (*c.* 271 mM), effectively did not reach the outflow – in contrast to chloride. Even after 577 days of APW_{OPC} injection, the hydroxide concentration never exceeded 0.002 mM, indicating a nearly complete consumption (pH only increased to *c.* 7.8). The conclusion from this is that OH^- was completely consumed by mineral dissolution reactions and surface de-protonation reactions (limited capacity).

The increase in ionic strength after 275 days of APW_{OPC} infiltration was mainly the result of a sulphate increase despite its lower concentration in APW_{OPC} compared to APW_{OPA}. Dissolution of gypsum is the most likely mechanism, and this process is fostered by precipitation of calcite that continuously removes Ca^{2+} from solution. Some pyrite dissolution may have occurred under the not strictly reducing conditions (but protected from the atmosphere), but there was no evidence of significant pyrite oxidation seen in the SEM (Fe-oxyhydroxides), nor for intermediate sulphur species such as thiosulphate (visible in ion chromatography). The extra cations to balance the increase in sulphate are generated by mineral dissolution (OH^- consumption), and this overall process is therefore limited to the reactive zone located very near the infiltration region.

The mass balance shows a loss of Na^+ from the core, released from the exchanger (Table 5). K^+ and Ca^{2+} were retained in the bentonite, by ion exchange for K^+, and probably also by calcite precipitation for Ca^{2+}. Dissolved Si and Al were enriched in the outflow during the high-pH infiltration.

Evolution of hydraulic conductivity. The hydraulic conductivity dropped twice during the high-pH infiltration phase: the first time directly after switching fluids and the second time after approximately 450 days (0.2 PV_{OPC}). The first phase of reduced hydraulic conductivity is interpreted as a transient phase (Dolder *et al.* 2014), while the second one seems to approach steady-state conditions. The Péclet number during the transient-flow phase indicated advection-dominated flow. It decreased thereafter to approximately 3.6, and reached a value of between 0.9 and 1.4 after about 650 days (0.24 PV_{OPC}), becoming more diffusion-dominated. Two main mechanisms for the reduction were proposed by Dolder *et al.* (2014): (1) mineral precipitation; and (2) a reduction in swelling pressure and resultant

Table 4. *Calculated saturation indices (SIs) of the outflow solution*

APW	Time (days)	PV (PV_{OPC})	Ionic strength (molal)	Charge balance error	Saturation indices							
					cc	gy	K-fsp	mont	qz	cristo	$SiO_{2amorph}$	Log ($pCO_{2(g)}$)
$APW_{Äspö}$			0.26	−1.5	−0.3	−0.8						−3.2
APW_{OPA}			0.23	2.1	−0.1	−0.3						−3.2
APW_{OPC}			0.28	7.3	1	−2.6	−10		−4.4	−5.2	−5.7	−12
	31	0.03 (0)	0.48	1.6	−2.8	−0.04	0.3	3	1.3	0.5	0	−1.1
	134	0.18 (0)	0.43	0.8	0.3	−0.06	3.8	7.3	1.3	0.5	0	−2.3
	191	0.22 (0.04)	0.39	5	0.8	−0.1	4.6	8.6	1.5	0.7	0.2	−1.7
	290	0.29 (0.11)	0.34	−0.1	0.8	−0.1	4.9	8.8	1.6	0.8	0.3	−2.1
	497	0.4 (0.21)	0.36	16.8	0.9	−0.2	5.7	9.2	1.7	0.9	0.5	−2.3
	690	0.43 (0.25)	0.34	7.2	1	−0.3	5.7	9	1.7	0.9	0.5	−2.5

APW, artificial porewater; PV, pore volume; cc, calcite; gy, gypsum; K-fsp, alkali feldspar; mont, montmorillonite; cristo, cristobalite; pCO_2, partial pressure of CO_2.

Table 5. *Ion-mass balance of the inflow and the outflow*

Ions	Total elements in infiltration fluids			Total elements in outflow	Difference between the inflow and outflow
	$APW_{Äspö}$	APW_{OPA}	APW_{OPC}		
Na^{+} (mmol)	2.56	1.4	1.35	6.93	−1.61
K^{+} (mmol)	0	0.02	2.12	0.06	2.08
Ca^{2+} (mmol)	1.58	0.1	0.02	0.25	1.44
Mg^{2+} (mmol)	0.05	0.08	0	0.15	−0.02
Al^{3+} (mmol)	–	–	3.5×10^{-4}	4.6×10^{-5}	4.6×10^{-5}
Si^{4+} (mmol)	–	–	7×10^{-4}	0.06	−0.06
SO_4^{2-} (mmol)	0.05	0.19	0.03	1.47	−1.2
Cl^{-} (mmol)	5.91	1.33	–	4.34	2.9
HCO_3^{-} (mmol)	7×10^{-3}	6×10^{-3}	3×10^{-3}	0.16	−0.15

–, not measured.

compaction due to the higher ionic strength of the APW_{OPC}. The sharp change in hydraulic conductivity after switching infiltration fluids, explained by the higher ionic strength (Dolder *et al.* 2014), had to be revised because the repeated removal of the experiment for CT scanning, including disconnection of the outflow capillary, induced errors in the measurements of the outflow. Mechanism (1) relates the drop in hydraulic conductivity to mineral precipitation in the filter, as well as in the bentonite. Mineral precipitation in the filter was recognized in the CT scans after 8 days of APW_{OPC} infiltration and had a diameter of approximately 12 mm in the last scan. Post-mortem analysis of the experiment identified these precipitates in the filter as calcite, with minor amounts of gypsum, talc and, possibly, some brucite. Massive precipitation of calcite and, possibly, an Mg-smectite was observed in the bentonite, identical to the high-density features detected in the CT. Mechanism (2) is based on the idea that an increase in ionic strength of the infiltration solution reduces the hydraulic conductivity in the confining pressure-constrained experimental set-up. APW_{OPC} has a slightly higher ionic strength compared to APW_{OPA} (Table 4). Not known in detail is how the high-pH, $K^{+}-Na^{+}-Ca^{2+}-OH^{-}$ fluid evolves in the bentonite. The possibilities range from locally increasing ionic strength due to gypsum dissolution to lowering it by consumption of the hydroxide ions.

Aqueous speciation modelling using PhreeqC was performed to identify possible processes. Directly at the interface, we deduced that the original high pH of the APW_{OPC} was at or near gypsum saturation. This led to a hypothetical pore fluid with an ionic strength of approximately 0.4 molal, which is clearly higher compared to the value of 0.28 molal for APW_{OPC}. This fluid would be oversaturated with respect to talc, calcite and cement phases, such as C–S–H and ettringite. A bit further away from the interface, the pH of the APW_{OPC} is expected to be buffered to approximately 7.8, as seen in the outflow (Fig. 3). At gypsum saturation, this would lead to a hypothetical fluid with an ionic strength of approximately 0.38 molal, induced mainly by the high sulphate concentration. This fluid would be oversaturated with respect to the Si–Al phases, but undersaturated with respect to the calcite and cement phases.

The calculations indicate that the hypothetical local pore fluids have a higher ionic strength compared to APW_{OPC} and APW_{OPA}. Karnland *et al.* (2006) observed in experiments, using fixed-volume cells and MX-80 bentonite, the following relationships for NaCl-based fluids: the higher the ionic strength of a porewater, the higher the hydraulic conductivity; and the lower the swelling pressure, the lower the density. Karnland *et al.* (2007) showed further for NaCl- and NaOH-based solutions that both had lower swelling pressures at higher ionic strengths. In the case of replacing a Cl^{-}-based solution with an OH^{-}-based solution at a constant ionic strength, the swelling pressure did drop drastically. Karnland *et al.* (2007) suggested that the dissolution of smectite and cristobalite mainly induced this effect.

For our experimental approach at constant total pressure, the expected behaviour is different and may be summarized as follows: an increased ionic strength reduced the diffuse layer volume, which allowed compaction if the volume was not fixed (Dolder *et al.* 2014). A reduction in the core volume was confirmed by CT measurements, and started 72 days after switching fluids (Fig. 5b). A compaction could be observed in the CT data, as well as in the post-mortem density measurements (Figs 4a & 7).

In summary, the drop in hydraulic conductivity after switching infiltration fluid was most likely to be dominated by extensive mineral precipitation in the filter region and in the first millimetres of bentonite, while the transient phase was linked to core compaction induced by gypsum dissolution, as well as by an increase in ionic strength of the pore water, while releasing water from the porosity.

Volume reduction and gypsum dissolution. The main period of core-volume reduction occurred during the transient-flow phase described in detail above. The propagation of the reduction in core diameter reached the end of the core simultaneously with the drop in hydraulic conductivity (Figs 2b & 4a), inferring that the core shrinkage and densification was related to this change in the porewater chemistry (ionic strength). The core shrinkage slowed down at the moment the hydraulic conductivity reached its lowest value after approximately 620 days (483 days of APW_{OPC} infiltration). This was the time at which the flow pattern changed from advection-dominated to diffusion-dominated. The absolute amount of shrinkage measured by CT was approximately 800 mm^3, which amounted to about 10% of the 9.3×10^4 mm^3 (9.3 ml) of porewater sampled during the transient-flow phase. The water ejection can explain a small fraction of the higher flow observed during this phase. It is argued that the core shrinkage is a combined effect of the above-described increase in ionic strength and the continuous dissolution of gypsum.

XRD measurements of the bentonite core showed much smaller quantities of gypsum in comparison to the pristine material, suggesting that significant quantities did dissolve. In order to estimate the quantities, a mass-balance calculation for sulphate was performed (Table 5). MX-80 bentonite contains 0.9 ± 0.3 wt% gypsum (Karnland 2010), which corresponds in our core to 1.27 g, or 7.36 mmol (gypsum, sulphate). In total, 0.27 mmol of sulphate were infiltrated and 1.47 mmol were measured in the outflow, which is an excess of 1.2 mmol (Table 5). If one now compares the maximum amount of sulphate originating from gypsum dissolution with the measured sulphate in the outflow, subtracting the inflows, one obtains a minimum amount of dissolved gypsum of approximately 19 wt%, subtracting sulphate in solution. Saturation indices (SI) of gypsum are shown in Table 4, and indicate that the outflow is initially at saturation and becomes undersaturated over time.

Dissolution of silicate minerals. The silicon concentration in the outflow doubled after switching to APW_{OPC} and tripled until the end of the experiment compared to the infiltration fluid (Fig. 3). The aluminium concentration doubled after changing fluids, but was still lower compared to APW_{OPC}. The modelling showed that the silicon concentration in the outflow was in equilibrium with amorphous SiO_2 during the APW_{OPA} infiltration (Table 4). The change to APW_{OPC} led to an oversaturation of amorphous silica. The fact that much more silicon was in solution relative to aluminium indicates that a pure SiO_2 phase, such as quartz, cristobalite, tridymite or amorphous silica, is involved, as observed by SEM. All these phases are distinctly oversaturated in the last aliquot sampled (Table 4). Preferential dissolution of cristobalite is in agreement with our XRD data, as well as with previous studies (Karnland 1997, 2004, 2010; Bouchet *et al.* 2004; Karnland *et al.* 2007; Fernández *et al.* 2013). The dissolution of quartz, as reported in previous studies (Fernández *et al.* 2006; Karnland *et al.* 2007), could not be confirmed. Amorphous silica would be difficult to detect with any analytical method, and its potential role remains uncertain.

Reaction zones. Two overlapping reaction zones could be distinguished in the bentonite by SEM EDX and CT measurements. One was characterized by strong MgO-enrichment in the clay matrix, the other by CaO-enrichment in the form of calcite precipitation (Fig. 10b, c).

The magnesium-enriched zone contained approximately 0.21 mmol of Mg^{2+} (3.4×10^{-3} mmol mm^{-3}), which was 0.15 mmol more compared to the amount in the same volume of pristine bentonite (*c.* 1.0×10^{-3} mmol mm^{-3}). The mass balance of the entire core showed a small net loss of 2.0×10^{-2} mmol of Mg^{2+} (Table 5), presumably displaced from the exchanger. Magnesium enrichment could only take place if an additional source was available. Two hypotheses are discussed: (1) infiltration of magnesium by the APW_{OPA}; or (2) dissolution of a magnesium phase. (1) In total, 8.0×10^{-2} mmol of magnesium were infiltrated during the equilibration phase (Table 5), explaining half of the enrichment. (2) The only major Mg-bearing mineral phase in MX-80 bentonite is montmorillonite. Dissolution of approximately 100 mm^3 of pure montmorillonite could explain the additional magnesium in the Mg-enrichment zone using amounts of magnesium of 7.4×10^{-2} mmol for Mg zone 1 and 0.14 mmol for Mg zone 2. A reduced core density and smaller core diameters in the inflow region, as well as smectite depletion directly at the inlet, support the dissolution hypothesis. It is concluded that the magnesium enrichment is a combination of both hypotheses. It seems that at high-pH conditions, the observed Mg phase was the only stable magnesium-bearing mineral phase. No magnesium-depletion zone was observed in the bentonite, which excludes diffusion of magnesium during APW_{OPC} infiltration that might have occurred by precipitation of an Mg phase. The magnesium enrichment took place in the clay matrix, but no distinct new mineral phases were observed at the given resolution.

A triangular $SiO_2-Al_2O_3+Fe_2O_3-MgO$ plot (Fig. 14) is used for further interpretation. Montmorillonite plots in a field with an average composition of 71 wt% SiO_2, 26 wt% $Al_2O_3+Fe_2O_3$ and 3 wt% MgO, in agreement with Fernández *et al.* (2006) and Karnland (2010). The Mg^{2+}-exchanged MX-80 smectite would not shift much in MgO direction and this composition is still far from the measured MgO-enriched zone: MgO zone 1 and 2 plot in a region of 61 wt% SiO_2, 20 wt% $Al_2O_3+Fe_2O_3$ and 19 wt% MgO. The measurements indicate the presence of newly formed Mg-rich mineral phases, such as saponite, M–S–H (Mg–Si-hydrate), hydrotalcite, talc or brucite (see the dotted grey lines in Fig. 14). We assume that the MgO-rich phase is related to a trioctahedral clay reflection observed in the XRD spectra, presumably saponite. The main chemical difference to montmorillonite is strong magnesium enrichment in the octahedral layer. Raman spectroscopy measurements indicate no major changes in the clay structure. M–S–H is amorphous with weak and broad humps in the XRD patterns (Brew & Glasser 2005). Brucite and talc were detected by XRD in small amounts in the inlet filter, but could not be identified in this zone of the bentonite. Several others observed similar reactions in high-pH experiments: Karnland (1997) observed an increased MgO content in the clay in MX-80 bentonite at ambient temperature; Ramirez *et al.* (2002*a*) observed the formation of a trioctahedral Mg-silicate phase and the dissolution of smectite at pH >12.6 and temperatures ≥35°C in FEBEX bentonite; Cuevas *et al.* (2006) showed the dissolution of montmorillonite and the formation of a new trioctahedral smectite, saponite–stevensite, in FEBEX bentonite at 35–90°C; Fernández *et al.* (2006) observed brucite precipitation in FEBEX bentonite, while montmorillonite dissolved at pH >13 and at ambient temperatures; Fernández *et al.* (2009) observed the formation of brucite interlayers in smectite (di-trioctahedral-chlorite-like phase without a continuous and ordered intercalation of brucite) in FEBEX bentonite at 60°C; Fernández *et al.* (2010) observed the formation of a trioctahedral sheet silicate, which they described as talc, in FEBEX bentonite at 60°C; and Fernández *et al.* (2013) observed the formation of a Mg-layer silicate phase (di-trioctahedral chlorites) in FEBEX bentonite at 90°C.

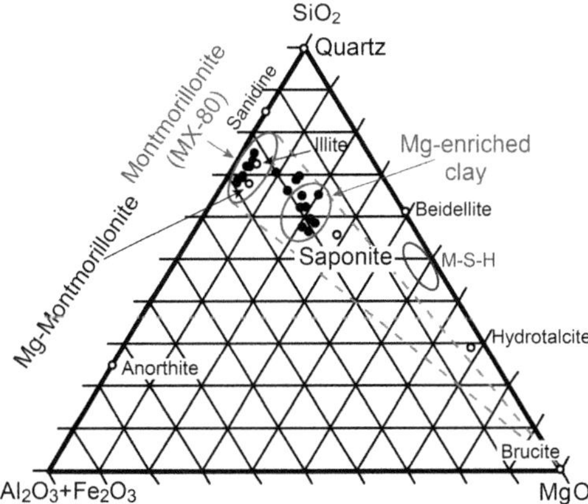

Fig. 14. A triangular $SiO_2-Al_2O_3+Fe_2O_3-MgO$ plot including an SEM-EDX point analysis of the MgO-enriched zone. Filled circles correspond to EDX measurements; and open circle and ellipses correspond to mineral compositions from the literature: MX-80 montmorillonite (Karnland 2010); Mg^{2+}-exchanged montmorillonite (Karnland *et al.* 2006); saponite (Duda & Rejl 1990); hydrotalcite (Anthony *et al.* 1995); and magnesium silicate hydrate gel (M–S–H) (Brew & Glasser 2005).

The CaO-enrichment propagated 7.7 mm into the bentonite core, characterized by intense calcite precipitation, verified by XRD and Raman spectroscopy measurements. SEM-EDX elemental spot analyses indicate a strong Ca-enrichment in this new phase, but, owing to the small grain size, other possible phases such as C–(A)–S–H cannot be excluded: however, neither XRD nor Raman spectroscopy measurements support this. A triangular $SiO_2-(Al_2O_3+Fe_2O_3)-CaO$ plot is shown in Figure 15, with analysis of the newly grown CaO phase. The data plot along a line, ending at the CaO–calcite edge. It can be concluded that the measurements are either pure calcite or clay intermixed with calcite.

The Ca mass balance shows a net gain in the core (Table 5). The Ca source for the calcite precipitation is the dissolution of gypsum, a displacement from the clay exchanger by K^+ and Na^+, and a minor contribution from the inflowing APW_{OPC}.

The bicarbonate mass balance shows a net loss of 0.15 mmol from the core, indicating dissolution of carbonates outside the precipitation zone (Table 5). The amount of infiltrated bicarbonate from the APWs is 1.6×10^{-2} mmol, with 3.0×10^{-3} mmol originating from the APW_{OPC}. This could induce the precipitation of 0.6 mm^3 of calcite, while the fraction originating from the APW_{OPC} contributes only 9.0×10^{-2} mm^3. A calcite content of 0.5–1.5 wt% in the two reaction zones, based on XRD reflection intensity data, corresponds to approximately 4 mm^3 of pure calcite. This indicates that an additional carbonate source is required, which may originate from the dissolution of carbonate compounds further away, reported by Karnland (2010) and associated back-diffusion. PhreeqC equilibrium

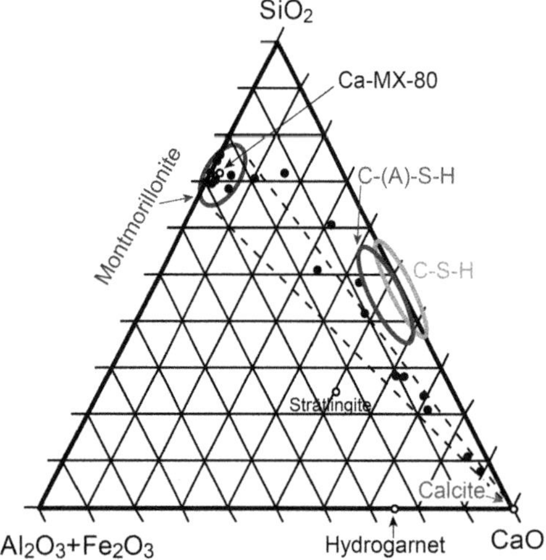

Fig. 15. A triangular $SiO_2-Al_2O_3+Fe_2O_3-CaO$ plot including an SEM-EDX point analysis of the newly formed CaO phase, as well as of the CaO-enriched regions. Filled circles correspond to EDX measurements; and open circle and ellipses to mineral compositions from the literature: MX-80 montmorillonite (Karnland 2010); Ca^{2+}-saturated MX-80 montmorillonite (Karnland *et al.* 2006); C–S–H (Lothenbach *et al.* 2008); C–(A)–S–H (Lothenbach & Winnefeld 2006; Lothenbach *et al.* 2008); and hydrogarnet (Matschei *et al.* 2007; Lothenbach *et al.* 2008).

calculations show that an increase in the pH of a clay porewater leads to calcite oversaturation.

The precipitation of Ca phases, like calcite, in high-pH environments are described in various studies: Sánchez *et al.* (2006) observed increased concentrations of calcite (<5%) in low- and high-temperature batch experiments using FEBEX bentonite; and Jenni *et al.* (2014) observed a CaO-enriched layer at a claystone–OPC interface, along with an elevated calcite concentration therein. No C–S–H precipitation was observed in our experiment in contrast to previous studies at ambient temperatures (Karnland 1997; Ramirez *et al.* 2002*a*, *b*; Bouchet *et al.* 2004; Mosser-Ruck & Cathelineau 2004; Fernández *et al.* 2006, 2010).

The evolution of the reaction plume showed a density increase in the inlet filter and the bentonite core over time (Fig. 16). Judging from its shape and its evolution, it is most likely that the imaged reaction plume in the CT scans represents parts of the calcite precipitation zone (mainly Ca zone 1 and 2) and the Mg-mineral precipitation zone. The first change was observed after 3 days of APW_{OPC} infiltration, with a small plug of most probably calcite forming in the filter. After 36 and 137 days, the zone kept growing mainly in the filter and it evolved into two ring structures. The ring structures were most probably induced by the grooved POM-adapter surface supporting the filter. This POM-adapter was produced by turning the piece in the workshop. After 483 days, there was clear evidence that the reaction plume had proceeded into the bentonite. After 588 days, a reaction plume 8 mm in thickness, with a diameter of 18 mm and a volume of 393 mm^3, was observed. The heterogeneous nature of the plume is mainly due to preferential flow paths. The position of the reaction plume seen in the CT scan close to the end of the experiment is in agreement with the post-mortem measured extension of the CaO–calcite-enrichment zone. Analyses of the growth of the reaction plume yielded an average growth rate of around 1.0 mm^3/day in the first 137 days, decreasing to approximately 0.6 mm^3/day thereafter. A comparison between volumetric flow and reaction progress shows a retardation of 34 during the first 137 days, and 20 by the end. This leads to the conclusion that the system did buffer the high-pH fluid by dissolution of cristobalite and montmorillonite (and gypsum), coupled with calcite and saponite precipitation.

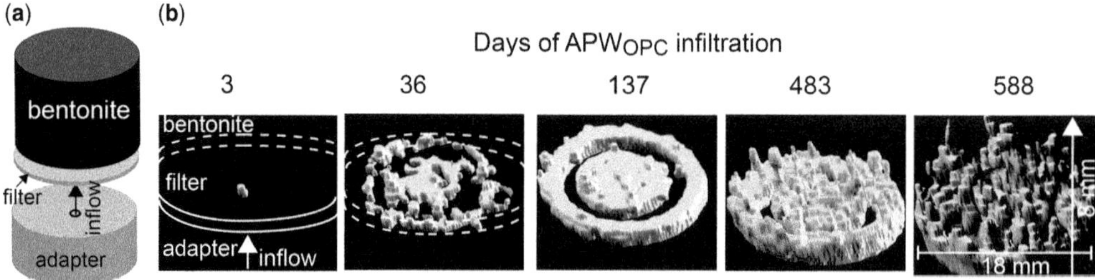

Fig. 16. Evolution of the reaction plume based on CT images and analysed by image processing: (**a**) sketch of the analysed section; and (**b**) zone of the reaction, consisting of an increased X-ray attenuation.

Conclusion

Infiltration of a high-pH pore fluid after 136 days of equilibration time led to a decrease in hydraulic conductivity from approximately 1.1×10^{-13} to approximately 4.2×10^{-15} m s^{-1}. The former advective dominated flow changed to diffusion-dominated flow with column Péclet numbers close to 1. This decrease resulted from a combination of mineral precipitation in the inlet filter and in the adjacent bentonite, as well as from an increase in ionic strength of the porewater mainly by extensive gypsum dissolution. The high-pH infiltration led to a core shrinkage of approximately 0.9 vol%, detected by CT measurements, induced by an increase in ionic strength with subsequent compaction and gypsum dissolution.

The outflow showed a decrease in all ion concentrations except for sulphate, silicon, aluminium and potassium, while pH increased just slightly from approximately 7.5 to 8. After switching to the high-pH fluid, chloride was replaced as main anion charge carrier by sulphate after 0.4 pore volumes (0.4 PV).

CT scans track a progressing hemispherical reaction plume from the filter into the first few millimetres of the bentonite sample, revealing a distinct zone of increased bulk density. This plume is characterized in the bentonite by two distinct, but overlapping, zones of MgO- and CaO-enrichment. The region enriched in Ca is related to intense calcite precipitation in the clay matrix. The MgO = enriched zone is most likely to be related to saponite, but the occurrence of other Mg phases cannot be excluded. Dissolution of cristobalite was observed in the same region, which is assumed to be the main source for elevated silicon concentrations in the outflow.

The experiment attests an effective buffering capacity for bentonite and a progressing coupled hydraulic–chemical sealing process. Also, the physical integrity of the interface region was preserved in this set-up with a total pressure boundary condition on the core sample. This is in contrast to some experiments carried out at under constant-volume condition, where shrinkage in bentonite led to the formation of preferential pathways and the breakthrough of the high-pH plume (e.g. Fernández *et al.* (2009)). Constant confining-pressure constraints may be more realistic for a repository environment due to the ability of the surrounding and unaltered bentonite to compensate volumetric changes at the interface.

Financial support from NAGRA is gratefully acknowledged. The Department of Forensic Medicine at the University of Bern is acknowledged for granting access and providing technical support for operating their CT scanner. The machine shop at our institute is acknowledged for constructing the infiltration apparatus, and our analytical laboratory performed the IC and ICP-OES analyses. Barbara Lothenbach (EMPA) kindly provided the modelled fluid composition for the OPC. Two anonymous reviewers – one in particular – significantly improved the clarity of our observations and discussions.

References

Adler, M. 2001. *Interaction of claystone and hyperalkaline solutions at 30°C: a combined experimental and modeling study*. PhD thesis, University of Bern, Switzerland.

Adler, M., Mäder, U. & Waber, H.N. 2001. Core infiltration experiment investigating high-pH alteration of low-permeability argillaceous rock at 30°C. *In*: Cidu, R. (ed.) *Proceedings WRI-10 (10th International Symposium on Water-Rock Interaction)*. Balkema, Villasimius, Italy, 1299–1302.

Alia, J.M., Edwards, H.G., Garcia-Navarro, F.J., Parras-Armenteros, J. & Sanchez-Jimenez, C.J. 1999. Application of FT-Raman spectroscopy to quality control in brick clays firing process. *Talanta*, **50**, 291–298.

Anthony, J.W., Bideaux, R.A., Bladh, K.W. & Nichols, M.C. 1995. *Handbook of Mineralogy*. Mineral Data Publishing, Tucson, AZ.

Bauer, A. & Berger, G. 1998. Kaolinite and smectite dissolution rate in high molar KOH solutions at 35°C and 80°C. *Applied Geochemistry*, **13**, 905–916, https://doi.org/10.1016/S0883-2927(98)00018-3

Bauer, A. & Velde, B. 1999. Smectite transformation in high molar KOH solutions. *Clay Minerals*, **34**, 259–273.

Blacksberg, J., Rossman, G.R. & Gleckler, A. 2010. Time-resolved Raman spectroscopy for in situ planetary mineralogy. *Applied Optics*, **49**, 4951–4962, https://doi.org/10.1364/ao.49.004951

Blanc, P., Lassin, A. & Piantone, P. 2007. *Thermoddem A Database Devoted to Waste Minerals*. BRGM (Bureau de Recherches Géologiques et Minières), Orléans, France, http://thermoddem.brgm.fr

Bouchet, A., Casagnabère, A. & Parneix, J.C. 2004. *Batch Experiments: Results on MX80*. (ANDRA) European contract FIKW-CT-2000-0028 ANDRA (National Agency for Radioactive Waste Management), Paris.

Bradbury, M.H. & Baeyens, B. 2009. Experimental and modelling studies on the pH buffering of MX-80 bentonite porewater. *Applied Geochemistry*, **24**, 419–425, https://doi.org/10.1016/j.apgeochem.2008.12.023

Brew, D.R.M. & Glasser, F.P. 2005. Synthesis and characterisation of magnesium silicate hydrate gels. *Cement and Concrete Research*, **35**, 85–98, https://doi.org/10.1016/j.cemconres.2004.06.022

Brindley, G.W. & Brown, G. 1980. *Crystal Structures of Clay Minerals and Their X-ray Identification*. Mineralogical Society, London.

Cuevas, J., Vigil de la Villa, R., Ramírez, S., Sánchez, L., Fernández, R. & Leguey, S. 2006. The alkaline reaction of FEBEX bentonite: a contribution to the study of the performance of bentonite/concrete engineered barrier systems. *Journal of Iberian Geology*, **32**, 151–174.

DAUZERES, A., LE BESCOP, P., SARDINI, P. & CAU DIT COUMES, C. 2010. Physico-chemical investigation of clayey/cement-based materials interaction in the context of geological waste disposal: experimental approach and results. *Cement and Concrete Research*, **40**, 1327–1340.

DE LA VILLA, R.V., CUEVAS, J., RAMIREZ, S. & LEGUEY, S. 2001. Zeolite formation during the alkaline reaction of bentonite. *European Journal of Mineralogy*, **13**, 635–644.

DE WINDT, L., PELLEGRINI, D. & VAN DER LEE, J. 2004. Coupled modeling of cement/claystone interactions and radionuclide migration. *Journal of Contaminant Hydrology*, **68**, 165–182, https://doi.org/10.1016/S0169-7722(03)00148-7

DOLDER, F., MÄDER, U., JENNI, A. & SCHWENDENER, N. 2014. Experimental characterization of cement–bentonite interaction using core infiltration techniques and 4D computed tomography. *Physics and Chemistry of the Earth, Parts A/B/C*, **70–71**, 104–113, https://doi.org/10.1016/j.pce.2013.11.002

DUDA, R. & REJL, L. 1990. *Minerals of the World*. Arch Cape Press, New York.

EBERL, D.D., VELDE, B. & MCCORMICK, T. 1993. Synthesis of illite-smectite from smectite at earth surface temperatures and high pH. *Clay Minerals*, **28**, 49–60.

FERNÁNDEZ, R., CUEVAS, J., SANCHEZ, L., DE LA VILLA, R.V. & LEGUEY, S. 2006. Reactivity of the cement-bentonite interface with alkaline solutions using transport cells. *Applied Geochemistry*, **21**, 977–992.

FERNÁNDEZ, R., MÄDER, U., RODRÍGUEZ, M., DE LA VILLA, R.V. & CUEVAS, J. 2009. Alteration of compacted bentonite by diffusion of highly alkaline solutions. *European Journal of Mineralogy*, **21**, 725–735, https://doi.org/10.1127/0935-1221/2009/0021-1947

FERNÁNDEZ, R., RODRÍGUEZ, M., DE LA VILLA, R.V. & CUEVAS, J. 2010. Geochemical constraints on the stability of zeolites and C-S-H in the high pH reaction of bentonite. *Geochimica et Cosmochimica Acta*, **74**, 890–906.

FERNÁNDEZ, R., MÄDER, U. & JENNI, A. 2011*a*. *Multi-Component Advective-Diffusive Transport Experiment in MX-80 Compacted Bentonite: Method and Results of 1st Phase of Experiment*. Internal NAGRA Working Report NAGRA (National Cooperative for the Disposal of Radioactive Waste), Wettingen, Switzerland.

FERNÁNDEZ, R., MÄDER, U. & STEEFEL, C. 2011*b*. Modelling of a bentonite column experiment with CrunchFlow including new clay-specific transport features. *Mineralogical Magazine*, **75**, 839.

FERNÁNDEZ, R., VIGIL DE LA VILLA, R., RUIZ, A.I., GARCÍA, R. & CUEVAS, J. 2013. Precipitation of chlorite-like structures during OPC porewater diffusion through compacted bentonite at 90°C. *Applied Clay Science*, **83–84**, 357–367, https://doi.org/10.1016/j.clay.2013.07.021

FERRAGE, E., LANSON, B., SAKHAROV, B.A. & DRITS, V.A. 2005. *Investigation of Smectite Hydration Properties by Modeling Experimental X-Ray Diffraction Patterns. Part I. Montmorillonite Hydration Properties*. Mineralogical Society of America, Washington, DC.

GAUCHER, E.C. & BLANC, P. 2006. Cement/clay interactions – a review: experiments, natural analogues, and modeling. *Waste Management*, **26**, 776–788, https://doi.org/10.1016/j.wasman.2006.01.027

HUMMEL, W., BERNER, U., CURTI, E., PEARSON, F.J. & THOENEN, T. 2002. *NAGRA/PSI Chemical Thermodynamic Data Base 01/01*. NAGRA Technical Report NTB 02-16 NAGRA (National Cooperative for the Disposal of Radioactive Waste), Wettingen, Switzerland/Universal Publishers), Parkland, FL.

JACQUES, D. 2009. *Benchmarking of the Cement Model and Detrimental Chemical Reactions Including Temperature Dependent Parameters. Project Near Surface Disposal of Category A Waste at Dessel*. NIRAS-MP5-03 DATA-LT(NF) Version 1.ONDRAF/NIRAS (Belgian Agency for Radioactive Waste and Enriched Fissile Materials), Brussels.

JENNI, A., MÄDER, U., LEROUGE, C., GABOREAU, S. & SCHWYN, B. 2014. In situ interaction between different concretes and Opalinus Clay. *Physics and Chemistry of the Earth, Parts A/B/C*, **70–71**, 71–83, https://doi.org/10.1016/j.pce.2013.11.004

KARNLAND, O. 1997. *Cement/Bentonite Interaction: Results from 16 Month Laboratory Tests*. SKB Technical Report 97-32. SKB (Swedish Nuclear Fuel and Waste Management), Stockholm.

KARNLAND, O. 2004. *Laboratory Experiments Concerning Compacted Bentonite Contacted to High-pH Solutions*. (ANDRA) European contract FIKW-CT-2000-0028 ANDRA (National Agency for Radioactive Waste Management), Paris.

KARNLAND, O. 2010. *Chemical and Mineralogical Characterization of the Bentonite Buffer for the Acceptance Control Procedure in a KBS-3 Repository*. SKB Technical Report 10-60. SKB (Swedish Nuclear Fuel and Waste Management), Stockholm.

KARNLAND, O., OLSSON, S. & NIELSSON, U. 2006. *Mineralogy and Sealing Properties of Various Bentonites and Smectite-Rich Clay Materials*. SKB Technical Report 06-30. SKB (Swedish Nuclear Fuel and Waste Management), Stockholm.

KARNLAND, O., OLSSON, S., NILSSON, U. & SELLIN, P. 2007. Experimentally determined swelling pressures and geochemical interactions of compacted Wyoming bentonite with highly alkaline solutions. *Physics and Chemistry of the Earth, Parts A/B/C*, **32**, 275–286.

KARNLAND, O., OLSSON, S. ET AL. 2009. *Long Term Test of Buffer Material at the Äspö Hard Rock Laboratory, LOT Project*. Final Report on the A2 Test Parcel. SKB Technical Report 09-29. SKB (Swedish Nuclear Fuel and Waste Management), Stockholm.

KOSAKOWSKI, G., BERNER, U., WIELAND, E., GLAUS, M. & DEGUELDRE, C. 2014. *Geochemical Evolution of the L/ILW Near-Field*. NAGRA Technical Report NTB 97-04. NAGRA, Wettingen, Switzerland.

KOŠAŘOVÁ, V., HRADIL, D., NĚMEC, I., BEZDIČKA, P. & KANICKÝ, V. 2013. Microanalysis of clay-based pigments in painted artworks by the means of Raman spectroscopy. *Journal of Raman Spectroscopy*, **44**, 1570–1577, https://doi.org/10.1002/jrs.4381

LANCASTER, J.L. & MARTINEZ, M.J. 2007. *Mango – Multi-Image Analysis GUI*. Research Research Imaging, University of Texas Health Science Center, San Antonio, TX.

LOTHENBACH, B. 2010. *Hydration Experiments of OPC, LAC and ESDRED Cements: 1 h to 3.5 Years*. Mont

Terri Technical Note (Internal Report) Mont Terri Project, St-Ursanne, Switzerland.

Lothenbach, B. & Winnefeld, F. 2006. Thermodynamic modelling of the hydration of Portland cement. *Cement and Concrete Research*, **36**, 209–226.

Lothenbach, B., Matschei, T., Möschner, G. & Glasser, F.P. 2008. Thermodynamic modelling of the effect of temperature on the hydration and porosity of Portland cement. *Cement and Concrete Research*, **38**, 1–18, https://doi.org/10.1016/j.cemconres.2007.08.017

Mäder, U. 2009. *Reference Pore Water for the Opalinus Clay and 'Brown Dogger' for the Provisional Safety-Analysis in the Framework of Sectoral Plan – Interim Results (SGT-ZE)*. NAGRA Working Report NAGRA (National Cooperative for the Disposal of Radioactive Waste), Wettingen, Switzerland.

Mäder, U., Waber, H.N. & Gautschi, A. 2004. New method for porewater extraction from claystone and determination of transport properties with results for Opalinus Clay (Switzerland). *In*: Wanty, R.B. & Seal, R.R. (eds) *Proceedings WRI-10 (11th International Symposium on Water–Rock Interaction)*. Balkema, Saratoga Springs, NY, 445–448.

Mäder, U., Fierz, T. *et al.* 2006. Interaction of hyperalkaline fluid with fractured rock: field and laboratory experiments of the HPF project (Grimsel Test Site, Switzerland). *Journal of Geochemical Exploration*, **90**, 68–94.

Mäder, U., Jenni, A., Fernández, R. & de Soto, I. 2012. Reactive transport in compacted bentonite: porosity concepts, experiments and applications. *Mineralogical Magazine*, **76**, 2052.

Matschei, T., Lothenbach, B. & Glasser, F.P. 2007. Thermodynamic properties of Portland cement hydrates in the system CaO-Al2O3-SiO2-CaSO4-CaCO3-H2O. *Cement and Concrete Research*, **37**, 1379–1410.

Mazurek, M., Waber, H.N., Mäder, U., de Haller, A. & Koroleva, M. 2013. *Geochemical Synthesis for the Effingen Member in Boreholes at Oftringen, Gösgen and Küttigen*. NAGRA Technical Report NTB 12-07. NAGRA (National Cooperative for the Disposal of Radioactive Waste), Wettingen, Switzerland.

Moore, D.M. & Reynolds, R.C. 1989. *X-Ray Diffraction and the Identification and Analysis of Clay Minerals*. Oxford University Press, New York.

Mosser-Ruck, R. & Cathelineau, M. 2004. Experimental transformation of Na,Ca-smectite under basic conditions at 150°C. *Applied Clay Science*, **26**, 259–273, https://doi.org/10.1016/j.clay.2003.12.011

Muurinen, A. & Lehikoinen, J. 1999. Porewater chemistry in compacted bentonite. *Engineering Geology*, **54**, 207–214.

Muurinen, A., Karnland, O. & Lehikoinen, J. 2004. Ion concentration caused by an external solution into the porewater of compacted bentonite. *Physics and Chemistry of the Earth, Parts A/B/C*, **29**, 119–127, https://doi.org/10.1016/j.pce.2003.11.004

NAGRA 2002. *Demonstration of Disposal Feasibility for Spent Fuel, Vitrified High-Level Waste and Long-Lived Intermediate-Level Waste (Entsorgungsnachweis)*. NAGRA Technical Report NTB 02-05. NAGRA (National Cooperative for the Disposal of Radioactive Waste), Wettingen, Switzerland.

Parkhurst, D.L. & Appelo, C.A.J. 1999. *User's Guide to PHREEQC (version 2) – A Computer Program for Speciation, Batch Reaction, One-Dimensional Transport, and Inverse Geochemical Calculations*. United States Geological Survey, Water-Resources Investigations Report, **99-4259**.

Parkhurst, D.L. & Appelo, C.A.J. 2013. *Description of Input and Examples for PHREEQC Version 3 – A Computer Program for Speciation, Batch Reaction, One-Dimensional Transport, and Inverse Geochemical Calculations*. United States Geological Survey, Techniques and Methods, **6-A43**.

Ramirez, S., Cuevas, J., de la Villa, R.V. & Leguey, S. 2002*a*. Hydrothermal alteration of La Serrata bentonite (Almeria, Spain) by alkaline solutions. *Applied Clay Science*, **21**, 257–269.

Ramirez, S., Cuevas, J., Petit, S., Righi, D. & Meunier, A. 2002*b*. Smectite reactivity in alkaline solutions. *Geologica Carpathica*, **53**, 87–92.

Sánchez, L., Cuevas, J., Ramírez, S., Riuiz De León, D., Fernández, R., Vigil Dela Villa, R. & Leguey, S. 2006. Reaction kinetics of FEBEX bentonite in hyperalkaline conditions resembling the cement–bentonite interface. *Applied Clay Science*, **33**, 125–141, https://doi.org/10.1016/j.clay.2006.04.008

Savage, D., Noy, D. & Mihara, M. 2002. Modelling the interaction of bentonite with hyperalkaline fluids. *Applied Geochemistry*, **17**, 207–223.

Schindelin, J., Arganda-Carreras, I. *et al.* 2012. Fiji: an open-source platform for biological-image analysis. *Nature Methods*, **9**, 676–682.

Thoenen, T. & Kulik, D. 2003. *NAGRA/PSI Chemical Thermodynamic Database 01/01 for the GEM-Selektor (V.2-PSI) Geochemical Modeling Code*. Paul Scherrer Institut (PSI), Villigen, Switzerland, http://les.web.psi.ch/Software/GEMS-PSI/doc/pdf/

Van Loon, L.R., Glaus, M.A. & Müler, W. 2007. Anion exclusion effects in compacted bentonites: towards a better understanding of anion diffusion. *Applied Geochemistry*, **22**, 2536–2552.

Watson, C., Hane, K., Savage, D., Benbow, S., Cuevas, J. & Fernández, R. 2009. Reaction and diffusion of cementitious water in bentonite: results of 'blind' modelling. *Applied Clay Science*, **45**, 54–69.

Wolery, T.J. 1992. *EQ3/6, A Software Package for Geochemical Modeling of Aqueous Systems: Package overview and Installation Guide (Version 7.0)*. Lawrence Livermore National Laboratory, University of California, Livermore, CA.

Chemical erosion of the bentonite buffer: do we observe it in nature?

HEINI M. REIJONEN* & NURIA MARCOS

Saanio & Riekkola Oy, Laulukuja 4, 00420 Helsinki, Finland

**Correspondence: heini.reijonen@sroy.fi*

Abstract: It is planned to use bentonites as buffer materials in various types of geological repositories designed to host radioactive waste. For the bentonite materials, the hydrological and hydrogeochemical conditions are two of the main environmental factors affecting their evolution, assessed for a period of up to 1 Ma. Since it has been observed in laboratory tests that smectite, in particular montmorillonite, forms colloids and disperses in very dilute conditions, a scenario for chemical erosion has been discussed at length, especially in relation to geological repositories located in future glaciated terrains and locations otherwise potentially hosting dilute groundwaters. General understanding, based on laboratory experiments, is that bentonite erosion does not occur when the total charge equivalent of cations in groundwater is higher than 4 mM. However, based on current knowledge, it seems that chemical erosion is not observed in repository-relevant natural systems where groundwater conditions are below the given limit. Further investigation is therefore suggested to provide a better scientific understanding of the mechanisms involved in stabilizing smectites in these natural systems.

Bentonite, containing mainly the mineral montmorillonite, is the reference buffer material in the KBS-3V (and KBS-3H) concept for geological disposal of spent nuclear fuel at Olkiluoto, Finland (Posiva 2012*a*, *b*). Montmorillonite is used in the literature both as a synonym for smectite and as a mineral name for a certain type of smectite. Here, 'montmorillonite' is used for the mineral and 'smectite' is used when discussing the mineral group. The longevity of montmorillonite, in particular, and smectites, in general, in current and evolving groundwater conditions is of interest when assessing the long-term performance of the bentonite buffer (cf. Fig. 1). The stability of bentonite is of interest in variable salinity ranges, but especially in relation to potential dilute water intrusion during glacial retreat or to continued dilution due to an extended temperate period (i.e. taking global warming into account). Traditionally, process understanding regarding long-term bentonite EBS behaviour has been acquired from studies performed on bentonite deposits (often called natural analogues: e.g. see Miller *et al.* 2000). To date, detailed behaviour of bentonite deposits in relation to surrounding groundwater systems has been little investigated (see the discussion in Laine & Karttunen 2010). Smectites occur within the crystalline basement in various geological environments: for example, fracture gouge, altered intrusions and some weathering profiles (cf. Marcos 2002; Vartiainen 2005; Laine & O'Brien 2008 for Finnish examples).

Background

Chemical erosion of bentonite is related to conditions where there is dilute enough groundwater to produce, and high enough flow to transport, bentonite colloids away from the repository via water-conducting features. The chemical erosion process is acknowledged in the current requirements set for the repository geochemical environment, especially on the total charge equivalent of cations in groundwater, which should remain above 4 mM (SKB 2011; Posiva 2012*a*). Chemical erosion is often discussed in relation to smectite dispersion only, and depositional processes are not accounted for. Requirements are mainly based on experimental information simulating simplified systems in artificial fractures (see the references given below for details).

The approach for solving this problem has focused on modelling exercises and experiments in the laboratory with bentonite materials (e.g. Neretnieks *et al.* 2009; Schatz *et al.* 2013). Laboratory tests are largely based on small-scale experimental set-ups, including compacted bentonite samples with variable compositions from purified homoionic Na- and Ca- montmorillonites to natural bentonites. Based on experiments (see Schatz *et al.* 2013; Erikson & Schatz 2015 and references therein), the process of chemical erosion is dependent on the initial composition of the smectite or bentonite material, density, aperture of the fracture, fracture inclination and flow rate.

Groundwater–buffer interaction may alter the exchangeable cation composition (ECC) of smectite during repository evolution. In addition, accessory minerals can dissolve and, along with buffer saturation, may affect the cation composition of smectites.

Modelling results of the chemical erosion process to date show discrepancies (see Schatz *et al.*

From: NORRIS, S., BRUNO, J., VAN GEET, M. & VERHOEF, E. (eds) 2017. *Radioactive Waste Confinement: Clays in Natural and Engineered Barriers*. Geological Society, London, Special Publications, **443**, 307–317.
First published online June 24, 2016, https://doi.org/10.1144/SP443.13

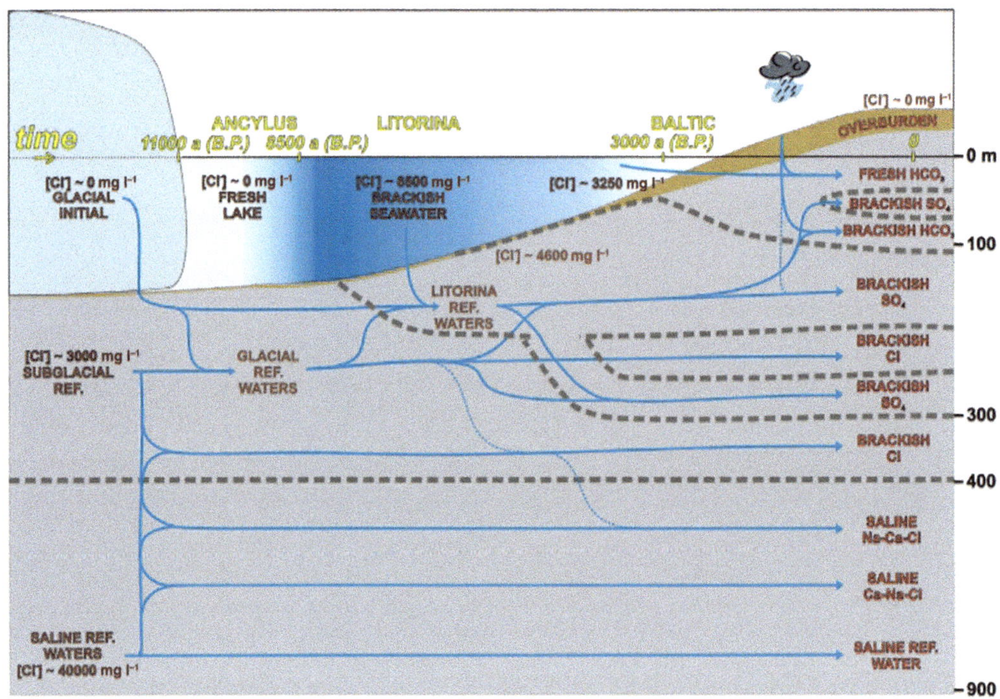

Fig. 1. Schematic representation of interpreted initial and boundary conditions at Olkiluoto since the last glacial period (redrawn from Posiva 2009). Potential salinity (as the Cl content) is shown for recharge waters (in the upper part along the time line) and bedrock groundwaters (on the left) at the initial stage of modelling. Reference waters correspond to figures 7–14 from Posiva (2013*a*). Saline and subglacial reference waters are contemporaneous initial waters (or end members) for the studied time period: that is, their formation dates are prior to the presented timescale. The generalized hydrochemical mixing hypothesis that is solved in detail, using initial waters with mass-balance calculations, is shown with blue arrows for the current groundwater types. Dashed lines between arrows imply minor mixing. Major groundwater types are bounded with grey dashed lines (Posiva 2013*a*).

2013) and, currently, the models are being developed further, along with an attempt to validate the models based on the experimental information (e.g. Moreno *et al.* 2015).

It should be noted that chemical erosion itself is not the only process that has a role in the long-term behaviour of bentonite in relation to mass transfer. Also, accessory minerals, such as quartz, that are of sufficient grain size and relatively insoluble may affect the system by forming filtering layers along the mass transfer path (e.g. fracture) (as suggested by Neretnieks *et al.* 2009). The consequent process of clogging of the water-conducting fractures may then occur, although the significance of this process in inhibiting erosion in unclear.

During the latest review on natural analogues relevant to Posiva's safety case, it was concluded that there are no dedicated natural analogue (NA) studies to support or disregard the chemical erosion process during potential dilute conditions during glacial meltwater intrusion at Olkiluoto (Posiva 2012*c*). The need for further research in relation to chemical erosion and development work has also been acknowledged elsewhere (e.g. see Reijonen & Alexander 2015 and references therein). There is great potential for gathering and interpreting existing information in the literature available on:

- the occurrence of dilute groundwater conditions at repository-relevant depths;
- bentonite/smectite occurrences/deposits where dilute water (e.g. meteoric) is in contact with the clay;
- other localities where smectite occurs in the crystalline basement.

Intrusion of dilute water

Buffering capacity of the bedrock and the nature of infiltrating fluids

In general, at Olkiluoto, the bedrock buffers both the redox conditions and the pH of the groundwaters (Posiva 2013*a*). Dilution of the groundwater can occur via two paths, either by prolonged infiltration of meteoric water during temperate climate conditions or during glaciation (retreat). It is assumed that during the latter, there may be increased flow in the conductive zones in the fractured rock (Posiva 2013*a*) and, hence, the time for water to react with the fracture minerals may be shorter.

Groundwater composition also affects the exchangeable cation composition (ECC) of montmorillonite. Arthur & Savage (2012) suggest that for Forsmark, the available evidence from glacial meltwaters does not support the promotion of chemical erosion. On the contrary, they suggest that the glacial meltwater composition would actually inhibit the erosion. This is based on the fact that the groundwater composition with $(Na^+)^2/Ca^{2+}$ activity ratios <0.05 (as seen in most Forsmark samples) would drive the clay exchanger systems towards a $>90\%$ Ca smectite end member, and thus maintain the bentonite in contact with groundwater in the stability field for gel formation instead of sol. The study of Arthur & Savage (2012) supports the view that the water–bentonite interface should be treated in more precise way in order to be able to describe the processes correctly at the conceptual level.

Local and regional considerations

Geochemical reactions do occur during filtration of the meteoric (or glacial melt) waters in the fractured bedrock system. Based on the current palaeohydrogeochemical knowledge from Olkiluoto (Fig. 1), it can be said that there is no evidence of glacial meltwaters penetrating below 300 m depth (Posiva 2013*a*). However, modelling results suggest that this could be the case in a few canister locations during the long-term evolution of the site (Posiva 2013*b*).

At other coastal and inland sites in Finland, more dilute conditions are observed (e.g. at Palmottu (Smellie *et al.* 2002), and at Hästholmen and Kivetty (Pitkänen *et al.* 1998, 2001)). At these sites, glacial groundwater signatures are present down to a depth of 400–500 m. In the future, when uplift continues at Olkiluoto, meteoric waters will continue infiltrating. This process will cause some dilution in the upper parts of the bedrock: however, its extent in the long term is not clear owing to model uncertainties. Saline groundwaters are present at depth throughout the Fennoscandian bedrock areas, but the depth where more saline conditions exist varies. For example, recently, deep geosphere groundwater chemistry has also been studied at the Outokumpu Deep Drill Hole Project (drill hole depth of 2500 m) (e.g. Kietäväinen *et al.* 2013). This study made at this inland site show no signs of glacial water intrusion. The total dissolved solids (TDS) of the groundwaters here are around $10\ g\ l^{-1}$ down to 1500 m, after which salinity increases. Natural analogue information on the processes near a terminating glacier of the Greenland Ice Sheet has been collected as a part of the Greenland Analogue Project (GAP), but the data are not yet public. This study will provide insight into the groundwater chemistry development near the terminus of a currently existing ice sheet.

Regional knowledge is important in order to understand the overall processes taking place in the crystalline basement areas. To assess the areal information on the groundwater compositions of these locations is beyond the scope of this work, but it has been noted that geological structural controls and local topography are the most important factors controlling the groundwater mixing processes. The only thing that can be concluded is that dilute water intrusion is a process that takes place in fractured bedrock, or any other permeable zone, when conditions favour it. Hence, estimating of the future behaviour of any site must first and foremost rely on an understanding of the site.

Natural analogues of smectite erosion

Bentonite deposits

Puura & Kirsimäe (2010) approached chemical erosion by discussing natural bentonite deposits, but made only cursory suggestions on potential study sites. Reijonen & Alexander (2015) noted that the erosional histories of known bentonite deposits that were impacted by the last glaciation or by meteoric waters could be of interest. Bentonite deposits could provide information, especially in relation to ECC development in contact with groundwater. Kuno *et al.* (2002) conducted a preliminary study of the colloidal dispersion of bentonite in the groundwater of a bentonite mine (Fig. 2), and a similar approach could be used at a more appropriate site and under more controlled conditions (Reijonen & Alexander 2015).

Surficial bentonite deposits are often vast, and erosional processes occurring are mostly controlled by meteoric water drainage, resulting in instability of the slopes. Thus, in most outcropping bentonites, the amount of historical erosion due to colloidal transport would be difficult to assess (Fig. 3).

Other smectite occurrences

As mentioned earlier, in addition to bentonites, smectites have been observed in various geological environments, including crystalline basement areas, such as in the Fennoscandian Shield. Smectites also occur in soils (e.g. Righi *et al.* 1997), as fracture- and fault-gouge-filling minerals (e.g. Pearson *et al.* 2003; Gehör 2007; Sandström *et al.* 2008; Koroleva *et al.* 2009), in general as an alteration product of,

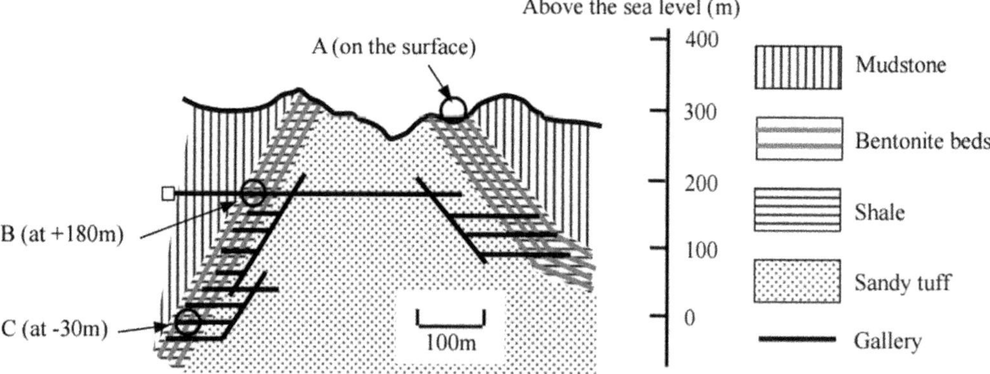

Fig. 2. Example of a situation where natural bentonite erosion could be studied. Schematic cross-section through the Tsukinuno bentonite mine, northern Japan. Locations of sampled groundwater (A–C) and their depths (above the sea level) are indicated (Kuno *et al.* 2002).

for examples, intrusive rocks (e.g. Laine & O'Brien 2008) and within bedrock weathering zones (e.g. Vartiainen 2005). In Table 1, a few examples of smectite occurrences in Fennoscandia are presented; smectite is widespread in many surficial and shallow bedrock environments, where formation of smectite is related to processes that have been active in timescales of hundreds of millions of years. It is worth noting that in crystalline basement areas (e.g. the Karelian craton, eastern Finland), intrusions of basic and ultrabasic rocks occur. In some cases, these intrusive rocks, with volumes usually comparable to a bentonite-filled repository tunnel, have altered to clays, including smectites, but to date these occurrences have not been studied in relation to deep geological repositories. However, several other examples will be discussed here: Hyrkkölä, Olkiluoto and other sites included in the Finnish site selection process (e.g. see McEwen & Äikäs 2000).

Hyrkkölä. A good example of a site where smectite has been observed as a fracture-filling mineral in water-conducting fractures is at Hyrkkölä, Finland (Fig. 4). Marcos (2002) reported smectite occurrences at the site. For example, at a depth of approximately 80 m, smectite is found at a location where current groundwater conditions are relatively dilute (*c.* 1.5 mM; TDS 116 mg l^{-1}: Marcos & Ahonen 1999) (Fig. 5; Table 2) and, hence, under the 4 mM limit of concern. Despite this, it seems that the smectite persists, so more detailed analyses of the smectite properties (e.g. ECC, smectite mineralogy and swelling properties) would be of great interest.

Olkiluoto and other repository (candidate) sites. The potential for using naturally occurring host-rock smectites to assess the longevity of industrial EBS bentonite has been previously raised by Apted *et al.* (2010) and Arthur & Savage (2012) when reviewing information on the Forsmark site for the Swedish regulator (SSM) (see also Savage 2012). Arthur & Savage (2012, p. 307) noted that:

> Site mineralogical data (SKB 2011) show that smectite and calcite occur at all depths in Forsmark fractures, with no evidence for removal/dissolution by previous glacial episodes. This natural analogue implies that these minerals may not have been dissolved/eroded during previous glacial episode.

Indeed, further investigation since has shown that the smectites occur throughout the host rock (Sandström *et al.* 2008). Dating of the fracture systems suggests that the smectites have been present since at least the Palaeozoic (i.e. 540–250 Ma BP), with the deep brines that are in contact with the smectites having been stable for at least several million years, if not longer (Smellie *et al.* 2008).

As discussed by Reijonen & Alexander (2015), montmorillonite has been reported from the fracture gouges sampled at Olkiluoto (e.g. Front *et al.* 1998), as well as in other Finnish site investigation reports (e.g. for Hästholmen (Gehör *et al.* 1997*b*; Front *et al.* 1999), Kivetty (Gehör *et al.* 1997*a*) and Romuvaara (Kärki *et al.* 1997)), suggesting smectite to be present at most depths.

The above-mentioned locations vary from inland (e.g. Romuvaara and Kivetty) to coastal (e.g. Olkiluoto and Forsmark). Deep groundwater systems in the Fennoscandian bedrock have been thoroughly studied in relation to site investigations, and these data provide both the salinity range for conditions where smectites occur and the spatial and temporal evolution of the groundwater salinities (e.g. see Posiva 2013*b*). As such, an assessment of likely

Fig. 3. Bentonite pillars protected by caprocks in North Dakota, USA (Bluemle 1991).

previous groundwater salinities at any given borehole depth can be conducted, meaning that by analysing the associated smectite minerals, in particular their paragenesis and mode of occurrence, their long-term stability could be assessed.

This would be of interest for understanding the situation at the Olkiluoto site, but, even more importantly, these sites with proven dilute deep groundwater conditions could be used as analogues for deep dilute water infiltration and interaction with smectites in general. In this regard, of great importance is that, at sites such as Kivetty and Romuvaara, smectites are present in conditions that show more dilute conditions than those noted in the national programme requirements (see earlier). Why these minerals remain here and in numerous surficial occurrences (e.g. weathering horizons) in seemingly too dilute conditions (see Table 2) would need further clarification. The mineralogical data currently available unfortunately does not allow a thorough assessment of this.

Other potential areas of research

In addition to the chemical erosion process, the formation of filter cakes was mentioned by Apted *et al.* (2010) as a topic that has not been backed up with any observations from nature. This process is

Table 1. *Examples of smectite occurrences in Fennoscandia*

Type	Locality	Geology	Smectite content	Age and genesis	Reference
Weathered profile	Mätäsaapa, Pelkosenniemi, Lapland, Finland	Weathering zone in crystalline basement (up to 20 m in depth), current overburden from 10 to 45 m	16–17 drill core samples contained 5–15% smectite. Smectite content in the fine fraction (<20 μm) is 30–40% higher than in bulk samples	Definite ages unknown, but ages between Precambrian and 360 Ma are likely.	Vartiainen (2005)
	Vitikko, Salla, Lapland, Finland	Weathered (originally albite carbonate rock) profile (2–6 m thick)	Large smectite content reported, in the fine fraction (<20 μm) over 50%	Related to kaolinization during warm and humid climate conditions	Vartiainen (2005)
	Vitsaoja, Salla, Lapland, Finland	Weathered zone of crystalline basement, current overburden 14–16 m	In some samples, up to 60% smectite		Vartiainen (2005)
	Lapioaapa, Rovaniemi mlk, Lapland, Finland	Weathered zone in the crystalline basement (quarzites, vulcanites and dolomite)	X-ray diffraction (XRD) results show smectite as a weathering product down to depths of 70 m. The current overburden is tens of metres		Vartiainen (1996)
Altered alkaline intrusion	Helsinki, Finland Sewage water tunnel site (tunnel collapse site due to smectite swelling)	Altered lamprophyre	Smectitization reported for a lamprophyre dyke	Age unknown, probably hydrothermal alteration of lamprophyre	Suominen (1997)
	Kaavi, Finland	Altered kimberlite	Thorough saponitization of kimberlite	Age of the alteration unknown	Laine & O'Brien (2008)
Fault	Stuoragurra, Norway	Post-glacial fault with clay gouges	Smectite as a fracture-filling mineral		Kukkonen *et al.* (2010)
Fractured crystalline rock	Hyrkkölä, Finland	Crystalline basement	Smectite as a fracture-filling mineral		Marcos (2002)
	Olkiluoto, Finland	Crystalline basement	Smectite as a fracture-filling mineral		Front *et al.* (1998)
	Hästholmen, Finland	Crystalline basement	Smectite as a fracture-filling mineral		Gehör *et al.* (1997*a*), Front *et al.* (1999)
	Kivetty, Finland	Crystalline basement	Smectite as a fracture-filling mineral		Gehör *et al.* (1997*b*)
	Romuvaara, Finland	Crystalline basement	Smectite as a fracture-filling mineral		Kärki *et al.* (1997)
	Forsmark, Sweden	Crystalline basement	Smectite as a fracture-filling mineral	Palaeozoic (likely)	Sandström *et al.* (2008)

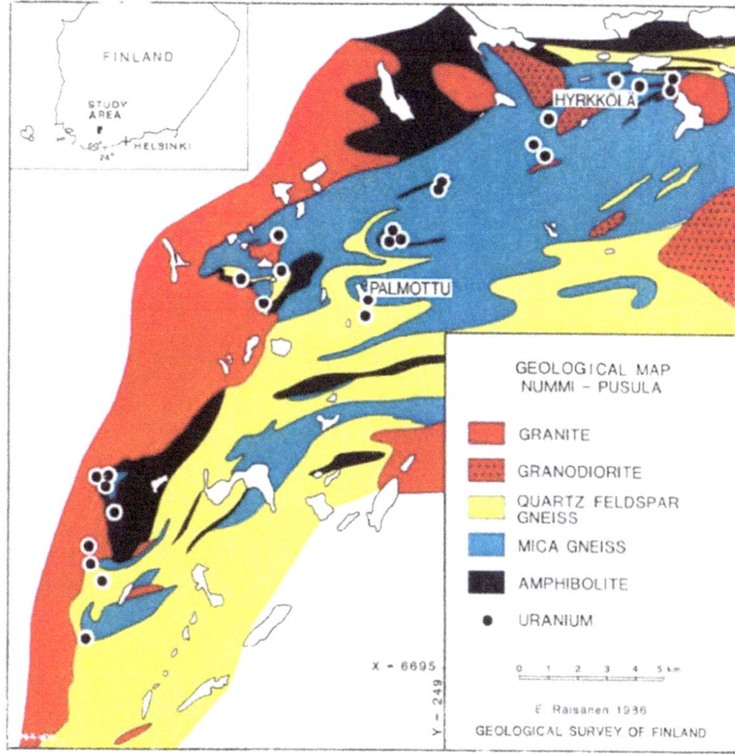

Fig. 4. Geological map of the Hyrkkölä area (Marcos & Ahonen 1999).

coupled with erosion, as when the smectite can be chemically eroded away in dilute enough conditions by the groundwater flow, other minerals initially bound in bentonite can remain in place/or build up in small aperture fractures This is because the grain size of these minerals (such as quartz) is larger and, owing to their other mineralogical properties, cannot be eroded in the repository conditions. This

Fig. 5. Smectite from the Hyrkkölä study area drill core (photographs by Nuria Marcos).

Table 2. *Groundwater compositions in connection with the occurrences of fracture smectite*

Locality	TDS ($mg\ l^{-1}$)	Total charge equivalent of cations (mM)	Cl ($mg\ l^{-1}$)	SO_4 ($mg\ l^{-1}$)	Na ($mg\ l^{-1}$)	K ($mg\ l^{-1}$)	Ca ($mg\ l^{-1}$)	Mg ($mg\ l^{-1}$)	HCO_3 ($mg\ l^{-1}$)	Fe ($mg\ l^{-1}$)	O ($mg\ l^{-1}$)	pH
Olkiluoto meteoric (PVP)	567	*c.* 7	60	48	24.6	6.7	92.3	15.6	291			
Grimsel water	83	*c.* 1	5.7	5.9	15.9	0.2	5.2	0.015	27.45	0.0002		9.6
Kivetty (minimum)	133	*c.* 0.8	1	0.1	4.7	0.2	11.2	0.9	82.4	0	0	7.2
Kivetty (maximum)	240	*c.* 3.6	5.8	3.5	27	2.4	33	8.6	158.6	1.4	0.4	8.5
Hyrkkölä	116	*c.* 1.5	3.8	10	7.1	2.8	14.4	4.3	57	0.05	4	6.6
Romuvaara (minimum)	62	*c.* 0.3	0.9	1	4.3	0.7	2.7	0.1	0	0	0	6.8
Romuvaara (maximum)	190	*c.* 3.6	7.4	6.5	43.2	3	23	5.3	123.3	3	1.1	9.3

Data for Hyrkkölä from Marcos & Ahonen (1999), Kivetty and Romuvaara from McEwen & Äikäs (2000). Olkiluoto reference compositions for Olkiluoto meteoric water and glacial meltwater (Grimsel) are given for comparison (Hellä *et al.* 2014).

phenomenon is mentioned, for example, by Schatz *et al.* (2013), who called for further experimental work to be undertaken as major uncertainties currently exist in the understanding of this process.

In addition to fracture smectites observed at a wide range of sites to date, post-glacial faults could provide useful targets for smectite stability research as smectite has been reported, for example, in the Stuoragurra Fault in northern Norway (Kukkonen *et al.* 2010 and references therein).

Discussion

The palaeohydrogeochemical evolution of the Olkiluoto site is well understood and supports the view that glacial meltwater has not intruded into the repository depths. However, current modelling of the groundwater processes during repository evolution cannot rule out the presence of dilute conditions in some waste canisters (Posiva 2013*b*). It is acknowledged that the models are far from perfect and, owing to uncertainties, very conservative assumptions have been employed.

Multiple lines of arguments are needed in order to assess the performance of a repository system. Knowledge that has been, and can still be, obtained by looking at the bentonite and smectite longevity in geological formations can help reduce the over-conservative treatment of repository processes and lead the way to more realistic assessments. Whilst many of the processes, such as ECC changes and the overall erosion pattern, can be seen in bentonite deposits themselves, fracture smectites and other smectite occurrences of relevance within the crystalline basement can provide information where bounding conditions are closer to the repository conditions (present and evolving). Regarding the Olkiluoto site, smectite occurrences above the assumed maximum depths of dilute water penetration support the likely longevity of smectite during future glaciations, while even more convincing support comes from other sites where the current groundwater salinity in smectite-hosting formations is below the requirements set for smectite stability.

Conclusion

Bentonite erosion in surficial natural deposits is mainly controlled by weathering and physical erosion, which makes observation of the process of chemical erosion within these difficult to assess.

However, smectite persists in widespread locations regardless of the presence of glacial meltwaters in the past, suggesting that the waters were not dilute and already buffered in the upper bedrock or that the limit set for the erosion is very conservative. Smectites are observed in localities where current groundwater composition is below the 4 mM limit given in the requirements. Many targets exist that could be investigated as potential analogues, although currently available information is not sufficient to assess their relevance with respect to repository conditions. Detailed studies would be required to assess the mode-of-occurrence of fracture smectites or other smectite bodies that occur in current and future repository-relevant hydrogeochemical and hydrogeological conditions.

The answer to the question, 'do we observe chemical erosion in nature?', would be 'no' – or 'unproven'. The natural analogues presented here do not support the widespread occurrence of chemical erosion and, thus, would support the stability of smectites in dilute conditions over geological timescales, something that cannot be investigated in the laboratory. However, further research is needed to provide a sound basis for this statement at the level of detail needed in the safety case.

The authors would like to thank Posiva, Finland for financial support.

References

APTED, M.J., ARTHUR, R., BENNETT, D., SAVAGE, D., SÄLLFORS, G. & WENNERSTRÖM, H. 2010. *Buffer Erosion: An Overview of Concepts and Potential Safety Consequences*. Research Report 2010:31. Swedish Radiation Safety Authority, Stockholm, Sweden.

ARTHUR, R. & SAVAGE, D. 2012. Equilibrium constraints on buffer erosion based on the chemistry and chemical evolution of glacial meltwaters. *In*: *Proceedings of the 5th International Meeting on Clays in Natural and Engineered Barriers for Radioactive Waste Confinement*, 22–25 October 2012, Montpellier, France, 306–307.

BLUEMLE, J.B. 1991. *The Face of North Dakota*. Revised edn. North Dakota Geological Survey, Educational Series, **21**.

ERIKSON, R. & SCHATZ, T. 2015. Rheological properties of clay material at the solid/solution interface formed under quasi free swelling conditions. *Applied Clay Science*, **108**, 12–18.

FRONT, K., PAULAMÄKI, S. & PAANANEN, M. 1998. *Updated Lithological Bedrock Model of the Olkiluoto Study Site, Eurajoki, Southwestern Finland*. Posiva Working Report 98-57. Posiva Oy, Helsinki, Finland [in Finnish with English abstract].

FRONT, K., PAULAMÄKI, S., AHOKAS, H. & ANTTILA, P. 1999. *Lithological and Structural Bedrock Model of the Hästholmen Study Site, Loviisa, SE Finland*. Posiva Report 99-31. Posiva Oy, Helsinki, Finland.

GEHÖR, S. 2007. *Mineralogical Characterization of Gouge Fillings in ONKALO Facility at Olkiluoto*. Posiva Working Report 2007-33. Posiva Oy, Olkiluoto, Finland.

GEHÖR, S., KÄRKI, A., SUOPERÄ, S. & TAIKINA-AHO, O. 1997*a*. *Kivetty, Äänekoski: Petrology and Low Temperature Minerals in the K1-K11 Drill Core Samples*.

Posiva Working Report 97-16. Posiva Oy, Helsinki, Finland.

GEHÖR, S., KÄRKI, A. & TAIKINA-AHO, O. 1997*b*. *Loviisa, Hästholmen, Petrology and Low Temperature Fracture Minerals in Drill Core Samples HH-KR1, HH-KR2 and HH-KR3*. Posiva Working Report 97-40. Posiva Oy, Helsinki, Finland.

HELLÄ, P., PITKÄNEN, P., LÖFMAN, J., PARTAMIES WERSIN, P. & VUORINEN, U. 2014. *Safety Case for the Disposal of Spent Nuclear Fuel at Olkiluoto – Definition of Reference and Bounding Groundwaters, Buffer and Backfill Porewaters*. Posiva 2014-4. Posiva Oy, Eurajoki, Finland.

KÄRKI, A., GEHÖR, S., SUOPERÄ, S. & TAIKINA-AHO, O. 1997. *Romuvaara, Kuhmo, Petrology and Low Temperature Fracture Minerals in the R0-R11 Drill Core Samples*. Posiva Working Report 97-19. Posiva Oy, Helsinki, Finland [in Finnish with English abstract].

KIETÄVÄINEN, R., AHONEN, L., KUKKONEN, I.T., HENDRIKSSON, N., NYYSSÖNEN, M. & ITÄVAARA, M. 2013. Characterisation and isotopic evolution of saline waters of the Outokumpu Deep Drill Hole, Finland – Implications for water origin and deep terrestrial biosphere. *Applied Geochemistry*, **32**, 37–51.

KOROLEVA, M., DE HALLER, A., MÄDER, U., WABER, H.N. & MAZUREK, M. 2009. *Technical Report TR-08-02: Borehole DGR-2: Pore-Water Investigations. Rock–Water Interaction*. Institute of Geological Sciences, University of Berne, Berne, Switzerland.

KUKKONEN, I., OLESEN, O., AKS, M.V.S. & PFDP WORKING GROUP 2010. Postglacial Faults in Fennoscandia: targets for scientific drilling. *GFF*, **132**, 71–81.

KUNO, Y., KAMEI, G & AND OHTANI, H. 2002. Natural colloids in groundwater from a bentonite mine – correlation between colloid generation and groundwater chemistry. *MRS Proceedings*, **713**, JJ8.4.1–JJ8.4.8.

LAINE, H. & KARTTUNEN, P. 2010. *Long-Term Stability of Bentonite – A Literature Review*. Posiva WR 2010-53. Posiva Oy, Eurajoki, Finland.

LAINE, H. & O'BRIEN, H. 2008. Alteration and primary kimberlite rock type classification for Lahtojoki kimberlite, Finland. Extended Abstract No. 9IKC-A-0034 presented at the 89th International Kimberlite Conference, 10–15 August 2008, Frankfurt, Germany.

MARCOS, N. 2002. *Lessons From Nature – The Behaviour of Technical Natural Barriers in the Geological Disposal of Spent Nuclear Fuel*. Acta Polytechinca Scandinavica, Civil Engineering and Building Construction Series. Finnish Academy of Technology, Helsinki, Finland.

MARCOS, N. & AHONEN, L. 1999. *New Data on the Hyrkkölä U-Cu Mineralization: the Behaviour of Native Copper in a Natural Environment*. Posiva 99-23. Posiva Oy, Helsinki, Finland.

MCEWEN, T. & ÄIKÄS, T. 2000. *The Site Selection Process for a Spent Fuel Repository in Finland – Summary Report*. Posiva 2000-15. Posiva Oy, Helsinki, Finland.

MILLER, W.M., ALEXANDER, W.R., CHAPMAN, N.A., MCKINLEY, I.G. & SMELLIE, J.A.T. (eds). 2000. *The Geological Disposal of Radioactive Wastes and Natural Analogues: lessons from Nature and Archaeology*. Waste Management Series, **2**. Elsevier Science, Oxford.

MORENO, L., NERETNIEKS, I. & LIU, L. 2015. Modelling of bentonite erosion applied to Schatz *et al.* experiments. *In*: MISSANA, T. (ed.) *Proceedings of the 3rd Annual Workshop of the EU BELBaR Project*. Ciemat, Madrid, Spain, 25–26.

NERETNIEKS, I., LIU, L. & MORENO, L. 2009. *Mechanisms and Models for Bentonite Erosion*. SKB TR-09-35. Swedish Nuclear Fuel and Waste Management (SKB), Stockholm, Sweden.

PEARSON, F.J., ARCOS, D. ET AL. 2003. *Mont Terri Project – Geochemistry of Water in the Opalinus Clay Formation at the Mont Terri Rock Laboratory*. Reports of the FOWG, Geology Series No. 5 – Berne 2003 Swisstopo, Berne, Switzerland.

PITKÄNEN, P., LUUKKONEN, A., RUOTSALAINEN, P., LEINO-FORSMAN, H. & VUORINEN, U. 1998. *Geochemical Modelling of Groundwater Evolution and Residence Time at the Kivetty Site*. Posiva 98-07. Posiva Oy, Helsinki, Finland.

PITKÄNEN, P., LUUKKONEN, A., RUOTSALAINEN, P., LEINO-FORSMAN, H. & VUORINEN, U. 2001. *Geochemical Modelling of Groundwater Evolution and Residence Time at the Hästholmen Site*. Posiva 2001-01. Posiva Oy, Eurajoki, Finland.

POSIVA 2009. *Olkiluoto Site Description 2008*. Posiva 2009-01. Posiva Oy, Eurajoki, Finland.

POSIVA 2012*a*. *Safety Case for the Disposal of Spent Nuclear Fuel at Olkiluoto – Design Basis 2012*. Posiva 2012-03. Posiva Oy, Eurajoki, Finland.

POSIVA 2012*b*. *Safety Case for the Disposal of Spent Nuclear Fuel at Olkiluoto – Description of the Disposal System 2012*. Posiva 2012-05. Posiva Oy, Eurajoki, Finland.

POSIVA 2012*c*. *Safety Case for the Disposal of Spent Nuclear Fuel at Olkiluoto – Complementary Considerations 2012*. Posiva 2012-11. Posiva Oy, Eurajoki, Finland.

POSIVA 2013*a*. *Olkiluoto Site Description 2011*. Posiva 2011-02. Posiva Oy, Eurajoki, Finland.

POSIVA 2013*b*. *Safety Case for the Disposal of Spent Nuclear Fuel at Olkiluoto – Performance Assessment 2012*. Posiva 2012-04. Posiva Oy, Eurajoki, Finland.

PUURA, E. & KIRSIMÄE, K. 2010. *Impact of the Changes in the Chemical Composition of Pore Water on Chemical and Physical Stability of Natural Clays – A Review of Natural Cases and Related Laboratory Experiments and the Ideas on Natural Analogues for Bentonite Erosion/Non-Erosion*. SKB TR-10-24. Swedish Nuclear Fuel and Waste Management (SKB), Stockholm, Sweden.

REIJONEN, H.M. & ALEXANDER, W.R. 2015. Bentonite analogue research related to geological disposal of radioactive waste: current status and future outlook. *Swiss Journal of Geosciences*, **108**, 101–110.

RIGHI, D., RÄISÄNEN, M.L. & GILLOT, F. 1997. Clay mineral transformations in podzolized tills in central Finland. *Clay Minerals*, **32**, 531–544.

SANDSTRÖM, B., TULLBORG, E.-L., SMELLIE, J.A.T., MACKENZIE, A.B. & SUKSI, J. 2008. *Fracture Mineralogy of the Forsmark Site: Final Report*. SKB R-08-102. Swedish Nuclear Fuel and Waste Management (SKB), Stockholm, Sweden.

SAVAGE, D. 2012. *Geochemical Constraints on Buffer Pore Water Evolution and Implications for Erosion*. SSM report 2012:61. Swedish Radiation Safety Authority (SSM).

SCHATZ, T., KANERVA, N., MARTIKAINEN, J., SANE, P. & KOSKINEN, K. 2013. *Buffer Erosion in Dilute*

Groundwater. Posiva WR-2012-44. Posiva Oy, Eurajoki, Finland.

SKB 2011. *Long-Term Safety for the Final Repository for Spent Nuclear Fuel at Forsmark Main Report of the SR-Site Project, Volume I*. SKB TR-11-01. Swedish Nuclear Fuel and Waste Management (SKB), Stockholm, Sweden.

SMELLIE, J.A.T., BLOMQVIST, R. *ET AL*. 2002. Palaeohydrogeological implications for long-term hydrochemical stability at Palmottu. *In*: *Natural Analogue Working Group Meeting*, March 1999, Strasbourg, France. European Commission, Luxembourg, 201–207.

SMELLIE, J.A.T., TULLBORG, E.-L., NILSSON, A.-C., SANDSTRÖM, B., WABER, H.N., GIMENO, M. & GASCOYNE, M. 2008. *Explorative Analysis of Major Components and Isotopes*. SDM-Site Forsmark. SKB R-08-84. Swedish Nuclear Fuel and Waste Management (SKB), Stockholm, Sweden.

SUOMINEN, V. 1997. Smectite alteration in a lamprophyre dyke; A case study. *GFF*, **119**, 149–150.

VARTIAINEN, R. 1996. *Rapakalliotutkimukset Rovaniemen MLKn Lapioaavalla 1995–1996*. M19/3611/-96/1/83. Geological Survey of Finland, Espoo, Finland [in Finnish with English abstract].

VARTIAINEN, R. 2005. *Teollisuusmineraali- ja rakennuskivitutkimukset Pohjois-Suomessa vuosina 2002–2004*. M 10.4/2005/4. Geological Survey of Finland, Espoo, Finland [in Finnish with English abstract].

Modelling of water and gas flow through an excavation damaged zone in the Callovo-Oxfordian argillites in the framework of a single porosity model

LUC-VINCENT BÉNET[1]*, ÉTIENNE BLAUD[1] & JACQUES WENDLING[2]

[1]*SOCOTEC SA/Projets Industriels/AME, 1 Avenue du Parc, 78640 Montigny-le-Bretonneux, France*

[2]*ANDRA, 1–7 rue Jean Monnet, Parc de la Croix-Blanche, 92298 Châtenay-Malabry Cedex, France*

**Correspondence: luc-vincent.benet@socotec.com*

Abstract: The excavation damaged zone (EDZ) of the Callovo-Oxfordian argillites can be regarded as a double-porosity media consisting of blocks of undisturbed argillites separated by cracks generated by shear or traction stresses. However, owing to hardware limitations, most of the hydraulic-gas simulations undertaken on large scales are based on an equivalent-porous-media model of the EDZ, where the fractures are not explicitly represented. The aim of this study is to improve the equivalent-porous-media model's consistency with the double-porosity flow behaviours shown by experimental data while keeping the same level of simplification. The new model has been developed through a performance assessment (PA)-like approach in two steps: a phenomenological model including explicit fractures has been designed in accordance with the actual geometrical and hydraulic understanding of the EDZ; then, the properties of the simplified model have been calibrated to comply with the hydraulic behaviour of the explicit-fracture model. This methodological approach has been used on an access gallery parallel to the main horizontal stress axis. The new equivalent-porous-media model differs from the previous one on the following three main points: the gallery EDZ has an anisotropic intrinsic permeability that is lower than previously; the same retention law is used as the undisturbed host rock; and there is a higher relative permeability for gas and water than previously.

Context

Since the end of the 1990s, ANDRA (the French National Radioactive Waste Management Agency) has been studying the feasibility of locating a repository in the Callovo-Oxfordian argillites on the borders of the Meuse and Haute-Marne departments at a depth of about 500 m. The approximately 150 m-thick Callovo-Oxfordian argillites layer is a natural barrier against radionuclide transfers from the repository to the biosphere, due mainly to the very low hydraulic conductivity of these clay-rich rocks. However, the rock will be disturbed by the excavation work. The excavation of a tunnel at the repository level results in de-confinement of the rock surrounding the excavation. The de-confinement modifies the natural stress field significantly and, consequently, a network of cracks forms over the so-called excavation damaged zone (EDZ).

The hydraulic properties are modified by the excavation work. Because the fractures are more permeable and less capillary active than undisturbed argillites, the EDZ offers a potential preferential pathway for gas flow and water flow. During the operational phase, the EDZ will have a key role in the de-saturation of the host rock around the excavation by facilitating vapour exchange with the ventilation airflow. After closure of the repository, the galleries and shafts are backfilled and sealed. Then, the EDZ could facilitate water and gas flow along galleries, and, consequently, increase the transport of soluble or gaseous radioactive species between sealing plugs. So the characterization of the hydraulic properties of the EDZ, and their implementation in a representative model, are important issues for safety assessments based on numerical simulations.

EDZ characterization

Work to characterize the EDZ has been performed in the underground rock laboratories (URLs) of Meuse/Haute-Marne over the last decade. Armand *et al.* (2014) described different methods that have been used in the Meuse/Haute-Marne URL to assess the extension of the excavation-induced fractured networks. They showed that microfractures

From: Norris, S., Bruno, J., Van Geet, M. & Verhoef, E. (eds) 2017. *Radioactive Waste Confinement: Clays in Natural and Engineered Barriers*. Geological Society, London, Special Publications, **443**, 319–332.
First published online June 24, 2016, https://doi.org/10.1144/SP443.6

extend beyond the fractured zone surrounding the excavation on the bases of hydraulic conductivity results. De La Vaissière *et al.* (2014) studied the effect of mechanical compression within the EDZ through the Compression of the Damaged Zone (CDZ) experiment, and observed self-sealing processes. Zhang (2011) also investigated the self-sealing of fractures in argillites by testing samples through a similar experimental approach, and found strong evidence for complete self-sealing within the time period of months to years.

The geometry of the EDZ is anisotropic and depends on the tunnel orientation. The EDZ cross-section has a horizontal oval shape around a tunnel parallel to the major stress axis (N155° E), and has an overall vertical shape around a tunnel parallel to the minor stress axis (N65° E). In the same direction and from a qualitative point of view, the data collected highlight the EDZ similarity in structure and in extension for excavations with diameters from 0.1 to 8 m. Most quantitative studies have been performed on tunnels 5–6 m in diameter orientated in the direction of the main horizontal stress. Cracks and fractures have been described in terms of extension, orientation, aperture, interconnection and spacing. Hydraulic conductivity measurements have been performed horizontally and vertically at different distances from the excavation through the wall and above the vault.

This characterization work results in an EDZ conceptualization in two distinct zones. The connected EDZ (EDZ-c) consists of a dense network of fractures and cracks of heterogeneous orientation generated by traction and shear stress. The disconnected EDZ (EDZ-d) that extends beyond the EDZ-c consists of a weakly connected network of homogeneously orientated microcracks.

Motivation

The EDZ can be regarded as a double-porosity media consisting of blocks of argillites crossed by more or less permeable fractures or cracks. However, most of the hydraulic-gas simulations that ANDRA has undertaken are based on an 'equivalent-porous-media' model, where fractures are not explicitly represented and physical parameters are uniform in the two subzones EDZ-c and EDZ-d. Indeed, this model is simple enough to deal with computing performance and modelling work issues, but the current version has to be improved since the knowledge on EDZ structure and hydraulic behaviour has dramatically increased over recent years. Since 2005, long-term pumping tests have been performed in the Meuse/Haute-Marne URL and have enhanced the accuracy of the undisturbed claystone permeability assessments (Vinsot *et al.* 2011). The hydraulic conductivity in the undisturbed claystone has been updated in ANDRA's 2009 model, but hydraulic properties of the EDZ had not changed since 2005 (ANDRA 2005). The aim of this study is to improve the consistency of the equivalent-porous-media model of the EDZ with the double-porosity flow behaviour.

Modelling approach

The different modelling methods available to build a model of a fractured media fall into two main categories: discrete fracture network (DFN) models; and continuum models (Wu *et al.* 2004). In an upscaling approach, Hawkins *et al.* (2011) developed a DFN of the EDZ and used it to define the EDZ permeabilities in the framework of a multiple interacting continua (MINC) model. The bulk properties from data reported by ANDRA (e.g. ANDRA 2005) were used for other physical parameters. In the DFN model, the EDZ-c is represented by small random fractures, and the EDZ-d is represented by inclined planes regularly spaced. Ababou *et al.* (2011) used the rock matrix, the 3D geometrical structure of EDZ-c microfissures and small fractures, and the EDZ-d subvertical fractures to develop a phenomenological model. They then perform a hydraulic homogenization to define an equivalent macropermeability tensor K_{ij} (xyz) distributed over the EDZ on a continuum model. The bulk properties of the claystone are based on the ANDRA (2005) hydraulic model. The resulting global tensor anisotropy ratio found in the EDZ is $3/8$.

Our equivalent-porous-media modelling work is based on a performance assessment (PA)-like approach. The new properties of the equivalent-porous-media model derive from the hydraulic behaviour of a phenomenological model that is designed to be consistent with geometrical and hydraulic data collected by *in situ* measurements. The phenomenological model used in this study is an explicit fracture model that includes full flow interactions between the rock matrix and the fractures by taking into account the geometrical continuity between the two media. The geometrical model of the fracture network is based on a deterministic approach (regular fracture pacing, orientation end aperture along the excavation). The equivalent-porous-media model is achieved in two steps: first, the flow behaviour is simulated by the phenomenological model; and, secondly, new retention functions, new intrinsic permeability and relative permeability functions for gas and water are defined in the frame of an equivalent-porous-media model to agree with the phenomenological model results regarding steady-state flows crossing the EDZ-c or the EDZ-d, longitudinally or transversally, over a large range of relative humidity.

Scope

This modelling work focuses on the EDZ hydraulic behaviour around tunnels parallel to the main horizontal stress axis, which includes storage cells for high- and intermediate-level radioactive waste (HLW and ILW) and some access galleries. These excavations generate an anisotropic extension of the EDZ, with the main extension being horizontal. The hydraulic properties of the EDZ have been evaluated for two excavation diameters: $D_{exc} = 7.5$ m, which is representative of an access gallery; and $D_{exc} = 0.7$ m, which is representative of a HLW storage cell. The modelling of transport properties in the framework of an equivalent-porous-media model is not addressed here. The evolution in time of the EDZ self-sealing is not taken into account in this modelling work.

Physical modelling framework

Basic definitions and assumptions

The physical model of the water and gas flow in porous media is based on generalized Darcy equations (Darcy equations for unsaturated flow). The Darcy velocity for gas and water depends on the gradient of the dynamic gas pressure, H_g, and the dynamic water pressure, H_w (or Darcy pressure), respectively:

$$U_w = -\frac{k_i\, k_r^w}{\mu_w}\, \nabla(H_w)$$

$$U_g = -\frac{k_i\, k_r^g}{\mu_g}\, \nabla(H_g)$$

where k_i (in m^2) is the intrinsic permeability, k_r is the relative permeability and U (in m s^{-1}) is the Darcy velocity. The Darcy velocity can also be written in terms of hydraulic conductivity, K_w (in m s^{-1}), and the Darcy pressure head in terms of the water level (m_{WL}) in saturated media. Thus:

$$K_w = \frac{k_i\, \rho_w\, g}{\mu_w}.$$

The water pressure, P_w, is composed of two terms: the Darcy or dynamic pressure, H_w; and the hydrostatic pressure, P_{stat}, due to the weight of the water:

$$P_w = H_w + P_{stat}.$$

Before excavation (i.e. *in situ* conditions), the water pressure, P_w, considered in the host rock at the repository level ($z = 0$ m in the model) is 5 MPa, and the initial Darcy pressure is defined as nil in the entire model (i.e. no water flux in the calculation domain). After excavation, the problem is solved by using the auxiliary variable h_w^z:

$$h_w^z = H_w^z + P_w^0.$$

In the gas, the dynamic pressure and the total pressure are considered equal because the gas weight is neglected:

$$H_g \approx P_g.$$

The capillary pressure, P_c, is defined as the difference between the water pressure, P_w, and the gas pressure, P_g:

$$P_c = P_w - P_g$$

and therefore is negative in unsaturated media and equal to zero in saturated media.

The gas and water mass balance depends on the water-content variation with time, the Darcy flow, the gas dissolution and transport in water, and the mass consumption or generation processes such as phase change or oxidation. The transport of different gaseous species is not taken into account here:

$$\omega \frac{\partial \rho_w S_w}{\partial t} + \nabla\, \rho_w U_w = Q_w$$

$$\omega \frac{\partial \rho_g S_g}{\partial t} + \nabla\, \rho_g U_g = Q_g - E_{g/w}$$

$$\omega \frac{\partial C_g S_w}{\partial t} + \nabla\, C_g U_w = \nabla\, d_w^e \nabla(C_g) + E_{g/w}$$

where S_w is the water saturation, $S_g = 1 - S_w$, C_g is the gas concentration in water, ω is the porosity, Q_w and Q_g are the water and gas sources, respectively, $E_{g/w}$ is the gas dissolution term, and d_w^e is the effective gas diffusivity in water.

Water flow in unsaturated media

The water-retention curve depends on the capillary pressure. The water-retention curve used was proposed by van Genuchten (1980):

$$S_w = \left[1 + \left(\frac{P_c}{P_r}\right)^n\right]^{-m}$$

where P_r is the reference pressure for a particular porous medium, n is the shape parameter and $m = 1 - 1/n$. Both parameters P_r and n have specific values that have to be defined for each porous material.

The relative permeability of the water, k_r^w, depends on variations in water saturation and, therefore, on capillary pressure variations. The formulation used has been established by combining the van Genuchten retention curve (van Genuchten 1980) with the model generalized by Mualem (1976):

$$k_r^w = \sqrt{S_w}[1 - (1 - S_w^{1/m})^m]^2$$

The gas phase is supposed to be saturated with water vapour, and its partial pressure depends on capillary pressure, $P_c(S_w)$, according to the Kelvin equation:

$$P_{vap}^g(T, S_w) = \exp\left(\frac{M_w P_c(S_w)}{\rho_w RT}\right) P_{vap}^{sat}(T)$$

where M_w is the molar mass of water, R is the ideal gas constant, T is the temperature (in Kelvins in the formula), and P_{vap}^{sat} is the vapour pressure at the state of saturation and without a capillary effect.

Gas flow in unsaturated media

As for the liquid phase, the relative permeability of the gas, k_r^g, is established by combining the van Genuchten retention curve (van Genuchten 1980) with the model generalized by Mualem (1976):

$$k_r^g = \sqrt{1 - S_w}[1 - S_w^{1/m}]^{2m}.$$

The gas should be composed of air and vapour during the ventilated operational phase. After the repository is closed, the ventilation stops. Then hydrogen formed by anaerobic corrosion will become the main gaseous species after the oxygen has been consumed. The hydrogen pressure will build up beyond the *in situ* water pressure (i.e. about 5 MPa) after thousands of years. The vapour pressure will remain relatively low at the *in situ* temperature (about 22°C) compared to the atmospheric pressure during the operational phase, or compared to the hydrogen pressure level reached after closure.

Hydraulic properties of the 2009 equivalent-porous-media model of the EDZ

The reference values of the 2009 set of hydraulic parameters are shown in Table 1 for undisturbed and damaged host rock. They have been defined regardless of the excavation diameter.

Explicit-fracture model

Introduction

The 'explicit-fracture' model is designed to perform numerical simulations of gas and water flows through the fracture network and inside the blocks of undisturbed rock between fractures in the EDZ for a 7.5 mm-diameter excavation (the reference gallery). The aim of these numerical simulations is to determine the global hydraulic properties of the EDZ-d and EDZ-c in radial and longitudinal directions with respect to a gallery orientated along the main *in situ* stress (N° 155E). The model consists of two different porous media with highly contrasting dimensions (fracture aperture v. the size of argillite block between fractures) and physical properties (permeability and water retention). This work has been performed in three steps: geometrical characterization of the EDZ based on geomechanical analyses; geometrical modelling (meshing work); and hydraulic modelling based on *in situ* measurements of permeability.

Characterization of fractured media

The geomechanical analysis of the EDZ results in a conceptual model on which the geometrical design of the explicit-fracture model is based. Data were mainly acquired from core samples extracted from the walls of galleries in the Bure URL (galleries GET and GCS: ANDRA 2013; Armand *et al.* 2014). The two galleries have the same orientation (N155° E) but a smaller diameter (about 5.3 m) than the reference gallery. According to the conceptual model, the EDZ is composed of two kinds of fractures: the subparallel fractures (Fbpa) generated by traction stress near the excavation, which are

Table 1. *Reference physical properties of argillite materials in the 2009 model of the EDZ*

Parameters	Symbols	Undisturbed rock	EDZ-c	EDZ-d
Porosity	ω (%)	18	20	18
Horizontal intrinsic permeability	k_i^{xy}(m^2)	2.3×10^{-20}	10^{-16}	10^{-18}
Vertical intrinsic permeability	k_i^z (m^2)	9.4×10^{-21}	10^{-16}	10^{-18}
Shape parameter (retention law)	n	1.49	1.5	1.5
Reference pressure head (retention law)	P_r (m$_{WL}$)	1500	200	800
Initial water saturation	S_w (%)	100	100	100

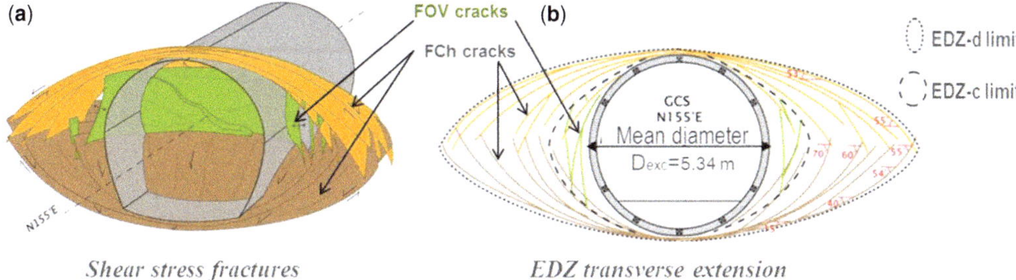

Fig. 1. (**a**) Three-dimensional view of the fracture generation at the excavation front and (**b**) cross-sectional view of the EDZ conceptualization into two subzones.

globally parallel to the wall; and the 'chevron' fractures (FCh) generated by shear stress, which run obliquely with respect to the longitudinal direction, either from the wall or from the floor and the roof (see the fractures in green, brown and orange, respectively, in Fig. 1a).

The rupture zone (i.e. the EDZ-c) is a network of connected fractures of heterogeneous orientation (see Fig. 2): the fracture planes in zone a are mainly composed of subparallel fractures (Fbpa) and subvertical fractures (FOV) from the wall. The rupture zone extends over an average thickness of 0.2 times the excavation diameter in the wall, and 0.1 times the diameter above and below the roof and the floor. The subparallel fracture average spacing in the radial direction is 0.1 m and the average aperture is about 1 mm. The average subvertical fracture spacing in the radial direction is 0.17 m and the average aperture of the fractures is less than the impregnation limit of 10 μm. Fracture apertures diminished with time due to self-sealing process.

The microcrack zone (i.e. the EDZ-d) is composed of FCh cracks of similar orientation that run overall horizontally from the roof and the floor deeply into the rock. This zone extends in an 'eye shape' from the excavation wall over an average distance from excavation wall of 0.8 times the excavation diameter. Thus, the fractures from the roof and the floor could possibly intersect each other at about half the gallery height. The vertical extensions of this zone above the roof and below the floor are not characterized. The average spacing in the radial direction at half the gallery height is about 0.45 m (see Fig. 2: the fracture planes in zone b). The aperture of the fractures is below the impregnation threshold of 10 μm. The transversal extensions of the connected and disconnected EDZ subzones are shown in Figure 1b.

Geometrical model

The 'explicit-fracture' model of the EDZ has been designed to be consistent with the geometrical

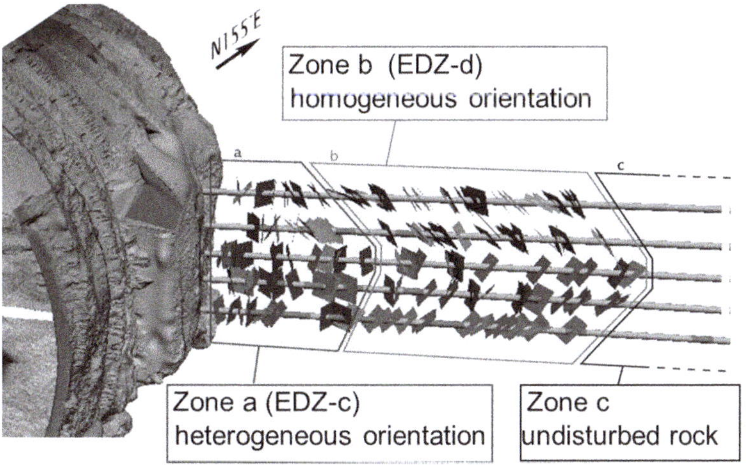

Fig. 2. Fracture planes in the lateral damaged zone.

data collected *in situ*: it has the same horizontal and vertical extensions relative to the excavation diameter as defined in the geomechanical conceptual model, and the same range of fracture spacing on the horizontal plane of symmetry. However, the geometrical model has been simplified given the complexity or the real EDZ. The gallery excavation is considered as circular. The EDZ-c network of connected and heterogeneously orientated fractures is simplified in a network of cylindrical (see the mesh in white in Fig. 3b) and transverse fractures (see the mesh in white in Fig. 3c) crossing each other; the fracture apertures are defined to be uniform in each zone. Moreover, transverse fractures have been artificially added to allow a degree of connectivity between the discrete fractures of the EDZ-d (see the mesh in grey in Fig. 3c). The connectivity level may vary, depending on the permeability of the transverse fractures, and can be neutralized by using the argillite permeability value in these fractures (see the hydraulic properties in Table 1).

For each fracture, one can define three kinds of fracture aperture: the geometrical aperture, which is characterized by impregnation tests, determines the water and gas content of fractures; the hydraulic aperture, which is lower than the geometrical one, determines the fracture permeability; and the mesh aperture, which must be large enough to minimize numerical problems.

Furthermore, in the EDZ-d, the subvertical fractures running from the roof and the floor cross each other at a plane located at half the height of the tunnel, instead of at two-thirds the height, as observed *in situ* (see Fig. 1a). This geometrical simplification allows us to introduce a horizontal symmetry plane. In the mesh, the EDZ-d fractures (see the mesh in grey in Fig. 3b) intersect this horizontal plane with an angle varying between 52.5° and 69° from the outer limit to the inner limit of the subzone, respectively, which is consistent with the EDZ conceptualization shown in Figure 1b.

The mesh represents a quarter of the EDZ over an 18.8 m-long stretch (see Fig. 3, which shows the explicit-fracture mesh of the EDZ-c fractures in white and the EDZ-d fractures in grey). It is composed of 66 570 cells, with 25 200 cells in the blocks of undisturbed rock, 26 470 cells in the EDZ-c fractures and 14 900 cells in the EDZ-d fractures. All of the mesh, the physical model and the solver algorithm were set up using the GIBIANE language. All the flow calculations were performed using Cast3M software (Cast3M 2010: http://www-cast3m.cea.fr/).

Hydraulic model

The reference values of the hydraulic properties are defined by assuming a uniform hydraulic aperture, e_h, of the fractures in each EDZ subzone and blocks of undisturbed rocks between fractures; the hydraulic aperture is defined as the effective aperture relating to head loss measurements. The fracture porous volume is deduced from the geometrical aperture, e_g, which, as an approximation, is considered to be twice the hydraulic aperture (Gentier *et al.* 2000):

$$\frac{e_g}{e_h} \cong 2.$$

The intrinsic permeability of a fracture is deduced from the hydraulic aperture by considering longitudinal flow through a fracture (Idel'cik 1979; Nicolas 2003):

$$k_i = \frac{e_h^2}{12}.$$

The water relative permeability in the fracture is defined as a linear function of the water saturation, by analogy to the linear variation of head loss with water-level variation along a vertical channel:

$$k_r = S_w.$$

This formula is also used to model water flowing transversally to the fractures in order to prevent an artificial barrier effect to water flow that is found using the reference law (see the earlier subsection

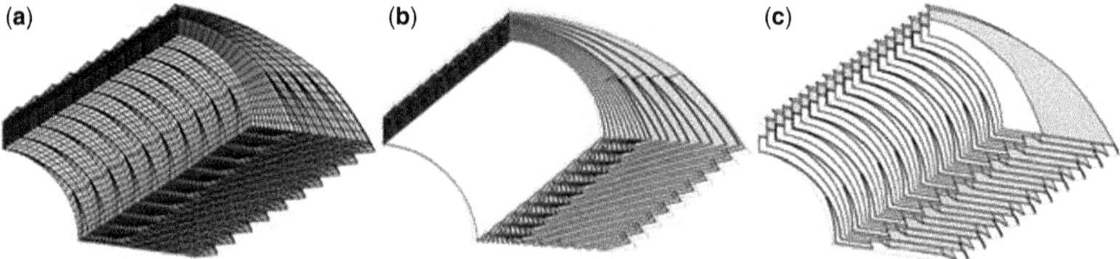

Fig. 3. (**a**) Explicit-fracture mesh with undisturbed argillite blocks, (**b**) longitudinal and oblique fractures and (**c**) transverse fractures; EDZ-c fractures are in white and the EDZ-d fractures are in grey.

on 'Gas flow in unsaturated media') in an unsaturated fracture. Indeed, the explicit-fracture mesh completely isolates the argillite blocks from each other in a geometrical point of view: however, this is not the case in a real DFN due to local fracture discontinuities or areas of zero aperture.

The reference pressure head, P_r, in a fracture is defined by using a simplified formulation of the Young–Laplace equation that maximizes the capillary water level in a thin vertical channel of uniform aperture, e_h:

$$P_r = \frac{1}{R_t}\frac{\gamma_w}{\rho_w g} = \frac{2\gamma_w}{e_h \rho_w g}$$

where $\gamma_w = 73 \times 10^{-3}\ \mathrm{N\,m^{-1}}$ is the surface tension of the water and R_t is the radius of the curvature of the water surface, which is maximized by the hydraulic half-aperture in the third part of the equation.

The hydraulic conductivity measurements are considered as representative of the longitudinal mean isotropic conductivity of an equivalent homogeneous medium at the scale of the measurements (about a few decimetres). The boreholes have been orientated perpendicular to the excavation surface for the purpose of intersecting the fractures (ANDRA 2013). Multipaker systems were used to perform gas permeability tests for hydraulics measurements. The permeability tests that are usually performed in the EDZ are uniform pressure tests or uniform flow rate tests. Most of the hydraulic measurements were performed before the crack self-sealing is complete.

The mean hydraulic conductivity per zone is deduced from the hydraulic conductivity measurements in three steps: first, a vertical profile above the vault and a horizontal profile in the wall have been interpolated from the measurements of the hydraulic conductivity (see Fig. 4, which shows the profiles (lines) and the measurements (dots)); secondly, the hydraulic conductivity variations in space $K_{\mathrm{EDZ}}^{d_r,\alpha}$ have been defined over an 'eye-shape' EDZ cross-section (see Fig. 5, which shows the hydraulic conductivity values) as a linear combination of vertical and horizontal profiles of conductivity; and, thirdly, the hydraulic conductivity has been averaged over the EDZ-d and EDZ-c, separately. The mean hydraulic conductivity derived from $K_{\mathrm{EDZ}}^{d_r,\alpha}$ (see the EDZ hydraulic conductivity in Table 2) is about half as much in the EDZ-c and 5 times lower in the EDZ-d compared with the 2009 reference values (deduced from the EDZ intrinsic permeability in Table 1):

$$K_{\mathrm{EDZ}}^{d_r,\alpha} = \frac{90^\circ - \alpha}{90^\circ} K_{\mathrm{wall}}^{d_x,\alpha=0^\circ} + \frac{\alpha}{90^\circ} K_{\mathrm{vault}}^{d_z,\alpha=90^\circ}$$

where $K_{\mathrm{wall}}^{d_x,\alpha=0^\circ}$ and $K_{\mathrm{vault}}^{d_z,\alpha=90^\circ}$ are, respectively, the horizontal and vertical profiles deduced from the hydraulic conductivity measurements, α is the angle to the horizontal, and d_x, d_z and d_r are, respectively, the distances to the excavation horizontally from the wall, vertically from the vault and radially.

It has to be noticed on Figure 4 that measurements of the hydraulic conductivity at about 0.1 diameters from the excavation are erratic, and there are few measurements below this distance, which induces an uncertainty on hydraulic conductivities in this zone.

The fracture aperture is calibrated on each subzone, so that the mean longitudinal hydraulic conductivity calculated with the explicit-fracture model is the same as the average of the hydraulic conductivity variations in space $K_{\mathrm{EDZ}}^{d_r,\alpha}$ derived from the measurements. This calibration work has been carried out through a series of numerical simulations of water flows crossing the fractures and the argillite blocks.

It has to be noticed that the fracture volume is less than 100th of the overall porous volume (see Table 2, which lists the fracture volume ratio in the EDZ-c and EDZ-d). Furthermore, the very low reference pressure in the fractures compared with the argillite (see Table 1) shows the high contrast in capillary behaviour between the two media.

The equivalent-porous-media model of the gallery

Calibration process

The geometrical and physical parameters of the equivalent-porous-media model (see Table 1) are modified in accordance with the geometry and hydraulic behaviours of the explicit-fracture model (see Table 2) so that, in each EDZ subzone, the transverse cross-sectional area, the gas and water volumes, and the transmissivities (i.e. the flow rate ($\mathrm{m^3\,s^{-1}}$) v. the dynamic pressure load, ΔH) have the same values as those calculated with the explicit-fracture model.

The explicit-fracture model provides a flow rate throughout a subzone by solving Darcy equations for steady flow, given a pressure load applied on the boundaries, either between the two cross-sectional limits for longitudinal flow or between inner and outer limits for transverse flow. Several calculation cases have been solved over a large range of capillary pressures for the following cases: EDZ connected or disconnected subzones; transverse or longitudinal flow; gas or water. In unsaturated conditions, a series of simulations of gas or water flow has been performed to cover a large range of capillary pressures in order to define the variation in the

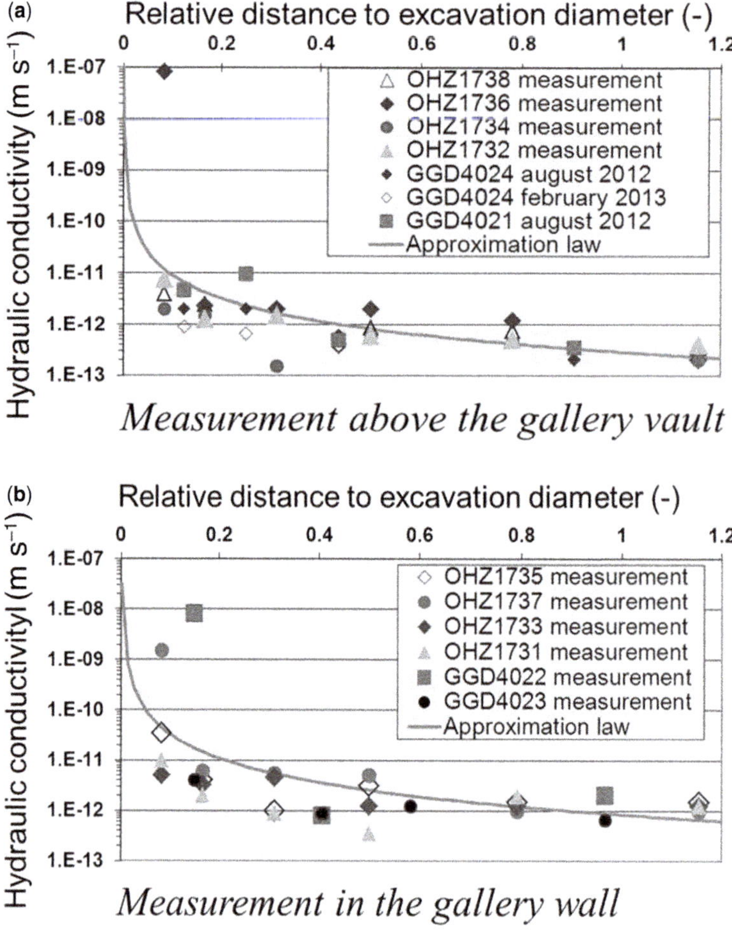

Fig. 4. Hydraulic conductivity measured (dots) through the EDZ (**a**) above the vault and (**b**) in the wall, and approximated as a function of the distance to the excavation (lines).

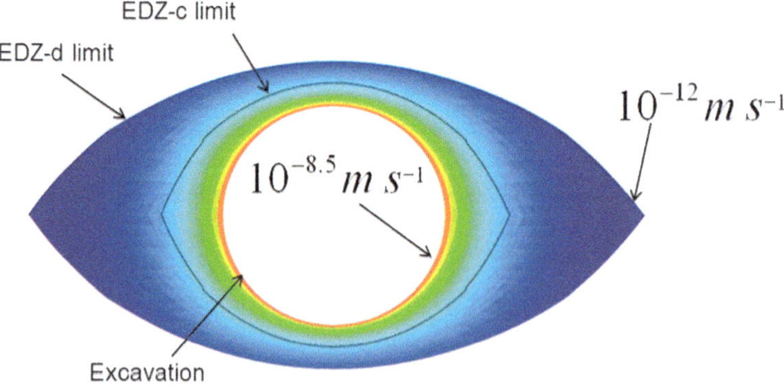

Fig. 5. Hydraulic conductivity approximated over an EDZ-c cross-section inside the line and over the EDZ-d (outside the line).

Table 2. *Physical and geometrical properties of the explicit-fracture model*

Fracture parameters	Symbols	EDZ-c	EDZ_d
Hydraulic aperture	e_h (μm)	2.45	1.1
Fracture volume ratio	$V\omega_F/V\omega_T$ (%)	0.67	0.18
Mean longitudinal hydraulic conductivity	K_L (m s^{-1})	2.1×10^{-10}	4.7×10^{-12}
Horizontal transverse spacing	(m)	0.1–0.15	0.45
Horizontal longitudinal spacing	(m)	0.33	1
Reference pressure head (retention law)	P_r (m_{WL})	3.44	13.27

water content and in the transmissivity as a function of capillary pressure or water saturation. Thus, the new parameter values of the equivalent-porous-media model depend on the geometrical properties and the hydraulic behaviour of the explicit-fracture model.

The thickness, E, and the outer radius, r_2, of a subzone are deduced from the cross-sectional area, A_t, of the explicit-fracture model, given the inner radius, r_1, of the subzone (i.e. excavation radius for the EDZ-c):

$$E = r_2 - r_1 = \sqrt{\frac{A_t + \pi r_1^2}{\pi}} - r_1.$$

The mean porosity, $\bar{\omega}$, is deduced from the local porosity, ω_i, in the fractures and the undisturbed rock:

$$\bar{\omega} = \frac{\sum_{i\in[1,N]} \omega_i \, \delta v_i}{\sum_{i\in[1,N]} \delta v_i}$$

where δv_i is the volume of the ith cell and N is the number of cells of the explicit-fracture mesh.

The mean water content, $\bar{\theta}_w$, is deduced from the local water saturation, $S_{w,i}$, in the fractures and the undisturbed rock:

$$\bar{\theta}_w = \frac{\sum_{i\in[1,N]} \omega_i \, S_{w,i} \delta v_i}{\sum_{i\in[1,N]} \delta v_i}.$$

The longitudinal effective hydraulic permeability is deduced from the longitudinal transmissivity, T_L (m^2 s^{-1}):

$$K_L = T_L \frac{L_G}{A_t}$$

where L_G is the length of the gallery stretch of the explicit-fracture model.

The transverse effective hydraulic conductivity is deduced from the transverse transmissivity, T_T, (m^2 s^{-1}), given the circular shape of subzone limit in the equivalent-porous-media model:

$$K_T = T_T \frac{\ln(r_2) - \ln(r_1)}{2\pi L_G}.$$

Physical properties

The updated values of the physical property parameters of the EDZ are given in Table 3. The new hydraulic model differs from the previous one in three main ways: it is less permeable, more capillary active and anisotropic in each subzone. The porosity and the retention law of the two EDZ porous media are the same as the undisturbed argillite because the fracture volume is negligible compared to the overall volumes of the porous media (<1% of the undisturbed host rock). The new model is more capillary active than that of the 2009 model (compare the reference pressure head, P_r, for the retention laws in the EDZ between the 2009 model (Table 1) and the new model (Table 3)), despite the very low reference pressures defined for the fractures in the phenomenological model (see P_r in Table 2). In addition,

Table 3. *Physical properties of the new equivalent-porous-media model of the reference gallery EDZ*

Parameters	Symbols	New EDZ-c	New EDZ_d
Porosity	ω (%)	18	18
Longitudinal intrinsic permeability	k_i^l (m^2)	2.1×10^{-17}	4.7×10^{-19}
Transverse intrinsic permeability	k_i^t (m^2)	5.2×10^{-18}	1.1×10^{-19}
Shape parameter (retention law)	n	1.5	1.5
Reference pressure head (retention law)	P_r (m_{WL})	1500	1500
Subzone thickness	E (m)	0.14 D_{exc}	0.27 D_{exc}

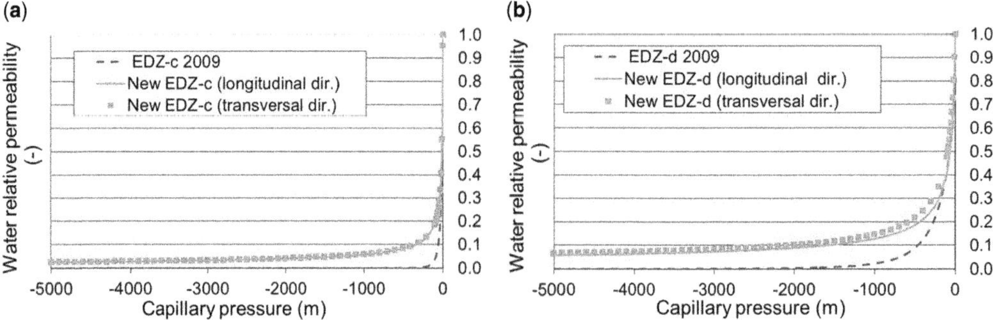

Fig. 6. Relative permeability v. capillary pressure for water flow in (**a**) the EDZ connected and (**b**) the EDZ disconnected reference gallery – comparison between the new model and the 2009 model.

the equivalent longitudinal intrinsic permeability is about 5 times lower in the EDZ-c and about half in the EDZ-d compared to the 2009 model. Moreover, the fracture network induced an anisotropic permeability, whereas the 2009 EDZ model is isotropic. The anisotropy is characterized by a lower permeability in the transverse direction than in the longitudinal direction in each subzone.

The new relative permeability laws for water and gas are compared with the 2009 ones in Figures 6 and 7, respectively. In the two subzones, and for both gas and water, the new relative permeability functions lie above those of the previous ones because of specific flow behaviour in the fracture network that is simulated by the phenomenological model. The gas relative permeability is higher because the fractures de-saturate much more than with the 2009 EDZ model assumptions. The water relative permeability is higher as a result of the linear law in water saturation used in the fractures. In the EDZ-d, a slight anisotropy has to be noted in the relative permeability laws for water and gas (see the differences between the dots and lines in Figs 6b & 7b).

Flow behaviour

The new model has been compared with the 2009 model through a two-dimensional (2D) simulation of de-saturation and re-saturation of the EDZ. The mesh and the zones of different materials are shown in Figure 8. The initial water saturation is 80% in the concrete liner and 100% in the host rock. The gallery is ventilated over 100 years and ventilation then stops when the gallery is closed. After closure of the gallery, no backfill material is considered. The gas pressure is supposed to be constant in time and equal to the atmospheric pressure. Gas transfers have not been modelled.

The evolution over time of water saturation is shown in Figure 9. During the ventilation period, water saturation decreases as a result of vapour exchange with the air. The dry ventilation air (about 47% RH) is modelled by imposing a capillary pressure of − 100 MPa on the inner boundary of the concrete liner. After the gallery is closed, the water saturation increases until the EDZ and the concrete liner are fully saturated. The new model predicts a lower de-saturation in the EDZ and a higher

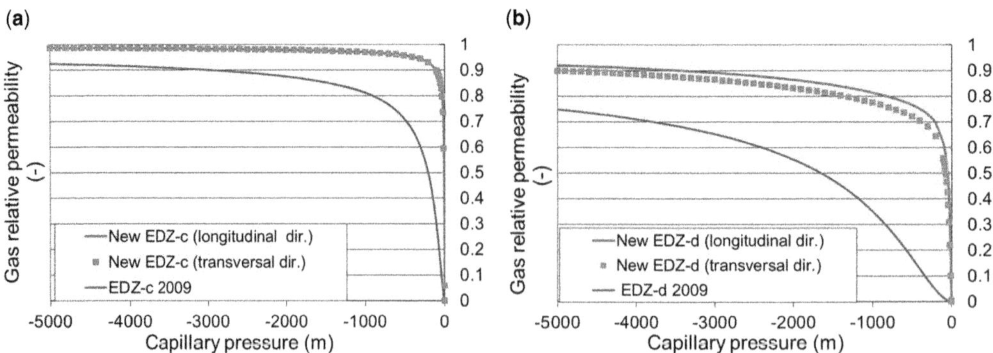

Fig. 7. Relative permeability v. capillary pressure for gas flow in the (**a**) EDZ connected and (**b**) the EDZ disconnected reference gallery – comparison between the new model and the 2009 model.

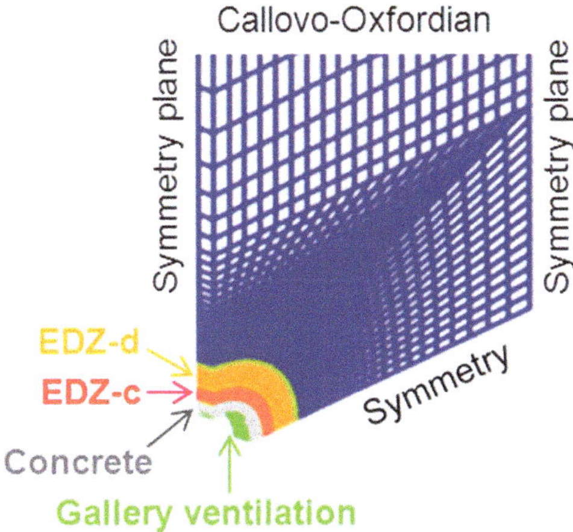

Fig. 8. View of the one-quarter gallery model; mesh, materials and boundary conditions.

de-saturation in the concrete liner at the end of the ventilation period. In the EDZ, the higher capillary forces limit the de-saturation and the porewater flow from the host rock toward the concrete liner. Consequently, ventilation dries the concrete liner much more during the ventilation period, and the re-saturation of the system is reached 100 years earlier by using the new model instead of the 2009 model. The two hydraulic models lead to similar phenomenological behaviours in this test case because the transverse water flow mainly depends on the least permeable material, which is the undisturbed host rock.

Sensitivity to the fracture-connection level in the EDZ-d

In the EDZ-d, the fractures have a great horizontal extension but a similar orientation. So, they have been considered as being disconnected from each other (see the fracture planes in zone c in Fig. 2). Nevertheless, a certain degree of connectivity could not be excluded. Therefore, some transverse fractures have been added at every 1 m in the mesh to take into account a possible interconnection between the EDZ-d oblique fractures. A sensitivity study of the hydraulic apertures of these fractures has been carried out to assess the importance of such an interconnection on longitudinal and transverse flow in saturated media. The reference hydraulic aperture (0.55 μm) of the transverse fractures is half that of the oblique fractures (1.1 μm). Figure 10 shows the longitudinal hydraulic conductivity (Fig. 10a) and the anisotropy factor of the transverse hydraulic conductivity (Fig. 10b) of transverse fracture hydraulic apertures between 0 and 1.1 μm. The comparison between those with and without transverse fractures (i.e. zero aperture) allows assessment of how the transverse fractures affect longitudinal and transverse flows in the reference calculation case (0.55 μm aperture): the transverse fractures induce a 45% increase in the longitudinal hydraulic conductivity and a 65% increase in the transverse hydraulic conductivity.

The equivalent-porous-media model of the EDZ around a HLW storage cell EDZ

Introduction

In ANDRA's project of a deep repository, the high-level waste (HLW) storage cells will be orientated according to the major stress axis in order to reduce the vertical extension of the EDZ for safety reasons. Drilling experiments carried out in the URL at Bure have highlighted the similarity in fracture structure and global extension of the EDZ over a range of excavation diameters from 0.1 to 8 m. Figure 11

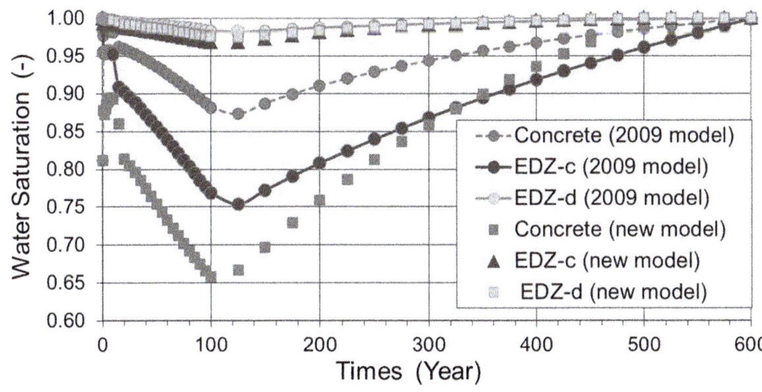

Fig. 9. Evolution with time of the water saturation in the concrete liner and in the EDZ calculated with the 2009 model and the new model.

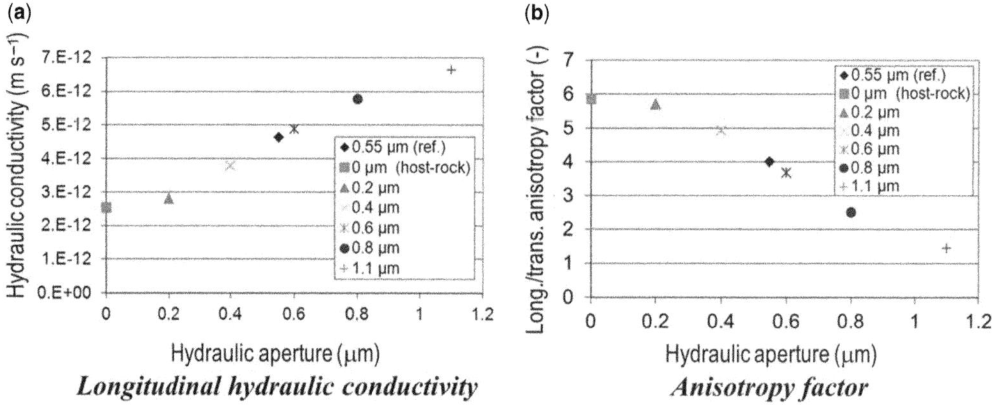

Fig. 10. Evolutions of (**a**) the longitudinal hydraulic conductivity and (**b**) the anisotropy factor in the transversal direction in the EDZ-d as a function of the transverse fracture hydraulic aperture.

illustrates these observations for excavations the size of a rock-bolt and a HLW storage cell: the EDZ extends in an 'eye shape', similar to the EDZ around the reference gallery.

However, hydraulic modelling of the EDZ of a HLW storage cell cannot be performed in the same way as for the reference gallery because characterization of the EDZ on a decimetre scale has not yet been undertaken. Neither geometrical data nor permeability data are available to allow a fracture network to de designed quantitatively. So, a qualitative approach has been adopted in order to extrapolate from the geometry of the gallery EDZ ($D_{\text{exc}} = 7.5$ m) to the scale of a storage cell ($D_{\text{exc}} = 0.7$ m) using a homothetic transformation.

Modelling hypotheses

The explicit-fracture model of the damaged zone around a HLW storage cell is based on two hypotheses: the fracture network is supposed to be homothetic to that around the gallery in both connected and disconnected subzones, except for the fracture apertures; and the fracture apertures are supposed to be the same as for the gallery, except in the case of the oblique fractures of the EDZ-d, which are supposed to have the same aperture as the EDZ-c fractures (2.45 μm). In fact, a visual analysis of Figure 11 suggests that, in the EDZ-d, the long-extension fractures generated by an excavation below the metre scale are thicker than those generated by an excavation at a half-decametre scale (i.e. an experimental gallery), probably because of smaller fracture extension. Figure 11 shows approximately the same 'eye-shape' EDZ, and the same ratio between the EDZ horizontal extension and the excavation diameter for a 5 cm and a 70 cm excavation diameter and also for the gallery. Furthermore, we observe approximately the same subvertical fracture number at the half height of the excavation (from 6 to 8) as in the EDZ-d gallery (see the oblique fractures in grey in Fig. 3b).

The equivalent-porous-media hydraulic properties

The longitudinal and transverse permeabilities have been deduced from the longitudinal and transverse

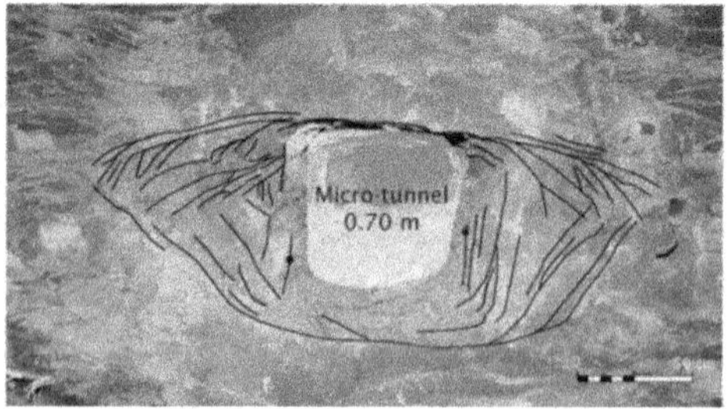

Fig. 11. Damaged zones around excavations 5 and 70 cm in diameter.

Table 4. *Updated values of physical property parameters for the EDZ of a HLW storage cell*

Parameters	Symbols	New EDZ-c	New EDZ-d
Porosity	ω (%)	18	18
Longitudinal intrinsic permeability	k_i^l (m^2)	2.07×10^{-16}	1.45×10^{-17}
Transverse intrinsic permeability	k_i^t (m^2)	6.27×10^{-17}	3.56×10^{-18}
Shape parameter (retention law)	n	1.5	1.5
Reference pressure head (retention law)	P_r (m_{WL})	1500	1500
Subzone thickness	E (m)	0.14 D_{exc}	0.27 D_{exc}

overall transmissivities calculated by the explicit-fracture model in the two subzones for saturated (see the intrinsic permeabilities in Table 4) and unsaturated media (see the relative permeabilities in Fig. 12). All other parameters have the same values as the reference gallery.

In the EDZ-c, the longitudinal intrinsic permeability is about 10 times higher than for the reference gallery. This result is inherent to the change in scale between the gallery and the storage cell, given that individual fracture permeabilities are the same in the two models whereas the fracture network becomes denser by changing the scale. In the EDZ-d, the longitudinal permeability is about 30 times higher than in the reference gallery. The intrinsic permeabilities are 3.3 and 4.1 times lower transversally in the EDZ-c and in the EDZ-d, respectively.

In unsaturated media, it has to be noticed that, in the EDZ-d, the relative permeability to gas is higher in the HLW storage cell model than in the reference gallery model because of lower capillary activity, given the larger aperture of the oblique fractures. The slight anisotropy of relative permeability shown in Figure 12b is partly induced by the fracture aperture difference between the oblique fractures (2.45 μm) and the transverse fractures (0.55 μm).

Conclusion

A hydraulic model of the EDZ has been developed through a PA-like approach in order to simulate flow behaviour in a fractured medium using an equivalent-porous-media model composed of two subzones of uniform physical parameters. This simplified model has been defined in two steps: first, the development of a realistic model, including an explicit representation of the fractures in the EDZ based on geomechanical and hydraulic measurements; and, secondly, the calibration of the equivalent-porous-media model in order to agree with the flow behaviour simulated using the explicit-fracture model developed in the first step.

This approach has been implemented to model the EDZ of two kinds of excavations parallel to the major stress axis: a 7.5 m-diameter excavation, representative of an access gallery; and a 0.7 m-diameter excavation, representative of a HLW storage cell. This modelling work results in specific physical properties of the EDZ, especially the following three points: (a) an anisotropic intrinsic permeability due to the fracture orientations; (b) the same retention law as the undisturbed host rock, because the fracture volume is negligible compared with the host-rock pore volume; and (c) relative permeability functions for gas and water that are higher

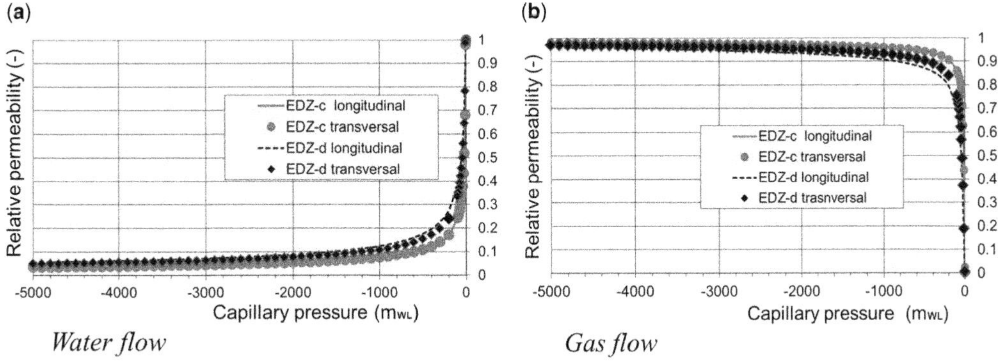

Fig. 12. Relative permeability v. capillary pressure for (**a**) water and (**b**) gas flows in the EDZ of a HLW storage cell (dots for radial flows and lines for longitudinal flows).

than previously, because the permeation behaviour under unsaturated conditions is very different between the fractures and the undisturbed rock.

Complementarily, a sensitivity study has been performed in order to determine the effect on transverse and longitudinal flows of the connectivity level inside the EDZ-d fracture network.

The consistency of these equivalent-porous-media models with the behaviour of the real EDZ could be improved if there were some additional experimental data: (a) regarding the gallery model, hydraulic measurements are few and erratic for distances less than a 10th of the diameter from the excavation; and (b) regarding the HLW storage cell model, no quantitative data are available as yet (qualitative observations have been used to define the EDZ around a storage cell as geometrically homothetic to the EDZ around a gallery.

Finally, this approach to modelling the EDZ could also be undertaken for excavations perpendicular to the major stress, including most of the access galleries to the ILW and HLW storage cells, or it could be applied to the modelling of transport phenomena, or to take into account the self-sealing of fracture apertures, assuming that adequate characterization data are available.

This work was funded by ANDRA, and the models presented are based on data provided by ANDRA.

References

ABABOU, R., CAÑAMON VALERA, I. & POUTREL, A. 2011. Macro-permeability distribution and anisotropy in a 3D fissured and fractured clay rock: 'Excavation Damaged Zone' around a cylindrical drift in Callovo-Oxfordian Argilite (Bure). *In*: ARANYOSSY, J.-F., FERNANDEZ, A.-M. ET AL. (eds) *Clays in Natural and Engineered Barriers for Radioactive Waste Confinement. Physics and Chemistry of the Earth, Parts A/B/C*, **36**, 1932–1948.

ANDRA 2005. *Dossier 2005 Argile: Phenomenological Evolution of a Geological Repository*, ANDRA Report C.RP.ADS.04/0025.B. ANDRA (National Agency for Radioactive Waste Management), Paris.

ANDRA 2013. *La zone endommagée initiale autour des ouvrages: Synthèse des travaux de caractérisation autour des ouvrages du laboratoire souterrain de Meuse/Haute-Marne*. Note technique CG.NT.AMFS. 13.0025. ANDRA (National Agency for Radioactive Waste Management), Paris.

ARMAND, G., LEVEAU, F. ET AL. 2014. *Geometry and properties of the excavation induced fractures at the Meuse/Haute-Marne URL drifts*. Rock Mechanics and Rock Engineering, **47**, 22–41, https://doi.org/10.1007/s00603-012-0339-6

DE LA VAISSIÈRE, R., MOREL, J. ET AL. 2014. Excavation-induced fractures network surrounding tunnel: properties and evolution under loading. *In*: NORRIS, S. & BRUNO, J. ET AL. (eds) *Clays in Natural and Engineered Barriers for Radioactive Waste Confinement*. Geological Society, London, Special Publications, **400**, 279–291, https://doi.org/10.1144/SP400.30

GENTIER, S., HOPKINS, D., RISS, J. 2000. Role of fracture geometry in the evolution of flow path under stress. *In*: FAYBISHENKO, B., WITHERSPOON, P.A. & BENSON, S.M. (eds) *Dynamics of Fluids in Fractured Rocks*. American Geophysical Union, Geophysical Monographs, **122**, 169–184.

HAWKINS, I.R., SWIFT, B.T., HOCH, A.R. & WENDLING, J. 2011. Comparing flows to a tunnel for single porosity, double porosity and discrete fracture representations of the EDZ. *In*: ARANYOSSY, J.-F., FERNANDEZ, A.-M. ET AL. (eds) *Clays in Natural and Engineered Barriers for Radioactive Waste Confinement. Physics and Chemistry of the Earth, Parts A/B/C*, **36**, 1990–2002.

IDEL'CIK, I.E. 1979. *Memento des pertes de charges*. [Handbook of Energy Losses.] Eyrolles, Paris [in French].

MUALEM, Y. 1976. A new model for predicting the hydraulic conductivity of unsaturated porous media. *Water Resources Research*, **12**, 513–522.

NICOLAS, M. 2003. *Ecoulements dans les milieux poreux. DEA Mécanique Energétique, Ecole Doctorale Mécanique énergétique et Modélisation*, Université de Provence, Marseille.

VAN GENUCHTEN, M.T. 1980. A closed-form equation for predicting the hydraulic conductivity of unsaturated soils. *Soil Science Society of America Journal*, **44**, 892–898.

VINSOT, A., DELAY, J., DE LA VAISSIÈRE, R., CRICHAUDET, M. 2011. Pumping tests in a low permeability rock: results and interpretation of a four-year long monitoring of water production flow rates in the Callovo-Oxfordian argillaceous rock. *In*: ARANYOSSY, J.-F., FERNANDEZ, A.-M. ET AL. (eds) *Clays in Natural and Engineered Barriers for Radioactive Waste Confinement. Physics and Chemistry of the Earth, Parts A/B/C*, **36**, 1679–1692.

WU, Y.-S., LIU, H.H. & BODVARSSON, G.S. 2004. A triple-continuum approach for modelling flow and transport in fractured rock. *Journal of Contaminant Hydrology*, **73**, 145–179.

ZHANG, CH.-L. 2011. Experimental evidence for the self-sealing of fractures in claystone. *In*: ARANYOSSY, J.-F., FERNANDEZ, A.-M. ET AL. (eds) *Clays in Natural and Engineered Barriers for Radioactive Waste Confinement. Physics and Chemistry of the Earth, Parts A/B/C*, **36**, 1972–1980.

The gas permeability, breakthrough behaviour and re-sealing ability of Czech Ca–Mg bentonite

JAN SMUTEK*, LUCIE HAUSMANNOVA & JIRI SVOBODA

Centre of Experimental Geotechnics, Faculty of Civil Engineering, Czech Technical University in Prague, Thakurova 7, 166 29 Prague 6, Czech Republic

**Correspondence: jan.smutek@fsv.cvut.cz*

Abstract: The study's objectives were to test the gas permeability of unsaturated compacted bentonite, the gas-breakthrough behaviour of saturated bentonite, and the re-sealing ability compared with the material's dry density, hydraulic conductivity and swelling pressure. During the project's pilot phase, a series of laboratory experiments were conducted on samples of two Czech Ca–Mg bentonites (Bentonite 75 and Cerny Vrch). The testing procedure, employing repeated hydration and gas-breakthrough test phases, simulated conditions within deep geological repositories for radioactive waste, wherein the buffer will be progressively hydrated from the surrounding host rock and loaded with pressure exerted by gases created within the repository.

It was found that the gas permeability of the bentonites tested was affected by dry density and, as with the two materials (FEBEX and MX-80) used for comparison, it correlated well with the accessible void ratio. Gas-breakthrough tests revealed correlations between time to breakthrough and both dry density and sample height. The recorded repeated test breakthrough times on the same samples varied, which implies that the pathways created are complex and differ for repeated gas-injection cycles. However, no systematic deterioration in sample properties was detected, thus indicating the extensive re-sealing ability of the materials tested.

Deep radioactive waste repository construction concepts currently envisage that swelling clays (bentonites) will be used in different forms (e.g. prefabricated blocks) in order to create a layer surrounding the container with the waste (a so-called buffer) and for the backfilling of the access galleries (e.g. a mixture of bentonite with crushed rock). The bentonite layers will be affected by heat loading from the container and by the groundwater present in the host rock. Furthermore, it is expected that gases will be produced within the repository following disposal consisting principally of hydrogen created by the corrosion processes impacting the container and water radiolysis. Theoretically, should the container become damaged, radionuclides could migrate from the repository into the environment, utilizing water or gases as the transport medium: hence, the main role of the clay layers is to retard radionuclide migration. The most important requirements with regard to the barriers, especially the buffer, consist of a hydraulic conductivity value lower than 10^{-12} m s^{-1}, and a swelling pressure value of between 1 and 10 MPa (according to POSIVA: Rautioaho & Korkiala-Tanttu 2009). These requirements are related particularly to the migration of water. The gases generated could negatively affect the performance of the buffer: therefore, it is important to gain a thorough understanding of the migration processes of gases within both partially and fully water-saturated bentonite. The Czech disposal concept considers the use of local bentonites for the barriers in the future geological repository. Hence, the aim of this investigation is to study the gas transport properties of compacted Czech bentonites.

Both laboratory and *in situ* gas injection tests can be performed in order to study gas-flow behaviour within bentonite. Pusch & Forsberg (1983) were amongst the first to conduct research on, and, subsequently, publish the results of the laboratory testing of, bentonite. This involved a series of gas-injection tests (GITs) on samples of compacted Mx-80 bentonite. Some years later, Gallé & Tanai (1998) conducted tests on saturated compacted clays (Fo–Ca clay) with high dry densities, and Horseman *et al.* (1999) began the study of the gas transport properties of saturated Mx-80 using helium injection testing methods. Subsequently, a large number of experiments were conducted on clays under both laboratory and *in situ* conditions. Recently, a great deal of work on bentonites was conducted as part of the European FORGE project (Shaw 2015). Similar gas-flow behaviour to that of compacted bentonites has been observed with respect to certain other fine-grained sedimentary rocks, and laboratory testing is also currently underway on samples of such rocks with a view to them providing host-rock formations for geological repositories: for example, Boom clay (Belgium), Opalinus claystone in Switzerland (Hildenbrand

From: NORRIS, S., BRUNO, J., VAN GEET, M. & VERHOEF, E. (eds) 2017. *Radioactive Waste Confinement: Clays in Natural and Engineered Barriers*. Geological Society, London, Special Publications, **443**, 333–348.
First published online June 22, 2016, https://doi.org/10.1144/SP443.5

et al. 2002) and Calovo-Oxfordian claystone (France: Cuss *et al.* 2014).

The most noteworthy tests on bentonite performed to date consist of those conducted on FEBEX bentonite (Villar *et al.* 2012) and Mx-80 bentonite (Graham & Harrington 2014). Results of gas injection tests performed on partially saturated bentonite by Villar *et al.* (2005) and Gutiérrez-Rodrigo *et al.* (2014) showed that gas permeability is affected significantly by the dry density and water content of the bentonite, and correlates well with the accessible void ratio (which is the inverse to the liquid degree of saturation). Observations from the gas injection testing of saturated bentonite and claystone revealed that advective gas flow occurs not via visco-capillary flow, but rather through a network of pressure-induced dilatant pathways (Graham *et al.* 2012). This theory is very well supported by numerous indirect observations and, recently, as the result of the use of direct detection methods (Harrington *et al.* 2012; Wiseall *et al.* 2015).

A typical gas injection test on saturated bentonite includes the recording of two threshold pressures: the first, known as the 'entry pressure', represents the pressure value at the time the injected gas first enters the sample; and, the second, the 'breakthrough pressure', corresponds to the pressure exerted at the moment the gas begins to flow through the sample. Following a gas migration event, saturated bentonite has the ability to self-seal the pathways created due to its high swelling capacity. Although there has been very significant progress in terms of understanding of the various processes involved (Shaw 2015), a number of questions remain relating to the transport of gases in saturated bentonite. Graham & Harrington (2014) pointed out uncertainties concerning breakthrough repeatability, as well as pathway stability. Dilatant pathways that develop within bentonite usually exhibit a complicated spatial–temporal evolution.

The primary objective of the research presented herein was to study the gas-breakthrough behaviour of saturated bentonite samples in comparison with the dry density, hydraulic conductivity and swelling pressure of the materials. The aim of the breakthrough tests was to create pathways for gas flow within water-saturated bentonite by means of the application of very high gas pressures that exceed the swelling pressure of bentonite. Moreover, it was intended that the repetition of water-saturation and gas-injection test cycles would allow the re-sealing properties of the material to be tested. Further, it was hoped that a comparison of the repeated gas-breakthrough tests using the same re-saturated samples would add to the existing knowledge of the repeatability of gas-pathway-formation processes. Finally, one of the secondary objectives of the project was to study the gas permeability of unsaturated samples.

This paper presents the objectives and testing procedures employed, and reveals the initial results of the research project that is being conducted for the Czech Science Foundation.

Material and testing equipment

Two types of Czech bentonite (Bentonite 75 (B75) and Cerny Vrch (CV)) were used for testing purposes in the first phases of the research project. Both of the bentonites were extracted from the same location, Cerny Vrch, in the NW of the Czech Republic. The Cerny Vrch area is rich in high-smectite-content (*c.* 60%) Ca–Mg bentonite, with notably high iron content in the octahedral layer of the smectite. Both materials are produced by the KERAMOST company in powder form. The difference between the two bentonites is that B75 (supplied in 2013) is industrially produced and may be influenced by the technological processes employed during production (mainly with concern to the addition of sodium). The Cerny Vrch bentonite, however, was specially prepared with a view to preparing a 'clean' material during the production process.

B75 2013 has a liquid limit of 171% and a specific density of 2.855 Mg m^{-3}; the Cerny Vrch material has a liquid limit of 130% and a specific density of 2.872 Mg m^{-3}. The initial water content of the powder materials is around 5% (B75) and 11% (CV). The saturated hydraulic conductivity (Fig. 1) and swelling pressure (Fig. 2) of both materials were compared with other bentonites that have been tested for deep geological repository purposes: MX-80 (Wyoming locality: Åkesson *et al.* 2010) and bentonite from Cabo de Gata, which was used in the FEBEX experiment (Villar 2002).

The samples were prepared by means of the direct compaction of the bentonite material into steel cylinders 30 mm in diameter. The maximum applied compaction pressure for the sample with the highest dry density (1.6 Mg m^{-3}) was 40 MPa. The samples were then placed between permeable plates (sintered steel) and enclosed within a stainless steel cell (Fig. 3), which takes the form of a constant volume vessel that can be connected either to a water permeameter or a gas-injection system. The water permeameters enable the saturation of the material, and the subsequent measurement of hydraulic conductivity and swelling pressure values. Neither confining nor backpressure was applied to the samples. The samples were saturated using distilled water pushed up from the bottom of the cells under a constant injection pressure of 1 MPa.

The gas properties of each sample were investigated in a number of stages employing two types

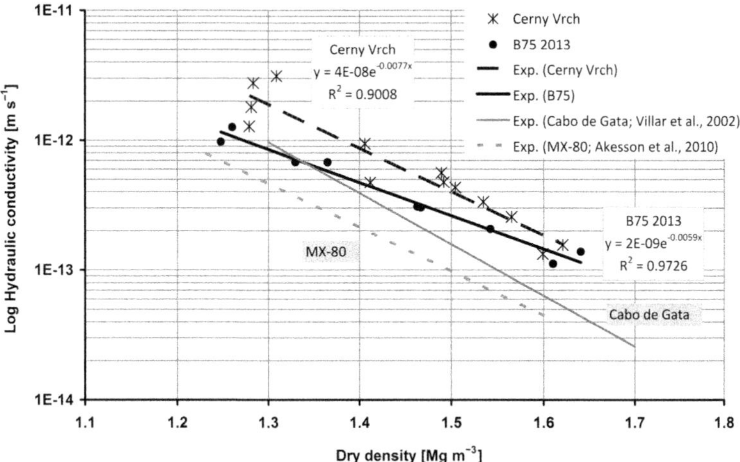

Fig. 1. Hydraulic conductivity of the tested materials in comparison with other bentonites.

of injection system, both of which employed dried air as the injection medium. The gas permeability of the unsaturated samples was tested following the connection of the cell to a mobile gas measuring station, which allowed for the setting and control of the upstream gas injection pressure; the downstream pressure was atmospheric. The maximum working pressure of the apparatus was 12 MPa and the range of the gas-flow meter was $0–20\ dm^3\ min^{-1}$ at STP. The set-up of the tests on the saturated samples (gas-breakthrough tests) is depicted in Figure 3.

All the various material characteristics investigated are dependent on the dry density of the material: therefore, samples with several differing dry density values were always tested. In addition, in order to allow the study of the influence of sample geometry, different sample heights (between 8 and 21 mm) were employed. The diameter of the samples was a constant 30 mm. The main physical properties of the samples used in the study are summarized in Table 1.

Test procedure and evaluation

The first sample testing stage included the performance of GITs on an unsaturated sample using the mobile measuring station. The test procedure consisted of an ascending cycle of step-by-step increases in gas pressure, followed by a corresponding descending cycle.

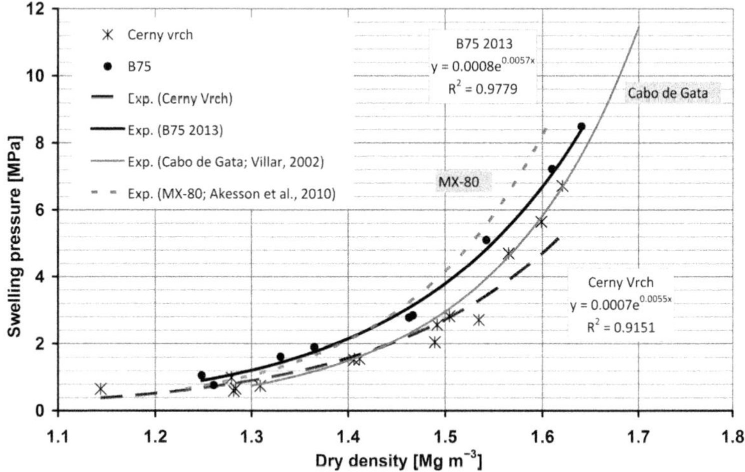

Fig. 2. Swelling pressure of the tested materials in comparison with other bentonites.

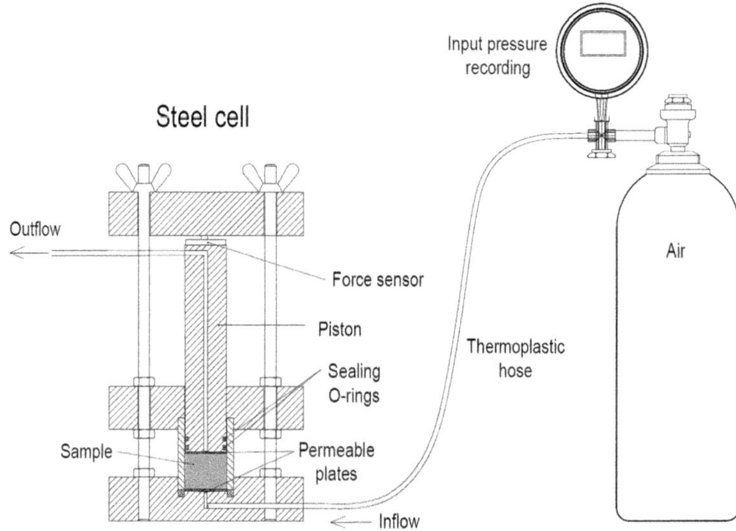

Fig. 3. Scheme of the set-up of the gas-breakthrough tests.

In order to obtain gas permeability directly, the samples would have to have been completely dry. Unfortunately, however, the drying of the samples in an oven was not possible as this procedure would have damaged the microstructure of the materials. Since the samples were not completely dry following compaction, the effective permeability, which can be calculated from the test (equation 1), consists of a multiple of the theoretical intrinsic permeability measured with the gas and relative gas permeability:

$$k_{\text{eff}} = k_{\text{ig}}\, k_{\text{rg}} = \frac{2Q\,\mu_{\text{g}}\,L\,P_0}{A\left(P_1^2 - P_2^2\right)} \quad (1)$$

where k_{eff} is the effective gas permeability (in m^2), k_{ig} is the intrinsic permeability measured with the gas (in m^2), k_{rg} is the relative gas permeability, Q is the gas-flow rate (in $m^3\ s^{-1}$), μ_g is the dynamic viscosity (in Pa s), L is the height of the sample (in m), P_0 is the reference (atmospheric) pressure (in Pa), P_1 is the gas upstream pressure (in Pa) and P_2 is the gas downstream pressure (atmospheric) (in Pa).

The data from the GITs was subjected to an analysis of whether or not measured effective permeability is affected by the Klinkenberg effect (Klinkenberg 1941), which expresses the extent of gas slippage that may occur in materials with small pores. Gas slippage occurs at low flowing pressures and results in an overestimation of

Table 1. *Geotechnical properties of samples*

Sample*	Height (mm)	Density ($Mg\ m^{-3}$)	Dry density ($Mg\ m^{-3}$)	Initial water content (%)	Porosity (n)	Void ratio (e)	Initial saturation (S_r)	Accessible void ratio ($e\ (1 - S_r)$)
CV_01	20.1	1.377	1.238	11.3	0.57	1.32	0.25	1.00
CV_02	20.3	1.779	1.599	11.3	0.44	0.80	0.41	0.47
B75_03	20.1	1.238	1.178	5.1	0.59	1.44	0.10	1.29
B75_04	20.3	1.398	1.330	5.1	0.54	1.16	0.13	1.01
B75_05	20.6	1.542	1.467	5.1	0.49	0.96	0.15	0.81
B75_06	20.8	1.692	1.610	5.1	0.44	0.78	0.19	0.64
B75_07	20.1	1.311	1.248	5.1	0.57	1.30	0.11	1.15
B75_09	8.7	1.690	1.606	5.2	0.44	0.79	0.19	0.64
B75_10	15.6	1.680	1.597	5.2	0.44	0.80	0.19	0.65
B75_11	8.5	1.333	1.267	5.2	0.56	1.26	0.12	1.12
B75_12	15.4	1.317	1.252	5.2	0.56	1.29	0.12	1.14

*Material: CV (Cerny Vrch); B75 (Bentonite 75).

measured apparent permeability. It can be detected by the performance of GITs at multiple pressure levels. Effective permeability is then plotted against the reciprocal of the mean gas pressure (the average of the upstream and downstream pressures). If the function (equation 2) of the fitted line has a positive gradient, the line intersects the y-axis at the Klinkenberg permeability value (absolute permeability equivalent to liquid). The gradient of the line is a multiple of Klinkenberg permeability and the slip factor:

$$k_{\text{eff}} = k_0\left(1 + \frac{\alpha}{P_{\text{mean}}}\right) \quad (2)$$

where k_0 is the Klinkenberg permeability (in m^2), α is the slip factor (in Pa) and P_{mean} is the mean pressure (in Pa).

Once the gas permeability tests had been completed, the samples were saturated by means of the water permeameter. The saturation pressure was kept constant at the same value of 1 MPa for all the samples. The later stages consisted of the performance of the gas-breakthrough test and cyclical re-saturation. The aim of the breakthrough test was to create pathways for gas migration and then to verify whether subsequent water injection sealed the sample (re-sealing).

Note: The term ‘self-sealing’ is used in this text to describe the capacity of bentonite to seal gas pathways without the additional hydration of the bentonite; ‘re-sealing’ is used in the context of when a sample is subjected to additional hydration following a gas injection test.

Each saturation phase was accompanied by the measurement of the absolute pressure, from which the swelling pressure could be calculated in the final stage of sample testing, and water inflow into the sample which provided data for the calculation of hydraulic conductivity. The gas-breakthrough test/re-sealing investigation cycle was repeated several times in all cases. The total testing period for one sample consisted of several months.

The pilot breakthrough tests involved a step-by-step increase in the injection gas pressure of 5–10 bar every 15 min up to a maximum pressure level of 12 MPa, which was then maintained until gas breakthrough was achieved. The second type of test consisted of the constant load test, wherein the saturated sample was loaded directly with 12 MPa of gas pressure: the time required to attain breakthrough was then recorded. The pressure value was chosen with regard to exceeding the theoretical swelling pressure of all the samples. The gas injection system included a cylinder containing dried air with a volume of 2 l (Fig. 3). Following the connection of the cylinder to the bottom part of the testing cell via a thermoplastic hose, the input valve was opened and the samples loaded with a gas pressure rate of 0.2 MPa s^{-1} up to the maximum value of 12 MPa. The input valve was kept open and the pressure in the cylinder was monitored. The outflow pressure corresponded to atmospheric pressure. The pressure in the input reservoir was recorded continuously using a digital manometer. The test was completed once gas breakthrough and pressure stabilization in the input reservoir had been achieved.

Once the cell had been connected to the water permeameter, the absolute pressure within the sample was measured continuously by means of an axial force sensor positioned between the piston and the upper steel plate (Fig. 3). The absolute pressure was partially influenced by the water injection pressure: therefore, it was necessary to record the value of the force used in the evaluation of the swelling pressure with an injection pressure of zero once the measured force had stabilized. This procedure was always applied at the very end of the sample-testing phase. As the absolute values of swelling pressure were not known during the performance of the tests, measured absolute pressure values were used for the interpretation of the results instead of those of the swelling pressure.

Results

Gas permeability tests on the unsaturated samples

Gas permeability tests were performed on unsaturated samples with a height of 20 mm and with different densities. The tests were conducted at multiple pressure levels in order to allow the study of the effects of gas pressure on the resulting effective permeability values. The maximum pressure values of the samples with lower dry densities were dependent on, and limited by, a maximum flow rate of 5 dm^3 min^{-1} at STP. The samples with the highest dry density values were loaded with pressures of up to 4 MPa. The results are summarized in Table 2. Figure 4 presents the results of effective permeability dependent upon injection pressure. The results revealed a clear dependence between the effective permeability and dry density of the samples. The calculated effective permeability range was within two orders of magnitude.

Effective permeability values were plotted against the reciprocal of the mean pressure, as depicted in Figure 5. The positive gradient of the trend lines for most of the samples implies the possible existence of the gas-slippage effect. The highest slip factor was recorded for the sample with the lowest dry density; the difference between the

Table 2. *Results of permeability testing*

Sample	Effective permeability (average) (m^2)	Effective permeability (Klinkenberg) (m^2)	Klinkenberg slip factor (bar)	Minimum injection pressure (bar)	Maximum injection pressure (bar)	Water permeability (m^2)	k_{ig} (m^2)	k_{rg}
CV_01	8.9×10^{-15}	–	*	4.8	9.4	3.6×10^{-19}	–	–
CV_02	4.9×10^{-16}	4.9×10^{-16}	0.1	9.1	43.2	1.3×10^{-20}	–	–
B75_03	2.4×10^{-14}	3.9×10^{-15}	15.5	1.8	7.0	–	3.4×10^{-14}	0.70
B75_04	4.4×10^{-15}	4.0×10^{-15}	0.4	1.9	14.8	6.6×10^{-20}	1.2×10^{-14}	0.37
B75_05	2.9×10^{-15}	2.6×10^{-15}	0.6	3.8	15.2	3.0×10^{-20}	4.7×10^{-15}	0.62
B75_06	5.8×10^{-16}	5.2×10^{-16}	1.1	4.0	39.5	1.0×10^{-20}	1.8×10^{-15}	0.33
B75_07	1.1×10^{-14}	9.1×10^{-15}	0.7	1.9	9.3	8.9×10^{-20}	2.1×10^{-14}	0.52

*Negative Klinkenberg slip factor obtained.

average value of effective permeability and the corrected Klinkenberg permeability is, in this case, close to one order of magnitude. However, in most cases, the effect was not so pronounced and the corrected Klinkenberg permeability was close to the average values of the full range of pressure levels. Average values of effective permeability were employed for the purposes of further interpretation.

Subsequently, the effective permeability results were compared with the accessible void ratio, as in studies by Villar *et al.* (2012) (Fig. 6). The accessible void ratio (sometimes referred to as accessible porosity) in connection with gas-flow testing indicates the ratio between gas accessible pore volume and the particle volume. Correlation with a high reliability coefficient was revealed when the average values of effective permeability were compared.

Water-saturation phase

Figure 7 shows the hydraulic conductivity and swelling pressure results from the tested samples following the initial saturation phase. No influence of sample height on these parameters was observed.

When the degree of saturation is considered equal to 0 and the relative gas permeability is equal to 1 (dry sample), and by substituting these values into equation (3), which expresses the experimentally determined correlation between the accessible void ratio and the effective permeability of the

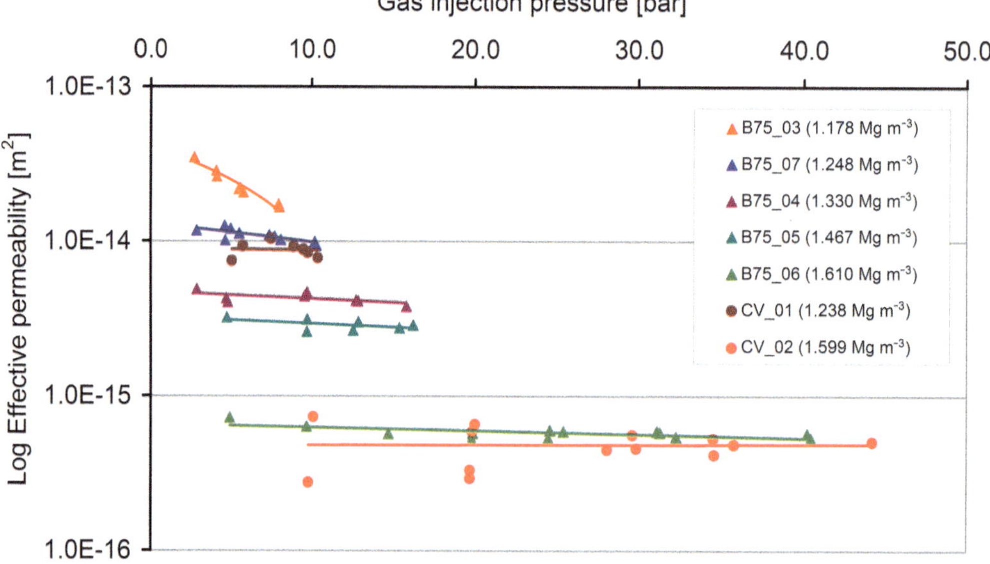

Fig. 4. Results of GITs on the unsaturated samples – dependence of effective permeability on injection pressure.

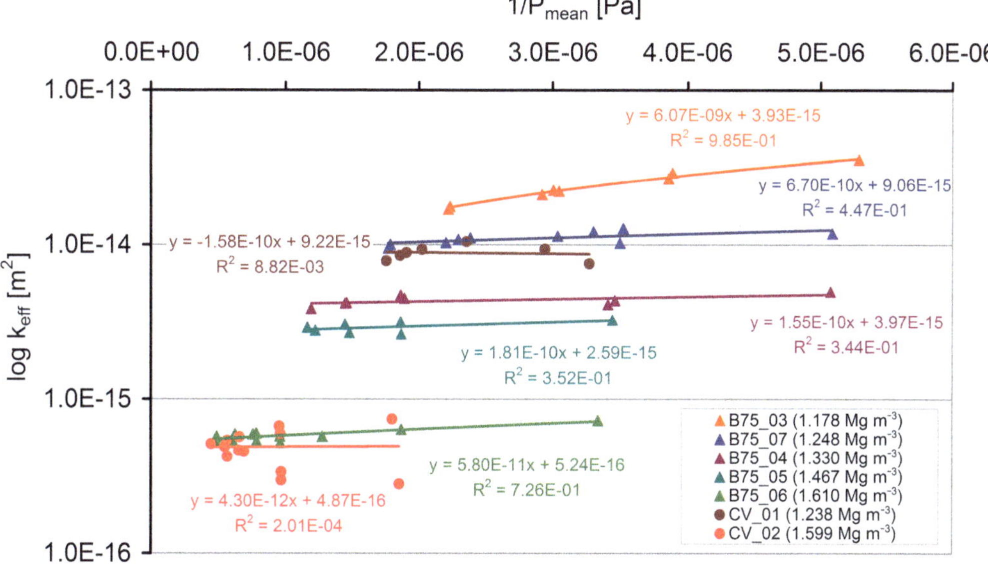

Fig. 5. Results of GITs on the unsaturated samples – Klinkenberg plot.

B75 material (Fig. 6), a formula for the calculation of intrinsic permeability is obtained (equation 4). Intrinsic permeability here is a function of the void ratio. The relative gas permeability of the tested samples can then be calculated as a ratio of effective permeability and intrinsic gas permeability. The results of these calculations are shown in Table 2 and, in Figure 8, they are compared with the intrinsic permeability values determined using water injection. The intrinsic permeability values determined for saturated bentonite using water as the permeating fluid were found to be around five orders of magnitude lower than those determined using a gas medium. The fitting functions of both exhibit very close exponents:

$$k_{ig}\,k_{rg} = 5.83 \times 10^{-15}[e(1 - S_r)]^{4.89} \qquad (3)$$

$$k_{ig} = 5.83 \times 10^{-15} e^{4.89}. \qquad (4)$$

Breakthrough tests on saturated samples

Initial pilot GITs with incremental pressure revealed that no breakthrough occurred during the ramp-up phase test lasting up to 6 h and ending at a pressure level of 12 MPa. These tests were performed on the

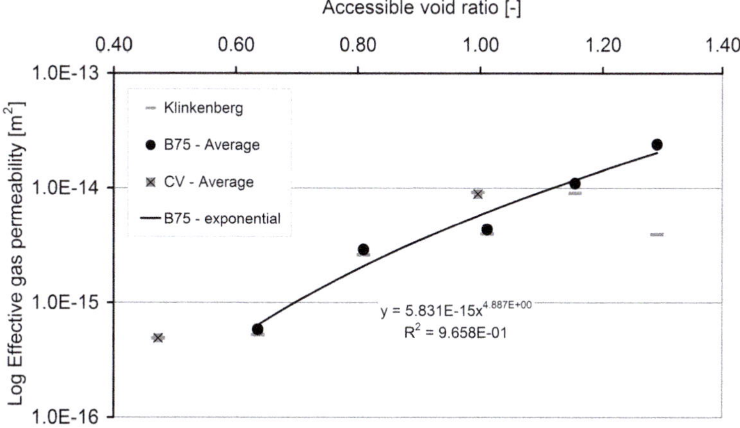

Fig. 6. Gas permeability as a function of the accessible void ratio.

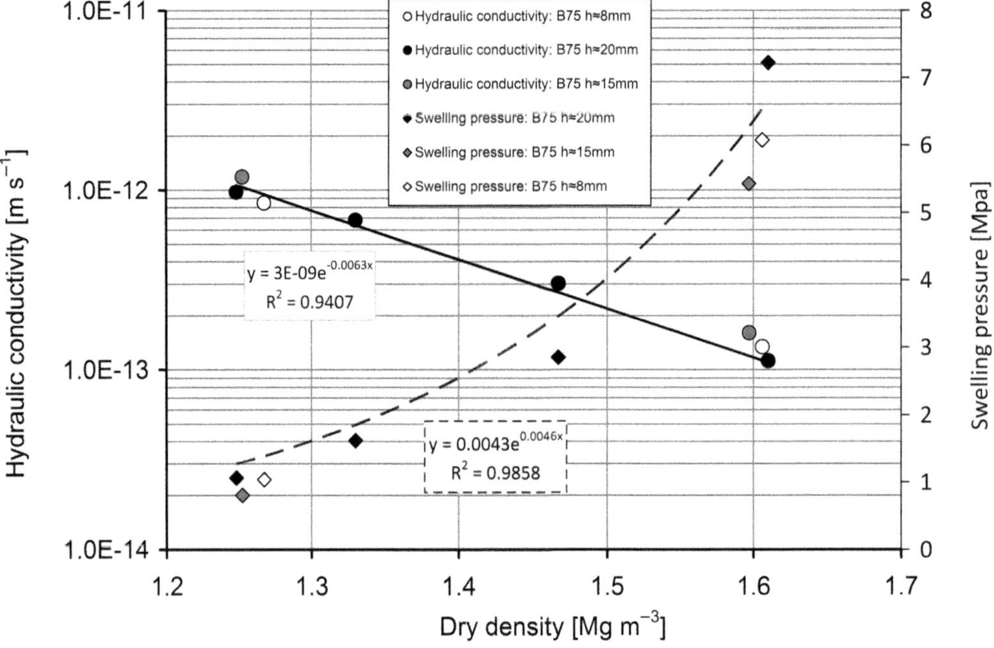

Fig. 7. Hydraulic conductivity and swelling pressure of the tested samples following the initial saturation phase.

CV, B75_04 and B75_05 samples following initial saturation. The results indicated that gas breakthrough was achieved no earlier than after several days of sample loading, with a gas pressure value exceeding that of the swelling pressure. Therefore, the gas-breakthrough test procedure was changed and all the samples were directly loaded with a gas pressure value of 12 MPa. The aim of the change in approach was to achieve breakthrough in a shorter time in order to be able to perform an extensive series of re-saturation and gas-breakthrough tests on the same samples, and thus to enable the study of re-sealing behaviour.

Figure 9 provides a comparison of the evolution of input gas pressure during repeated GITs on samples with different dry densities. Following the loading of the samples with gas pressure, no pressure decrease or outflow from the cell was detected up to the time that a breakthrough event occurred. Breakthrough was accompanied by a rapid drop in

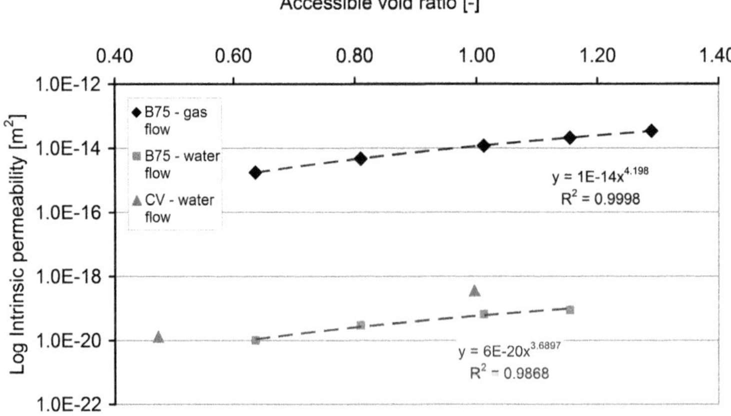

Fig. 8. Comparison of intrinsic permeability from saturated water flow and unsaturated gas flow.

input pressure, which usually occurred between a time period of 15 and 130 h. A number of typical input-pressure evolutions following a breakthrough event were detected (Fig. 9). The test equipment was not equipped with a shut-off valve that would automatically close the injection system following breakthrough. As a result, most of the tests ended with the total volume of compressed air in the cylinder passing through the samples (240 l of air at STP). A number of tests were characterized by a sudden cessation of outflow from the sample and a decrease in input pressure of varied time duration. Such events, which were interpreted as the closure of gas pathways, were discovered to be temporary in some cases and permanent in others.

Complete pathway closure was recorded on a regular basis with regard to just one sample (B75_07) with a low dry density and a maximum height, and only once with concern to the B75_04 sample. Such behaviour would tend to signify the self-sealing ability of the bentonite. However, residual pressure values were found to be even slightly above the swelling pressures of the samples. In total, two tests (samples B75_04, 05 during GIT_02) exhibited longer temporary pathway closure times, lasting between 10 and 15 h, which was then followed by a second breakthrough. It is conjectured that such exceptional behaviour was influenced by the fact that the first GIT was performed using a different experimental procedure. All the samples with a dry density value of 1.6 Mg m^{-3} showed a certain tendency to closure. Once the pressure decreased to a level of between 7 and 10 MPa, a short temporary closure occurred that lasted for between 15 and 60 min; however, this closure was incomplete (a low-pressure decrease and a slight, but notable, air outflow from the cell continued to be observed). For reasons of comparison, the range of values for the swelling pressures with concern to these samples was between 5 and 7 MPa.

The testing of all the other samples ended with the complete emptying of the air-input reservoir: however, both the flow-rate and the outflow evolution were observed to differ. The minimum registered time for the complete emptying of the reservoir was 2 h. A number of other tests were characterized by a gentle decrease in pressure, lasting up to 20 h.

Breakthrough time was found to be clearly dependent on dry density (Fig. 10), and the corresponding hydraulic conductivity and swelling pressure, as well as on the height of the sample (Figs 11 & 12). It was also found that the breakthrough time increased exponentially with increasing dry density. The relationship between the height of the sample and the breakthrough time, with regard to the high dry density samples, can be expressed as an exponential function (Fig. 11), while the low dry density samples appear to be linearly dependent (Fig. 12).

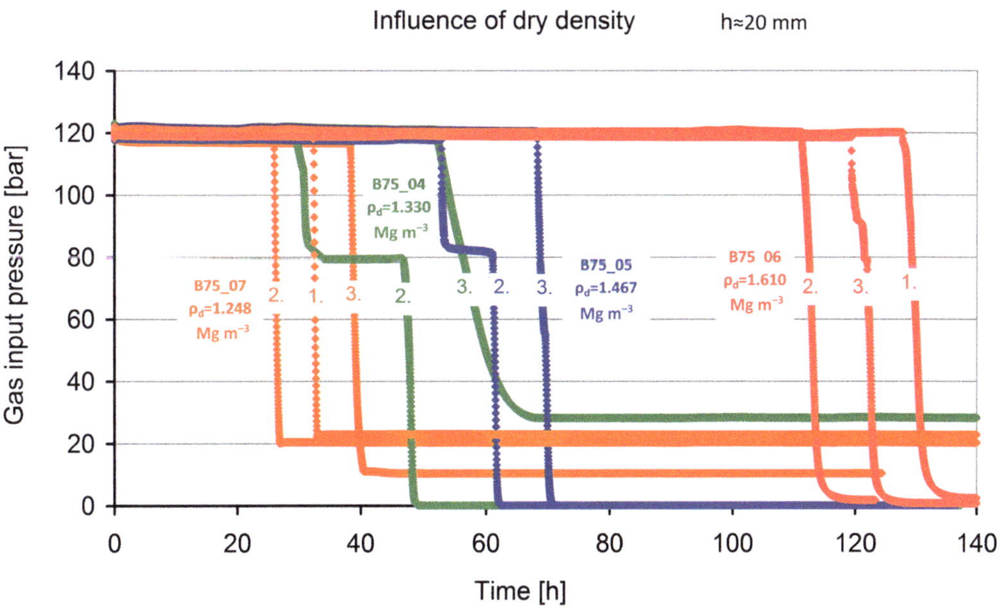

Fig. 9. Comparison of repeated gas-breakthrough tests for samples of different dry density and the same height of 20 mm.

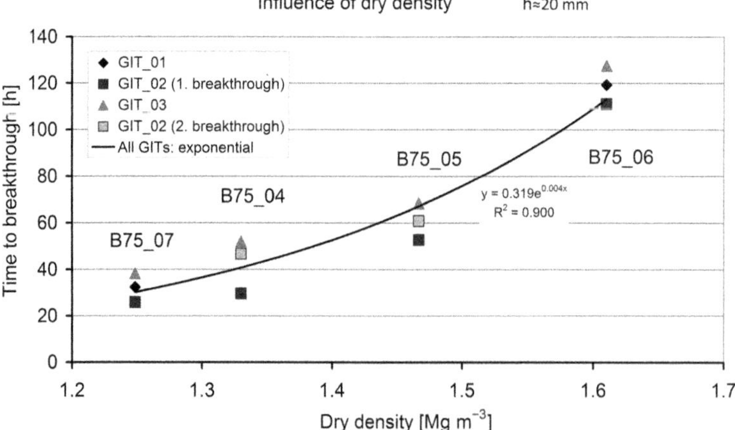

Fig. 10. Comparison of the breakthrough time for samples of different dry density and the same height of 20 mm.

Re-sealing investigation

The repeated cyclical saturation and gas-breakthrough tests were carried out in order to investigate the re-sealing properties of the bentonite materials. Figure 13 shows the typical complete testing cycle for one sample. The starting point on the graph corresponds to the moment at which the measurement of the water-flow rate commenced; the flow rate was not measured during the initial saturation time period. The performance of the GIT usually led to a decrease in the measured absolute pressure. Following the commencement of re-saturation, absolute pressure increased up to stabilization over a period in the range of several days. A much longer period was needed for the stabilization of the water-flow rate; the re-saturation phases lasted for a period of up to 2 months.

It was found to be typical for all the samples that, in the second saturation phase (following GIT_01), saturated hydraulic conductivity values were lower than following the initial saturation. A comparison of the hydraulic conductivity values determined for all the saturation phases is provided in Figure 14, while Figure 15 depicts the decrease in hydraulic conductivity following the first saturation phase. Subsequent to the first decrease in hydraulic conductivity, which appeared to be dependent on dry density, the following GITs did not substantially influence hydraulic conductivity in the next saturation phases. This observation clearly demonstrates the re-sealing ability of the samples. A possible

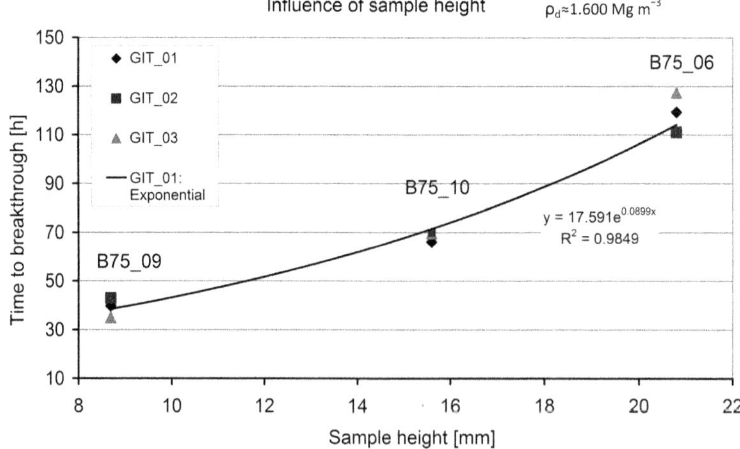

Fig. 11. Comparison of the breakthrough tests for samples with different heights and a dry density of 1.60 Mg m^{-3}.

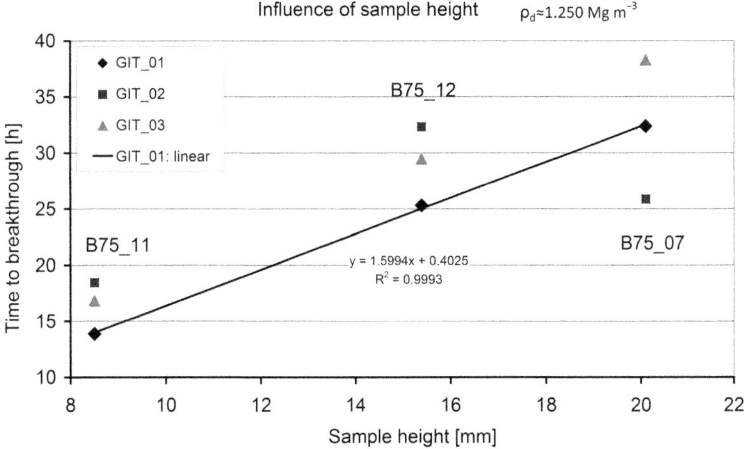

Fig. 12. Comparison of the breakthrough tests on samples of different height and a dry density of 1.25 Mg m^{-3}.

explanation for the first hydraulic conductivity decrease may lie in the fact that the sample was compacted by applied gas-injection pressure. If this were the case, the decrease in hydraulic conductivity would lead to a theoretical increase in dry density calculated using the relationship between dry density and a decrease in hydraulic conductivity following GIT_01 (Fig. 15) of less than 10%. A similar comparison was carried out with concern to absolute pressures (Figs 16 & 17). Absolute pressure was also found to slightly decrease: however, the changes were not so significant (< 10%) and were not related to dry density or sample height.

A comparison of breakthrough times for repeated GITs on identical samples is provided in Figures 9–12. The breakthrough times for repeated tests were not identical: however, they did occur within certain limited time ranges. The data were analysed in order to discover systematic changes in breakthrough times that might be supported by other observations and theories. With regard to the data gathered to date, none of the theories worked for all the samples. With concern to the low dry density samples (Fig. 12), it appeared that GIT_01 led to the compaction of the samples, as a result of which the breakthrough time in GIT_02 was extended. This theory seems to concur with decreased hydraulic conductivity observations: however, this was observed in connection with only two of three samples. A further common observation was detected regarding samples with a height of 20 mm (Fig. 10) – for all four samples, the recorded breakthrough time was

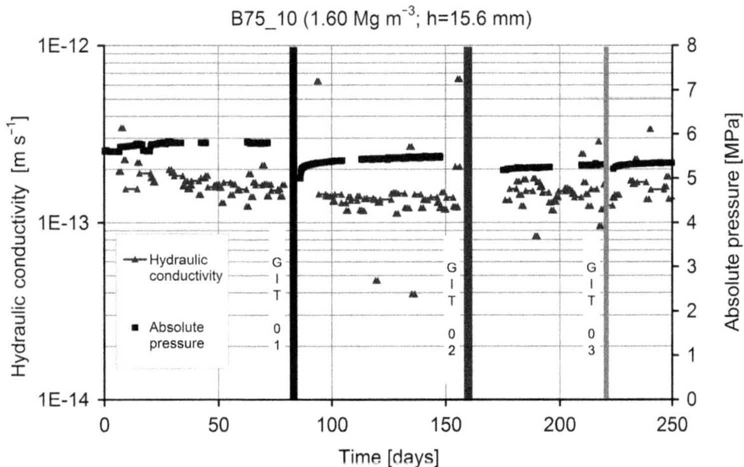

Fig. 13. The hydraulic conductivity and absolute pressure of one sample during the whole of the testing cycle.

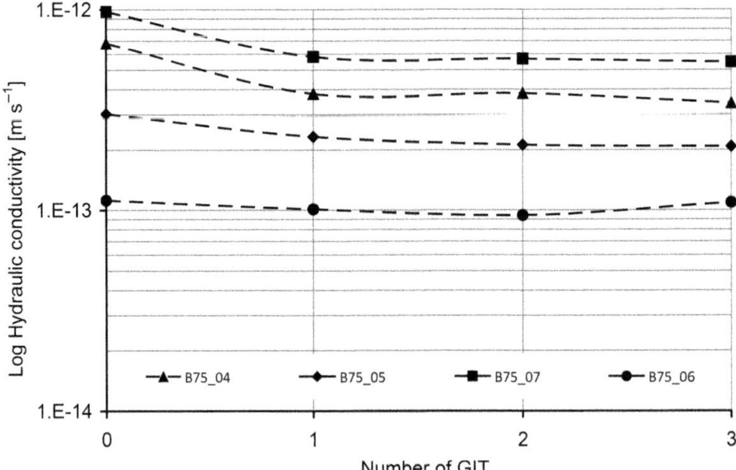

Fig. 14. Evolution of the saturated hydraulic conductivity of samples with different dry densities.

extended between GIT_02 and GIT_03: however, this observation was not supported by changes in hydraulic conductivity or absolute pressure. Generally, no systematic breakthrough time changes were recorded that would imply the progressive deterioration of the samples. This would appear to provide evidence of the material's strong ability to re-seal.

Discussion

Gas permeability tests

The results of the gas permeability tests can be compared with those performed on FEBEX bentonite in the context of the FEBEX (ENRESA 2000) and FORGE projects (Villar *et al.* 2012), as well as on MX-80 bentonite (Gutiérrez-Rodrigo *et al.* 2014). All the tests revealed the clear influence of dry density on effective gas permeability. An increase in density from 1.2 to 1.6 Mg m^{-3} led to a decrease in gas permeability of two orders of magnitude ($10^{-14} \div 10^{-16}$ m^2). For the purposes of comparison, tests on a sample series of FEBEX and MX-80 samples (Gutiérrez-Rodrigo *et al.* 2014) showed that an increase in dry density from 1.5 to 1.8 Mg m^{-3} led to a decrease of three orders of magnitude for similar water content levels. The water content of these two sample series was 16 ÷ 22% for

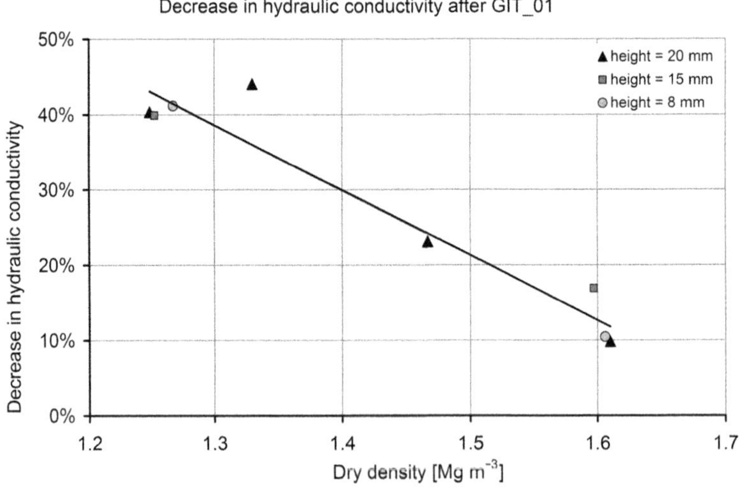

Fig. 15. Decrease in hydraulic conductivity following GIT_01 related to dry density – for samples with different heights.

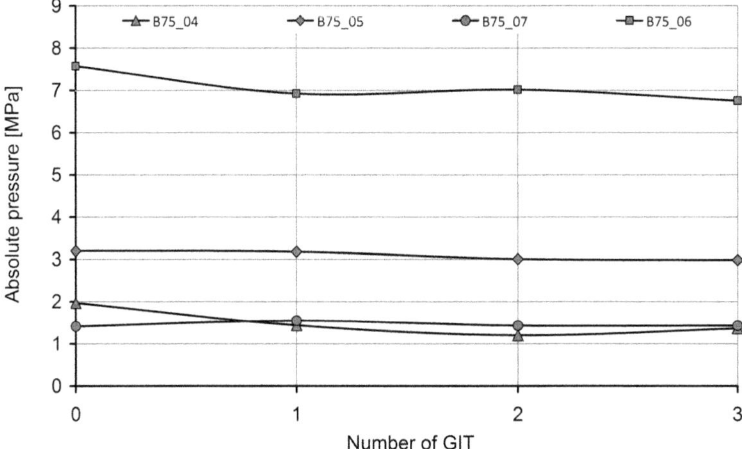

Fig. 16. Evolution of the absolute pressure of samples with different dry densities and a height of 20 mm.

FEBEX and 9 ÷ 20% for MX-80. As all the B75 samples had the same water content (corresponding to the hygroscopic water content of powder bentonite: i.e. 5% for B75), the effect of water content was not studied. The detected gas permeability values of B75 were found to be lower than those of FEBEX and MX-80 for the corresponding dry density values. For example, for a dry density value of 1.6 Mg m^{-3} and hydroscopic water content (both materials: i.e. B75, 5%; MX-80, 9%), the effective gas permeability of MX-80 was one order of magnitude higher than B75; and the values determined for FEBEX were found to be even higher. These differences could be explained by means of the different grain-size distribution of the compared bentonites.

It was found during the tests that gas permeability was affected by effective gas pressure. Within the range of applied pressure values (from 0.3 to 4 MPa), the effective permeability values for most of the samples slightly decreased with increasing injection pressure. This behaviour could be explained by the occurrence of gas slippage (the Klinkenberg effect): however, in theory, slippage should take place during periods of low gas-flow pressures (close to atmospheric pressure). For purposes of comparison, the series of tests on FEBEX bentonite performed as part of the FORGE project (Villar *et al.* 2012) revealed that the Klinkenberg effect was not particularly significant: for applied pressure values of up to 1 MPa, the Klinkenberg

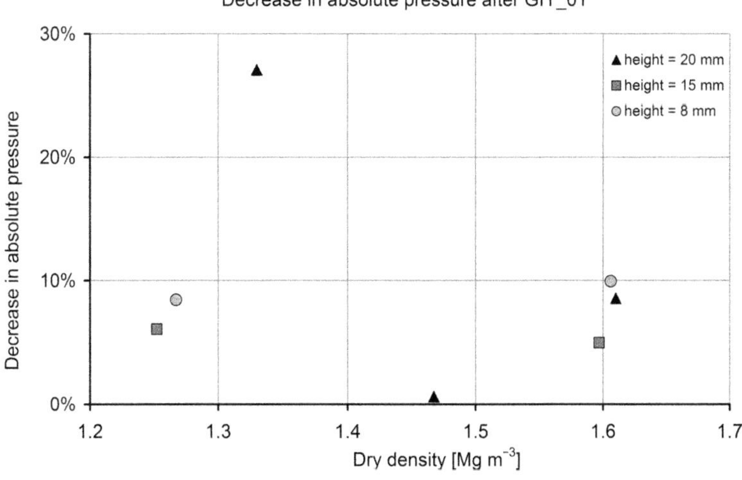

Fig. 17. Decrease in absolute pressure following GIT_01 related to dry density – for samples with different heights.

plot slopes were sometimes even seen to be negative. However, it was found that gas permeability tended to be slightly lower when effective pressure was increased (by means of an increase in confining pressure or the lowering of back-pressure). As far as the set-up used in the tests performed as part of this study are concerned, in which no confining pressure was applied on the samples and the residual back-pressure was merely atmospheric, the changes in effective permeability may well have been affected by axial pressure acting upon the samples. However, these changes were not particularly significant and the average permeability values were later used for further analysis.

The flow of air in porous media requires that the gas phase is continuous. As water content increases, the water-filled pores block the flow of gas, thus leading to a decrease in gas permeability. Ideally, intrinsic permeability is dependent solely on soil structure, and it should exhibit the same value for both gas and liquid flow. However, water reacts with the clay minerals within expansive clays, which leads to swelling. As a result, a reduction in pore space and the modification of pore-size distribution occur. Thus, values of intrinsic permeability determined by gas tests in dry materials and by water flow in saturated materials differ. During the testing of FEBEX bentonite, it was found that this difference can be up to eight orders of magnitude (ENRESA 2000).

In a similar way as was discovered during the testing of FEBEX and MX-80 bentonites (Villar *et al.* 2005), effective gas permeability correlated best with the accessible void ratio of B75. Based on this correlation, the function for the calculation of intrinsic permeability from the gas tests was determined. With respect to B75 bentonite, the detected difference between intrinsic permeability values obtained using water and gas was almost six orders of magnitude. The results are compared with the other bentonites in Figure 18, in which intrinsic permeability is plotted as a function of the accessible void ratio. It can be seen that the saturated water permeability fitting curves for the FEBEX and B75 bentonites are very close. The FEBEX samples series used for purposes of comparison had dry density values from 1.4 to 1.8 Mg m^{-3} and a water content from 16 to 22%: therefore, they had a much lower accessible void ratio. It can also be seen that, for higher water content values, intrinsic permeability decreases more rapidly with decreasing accessible void ratio following a power function.

Gas-breakthrough tests

The aim of the breakthrough tests was not to determine breakthrough pressure values, but rather to measure and compare the time to breakthrough for samples loaded with a constant gas pressure of 12 MPa. As with the dependency of hydraulic conductivity and swelling pressure on dry density, a clear correlation was found between dry density and breakthrough time for samples with the same height.

For those samples with a low dry density value of 1.25 Mg m^{-3}, a clear linear dependence was determined between sample height and breakthrough time (Fig. 12), while the breakthrough time of samples with a high dry density value of 1.6 Mg m^{-3} was found to be exponentially dependent on sample height (Fig. 11). The linear dependence could be explained by the fact that the gas developed pathways through the samples via the path of least

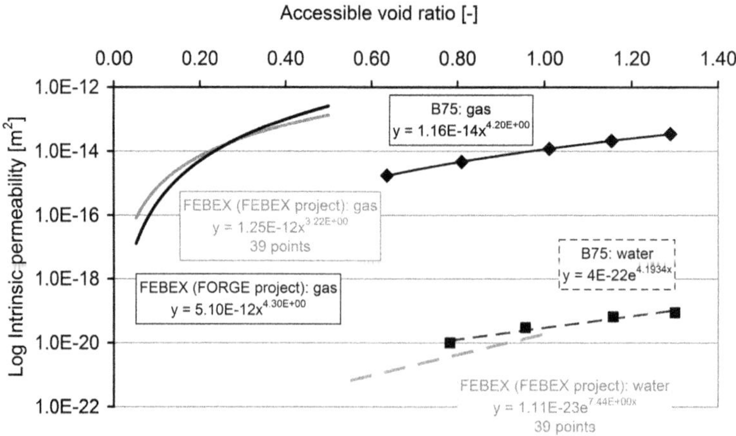

Fig. 18. Comparison of intrinsic permeability for B75 and FEBEX bentonite obtained from the unsaturated gas-flow tests and the saturated water-flow tests.

resistance (i.e. along the interface between the bentonite and the cell walls). It seems reasonable that this might occur with regard to samples with low dry densities since they exert significantly lower pressures on the cell walls (swelling pressure) in comparison with the applied gas-injection pressure. Conversely, the exponential relationship with regard to the high dry density samples could imply that a complex system of pathways developed through the sample matrices, in a similar way to that described in other studies (Hildenbrand *et al.* 2002).

The detected decreases in hydraulic conductivity following the first GIT are assumed to have been caused by the partial compaction of the samples and the redistribution of mass within the samples during the gas-injection phases. One interpretation is that the application of gas-injection pressure leads to the compaction of the samples; following the release of gas pressure, the void space created by compaction is filled (by means of swelling) with material that has a lower dry density than the rest of the sample, the dry density of which increased. As a result, hydraulic conductivity values are seen to be lower in the subsequent saturation phases. This interpretation is further supported by the fact that the change in hydraulic conductivity is dependent on dry density. The highest changes (i.e. of 40%) were detected for the 1.25 Mg cm^{-3} samples: these samples were particularly vulnerable to compaction. Absolute pressure returned to values close to initial values, with a final decrease of up to 10%. Following these significant changes, brought about by the first GIT, values did not change significantly during the subsequent loading cycles. After comparing the repeated breakthrough times, no systematic theory-supported change was determined using the data collected to date. Gas breakthrough appears to occur randomly within a certain time range, which would seem to imply that the pathways created are different for repeated GITs.

Tests performed elsewhere focusing on the detection of breakthrough pressure found that repeated gas injection influences breakthrough-pressure values. Graham & Harrington (2014), using observations from Harrington & Horseman (2003), concluded that when gas injection is halted, gas pathways collapse in response to declining pressure: subsequent gas injection then exhibits a notable reduction in breakthrough pressure. This behaviour, known as the 'memory effect', is understood to be caused by residual gas that is left in localized regions of the bentonite. The memory effect was seen to decline significantly following the application of an additional hydration phase. With regard to the tests performed in this project, it is thought that the deliberately long hydration phases applied ensured the full saturation of the samples. The results of the repeated test cycles, which showed that the time to breakthrough is not extended systematically, demonstrated the resealing capacity of the tested bentonite samples.

A comparison of pressure evolution following breakthrough events led to a number of interesting observations. The samples with high dry density values all exhibited a certain tendency towards pathway closure that occurred when the pressure dropped to a level of between 7 and 10 MPa: these values were slightly higher than the detected swelling pressure values (5–7 MPa). Once the described temporary closure ceased, the input gas pressure dropped very rapidly and the tests ended with the complete emptying of the input reservoir. Other behaviour, such as complete pathway closure and a long-term gradual decrease in pressure, was recorded for just one sample: therefore, it is difficult to draw conclusions from the data provided by this one single event.

In order to come to any firm final conclusions, it will be necessary to gather data from a greater number of samples. The subsequent phases of this ongoing research project will include the testing of both homogeneous samples of other Czech bentonites and a series of inhomogeneous samples prepared from a mixture of pellets and powder.

Conclusions

During the first year of the research project, an extensive series of laboratory experiments were conducted on samples of two Czech Ca–Mg bentonites (Bentonite 75 and Cerny Vrch). The first objective was to test the gas permeability of unsaturated samples, and the second was to study the gas-breakthrough behaviour of saturated bentonite and its potential re-sealing ability.

The results of the tests on unsaturated bentonite were compared with those on two other materials studied elsewhere in the context of the development of deep geological repositories for radioactive waste (FEBEX and MX-80). The gas permeability of B75 bentonite was found to be affected by dry density and, in a similar way as the two materials used for the purposes of comparison, the best correlation was determined as that with the accessible void ratio.

The gas-breakthrough tests revealed correlations between time to breakthrough and both dry density (and corresponding hydraulic conductivity and swelling pressure) and sample height. This observation implies that, in low dry density conditions, gases may develop pathways along the interface between the bentonite and the cell walls. However, with concern to the high dry density samples, it appears that more complicated pathways develop throughout the sample matrix.

The testing procedure, involving repeated hydration and gas injection phases, simulated geological repository conditions, wherein the buffer will be progressively hydrated by water from the surrounding host rock. The breakthrough times recorded of repeated testing on the same samples, while not identical, did occur within certain limited time ranges: moreover, there was no clear link to the other properties monitored (hydraulic conductivity and swelling pressure). This would imply that the pathways created are rather complex and differ for repeated GITs. Generally, no systematic changes in breakthrough time were recorded that would imply the progressive deterioration of the samples, thus providing evidence of the re-sealing ability of the tested materials.

This research was supported by the Czech Science Foundation as part of the 'Gas migration through compacted smectitic clays' project (GA14-19655S).

References

ÅKESSON, M., BÖRGESSON, L. & KRISTENSSON, O. 2010. *SR-Site Data Report THM Modelling of Buffer, Backfill and other System Components*. SKB TR-10-44. Svensk Kärnbränslehantering AB, Stockholm, http://www.skb.se/publication/2095119/TR-10-44.pdf

CUSS, R.J., HARRINGTON, J.F., GIOT, R. & AUVRAY, C. 2014. Experimental observations of mechanical dilation at the onset of gas flow in Callovo-Oxfordian claystone. *In*: NORRIS, S., BRUNO, J. ET AL. (eds) *Clays in Natural and Engineered Barriers for Radioactive Waste Confinement*. Geological Society, London, Special Publications, **400**, 507–519, https://doi.org/10.1144/SP400.26

ENRESA 2000. *FEBEX Project Full-Scale Engineered Barriers Experiment for a Deep Geological Repository for High Level Radioactive Waste in Crystalline Host Rock*. Final Report ENRESA (National Radioactive Waste Company), Madrid, http://www.enresa.es/files/multimedios/PT01-00.pdf

GALLÉ, C. & TANAI, K. 1998. Evaluation of gas transport properties of backfill materials for waste disposal: H2 migration experiments in compacted Fo-Ca Clay. *Clays and Clay Minerals*, **46**, 498–508. https://doi.org/10.1346/CCMN.1998.0460503

GRAHAM, C.C. & HARRINGTON, J.F. 2014. *Final Report of FORGE 3.2.1: Key Gas Migration Processes in Compact Bentonite*. FORGE Report D3.33 (British Geological Survey Internal Report CR/14/064) British Geological Survey, Keyworth, Nottingham, https://www.bgs.ac.uk/forge/docs/reports/D3.33.pdf

GRAHAM, C.C., HARRINGTON, J.F., CUSS, R.J. & SELLIN, P. 2012. Gas migration experiments in Bentonite: implications for numerical modelling. *Mineralogical Magazine*, **76**, 3279–3292, https://doi.org/10.1180/minmag.2012.076.8.41

GUTIÉRREZ-RODRIGO, V., VILLAR, M.V., MARTÍN, P.L. & ROMERO, F.J. 2014. Gas transport properties of compacted bentonite. *In*: KHALILI, N., RUSSELL, A.R. & KHOSHGHALB, A. (eds) *Unsaturated Soils: Research & Applications*. Taylor & Francis, London, 1735–1740, https://doi.org/10.1201/b17034-253

HARRINGTON, J.F. & HORSEMAN, S.T. 2003. *Gas Migration in KBS-3 Buffer Bentonite, Sensitivity of Test Parameters to Experimental Boundary Conditions*. SKB Technical Report TR-03-02. Svensk Kärnbränslehantering AB, Stockholm, http://www.skb.com/publication/20108/TR-03-02.pdf

HARRINGTON, J.F., MILODOWSKI, A.E., GRAHAM, C.C., RUSHTON, J.C. & CUSS, R.J. 2012. Evidence for gas-induced pathways in Clay using a nanoparticle injection technique. *Mineralogical Magazine*, **76**, 3327–3336, https://doi.org/10.1180/minmag.2012.076.8.45

HILDENBRAND, A., SCHLÖMER, S. & KROOSS, B.M. 2002. Gas breakthrough experiments on fine-grained sedimentary rocks. *Geofluids*, **2**, 3–23, https://doi.org/10.1046/j.1468-8123.2002.00031.x

HORSEMAN, S.T., HARRINGTON, J.F. & SELLIN, P. 1999. Gas migration in clay barriers. *Engineering Geology*, **54**, 139–149, https://doi.org/10.1016/S0013-7952(99)00069-1

KLINKENBERG, L.J. 1941. The permeability of porous media to liquids and gases. *In*: *API Drilling and Productions Practices*. American Petroleum Institute, New York, 200–213.

PUSCH, R. & FORSBERG, T. 1983. *Gas Migration Through Bentonite Clay*. SKB Technical Report 83-71. Svensk Kärnbränslehantering AB, Stockholm, http://www.skb.se/upload/publications/pdf/TR83-71webb.pdf

RAUTIOAHO, E. & KORKIALA-TANTTU, L. 2009. *Bentomap: Survey of Bentonite and Tunnel Backfill Knowledge – State-of-the-Art*. VTT Working Papers, **133**. VTT Technical Research Centre of Finland, http://www.vtt.fi/inf/pdf/workingpapers/2009/W133.pdf

SHAW, R.P. 2015. The Fate of Repository Gases (FORGE) project. *In*: SHAW, R.P. (ed.) *Gas Generation and Migration in Deep Geological Radioactive Waste Repositories*. Geological Society, London, Special Publications, **415**, 1–7, https://doi.org/10.1144/SP415.17

VILLAR, M.V. 2002. *Thermo-Hydro-Mechanical Characterisation of a Bentonite from Cabo de Gata: A Study Applied to the Use of Bentonite as Sealing Material in High Level Radioactive Waste Repositories*. ENRESA (National Radioactive Waste Company), Madrid, http://www.enresa.es/files/multimedios/PT04-02.pdf

VILLAR, M.V., ROMERO, E. & LLORET, A. 2005. Thermo-mechanical and geochemical effects on the permeability of high-density clays. *In*: ALONSO, E.E. & LEDESMA, A. (eds) *Advances in Understanding Engineered Clay Barriers*. Taylor & Francis, London, 177–191.

VILLAR, M.V., GUTIÉRREZ-RODRIGO, V., MARTIN, P.L. & BARCALA, J.M. 2012. *Results of the Tests on Bentonite (Part 2)*. FORGE Report D3.27 (CIEMAT Technical Report CIEMAT/DMA/2G207/07/12) http://www.bgs.ac.uk/forge/docs/reports/D3.27.pdf

WISEALL, A.C., CUSS, R.J., GRAHAM, C.C. & HARRINGTON, J.F. 2015. The visualization of flow paths in experimental studies of clay-rich materials. *Mineralogical Magazine*, **79**, 1335–1342, https://doi.org/10.1180/minmag.2015.079.06.09

Measuring diffusion coefficients of dissolved He and Ar in three potential clay host formations: Boom Clay, Callovo-Oxfordian Clay and Opalinus Clay

E. JACOPS[1,2,3]*, N. MAES[1], C. BRUGGEMAN[1] & A. GRADE[1]

[1]*Belgian Nuclear Research Centre (SCK•CEN), Waste & Disposal Expert Group, Boeretang 200, B-2400 Mol, Belgium*

[2]*KU Leuven, Department of Earth & Environmental Sciences, Celestijnenlaan 200E, B-3001 Heverlee, Belgium*

[3]*Energy and Mineral Resources Group (EMG), Institute of Geology and Geochemistry of Petroleum and Coal, Lochnerstrasse 4-20, Haus B, D-52056 Aachen, Germany*

**Correspondence: ejacops@sckcen.be*

Abstract: For the long-term management of high- and intermediate-level radioactive waste and/or spent fuel, many countries prefer disposal in a geological repository. In Belgium, Switzerland and France, argillaceous formations are being explored as potential host formations for this purpose. In this context, knowledge of the diffusion coefficient of He is relevant within two research areas: first, diffusion coefficients are used in safety calculations to evaluate the balance between gas generation (mainly H_2) and gas dissipation; and, second, the diffusion coefficients of He and Ar are needed in the diffusion models of natural tracers. Owing to the lack of data on the diffusion coefficients of He and Ar for the different clay host formations, diffusion experiments with dissolved He and Ar were performed on Boom Clay, Opalinus Clay and Callovo-Oxfordian Clay. Samples were confined in a diffusion cell, and diffusion coefficients were measured by using the double through-diffusion technique. The diffusion coefficients (D_p or D_{pore}) for He in Boom Clay, Callovo-Oxfordian Clay and Opalinus Clay are 12.6×10^{-10}, 4.51×10^{-10} and 7.13×10^{-10} m^2 s^{-1}, respectively. The diffusion coefficients for Ar in Boom Clay, Callovo-Oxfordian Clay and Opalinus Clay are 18.6×10^{-11}, 4.06×10^{-11} and 3.65×10^{-11} m^2 s^{-1}, respectively.

The preferred option for the long-term management of high- and intermediate-level radioactive waste and/or spent fuel, adopted by many countries, is disposal in a geological repository. In Belgium, Switzerland and France, argillaceous formations are being explored as potential host formations for this purpose. The clays under consideration have a high sorption capacity for many radionuclides (Maes *et al.* 2004, 2008; Altmann *et al.* 2012), low hydraulic conductivity (Enssle *et al.* 2011; Yu *et al.* 2013) and self-sealing properties (Van Geet *et al.* 2008; Zhang 2013).

In Belgium, no formal decision has yet been taken on a host formation but, for research and development (R&D) purposes, the Belgian radioactive waste management organization (ONDRAF/NIRAS) considers Boom Clay (BC) as a potential host formation for a geological disposal facility.

In France, the National Agency for Radioactive Waste Management (ANDRA) selected the Callovo-Oxfordian Clay Formation (COx) in the east of France as a potential host formation (Labalette *et al.* 2010), and, in Switzerland, the National Cooperative for the Disposal of Radioactive Waste (NAGRA) proposed the Jurassic Opalinus Clay (OPA).

Within a geological nuclear waste repository, the production of gas is unavoidable. Gas is produced by different mechanisms: anaerobic corrosion of metals in waste and packaging; radiolysis of water and organic materials in the packages; and microbial degradation of various organic wastes. Anaerobic corrosion and radiolysis yield mainly hydrogen, while microbial degradation produces methane and carbon dioxide (Rodwell *et al.* 1999; Perko & Weetjens 2011). In a geological repository, anaerobic corrosion is considered to be the dominant gas production mechanism and, consequently, hydrogen will be the main gas component.

The generated gas will dissolve in the porewater and will migrate away from the repository by diffusion as dissolved species. If the rate of gas generation is larger than the diffusive flux into the clay, then the porewater within the disposal gallery will become oversaturated and a free gas phase will form. In order to compute a comprehensive and

From: Norris, S., Bruno, J., Van Geet, M. & Verhoef, E. (eds) 2017. *Radioactive Waste Confinement: Clays in Natural and Engineered Barriers*. Geological Society, London, Special Publications, **443**, 349–360.
First published online June 22, 2016, https://doi.org/10.1144/SP443.1

reliable balance between gas generation v. gas dissipation through engineered barriers and a host formation for a particular waste form and/or disposal concept, good estimates for diffusion coefficients of dissolved gases are essential. However, measuring the diffusion coefficient of hydrogen with a high degree of accuracy is difficult as hydrogen–air mixtures have a low explosion limit and H_2 easily leaks from the experimental set-up. But the major problem is the microbial conversion of H_2 into CH_4 by methanogenic bacteria (Jacops *et al.* 2015). Owing to these experimental difficulties and the current lack of accurate diffusion data for hydrogen, the diffusion coefficient of helium is often used as a surrogate for the diffusion coefficient of hydrogen.

In clay-rich formations, transport properties in porewater can be constrained from natural tracer profiles, which can be considered as long-term and large-scale natural experiments. Information obtained from laboratory-scale experiments and from experiments in underground research laboratories is restricted to spatial and temporal scales of typically 0.01–1 m and 1 year. In contrast, information provided by natural tracer profiles considers scales of tens to hundreds of metres and 0.1–1 myr. This information is very important for supporting the safety assessment of a deep geological repository and for upscaling from the laboratory scale to the repository scale. Moreover, natural tracer profiles represent independent evidence for system understanding. Natural tracer profiles have been determined for anions (Cl^-, Br^-, I^-), water isotopes ($\delta^{18}O$, δ^2H) and noble gases (mainly He, but also Ar) (Battani *et al.* 2011; Mazurek *et al.* 2011; Rebeix *et al.* 2014).

As natural tracers are present in deep clay layers for millions of years and diffusion is considered to be the dominant transport mechanism, it should be possible to represent these natural tracer profiles by diffusion models. For these models, reliable diffusion coefficients are required. Diffusion coefficients for anions and water isotopes are widely available, but the availability of reliable diffusion coefficients for He and Ar is limited.

Thus, on the one hand, reliable diffusion coefficients for He are needed for safety calculations (evaluation of gas generation v. gas dissipation – with the diffusion coefficient for He being used as an approximation of hydrogen). Yet, on the other hand, diffusion coefficients of both He and Ar are needed for use in the diffusive models of natural tracers. Owing to the lack of data on the diffusion coefficients of He and Ar for the different clay host formations, SCK•CEN started an experimental programme, in collaboration with NIRAS/ONDRAF, ANDRA and the University of Bern, to measure the diffusion coefficients of He and Ar with a newly developed set-up, as described in Jacops *et al.* (2013*b*).

Materials and methods

Clay samples

The Boom Clay is a highly plastic Tertiary argillaceous formation in NE Belgium, deposited at around 29–32 Ma. The underground research laboratory (URL) 'HADES', where experimental research on geological disposal for high-level and/or long-lived radioactive waste is performed, is located near the town of Mol at a depth of 225 m. The Boom Clay comprises clay minerals (up to 60%, dominated by illite, mixed layered illite–smectite, kaolinite and traces of chlorite), as well as quartz, K-feldspar, Na-plagioclase, pyrite and carbonates (Honty & De Craen 2012). The Boom Clay sample (O/N-Mol-1 core 84b) used in this study was cored at a depth of between 233.02 and 233.12 m below the drilling table, with the drilling table located at 29.73 m TAW (Tweede Algemene Waterpassing). The sample axis is orientated perpendicular to the bedding plane. The samples were cored in 1997 and have been stored under anoxic conditions (vacuum packed in aluminium-coated PE-foil) at 4°C. Prior to their use, they were visually inspected for signs of oxidation, which is the main threat to quality.

The Callovo-Oxfordian Clay is an indurated, Middle Jurassic mudstone from a marine origin that was deposited 150–160 myr ago. It has been studied by ANDRA over an area of 250 km^2. In this area, ANDRA had performed several drilling campaigns and constructed a URL at a depth of 490 m in the COx layer. The thickness of the COx formation is about 135 m at the URL. The COx consists mainly of clay minerals (illite, interstratified smectite/illite and others), calcite, dolomite/ankerite, quartz, feldspars and minor amounts of accessory phases, such as pyrite (Gaucher *et al.* 2004). The used sample (EST 49109: depth of 478.52 m) was taken from the OHZ6560 borehole, which had been drilled from a technical gallery at the URL. The sample axis is orientated perpendicular to the bedding plane.

The Opalinus Clay is a fine-grained sedimentary rock that was deposited 172 myr ago. The Opalinus Clay consists mainly of clay minerals (illite, illite/smectite, kaolinite and others), calcite, dolomite/ankerite, quartz, feldspars, pyrite and organic matter. Research on the Opalinus Clay is mainly performed in the Mont Terri URL, located in the canton Jura. The sample was, however, taken from the Schlattingen borehole (NE of Switzerland, canton Thurgau), at a depth of 860.32 m below surface. As the burial history for the Mont Terri site and

Schlattingen is different (with deeper burial in Schlattingen) (Mazurek *et al.* 2006), the physical properties of the Opalinus Clay are different at both locations. The sample axis is orientated perpendicular to the bedding plane. At the time at which the diffusion experiments were performed, there was no information available on the porewater composition of the samples from the Schlattingen borehole. Therefore, artifical porewater, relevant for the Opalinus Clay in Mont Terri, was selected.

Sample preparation

The Boom Clay sample (diameter 80 mm, length 30 mm) was loaded into the diffusion cell with a hydraulic press. This was achieved by placing a sharp cutting edge in front of the cell, which trims off the outer shell of the plastic clay sample; no resin was needed. Confinement (constant volume) was achieved by sealing the diffusion cell (welding). Owing to the high plasticity (Van Geet *et al.* 2008), perfect sealing between clay and cell was achieved. Porous stainless steel filter plates (diameter 80 mm, thickness 2 mm, porosity 40%) were used on both sides. This provided good contact between the gas-saturated water and the clay.

As Callovo-Oxfordian Clay and Opalinus Clay contain lower proportions of swelling clay minerals, their swelling capacity/plasticity is much lower compared to the Boom Clay. Therefore, the samples had to be embedded in a resin in order to seal the interface clay sample cell. First, cylinders (80 mm in diameter and 35 mm in height for Opalinus Clay, and 70 mm in diameter and 40 mm in height for the Callovo-Oxfordian Clay) were machined on a lathe. Next, the samples were placed in the diffusion cell and the remaining void (thickness 5 mm) was filled with resin (Sikadur 52 Injection Normal). After curing, the cell was machined again: the upper and lower portions of the cylinder were removed, leaving cylinders 25 mm high for the Opalinus Clay and 30 mm high for the Callovo-Oxfordian Clay. A Teflon filter (with a diameter the same as the diameter of the sample, thickness 2 mm) was placed in the upper and lower flange, and next the flanges were welded to the diffusion cell. By welding the cell and flanges, confinement (constant volume) was achieved. As the flanges were provided with a circulation loop, good contact between the water, which contains the dissolved gas, and the clay sample was ensured.

General description of the methodology

The methodology is described in detail in Jacops *et al.* (2013*b*) and only the main aspects are repeated here. The principal idea consists of a double through-diffusion test (Shackelford 1991) with two water reservoirs with dissolved gases placed on opposite sides of a saturated test core (Fig. 1).

Prior to the diffusion experiment, the hydraulic conductivity (K_v) is measured in a permeameter set-up. This measurement ensures full saturation of the sample and verifies proper sealing of the interfaces. The method used to measure hydraulic conductivity is described by Wemaere *et al.* (2008) and reference values for K_v are summarized in Table 1.

Next, both vessels of the set-up were filled with approximately 500 ml of oxygen-free porewater (the composition for each sample is given in Table 2) and 500 ml of gas at 1 MPa pressure. Once the test started, sampling occurred on a regular basis (generally once per week for Boom Clay, and once every 2 weeks for Callovo-Oxfordian and Opalinus Clay) until approximately 10 data points were obtained. The gas composition was analysed with a CP4900 micro GC (VARIAN, Palo Alto, CA, USA) and a Compact CGC-4 GC (Interscience, Breda, The Netherlands).

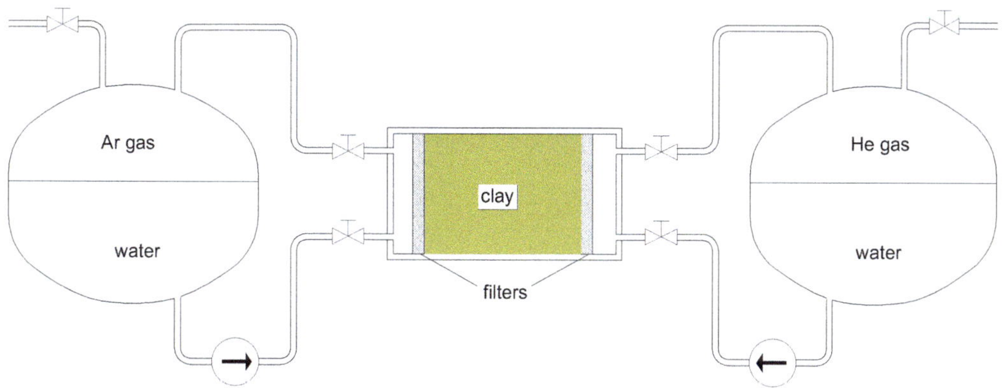

Fig. 1. Schematic design of the set-up to measure the diffusion of dissolved gases.

Table 1. *Overview of the parameter porosity used in the transport model and reference hydraulic conductivity*

	Reference hydraulic conductivity (K_v) ($m^2\ s^{-1}$)	Porosity (%)
Boom Clay	1.5×10^{-12}–8×10^{-12}*	37†
Callovo-Oxfordian Clay	5×10^{-14}–5×10^{-13}‡	18§
Opalinus Clay	6×10^{-15}–3×10^{-14}†	9.6‖

*Yu *et al.* (2013) – values measured orthogonal to bedding planes.
† Bruggeman *et al.* (2009).
‡Mazurek *et al.* (2011) – for Opalinus Clay, K_v values for samples from Schlattingen are not available. Therefore, values for samples taken in Benken (a town in North Switzerland – the nearest available location) are given.
§Jacquier *et al.* (2013).
‖Jacops *et al.* (2013*a*).

The experiment was performed in a temperature-controlled room (21 ± 2°C).

The diffusive transport model

The tests were interpreted with a simple diffusive transport model that represents the transport equation in a one-dimensional (1D) geometry, and which is based on Fick's first and second laws of diffusive transport in porous media. The use of Fick's laws to calculate diffusion coefficients is generally accepted. More detailed information on the used model can be found in Jacops *et al.* (2013*b*).

The diffusion–accessible porosity was set at a fixed value. For Boom Clay and Callovo-Oxfordian Clay, porosity was set at 37 and 18%, respectively. Both values are considered to be reference values for these materials (Bruggeman *et al.* 2009; Jacquier *et al.* 2013). For Opalinus Clay, no reference value for samples from the Schlattingen borehole is available, so the porosity was set at 9.6% (a value measured experimentally by drying the sample). The modelling accounts for a reduction in pressure in both vessels caused by sampling. All calculations were performed with COMSOL Multiphysics® version 3.5a, Earth science module. The diffusion coefficients are obtained by using a least-squares fitting procedure to the experimental data with the MATLAB Optimization Toolbox. This toolbox uses a Levenberg–Marquardt method on a Trust region, which is specifically optimized for a least-squares problems. In fact, the model calculates the best fit for two products: $D_{pore}\eta(=D_{eff})$; and $R\eta$.

D_{pore} denotes the pore diffusion coefficient (also written as D_p), D_{eff} is the effective diffusion coefficient, R is the retardation factor and η is the diffusion accessible porosity. As η is set to a fixed value and R is set = 1 (no retardation), the value for D_{pore} is readily obtained. The porosity used for each clay is given in Table 1.

The relationship between D_{app} (the apparent diffusion coefficient), D_{pore} and D_{eff} is described by equation (1):

$$D_{app} = \frac{D_{pore}}{R} = \frac{D_{eff}}{\eta R} \quad (1)$$

The definitions of D_{pore}, D_{eff} and D_{app} can be found in (Grathwohl 1998).

Table 2. *Composition of the artificial porewaters*

Chemical component ($mg\ l^{-1}$)	Boom Clay*	Callovo-Oxfordian Clay†	Opalinus Clay‡
$NaHCO_3$	1176	216	50
NaCl		1059	6132
Na_2SO_4		2216	1632
$CaCl_2.2H_2O$		947	1020
$MgCl_2.6H_2O$		911	1004
$SrCl_2.6H_2O$			84
KCl		78	60

*De Craen *et al.* (2004).
†Savoye *et al.* (2010, 2012).
‡Pearson *et al.* (2002).

Results

The measured hydraulic conductivities (K_v) for Boom Clay and Callovo-Oxfordian Clay are reported in Table 3. For Opalinus Clay, the time of measurement for the sample was only 18 days. Based on our experience with this kind of measurement, we must acknowledge that this is too short a period for samples with a low ($<10^{-12}\ m^2\ s^{-1}$) hydraulic conductivity. Owing to the short duration, the obtained result, $K_v = 1.2 \times 10^{-13}\ m\ s^{-1}$, is probably biased by artefacts and not fully representative of the sample. However, the main goal of the measurement was to verify whether good sealing

Table 3. *Overview of obtained diffusion coefficients for He and Ar, and measured hydraulic conductivity*

	D_p (He) (10^{-10} m^2 s^{-1})	D_p (He) 95% confidence interval (10^{-10} m^2 s^{-1})	D_p (Ar) (10^{-11} m^2 s^{-1})	D_p (Ar) 95% confidence interval (10^{-11} m^2 s^{-1})	D_{eff}* (He) (10^{-11} m^2 s^{-1})	D_{eff}* (Ar) (10^{-11} m^2 s^{-1})	K (10^{-12} m^2 s^{-1})
Boom Clay (BC)	12.6	12.2–13.1	18.6	18.0–19.1	46.7	6.87	3.30
Callovo-Oxfordian Clay	4.51	4.42–4.61	4.06	3.96–4.15	8.13	0.73	0.34
Opalinus Clay (OPA)	7.13	6.82–7.45	3.65	3.50–3.81	6.84	0.35	0.12

All measurements took place perpendicular to the bedding plane.
*$D_{eff} = D_p \times \eta$, and η is 37 (BC), 18 (Jacquier *et al.* 2013) and 9.6 (OPA), respectively.

was obtained between the clay, resin and the cell. As the measured hydraulic fluxes were low enough, the interface clay–resin–cell is considered to be tight.

Good sealing was also achieved for the Boom Clay and Callovo-Oxfordian Clay samples.

Based on the concentration profile recorded in the low-concentration compartment and the diffusive model, the diffusion coefficients for He and Ar could be determined (Fig. 2).

An overview of the measured diffusion coefficients for He and Ar, and the measured hydraulic conductivity, is given in Table 3.

An overview of the mineralogical composition and measured porosity for the Boom Clay and Opalinus Clay sample is given in Table 4. The methodology used to obtain these parameters is described by Wersin *et al.* (2013) for Opalinus Clay, and by Zeelmaekers *et al.* (2015) and Aertsens (2011) for Boom Clay. As the sample Callovo-Oxfordian Clay is still in use, characterization has not yet been carried out. The major geochemical properties and a general mineralogical composition can be found in Gaucher *et al.* (2004).

Discussion

Diffusion coefficients for He and Ar in water-saturated shales have been obtained for three different potential host formations. In the literature, diffusion coefficients for He have been reported for both Callovo-Oxfordian Clay (Rebour *et al.* 1997; Bigler *et al.* 2005) and Opalinus Clay (Gómez-Hernández 2000; Rubel *et al.* 2002). An overview of available data is given in Table 5. For argon, no data have been found in literature. A detailed discussion on the techniques used, their advantages, disadvantages and pitfalls, can be found in Jacops *et al.* (2013*b*).

As shown in Table 5, the diffusion coefficient of He for Opalinus Clay has been reported by Gómez-Hernández (2000) and Rubel *et al.* (2002). Gómez-Hernández (2000) performed an *in situ* in- and out-diffusion experiment with He parallel to the bedding plane on Opalinus Clay in the Mont Terri URL. The best fit for D_{app}(He), 7.0×10^{-10} m^2 s^{-1}, was obtained from an out-diffusion experiment (Gómez-Hernández 2000) but a porosity of 30% was used, which is twice the typical porosity values of 12–18% reported for Opalinus Clay (Mazurek *et al.* 2011). This might be explained by the fact that, during the drilling of the experimental borehole, an excavation damage zone (EDZ) was created, leading to a higher permeability and increased porosity – as discussed by Gómez-Hernández (2000). If the higher porosity is, indeed, related to the creation of an EDZ, the measured diffusion coefficient might be not relevant for diffusion under undisturbed conditions. Rubel *et al.* (2002) fitted D_{app} for He ($D_{app} = 3.5 \times 10^{-11}$ m^2 s^{-1}) in Opalinus Clay based on the natural helium concentration profile measured at Mont Terri, Switzerland. However, according to Mazurek *et al.* (2011), this D_{app} value is probably too small because of the use of an overly simplified model – in particular, the unsubstantiated assumption of a steady-state situation in which out-diffusion of He is balanced by *in situ* production.

The values reported by Gómez-Hernández (2000) and Rubel *et al.* (2002) differ by more than one order of magnitude, which can partly be explained by the difference in orientation with respect to bedding plane. Even when taking into account the anisotropy ratio for diffusion (around 4 for HTO in Opalinus Clay at Mont Terri: Van Loon *et al.* 2004), the difference between the reported values of Gómez-Hernández (2000) and Rubel *et al.* (2002) remains high. Owing to the technique used, the experimental conditions and the parameters that have been used to calculate the diffusion coefficient, the accuracy of the reported values is questionable. Moreover, one should consider that the reported values are relevant for the Opalinus Clay in Mont Terri, whereas our sample originates from Schlattingen. Therefore, no direct comparison can be made between the values reported in the literature and the values reported in this work.

For the Callovo-Oxfordian Clay, Bigler *et al.* (2005) performed an ^{4}He outgassing experiment

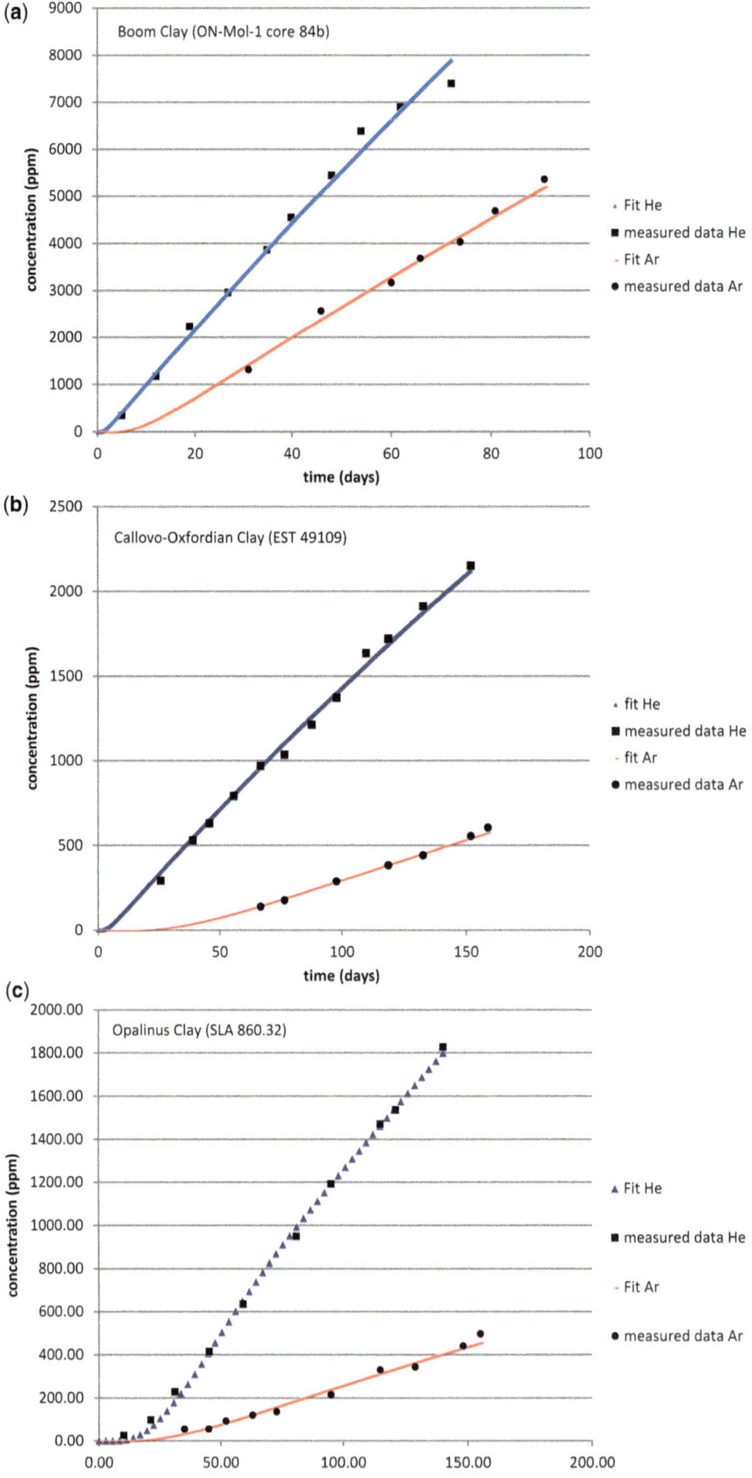

Fig. 2. Best fit for He and Ar for (**a**) Boom Clay, (**b**) Callovo-Oxfordian Clay and (**c**) Opalinus Clay.

Table 4. *Mineralogical composition, density and measured physical porosity of the Boom Clay and Opalinus Clay samples*

	Quartz (wt%)	*K*-feldspar (wt%)	Plagioclase (wt%)	Calcite (wt%)	Dolomite/ankerite (wt%)	Siderite (wt%)	Pyrite (wt%)	Gypsum (wt%)	C(org) (wt%)	Clay minerals (wt%)	Bulk dry density (g cm^{-3})	Grain density (g cm^{-3})	Physical porosity (vol.%)
Opalinus Clay (SLA 860.32)	28	3	b.d.	8	b.d.	b.d.	0.6		1.0	59	2.446	2.705	9.6
Boom Clay (ON-Mol-1 core 84b)	31	8	3	0.2	–	–	2	0.6	–	53	2.023	–	40

b.d., below detection.

Table 5. *Summary of diffusion coefficients for He in different clays*

Formation	Reference	Type	Orientation with respect to the bedding plane	Measured D ($m^2 s^{-1}$)	η (%)	T (°C)
Opalinus Clay	Gómez-Hernández (2000)	D_{app}	$\parallel$	7.0×10^{-10}	30	
Opalinus Clay	Rubel *et al.* (2002)	D_{app}	$\perp$	$3.5(\pm 1.3) \times 10^{-11}$		Room temperature
Opalinus Clay	This work	D_p	$\perp$	7.13×10^{-10}	9.6	21 ± 2
COx	Bigler *et al.* (2005)	D_p	Sphere	$7.5 \times 10^{-10} \pm 20\%$	16.3	20
COx	Bigler *et al.* (2005)	D_p	$\perp$	2.41×10^{-10}	15.4	20
COx	Rebour *et al.* (1997)	D_{app}	$\perp$	$5 \pm 1 \times 10^{-11} \pm 20\%$	23	50
COx	This work	D_p	$\perp$	4.38×10^{-10}	18	21 ± 2

The diffusion coefficients (D_{app} or D_p) refer to those used in the publications.

with a spherical sample. With the best fit between the experimental results and the analytical solution, $D_p = 7.5 \times 10^{-10}\ m^2\ s^{-1}$ was obtained with an uncertainty of 20%. The diffusion coefficient obtained for the spherical sample was actually a mixed diffusion coefficient with respect to bedding-plane orientation. In addition, the sample was not a perfect sphere and might have been disturbed by cutting. Therefore, it is reported that this value has to be considered as a maximum value, affected by experimental artefacts (Bigler *et al.* 2005). The latter statement is also supported by the authors of this manuscript. Bigler *et al.* (2005) also modelled an *in situ* pore diffusion coefficient based on the natural He profile in the Callovo-Oxfordian Clay: $D_p = 2.41 \times 10^{-10}\ m^2\ s^{-1}$, with a range of uncertainty from 0.8×10^{-10} to $7.2 \times 10^{-10}\ m^2\ s^{-1}$. This range of uncertainty is almost one order of magnitude, and the experimentally obtained value (from the spherical sample, $D_p = 7.5 \times 10^{-10}\ m^2\ s^{-1}$) reported in the same study does not fall within this range. This discrepancy could be explained by the disturbed condition of the spherical sample, leading to an increased and mixed (with respect to bedding plane) diffusion coefficient. As discussed above, the value $D_p = 7.5 \times 10^{-10}\ m^2\ s^{-1}$ has to be considered as a maximum value.

Rebour *et al.* (1997) used a through diffusion set-up. Despite the fact that this set-up could be used for different gases, only data for He diffusion on Callovo-Oxfordian Clay were reported ($D_{app} = 5 \pm 1 \times 10^{-11}\ m^2\ s^{-1}$). According to Bigler *et al.* (2005), the interpretation of the data suffered from complications such as anisotropy effects, which were not taken into account, and the measured porosity (23%) did not correspond to the porosity value needed to obtain a good fit (16%). One should also consider the origin of the sample: the only information given by Rebour *et al.* (1997) was that the sample was cored from Callovo-Oxfordian sediments from the NE of the Paris Basin at a depth of about 400 m. Given the lithostratigrapy of the Oxfordian (Gaucher *et al.* 2004), the sample used in Rebour *et al.* (1997) can be located in the calcareous Oxfordian, whereas the sample used in our work is located in the argillaceous Callovian–Oxfordian. Descostes *et al.* (2008) indicated that transport (and, thus, diffusion) properties are different for the different lithofacies of the Oxfordian, leading to, for instance, differences in the diffusion coefficient for HTO. For samples cored at a depth of 400 m, Descostes *et al.* (2008) reported an average D_{eff} (HTO) of $3.3 \times 10^{-12}\ m^2\ s^{-1}$, with an average porosity of 3.8%, which differs considerably from the data obtained from a sample at a depth of 376 m: average D_{eff} (HTO) of $22 \times 10^{-12}\ m^2\ s^{-1}$, with an average porosity of 19.5%. Given this information, the authors believe that no direct comparison can be made between the experimental results of Rebour *et al.* (1997) and this work.

Also, the diffusion coefficients reported for the Callovo-Oxfordian Clay by Rebour *et al.* (1997) and Bigler *et al.* (2005) differ by more than one order of magnitude. The value obtained in this work corresponds best to the values reported by Bigler *et al.* (2005): it is in-between the maximum value, which is obtained from outgassing of the spherical sample, and the value obtained from the natural tracer profile.

Currently, ANDRA (J. Talandier pers. comm. 16 January 2015; ANDRA 2005) considers a diffusion coefficient (D_p) of $4 \times 10^{-10}\ m^2\ s^{-1}$ for hydrogen (porosity 15%), based on values obtained for tritiated water (HTO). This value is in good agreement with the D_p obtained for He in this work.

As discussed earlier, the diffusion experiment with Opalinus Clay was performed with porewater relevant for Opalinus Clay in Mont Terri (low-salinity Pearson water). In the meantime, the porewater composition of the Opalinus Clay in

Table 6. *D_{eff} for HTO, calculated geometric factors* (G) *for tritiated water, He and Ar, and their ratio He/HTO and Ar/HTO (all for diffusion perpendicular to the bedding plane)*

	D_{eff} HTO (10^{-11} m^2 s^{-1})	*G* (He)	*G* (Ar)	*G* (HTO)†	*G* (He)/*G* (HTO)	*G* (Ar)/*G* (HTO)
Boom Clay	18‡	6	13	5	1.3	2.9
Callovo-Oxfordian Clay	2.61*	16	60	17	0.92	3.4
Opalinus Clay	1.37‡	10	67	15	0.67	4.4

*Mazurek *et al.* (2011); Savoye *et al.* (2012).
†For calculating the *G* of HTO, the porosity value that is derived from the HTO diffusion experiment is used (21% for COx).
‡Measured experimentally.

Schlattingen is now available (Mazurek *et al.* 2015). When comparing the composition of both porewaters, the porewater at Schlattingen has a higher salinity compared to the low-salinity Pearson water. Owing to this increase in salinity, the ionic strength also increased from 0.16 to 0.24 M. The effect of the increased ionic strength of the porewater on diffusion in Opalinus Clay was investigated and reported by Van Loon (2014). Diffusion experiments with HTO on samples cored in the Mont Terri URL were performed using porewaters with varying ionic strengths (0.17, 0.39 and 1.07 M, respectively). No significant effect of a variation in ionic strength on the diffusion parameters was observed. As noble gases are expected to have diffusive behaviour similar to that of HTO (they, too, are not affected by double-layer effects), the difference in composition of the used artificial porewater and the actual porewater should have no influence on the measured diffusion coefficients of He and Ar.

For Boom Clay, a fixed porosity of 37% was used to calculate the diffusion coefficients. As the measured physical porosity of the sample is 40%, a sensitivity analysis was performed. Calculations of D_{eff} with porosities of both 37 and 40% indicate no significant difference: only 1% for Ar and 0.2% for He. These small differences fit with the 95% confidence interval of the calculated D_{eff}. Therefore, the use of a standard porosity value of

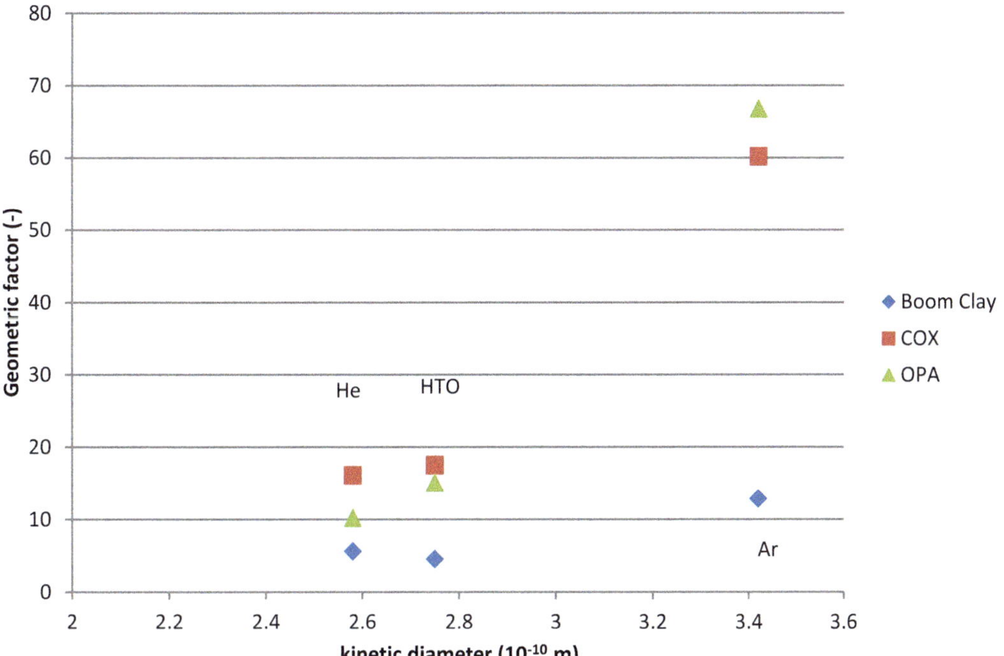

Fig. 3. Size of the diffusing gas molecule represented by the gas kinetic diameter (values taken from Hirschfelder *et al.* 1954) v. the geometric factor for three potential host formations.

37% for Boom Clay is considered to be a correct approximation.

In order to compare the diffusive behaviour of gases to the diffusive behaviour of water, diffusion coefficients for HTO, which are widely available, are given in Table 6.

Diffusion processes are influenced by the pore-network geometry (including, amongst others, shape, connectivity and orientation of the pores). The factor that describes this influence is called the geometric factor (G) (Bourg & Tournassat 2015).

For the three clays, the G for He, Ar and HTO is calculated according to:

$$G = \frac{\tau^2}{\delta} = \eta \frac{D_0}{D_{eff}} \qquad (2)$$

where τ^2 is the tortuosity ($-$), δ is the constrictivity ($-$), η is the diffusion accessible porosity (see Table 1) and D_0 is the molecular diffusion coefficient in pure water (in $m^2\ s^{-1}$).

Values for D_0 (He) of $7.28 \times 10^{-9}\ m^2\ s^{-1}$, D_0 (Ar) of $2.44 \times 10^{-9}\ m^2\ s^{-1}$ and D_0 (HTO) of $2.22 \times 10^{-9}\ m^2\ s^{-1}$ can be found in Boudreau (1997).

The tortuosity ($\tau = l_e/l$) is the ratio of the effective travelled distance (l_e) v. the distance between the start and end point (l), and constrictivity accounts for pore narrowing and widening (Collin & Rasmuson 1988). As stated above, the geometric factor is often seen as a characteristic of the porous medium (Grathwohl 1998) and provides information on the diffusion pathway geometry.

For all three molecules (He, Ar and HTO), the G values for Boom Clay are lower than for Callovo-Oxfordian Clay and Opalinus Clay (Table 6). When comparing the G ratio for He and HTO for the three host formations, it is observed that this ratio is rather similar for all three clays, which indicates that the diffusive behaviour of He and HTO is also similar. As the ratio is around 1, the pathway followed by He is rather similar to the pathway followed by HTO. However, the ratio G (Ar)/G (HTO) is >1, which means that Ar is more hindered and follows a longer diffusive pathway compared to water.

When comparing the diffusion coefficients measured for the three host formations to the size of the diffusing molecules (kinetic diameters: He $= 2.58 \times 10^{-10}$ m; HTO $= 2.75 \times 10^{-10}$ m; Ar $= 3.42 \times 10^{-10}$ m: Hirschfelder *et al.* 1954), it is evident that D_{eff} decreases with increasing size of the molecule. As D_{eff} and G are inversely proportional, G increases with kinetic diameter (Fig. 3). Therefore, the geometric factor (G) cannot be considered as a constant value, but depends on the diffusing molecule and the pore structure.

Conclusion

By using the set-up designed by SCK•CEN, diffusion coefficients for both He and Ar have been obtained with a small uncertainty for three potential host formations. The diffusion coefficients (D_p) for He in Boom Clay, Callovo-Oxfordian Clay and Opalinus Clay are 12.6×10^{-10}, 4.51×10^{-10} and $7.13 \times 10^{-10}\ m^2\ s^{-1}$, respectively. The diffusion coefficients (D_p) for Ar in Boom Clay, Callovo-Oxfordian Clay and Opalinus Clay are 18.6×10^{-11}, 4.06×10^{-11} and $3.65 \times 10^{-11}\ m^2\ s^{-1}$, respectively. The diffusion coefficients measured can be used in safety assessments (He as an approximation for H_2) and in the diffusive transport models of natural tracers (He and Ar).

The work on Boom Clay was performed in close co-operation with, and with the financial support of, ONDRAF/NIRAS, the Belgian Agency for Radioactive Waste and Fissile Materials, as part of the programme on geological disposal of high-level/long-lived radioactive waste that is carried out by ONDRAF/NIRAS. The work on Callovo-Oxfordian Clay was performed with the financial support of ANDRA, the French Agency for Nuclear Waste. The work on Opalinus Clay was partly funded by NAGRA. This work has been performed with the support of Tom Maes, Louis van Ravestyn, Serge Labat, Joan Govaerts, Eef Weetjens, Xavier Sillen, Jean Talandier and Martin Mazurek.

References

Aertsens, M. 2011. *Migration in Clay: Experiments and Models*. SCK-CEN ER-165. SCK•CEN, Mol, Belgium.

Altmann, S., Tournassat, C., Goutelard, F., Parneix, J., Gimmi, T. & Maes, N. 2012. Diffusion-driven transport in clayrock formations. *Applied Geochemistry*, **27**, 463–478, https://doi.org/10.1016/j.apgeochem.2011.09.015

ANDRA 2005. *La production et le transfert de gaz dans le stockage et la couche du Callovo-Oxfordien*, CNTASCM030042. ANDRA (National Agency for Radioactive Waste Management), Paris.

Battani, A., Smith, T., Robinet, J.C., Brulhet, J., Lavielle, B. & Coelho, D. 2011. Contribution of logging tools to understanding helium porewater data across the Mesozoic sequence of the East of the Paris Basin. *Geochimica et Cosmochimica Acta*, **75**, 7566–7584, https://doi.org/10.1016/j.gca.2011.09.032

Bigler, T., Ilhy, B., Lehmann, B. & Waber, H. 2005. *Helium Production and Transport in the Low-Permeability Callovo-Oxfordian Shale at the Site Meuse/Haute Marne*. NAGRA (National Cooperative for the Disposal of Radioactive Waste), Wettingen, Switzerland.

Boudreau, B. 1997. *Diagenetic Models and their Implementation*. Springer, Berlin.

Bourg, I.C. & Tournassat, C. 2015. Self-diffusion of water and ions in clay barriers. *In*: Tournassat, C., Steefal, C.I., Bourg, I.C. & Bergaya, F. (eds)

Natural and Engineered Clay Barriers. Developments in Clay Science, **6**. Elsevier, Amsterdam, 189–226.

Bruggeman, C., Maes, N., Aertsens, M. & De Canniere, P. 2009. *Tritiated Water Retention and Migration Behaviour in Boom Clay*. SFC1 Level 5 Report: First Full Draft – Status 2009. Report SCK-CEN ER-248. SCK•CEN, Mol, Belgium.

Collin, M. & Rasmuson, A. 1988. A comparison of gas diffusivity models for unsaturated porous media. *Soil Science Society of America Journal*, **52**, 1559–1565, https://doi.org/10.2136/sssaj1988.03615995005200060007x

De Craen, M., Wang, L., Van Geet, M. & Moors, H. 2004. *Geochemistry of Boom Clay Pore Water at the Mol Site*. Report SCK-CEN BLG-990. SCK•CEN, Mol, Belgium.

Descostes, M., Blin, V. et al. 2008. Diffusion of anionic species in Callovo-Oxfordian argillites and Oxfordian limestones (Meuse/Haute–Marne, France). *Applied Geochemistry*, **23**, 655–677, https://doi.org/10.1016/j.apgeochem.2007.11.003

Enssle, C.P., Cruchaudet, M., Croisé, J. & Brommundt, J. 2011. Determination of the permeability of the Callovo-Oxfordian clay at the metre to decametre scale. *Physics and Chemistry of the Earth, Parts A/B/C*, **36**, 1669–1678, https://doi.org/10.1016/j.pce.2011.07.031

Gaucher, E., Robelin, C. et al. 2004. ANDRA underground research laboratory: interpretation of the mineralogical and geochemical data acquired in the Callovian–Oxfordian formation by investigative drilling. *Physics and Chemistry of the Earth, Parts A/B/C*, **29**, 55–77, https://doi.org/10.1016/j.pce.2003.11.006

Gómez-Hernández, J.J. 2000. *FM-C Experiment: Part A) Effective Diffusivity and Accessible Porosity Derived from in-situ He-4 Tests. Part B) Prediction of HE-3 Concentration in a Cross-Hole Experiment*. Technical Note 2000-40. Mont Terri Project.

Grathwohl, P. 1998. *Diffusion in Natural Porous Media: Contaminant Transport, Sorption/Desorption and Dissolution Kinetics*. Kluwer Academic, Dordrecht.

Hirschfelder, J., Curtiss, C. & Bird, R. 1954. *Molecular Theory of Gases and Liquids*. Wiley, New York.

Honty, M. & De Craen, M. 2012. *Boom Clay Mineralogy – Qualitative and Quantitative Aspects*. SCK-CEN ER-194. SCK•CEN, Mol, Belgium.

Jacops, E., Grade, A., Govaerts, J. & Maes, N. 2013*a*. *Measuring the Diffusion Coefficient of He in Opalinus Clay*. SCK-CEN ER-238. SCK•CEN, Mol, Belgium.

Jacops, E., Volckaert, G., Maes, N., Weetjens, E. & Govaerts, J. 2013*b*. Determination of gas diffusion coefficients in saturated porous media: He and CH4 diffusion in Boom Clay. *Applied Clay Science*, **83–84**, 217–223, https://doi.org/10.1016/j.clay.2013.08.047

Jacops, E., Wouters, K. et al. 2015. Measuring the effective diffusion coefficient of dissolved hydrogen in saturated Boom Clay. *Applied Geochemistry*, **61**, 175–184, https://doi.org/10.1016/j.apgeochem.2015.05.022

Jacquier, P., Hainos, D., Robinet, J., Herbette, M., Grenut, B., Bouchet, A. & Ferry, C. 2013. The influence of mineral variability of Callovo-Oxfordian clay rocks on radionuclide transfer properties. *Applied Clay Science*, **83–84**, 129–136, https://doi.org/10.1016/j.clay.2013.07.010

Labalette, T., Landais, P., Farin, S. & Ouzounian, G. 2010. Site selection for a geological repository in France. Paper presented at WM2010 Conference, 7–11 March 2010, Phoenix, AZ.

Maes, N., Wang, L. et al. 2004. *Migration Case Study: Transport of Radionuclides in a Reducing Clay Sediment (TRANCOM 2)*. Contract No. FIKW-CT-2000-00008. Report EUR-21022. European Commission, Luxembourg.

Maes, N., Salah, S., Jacques, D., Aertsens, M., Van Gompel, M., De Canniere, P. & Velitchkova, N. 2008. Retention of Cs in Boom Clay: comparison of data from batch sorption tests and diffusion experiments on intact clay cores. *Physics and Chemistry of the Earth*, **33**, S149–S155, https://doi.org/10.1016/j.pce.2008.10.002

Mazurek, M., Hurford, A.J. & Leu, W. 2006. Unravelling the multi-stage burial history of the Swiss Molasse Basin: integration of apatite fission track, vitrinite reflectance and biomarker isomerisation analysis. *Basin Research*, **18**, 27–50, https://doi.org/10.1111/j.1365-2117.2006.00286.x

Mazurek, M., Alt-Epping, P. et al. 2011. Natural tracer profiles across argillaceous formations. *Applied Geochemistry*, **26**, 1035–1064, https://doi.org/10.1016/j.apgeochem.2011.03.124

Mazurek, M., Oyama, T., Wersin, P. & Alt-Epping, P. 2015. Pore-water squeezing from indurated shales. *Chemical Geology*, **400**, 106–121, https://doi.org/10.1016/j.chemgeo.2015.02.008

Pearson, F., Arcos, D. et al. 2002. *Geochemistry of Water in the Opalinus Clay Formation at the Mont Terri Laboratory*. Technical Report 2003-03. Mont Terri Project.

Perko, J. & Weetjens, E. 2011. Thermohydraulic analysis of gas generation in a disposal facility for vitrified high-level radioactive waste in Boom Clay. *Nuclear Technology*, **174**, 401–410.

Rebeix, R., Le Gal La Salle, C. et al. 2014. Chlorine transport processes through a 2000 m aquifer/aquitard system. *Marine and Petroleum Geology*, **53**, 102–116, https://doi.org/10.1016/j.marpetgeo.2013.12.013

Rebour, V., Billiotte, J., Deveughele, M., Jambon, A. & le Guen, C. 1997. Molecular diffusion in water-saturated rocks: a new experimental method. *Journal of Contaminant Hydrology*, **28**, 71–93.

Rodwell, W., Harris, W., Horseman, S., Lalieux, P., Müller, W., Ortiz, L. & Pruess, K. 1999. *Gas Migration and Two-Phase Flow through Engineered and Geological Barriers for a Deep Repository for Radioactive Waste*. A Joint EC/NEA Status Report EUR19122. European Commission, Luxembourg.

Rubel, A., Sonntag, C., Lippmann, J., Pearson, F. & Gautschi, A. 2002. Solute transport in formations of very low permeability: profiles of stable isotope and dissolved noble gas contents of pore water in the Opalinus Clay, Mont Terri, Switzerland. *Geochimica et Cosmochimica Acta*, **66**, 1311–1321, https://doi.org/10.1016/s0016-7037(01)00859-6

Savoye, S., Page, J., Puente, C., Imbert, C. & Coelho, D. 2010. New experimental approach for studying diffusion through an intact and unsaturated medium: a Case Study with Callovo-Oxfordian Argillite.

Environmental Science & Technology, **44**, 3698–3704, https://doi.org/10.1021/es903738t

Savoye, S., Frasca, B., Grenut, B. & Fayette, A. 2012. How mobile is iodide in the Callovo–Oxfordian claystones under experimental conditions close to the in situ ones? *Journal of Contaminant Hydrology*, **142–143**, 82–92, https://doi.org/10.1016/j.jconhyd.2012.10.003

Shackelford, C. 1991. Laboratory diffusion testing for waste disposal: a review. *Journal of Contaminant Hydrology*, **7**, 177–217, https://doi.org/10.1016/0169-7722(91)90028-Y

Van Geet, M., Bastiaens, W. & Ortiz, L. 2008. Self-sealing capacity of argillaceous rocks: review of laboratory results obtained from the SELFRAC project. *Physics and Chemistry of the Earth, Parts A/B/C*, **33**, (Suppl. 1), S396–S406.

Van Loon, L. 2014. *Effective Diffusion Coefficients and Porosity Values for Argillaceous Rocks and Bentonite: Measured and Estimated Values for the Provisional Safety Analyses for SGT-E2*. Report 12-03. NAGRA (National Cooperative for the Disposal of Radioactive Waste), Wettingen, Switzerland.

Van Loon, L.R., Soler, J.M., Muller, W. & Bradbury, M.H. 2004. Anisotropic diffusion in layered argillaceous rocks: a case study with opalinus clay. *Environmental Science & Technology*, **38**, 5721–5728, https://doi.org/10.1021/es049937g

Wemaere, I., Marivoet, J. & Labat, S. 2008. Hydraulic conductivity variability of the Boom Clay in north-east Belgium based on four core drilled boreholes. *Physics and Chemistry of the Earth, Parts A/B/C*, **33**, (Suppl. 1), S24–S36, https://doi.org/10.1016/j.pce.2008.10.051

Wersin, P., Mazurek, M., Waber, H.N., Mäder, U.K., Gimmi, T., Rufer, D. & De Haller, A. 2013. *Rock and Porewater Characterisation on Drillcores from the Schlattingen Borehole*. NAGRA Arbeitsbericht NAB 12-54. NAGRA (National Cooperative for the Disposal of Radioactive Waste), Wettingen, Switzerland.

Yu, L., Rogiers, B., Gedeon, M., Marivoet, J., De Craen, M. & Mallants, D. 2013. A critical review of laboratory and in-situ hydraulic conductivity measurements for the Boom Clay in Belgium. *Applied Clay Science*, **75–76**, 1–12, https://doi.org/10.1016/j.clay.2013.02.018

Zeelmaekers, E., Honty, M. et al. 2015. Qualitative and quantitative mineralogical composition of the Rupelian Boom Clay in Belgium. *Clay Minerals*, **50**, 249–272, https://doi.org/10.1180/claymin.2015.050.2.08

Zhang, C.-L. 2013. Sealing of fractures in claystone. *Journal of Rock Mechanics and Geotechnical Engineering*, **5**, 214–220, https://doi.org/10.1016/j.jrmge.2013.04.001

Index

Page numbers in *italics* refer to Figures. Page numbers in **bold** refer to Tables.